J. Thewlis DSc FInstP

Concise
Dictionary of
Physics
and related subjects

Second Edition
Revised & Enlarged

 PERGAMON PRESS OXFORD · NEW YORK · TORONTO · SYDNEY · PARIS · FRANKFURT

U.K.	Pergamon Press Ltd., Headington Hill Hall, Oxford OX3 0BW, England
U.S.A.	Pergamon Press Inc., Maxwell House, Fairview Park, Elmsford, New York 10523, U.S.A.
CANADA	Pergamon of Canada, Suite 104, 150 Consumers Road, Willowdale, Ontario M2J 1P9, Canada
AUSTRALIA	Pergamon Press (Aust.) Pty. Ltd., P.O. Box 544, Potts Point, N.S.W. 2011, Australia
FRANCE	Pergamon Press SARL, 24 rue des Ecoles, 75240 Paris, Cedex 05, France
FEDERAL REPUBLIC OF GERMANY	Pergamon Press GmbH, 6242 Kronberg-Taunus, Pferdstrasse 1, Federal Republic of Germany

First edition 1973

Reprinted with corrections 1974

Second edition revised and enlarged, 1979

British Library Cataloguing in Publication Data
Thewlis, James
Concise Dictionary of Physics and Related Subjects
Second ed., revised and enlarged.
I. Physics Dictionaries
530'.03   QC5   79-40209

ISBN 0 08 023048 2

Printed Offset Litho in Great Britain by Cox & Wyman Ltd, Fakenham, Norfolk

Concise
Dictionary of
Physics

Second Edition

Contents

Foreword to First Edition

In compiling this *Concise Dictionary of Physics* I have relied on the *Encyclopaedic Dictionary of Physics* to a considerable extent. However, since the new Dictionary consists of short definitions of terms which are each restricted (with some exceptions) to one concept, a new approach has had to be made to the choice and treatment of individual entries. The *Concise Dictionary* is, therefore, a new work in everything but scope, and even here there are noticeable differences between it and the *Encyclopaedic Dictionary,* if only because Physics does not stand still.

The scope of the work again covers not only Physics proper, but to a greater or lesser extent such related subjects as Astronomy, Astrophysics, Aerodynamics, Biophysics, Crystallography, Geophysics, Hydraulics, Mathematics, Medical Physics, Meteorology, Metrology, Photography, Physical Chemistry, Physical Metallurgy and so on.

It is hoped that a wide range of readers will find the Dictionary useful, from academic and industrial research physicists who may be interested in some branch of Physics removed from their own immediate specialities, through science teachers in Universities, Colleges or Schools, to students in Colleges and Universities and senior students in Schools, who may or may not be specializing in Physics, but who will find that the short summaries presented can save their consulting various dispersed sources for particular pieces of information.

Another kind of reader to whom the Dictionary should clearly be useful is the professional worker in some branch of science, such as one of those listed above, which is related to Physics. But there is also a large group of people with broader interests who will, I am sure, find the Dictionary a handy book of reference. I refer to managers and administrators in Government and Industry, who employ scientists and need to know the significance of what they read and hear; to politicians; to educationalists and teachers in general; to scientific editors; to writers and journalists; and of course to interested laymen of various kinds.

All this means naturally that not every entry in the Dictionary is suitable for every reader; but even the more specialized definitions will usually be found to contain an introductory sentence or so which serves to indicate the general nature of the term being defined.

To assist the reader cross-references have been given from individual terms to related terms; and the terms mentioned in the text are italicized where such cross-references have been made or where the reader might in any case be expected to look for them.

Terms which have their dictionary meaning or which are self-evident are not defined unless there is a special reason. The natural order of words in the terms has been adopted where possible, but there are many exceptions to this, for example, where groups of related terms are entered in close proximity. Thus *Emission spectrum* will be found under *Spectrum, emission,* with other related terms. One or two important terms are, however, entered in more than one place.

The order of the entries is strictly alphabetical and the use of apostrophes has been avoided for simplification, e.g. *Boyle law,* not *Boyle's law.* This strict alphabetical arrangement may occasionally lead to the inclusion of a term among a series of others to which it is not related (e.g. *Archimedes principle* and *Arctic air* come in the middle of terms dealing with *Arc*), but experience has shown that worse problems arise when one departs from the strict alphabetical order. The great advantage of the arrangement adopted is that there are no rules to remember, so that the reader knows precisely where to look for a given term.

This Dictionary is published at a time when a change to SI units is taking place in Schools, Colleges and Universities, and in Science and Technology in general. Even so, CGS units (and some units such as the electron volt, which are neither CGS nor SI) already form part of the literature and are likely to survive for some time to come. CGS and SI equivalents are, therefore, included in the definitions as appropriate. In addition, conversion tables are included in one of the Appendices, which give the values of CGS and related units, and of Imperial units, in terms of SI units.

The symbols are explained in each definition; but, apart from the general physical constants, are not necessarily the same in different definitions. Indeed a moment's reflection will show that this would be impossible. The usual convention has been adopted by which, in writing symbols for chemical elements or nuclides, the numerals attached have the following significance: upper left—mass number; lower left—

nuclear charge; lower right—number of atoms in the molecule. Thus $^{7}_{3}\text{Li}$ represents a lithium nucleus of mass 7, and $^{2}_{1}\text{H}_2$ or D_2 represents a molecule of deuterium. By extension $^{1}_{0}\text{n}$ represents a neutron. Where the state of ionization is needed it is given in the upper right-hand position.

In conclusion, I should like to acknowledge the help and encouragement given to me by the publishers, first by Mr. Robert Maxwell and later by Mr. Eric Buckley; and by the staff of Pergamon Press.

Finally I should like to say how much I owe to my wife, who has not only typed the script of the whole Dictionary (some parts more than once!) but has been of invaluable assistance in assembling and collating the individual entries.

Ardentinny,   J. Thewlis
Argyll

Introduction to Second Edition

The foreword to the original edition of the *Concise Dictionary of Physics* will serve the reader as a statement of the scope and plan of the present edition. In this edition an attempt has been made to fill gaps in the original edition, to modify or re-write entries in that edition which in the light of later work seemed incomplete or inaccurate, and to supply a number of new entries mostly dictated by recent discoveries.

Of the three original Appendices only the first, the Periodic Table and Symbols of the elements, remains untouched. The second, dealing with conversion tables, has been replaced by a fairly comprehensive table giving SI units, conversions and equivalents; and the third, giving values of some Fundamental Physical Constants, has been brought up to date by the inclusion of the latest recommended values.

Throughout the Dictionary the numerical values of physical quantities, ranging from the fundamental physical constants to experimentally determined values, have been examined and replaced by the latest values as necessary, having due regard to SI units.

I should again like to acknowledge the help and encouragement given to me by the publishers, notably Mr. Robert Maxwell and Mr. Alan Steel, and by members of the staff of Pergamon Press.

Finally I should once more like to thank my wife for typing and collating the revisions and additions, and for her invaluable assistance in general.

Ardentinny   J. Thewlis
Argyll

a

AB-. A prefix used to denote CGS electromagnetic units. See: *Units, CGS electromagnetic (e.m.u.).*

ABAMPERE. The absolute ampere, i.e. the CGS unit of current. One ampere is equal to 10^{-1} abamperes. See also: *Ampere, international. Biot.*

ABBÉ NUMBER. The reciprocal of the dispersive power for optical glass. See also: *Dispersive power, optical. Constringence.*

ABBÉ REFRACTOMETER. An instrument, employing the principle of total internal reflection, for the direct determination of the refractive index of a small quantity of liquid.

ABERRATION OF AN OPTICAL SYSTEM. Any failure of an optical system to form a point image of a point object. See also: *Astigmatism. Caustic. Coma (optics). Chromatic aberration. Spherical aberration.*

ABERRATION OF LIGHT. The alteration of the apparent position of a star or planet as a consequence of the Earth's motion and the finite velocity of light. See also: *Annual aberration. Diurnal aberration.*

ABLATION. A term originally describing the erosion and disintegration of meteors entering the Earth's atmosphere. Its meaning has now been extended to include the erosion of the outer surfaces of space vehicles, missiles, etc., on re-entering the Earth's atmosphere.

ABLATION MATERIALS. Protective materials, typically organic plastics, applied to the outer surfaces of space vehicles, missiles, etc., which are designed to undergo slow erosion on re-entering the Earth's atmosphere, thus affording protection from thermal damage to the space vehicle or missile itself.

ABNEY LAW. States that: if a spectral colour is desaturated by adding white light, its hue shifts towards the red end of the spectrum if the wavelength is below 5700 Å, and towards the blue if it is above. (*Note:* 1 Å is equal to 10^{-10} m.)

ABNEY MOUNTING. A modification of the Rowland mounting of a concave diffraction grating. See also: *Rowland circle. Rowland mounting.*

ABNORMAL SERIES. See: *Anomalous series.*

ABOHM. The absolute ohm, i.e. the CGS unit of resistance. One ohm is equal to 10^9 abohms. See also: *Ohm, international.*

ABSOLUTE PERMITTIVITY: PERMITTIVITY. Of a medium: the electric flux density or displacement produced in a medium by unit electric force. The term is often used to mean *relative permittivity.* See also: *Dielectric, refractive index of. Permittivity, relative.*

ABSOLUTE ZERO. The least possible temperature that could theoretically exist, according to the first and second laws of thermodynamics. At this temperature thermal energy is nil, but zero-point energy, attributed to the atoms by quantum mechanics, persists. See: *Temperature scale, international practical (IPTS).*

ABSORBANCE. The common logarithm of the absorptance. Also known as the optical extinction or optical density.

ABSORBED CHARGE. In a capacitor: the slow increase in charge which arises from the (slow) orientation of permanent dipolar molecules when the potential difference across a charged capacitor is maintained.

ABSORPTANCE. The ratio of the radiant flux absorbed by a body to the incident flux it receives. It is also known as the *absorptivity.*

ABSORPTION BAND. A band observed in an optical spectrum, arising from the excitation of the molecules of the absorbing medium to a number of excited states, with a consequent reduction of transmitted energy. See also: *X-ray absorption edge.*

ABSORPTION CELL. A vessel, usually with two transparent parallel walls, for containing a gas or liquid for which the absorption of radiation is to be measured. For the visible region glass walls are used, fused silica is commonly used for the ultraviolet region, and sodium chloride for the infrared. See also: *Beer law.*

ABSORPTION, CHEMICAL. The more or less uniform penetration of one substance into another across either a liquid or solid surface, involving a force of attraction analogous to valency. The process is not reversible. Alternative name: *Chemisorption*.

ABSORPTION COEFFICIENT FOR RADIATION. The quantity, α, in the expression $\alpha \Delta x$ for the fraction of a particular type of radiation absorbed in the passage of a parallel beam of the radiation through a thin layer of matter of thickness Δx. It is the absorption coefficient of that matter for that type of radiation. It is a function of the energy of the radiation and the nature of the matter, and, according as the thickness is expressed in terms of length, mass per unit area, moles per unit area, or atoms per unit area, the absorption coefficient is termed the *linear*, *mass*, *molar* or *atomic* absorption coefficient respectively. The absorption coefficient, α, occurs exponentially in the expression for the fraction of radiation transmitted $I = I_0 e^{-\alpha x}$ where I_0 refers to the incident and I to the transmitted radiation, and x is the thickness of matter traversed expressed in one of the above ways. See also: *Attenuation coefficient for radiation*.

ABSORPTION EDGE. See: *X-ray absorption edge*.

ABSORPTION INDEX. See: *Refractive index, complex*.

ABSORPTION LIMIT. See: *X-ray absorption edge*.

ABSORPTION OF RADIATION. (1) The transfer of energy to matter from radiation passing through the matter. (2) The disappearance in matter of an incident particle as a free particle, with or without the subsequent emission of other (possibly similar) particles. See also: *Absorption coefficient for radiation. Bouger–Lambert law of absorption. Photoelectric absorption*.

ABSORPTION, PHYSICAL. The more or less uniform penetration of one substance into another across either a liquid or solid surface, involving physical forces only. The process is reversible.

ABSORPTIVITY. See: *Absorptance*.

ABUNDANCE, CHEMICAL. Of an element: the fractional amount of that element (in terms of mass or number of atoms) present in a given environment, e.g. in the Earth (*terrestrial abundance*) or in the Universe (*cosmic abundance*).

ABUNDANCE, ISOTOPIC. Of a particular isotope in a mixture of isotopes of the same element: the fractional amount of that isotope. When expressed as a percentage it is known as the *relative abundance*.

ABUNDANCE, NATURAL. The isotopic abundance as found in nature. It is not necessarily constant.

ABUNDANCE RATIO. Of a particular isotope in a given material: the ratio of the number of atoms of that isotope to the number of atoms of another isotope of the same element present in the material.

ABUNDANT NUMBER. See: *Perfect number*.

ABVOLT. The absolute volt, i.e. the CGS unit of e.m.f. One volt is equal to 10^8 abvolts. See also: *Volt, international*.

ACCELERATING TUBE. The evacuated tube through which ions are accelerated in a Van de Graaff generator, Cockcroft–Walton apparatus, etc. It consists of a series of insulating rings sealed between conducting rings (the "electrodes") and is to be distinguished from the drift tubes or waveguide sections (a series of tuned radio-frequency cavities) used in a linear accelerator. See also: *Particle accelerator*.

ACCELERATION. The rate of change of velocity with time.

ACCELERATION DUE TO GRAVITY. See: *Gravitational acceleration*.

ACCELERATOR (PHOSPHOR). See: *Activator, optical*.

ACCEPTOR. An impurity or imperfection in a semiconductor which accepts electrons excited from the valence band, leading to hole conduction. The energy required to raise an electron from the valence band to the acceptor state is the ionization energy of the acceptor atom. See also: *Donor*.

ACCEPTOR LEVEL. An energy level associated with the acceptor state, commonly measured from the top of the uppermost filled energy band.

ACCIDENTAL COUNTS. In a coincidence counting system: the counts arising from effects unrelated to that under investigation.

ACCOMMODATION COEFFICIENT. A measure of the extent to which gas molecules leaving the surface of a solid in an atmosphere of the gas are in thermal equilibrium with it. It is defined as

$$a = \frac{T_i - T_r}{T_i - T_w}$$

where T_i is the temperature associated with the molecules incident on the surface, T_r is that associated with the molecules leaving the surface, and T_w is the temperature of the surface itself.

2

ACCOMMODATION TIME. Of an electrical discharge: the time that elapses from the production of the first electron to the establishment of the discharging condition (*ignition*).

ACCUMULATOR. An electrochemical cell in which storage is achieved and which is capable of being recharged when discharged, in contrast to a primary cell which has a limited life. The three main forms are the lead–acid, nickel–cadmium and nickel–iron cells. The accumulator is also known as a *storage cell* or *secondary cell*. See also: *Primary cell*.

ACCUMULATOR, EFFICIENCY OF. The ratio of the electrical energy obtained from the accumulator during its discharge to that supplied during charging. The energy is taken between specified limits of charge and discharge, and is usually expressed as a percentage.

ACCURACY. The reliability of a measurement, observation, etc., usually expressed as its error (i.e. its deviation from the "true" value). See also: Entries under *Error*. *Precision*.

ACHROMATIC COLOUR. One which has no hue, i.e. a grey.

ACHROMATIC SENSATION. One which gives merely a sensation of grey.

ACID. A substance having a tendency to lose a proton in a chemical reaction. See also: *Base*.

ACKERET THEORY. States that the disturbance due to an aerofoil in a supersonic air stream consists of two plane waves, occurring at the leading and trailing edges respectively, which are propagated outwards like sound waves. The theory leads to expressions for the pressures on the upper and lower surfaces of the aerofoil and to values for the lift, drag and momentum coefficients.

ACOUSTIC. See also under *Sound*.

ACOUSTIC ARRAY, DIRECTIONAL. An arrangement of acoustic sources or receivers which is designed to have directional properties. The term may be extended to include electrical signal-processing operations which influence the directional properties, e.g. a multiplicative array. See also: *Array, scanning of*.

ACOUSTIC ARRAY, DIRECTIONAL FUNCTION OF. (1) For a transmitting array: a function relating the strength of the acoustic signal at a distant receiving point to the bearing angle, as the array is rotated. (2) For a receiving array: a function expressing the electrical output amplitude from a fixed distant source, as the array is rotated.

ACOUSTIC ARRAY, MULTIPLICATIVE. A receiving array which is split into two parts, the signal voltages obtained from them being multiplied together. Also known as a *correlation array*.

ACOUSTIC CAPACITANCE. The negative imaginary part of acoustic impedance.

ACOUSTIC EMISSION FROM MATERIALS. The generation of sound when certain materials are mechanically deformed. In most cases the amplitude of the sound is too low, or the frequency too high, for it to be heard without special apparatus. The main causes are twinning (as in the well-known *cry of tin*), the sudden reorientation of large grains in polycrystalline material, the martensitic transformation, and slip in general. The effect has been applied in a variety of ways, from the warning of rock bursts in mines to studying the behaviour of nuclear reactor pressure vessels. See also: *Singing sands*.

ACOUSTIC FILTER. A device which filters out or attenuates certain frequencies initially present in a sound wave. The filters may consist of tubes, ducts or acoustic diffraction gratings, but the term is usually restricted to tubes with expanded or contracted sections, branch pipes or cavities, or diaphragms.

ACOUSTIC GRATING. A diffraction grating for the production of sound spectra. Both transmission and reflection types, including concave gratings, are used. With a frequency of 5000 Hz the grating spacing is about 10 cm.

ACOUSTIC IMPEDANCE. At a surface: the complex ratio of sound pressure averaged over the surface to the flux, or volume velocity, through it. It may be written $Z_A = P/Sv$, where P is the complex pressure, S is the area in question and v is the volume velocity. The units are the *acoustic ohm* (dyne s cm^{-5}) or the *SI acoustic ohm* (N s m^{-5}). See also: *Impedance. Mechanical impedance*.

ACOUSTIC IMPEDANCE, CHARACTERISTIC. A special case of the specific acoustic impedance for a plane wave. It is equal to the product of the mean density of the medium and the velocity of sound in the medium. It is sometimes known as the *intrinsic resistance*.

ACOUSTIC IMPEDANCE, SPECIFIC. The ratio of the (complex) sound pressure to the particle velocity, i.e. to the velocity of an infinitesimal part of the medium.

ACOUSTIC MODEL OF PARTICLE SCATTERING. An early form of the optical model, involving sound instead of light. See also: *Optical model of particle scattering*.

ACOUSTIC NOISE IN THE SEA. Underwater acoustic noise. It may be due to thermal agitation,

3

breaking waves, fish and other marine animals, the rolling of shingle, ships' propellers, etc. Its importance falls off with the operational frequency of the detection system and is unimportant in sonar systems working at frequencies above 200 kHz.

ACOUSTICS. The science dealing with the production, behaviour and detection of sound.

ACOUSTO-ELECTRIC EFFECTS. Phenomena which arise from the interaction of free electrons and holes with acoustic waves.

ACTINIC LIGHT. Light of a wavelength capable of producing a photographic effect.

ACTINIC VALUE. Of light of a given wavelength: the density of the image produced in a photographic emulsion under specified conditions.

ACTINIDE ELEMENTS. Members of the second rare earth series, the actinide series, named by analogy with the lanthanide series. The actinide elements comprise elements with atomic numbers 90 to 103. See also: *Rare earths. Appendix I.*

ACTINIUM SERIES. The radioactive series of elements beginning with ^{235}U and ending with ^{207}Pb. It is one of the three naturally occurring radioactive series, the others being the thorium series and the uranium series, which are separately defined. The series is also known as the *actino-uranium series*, or the *4n + 3 series* (since the atomic number of each member of the series can be expressed in this way).

ACTINOMETER. A radiometer used mainly in meteorology to measure solar and terrestrial radiation. An instrument designed to measure the intensity of the direct solar beam is known as a *pyrheliometer*. One designed to measure the total radiation falling on a surface is known as a *pyranometer*.

ACTINOMETRY. (1) The measurement of the light absorbed by a system undergoing photochemical change. Often extended to include the measurement of radiant energy absorption in general. (2) The measurement of the energy of solar radiation.

ACTINON. See: *Radioactive emanation: Emanation.*

ACTINO-URANIUM SERIES. See: *Actinium series.*

ACTION. Of a conservative dynamical system: the space integral of the total momentum of the system, i.e.

$$\int_{P_1}^{P_2} \Sigma_i m_i \frac{d\mathbf{r}_i}{dt} \cdot d\mathbf{r}_i$$

where m_i is the mass and $\mathbf{r}_i$ the position vector of the ith particle, t is time, and the system is assumed to pass from configuration P_1 to P_2. This expression

reduces to $2 \int_{t_1}^{t_2} E \, dt$ where t_1 and t_2 are the times at which the system has the configuration P_1 and P_2 respectively, E being the total energy of the system. See also: *Least action.*

ACTION AT A DISTANCE. The problem of the influence of one body upon another, with which it is not in contact. Contiguous or contact forces, such as tensions and pressures in electric and magnetic fields, were invoked at one time to resolve this problem, but the concept of action at a distance has now been modified into an interaction through some intermediate field, such as the electromagnetic field of Maxwell or the gravitational field of Einstein. A similar attitude exists in the quantum treatment of interaction.

ACTION POTENTIAL. The wave of depolarization travelling down biological cells, which is associated with signals carried by nerves or with the contraction of muscles. Also known as a *spike*. After the passage of an action potential down a nerve, the nerve is for a short time quite unexcitable. This time is the *absolute refractory period* of the nerve. The nerve then shows a reduced excitability for a further period, the *relative refractory period*, before its initial excitability returns. See also: *Resting potential.*

ACTION SPECTRUM. A plot against wavelength of the ability of a given type of radiation to produce a particular biological effect. This ability is expressed as the reciprocal of the energy required to produce the effect in question and the action spectrum corresponds more or less closely to the relevant absorption spectrum.

ACTIVATED DIFFUSION. The migration of ions, atoms or lattice defects across a potential barrier in a solid.

ACTIVATION ANALYSIS. See: *Radioactivation analysis.*

ACTIVATION ENERGY. The energy that must be acquired by an atomic or molecular system, or part of such a system, to enable a particular process to occur. Examples are: the energy needed by a molecule to take part in a chemical reaction, the energy needed to produce annealing or creep in crystalline materials, the energy needed to excite an electron in the production of luminescence, the energy required to raise an electron to the conduction band, or the energy needed to allow a lattice defect to move to a neighbouring site.

ACTIVATION, RADIOACTIVE. The process of

inducing radioactivity by irradiation.

ACTIVATOR, OPTICAL. An activating agent which is added in minute quantities to a phosphor to produce luminescence of a required colour. Alternative name: *Accelerator*.

ACTIVE COMPONENT. In electrical engineering: that component of the vector representing an alternating quantity which is in phase with some reference vector, e.g. the active component of the current, or *active current* (the component in phase with the voltage), the *active voltage* (the component in phase with the current), the active *volt-amperes* (the product of the active voltage and the current, or of the active current and the voltage, equal to the power in the circuit).

ACTIVE CURRENT. See: *Active component*.

ACTIVE DAYS: DISTURBED DAYS. Five days selected each month as international magnetic "disturbed" or "active" days, on the basis of many magnetic observations. See also: *Quiet days: Calm days*.

ACTIVE NITROGEN. A modification of ordinary nitrogen produced by passing various types of discharge through N_2. It exhibits a wide range of interesting physical and chemical properties, including significant changes in the optical spectrum. Active nitrogen emits a golden yellow glow (the *Lewis–Rayleigh afterglow*) which is believed to be associated with the recombination of nitrogen atoms.

ACTIVE VOLTAGE. See: *Active component*.

ACTIVE VOLT-AMPERES. See: *Active component*.

ACTIVITY. (1) Of a radioactive substance: the number of nuclear disintegrations occurring in a given quantity of substance per unit time. See also: *Becquerel. Curie. Radioactive decay constant: Radioactive disintegration constant*. (2) In a chemical reaction: the ideal or thermodynamic concentration of a substance, the substitution of which for the true concentration permits the application of the law of mass action. (3) Optical: the property of rotating the plane of polarization of light. It may arise from the asymmetric arrangement of molecules in a crystal or from the asymmetric arrangement of atoms in a molecule.

ACTIVITY COEFFICIENT. The ratio of the chemical activity to the true concentration.

ACTIVITY, SPECIFIC. Of a specified radioactive material or nuclide: the activity per unit mass.

ACTUATOR DISK. A simplified concept of a propeller in which it is assumed that the number of blades is infinite and that the thrust is uniformly distributed over the disk.

ADAPTATION (VISION). The change in the sensitivity of the eye after prolonged exposure to light or dark. *Dark adaptation* occurs when the subject moves from a bright to a dark environment, and is measured in minutes. *Light adaptation* occurs when the subject moves in the opposite sense, and is measured in seconds.

ADAPTIVE CONTROL. The automatic control of a process (e.g. the flight of an aircraft) in such a way that the process responds to changes in the operating environment (e.g. the autopilot of an aircraft would take account of change in altitude and speed).

ADAPTOMETER. An instrument for measuring the course of adaptation of the eye.

ADDITIVE COLOUR PROCESS. See: *Photography, colour*.

ADHERENCE NUMBER. A measure of the adhesive properties of solid particles, based on the number of particles adhering to a plate under specified conditions.

ADHESION, WORK OF. The work required to pull apart unit area of the interface between two solids, two liquids, or a liquid and a solid.

ADIABATIC APPROXIMATION. An approximation used when the Hamiltonian of a quantum system varies slowly with time.

ADIABATIC CHANGE. (1) Any process taking place without gain or loss of heat. (2) A reversible thermodynamical change of a system carried out with no change of entropy, i.e. without any heat entering or leaving the system. See also: *Polytropic change or process*.

ADIABATIC CHART. A chart showing the variations of a pair of thermodynamic variables during an adiabatic change. Also called an *isentropic chart*.

ADIABATIC DEMAGNETIZATION, COOLING BY. (1) Of a paramagnetic salt: the most common method of producing temperatures below 1 K. It involves the isothermal magnetization of the salt, followed by adiabatic demagnetization, and temperatures as low as about 10^{-3} K may be obtained in this way. (2) Of a substance with nuclear moments: a method involving the isothermal magnetic alignment of the nuclear spins at temperatures of about 10^{-2} K, followed by adiabatic

5

demagnetization. The production of temperatures as low as about 10^{-6} K should be feasible.

ADIABATIC LAPSE RATE. See: *Lapse rate.*

ADIABATIC SATURATION TEMPERATURE. Of a gas under specified initial conditions of temperature and humidity: the temperature that the gas would attain if it were saturated with vapour under adiabatic conditions. For the special case of the air–water system this temperature is equal to the wet-bulb temperature.

ADMITTANCE. The reciprocal of impedance. The real part is called the *conductance* and the imaginary part the *susceptance.*

ADMITTANCE, IMAGE. The reciprocal of the image impedance.

ADSORBATE. The film of adsorbed material present on an adsorbing surface, i.e. on the *adsorbent.* See also: *Adsorption.*

ADSORPTION. Absorption when only the surface functions as the absorbing medium. The material adsorbed (the *adsorbate*) is present as a film, which may be only one molecule thick, on the surface of the adsorbent.

ADSORPTION INDICATOR. A compound capable of revealing the end-point in a precipitation titration by changing the colour of the precipitate.

ADSORPTION ISOBAR. The relation between the quantity of gas adsorbed at a solid surface to the temperature, at constant pressure.

ADSORPTION ISOTHERM. The relation between the quantity of gas adsorbed at a solid surface to the pressure of the gas, at constant temperature.

ADVANTAGE FACTOR. In a nuclear reactor: the ratio of a specified radiation quantity at a position where an enhanced effect is produced to the value of the same radiation quantity at some reference position.

ADVECTION. Refers (mainly in meteorology) to the change in any property (e.g. temperature) of a given element of air which is brought about by the motion of the air in a gradient of the property concerned. It is usually limited to the change arising from the horizontal component of the motion only.

AEOLIAN TONE. A musical note set up when air flows past a fixed thin wire. It arises from the vortices formed in the wake of the air stream. See also: *Noise, aerodynamic.*

AELOTROPIC. Having different properties in different directions. The term usually refers to a crystal and is applied even if only one anisotropic property is exhibited.

AERIAL. A conductor or series of conductors, usually in the form of a wire or wires, erected in an elevated position and used for the transmission or reception of radio waves. Also known as *antenna.*

AERIAL, ADCOCK. A directional receiving aerial consisting of two spaced vertical dipoles connected to respond only to vertically polarized waves.

AERIAL ARRAY. A system of two or more spatially arranged, coupled aerials, which is designed to have particular directional radiating or receiving properties. See also: *Array, scanning of.*

AERIAL ARRAY, LINEAR. An aerial array in which the constituent elements, generally identical, are arranged along a straight line and separated by equal distances of a given fraction of a wavelength. It is designed to have highly directional properties.

AERIAL ARRAY, NULL DIRECTION OF. That direction in which the power radiated or received in unit solid angle is zero.

AERIAL, BEVERAGE. An aerial consisting of a long conductor mounted horizontally above the earth and running in the direction of arrival of the incoming waves. Also called *wave aerial.*

AERIAL, BICONICAL. An aerial formed by two conical conductors, having a common axis and vertex, and excited at the vertex.

AERIAL, DIPOLE. Usually refers to a straight conductor, of overall length one half-wavelength, and fed in the centre. Other lengths may, however, be used and a *full-wave* dipole is not uncommon.

AERIAL, DIRECTIVE PATTERN OF. See: *Aerial, polar diagram of.*

AERIAL, DIRECTIVITY OF. The ratio of the radiation intensity of an aerial, measured in the peak direction, to the radiation intensity of a standard uniform radiator, when both are subject to the same input power.

AERIAL, EFFECTIVE HEIGHT OF. Of a receiving aerial: the ratio of the total voltage induced at the terminals to the field strength (in volts per unit length) incident on the aerial.

AERIAL, FOLDED-DIPOLE. A dipole antenna folded back on itself to form a narrow rectangular loop one half-wavelength long.

AERIAL, FRAME. See: *Aerial, loop or frame.*

AERIAL GAIN. Of a directional aerial: the ratio, usually expressed in decibels, of the power supplied to the aerial for the production of a given field strength to that required for the production of the same field strength in the same direction by a non-directional comparison aerial.

AERIAL, HELICAL. An aerial in the form of a loosely wound helix in which the circumference of one turn is approximately equal to the wavelength. It has appreciable gain with the main lobe directed along the axis of the helix.

AERIAL, HORN. An open extension of a waveguide in which some degree of matching between the waves in the guide and those in space is obtained by suitably shaping the region of transition between the guide and the aerial aperture.

AERIAL IMPEDANCE. (1) Between two points in an aerial system: the self-impedance that would be offered to the voltage supply feeding the system if that supply were to be fed at the two points in question. (2) Of one point in an aerial system: the self-impedance that would be offered if the system were to be open-circuited at that point and the voltage source inserted. At the resonant frequency of the system the aerial impedance is purely resistive.

AERIAL, INDUCTION FIELD OF. The electromagnetic field distribution which is localized round an aerial, and which makes no contribution to the average power flowing into or out of the aerial. The magnetic induction field varies as the inverse square and the electric induction field as the inverse cube of the distance from the aerial.

AERIAL, LOOP OR FRAME. An aerial consisting of one or more rings, squares or rectangles of wire, the plane of the loops being in the direction of maximum transmission or reception. Of main interest are electrically small loops operated at high frequency.

AERIAL NOISE. A noise component (analogous to thermal noise) of the output from an aerial or aerial system which is directed at the open sky or at objects which absorb radio waves. Its magnitude is given by the *aerial temperature*, defined as the temperature of a black-body enclosure which, if it completely surrounded the aerial system and were in thermodynamic equilibrium with it, would produce the same amount of noise.

AERIAL, PARABOLIC. A directional aerial employing a reflecting parabolic mirror. It is fed (by means of dipoles, waveguide feed systems, or horns) at a point approximately at the focus of the reflector for a paraboloid of rotation, or on the focal line of the reflector for a parabolic cylinder.

AERIAL, PARASITIC. An element of an aerial array excited not by the primary source but by the field produced by other "driven" elements of the array.

AERIAL, POLAR DIAGRAM OF. The plot of the radiation intensity received or transmitted by an aerial as a function of direction in a given plane. Where the mapping is carried out in three dimensions the diagram is known as the *radiation pattern* or *directive pattern* of the aerial.

AERIAL, RADIATION PATTERN OF. See: *Aerial, polar diagram of.*

AERIAL, RHOMBIC. An aerial which has four wire elements arranged along the sides of a rhombus. The aerial is fed at one corner and the opposite corner is terminated in the resistance required to prevent reflections travelling back towards the feed.

AERIAL, SLOT. A radiating element formed by cutting a hole or slot in the wall of a waveguide or cavity resonator.

AERIAL TEMPERATURE. See: *Aerial noise.*

AERIAL, TURNSTILE. (1) An aerial comprising four vertical radiators arranged to intersect a horizontal plane at the four corners of a square. (2) An aerial consisting of two horizontal half-wave dipoles arranged at right angles and fed in phase quadrature.

AERIAL, VEE. An aerial consisting of a pair of equal straight conductors set at an angle to each other and fed at the junction between them.

AERIAL, WAVE. See: *Aerial, Beverage.*

AERIAL, YAGI. An aerial array containing one radiating element and a number of parasitic ones.

AERODYNAMIC CENTRE. The position within an aircraft or missile at which a change in the angle of incidence produces no change in the aerodynamical pitching moment. It is the point at which the additional forces induced by a change of incidence may be considered to be located. From the position of the vehicle's centre of gravity with respect to this point, the longitudinal stability or *static margin* of the vehicle, may be assessed.

AERODYNAMIC HEATING. The transfer of heat to a solid body immersed within, and moving relative to, a fluid, often air.

AERODYNAMIC NOISE. See: *Turbulence, sound from.*

AERODYNAMIC RANGE. An arrangement for observing the motion of a projectile, aircraft model or missile model in free flight. It is complementary to a wind tunnel.

AERODYNAMICS. The study of air in motion; by generalization, the science of the dynamics of gases. It deals largely with the disturbance produced by the relative motion between a solid and air or other gas in contact with it, and with the reactions experienced by the solid.

AERODYNAMICS, INTERNAL. The study of the flow of air or gas through compressors, ducts, fans, orifices, turbines, valves, etc.

AERODYNAMIC TIME. A unit of time, known as the *airsec,* formerly used in aerodynamic investigations of stability, and defined as $\dfrac{W/S}{\varrho g V}$ where W is the all-up weight of an aircraft, S the gross wing area, ϱ the density of the air, V the air speed and g the gravitational acceleration.

AERODYNE. See: *Aircraft.*

AEROELASTICITY. The science concerned with the dynamics of a system governed by inertial (and to some extent gravitational), aerodynamic and elastic forces. Its principal application lies in the field of aeronautics, but it is also of considerable significance in other fields, e.g. the stability of suspension bridges, the oscillation of towers and high buildings.

AEROFOIL. In aeronautics: a lifting wing or, more specifically, the shape of an appropriate section through such a wing.

AEROFOIL, INCIDENCE OF. The angle between the chord line of an aerofoil and the relative airflow.

AEROFOILS, CASCADE OF. A set (finite or infinite) of aerofoils arranged in rows or tiers, usually of constant chord, with constant *pitch* (distance between neighbouring members), whose purpose is to deflect a uniform airstream through a constant angle. By extension, sections of a set of turbine blades, taken at constant radius, can be developed into sections of an infinite cascade.

AEROFOIL, SPAN OF. The length along a specified line measured normal to the main air flow. See also: *Span.*

AEROFOILS, STAGGER OF. Of two or more aerofoils arranged in cascade (e.g. in a biplane system): the angle (in a section normal to the span) between a line joining the leading edges and the normal to a chord line. It is positive if the upper aerofoil ("up" being taken as the direction of lift) leads the lower.

AEROFOIL, WEDGE. A supersonic aerofoil section of small thickness ratio (5% or less) with plane surfaces instead of curved. It is usually wedge-shaped at both leading and trailing edges, sometimes with a parallel mid-portion, and is then known as a *double-wedge aerofoil.*

AEROGEL. A gel in which the dispersing medium is a gas.

AEROLITE. See: *Meteorite.*

AEROLOGY. The study of the atmosphere in three dimensions as contrasted with the study of the atmosphere from observations made only at the Earth's surface.

AERONOMY. The study of the physics of upper atmospheric phenomena, including the effects of solar radiation.

AEROPHYSICS. The study of the properties of gases at very high Mach numbers.

AEROSOL. A relatively stable suspension of solid particles or liquid droplets dispersed in a gas.

AEROSOL, RADIOACTIVE. A suspension in air of liquid or solid particles containing radioactive substances. The particle size may vary from 10^{-3} to 50 μm.

AEROSTAT. See: *Aircraft.*

AEROSTATICS. The study of gases at rest in equilibrium. An important application is to the equilibrium of balloons and airships.

AFFINE FIELD THEORIES, RELATIVISTIC. Theories which attempt to extend the general theory of relativity to explain the origin of the electromagnetic field and the meson field.

AFFINE SPACE. An n-dimensional space with particular properties associated with the transformation of coordinates. It finds a particular application in the general theory of relativity.

AFFINITY, CHEMICAL. Of a chemical or electrochemical reaction: the decrease in free energy accompanying the reaction.

AFTER-GLOW. The light emitted by a phosphor after the excitation is removed. See also: *Phosphorescence.*

AFTER-GLOW, LEWIS–RAYLEIGH. See: *Active nitrogen.*

AFTER-IMAGE. The image seen after the retina, or a portion of it, has been fatigued by exposure to intense light or to a continued fixed light stimulus. It may be positive or negative and sometimes exhibits colours complementary to the original.

AGE DETERMINATION BY RADIOACTIVITY. The determination of the time since a given radioactive nuclide was first formed, usually estimated from the ratio of the amounts of parent and daughter products, as in the case of potassium-argon dating and uranium series dating. In carbon dating the actual amount of ^{14}C present is determined, on the assumption that the decay rate has remained constant under terrestrial conditions during geological time. The technique has been applied to estimate the age of minerals, geological formations, meteorites, archaeological objects, etc., and has been extended to include the examination of fission tracks in minerals, which have resulted from spontaneous fission of uranium. See also: *Archaeomagnetism. Thermoluminescent dating.*

AGE HARDENING: PRECIPITATION HARDENING. Of an alloy: the hardening, by ageing at room temperature or above, of a supersaturated alloy quenched from a high temperature. The effect is due to the precipitation of new phases from the supersaturated matrix, usually in the form of platelets, known as *Guinier–Preston zones.*

AGEING OF MAGNETIC MATERIALS. The change of magnetic properties of a metastable alloy with time. Artificial ageing may be induced by heating at an elevated temperature.

AGE OF NEUTRONS. See: *Fermi age.*

AGGLUTINATION. The coalescing of small suspended particles into large masses which will form a precipitate.

AGGREGATE RECOIL. The ejection of radioactive atoms from the surface of a radioactive material. It occurs principally in α-particle disintegration where kinetic energy is transferred from the α-particle and the recoiling daughter nucleus to the parent atom, which may occasionally be ejected from the surface. It is observed as a migration of a fraction of the radioactivity on to nearby surfaces.

AGGREGATION. The gathering of particles, especially in the sense of their formation into larger entities or aggregates.

AGGREGATION, STATE OF. Formerly used to denote a separate physical condition with the classification into solids, liquids and gases. This is no longer sufficient and it can be considered that any discontinuity in a normally well-defined physical property marks some change in the state of aggregation. For example, solids may be divided into crystalline and amorphous; and superfluids and plasmas may be regarded as new states of aggregation.

AGREEMENT RESIDUAL. In crystal structure determination: a quantity expressing the extent to which the calculated structure amplitudes for a proposed structure disagree with those observed. It is given by the sum of the differences between the observed and calculated structure amplitudes, for all observed reflections, expressed as a fraction of the sum of the observed amplitudes. It is sometimes known as the *reliability factor.*

AIR COIL. A self-supporting coil of wire in which the core is free of any other material. It may be used, for example, as an impedance.

AIR-CONDITIONING. The control of the temperature, moisture content and purity of the air by ventilation, heating and cooling and adjustment of humidity.

AIRCRAFT. Any form of flying machine or vehicle supported by air. An *aerostat* is supported by its own buoyancy; an *aerodyne* by wings.

AIR ENGINE. A heat engine depending for its operation on the alternate expansion and contraction of a given mass of air.

AIR-EQUIVALENT MATERIAL. A material suitable for the walls of ionization chambers, particularly for standardizing measurements, having substantially the same effective atomic number as air. A suitable material is impregnated paper or light plastic, with graphite as the conducting material. Also known as *air-wall material.* See also: *Ionization chamber.*

AIRGLOW. An emission from the upper terrestrial atmosphere which, together with diffused light from the Sun and stars, contributes to the light of the night sky. It exhibits diurnal, annual and solar variations. The airglow emitted at twilight arises from the illumination of the upper layers of the atmosphere by the Sun and is known as the *twilight airglow.* See also: *Twilight phenomena.*

AIR MASS (METEOROLOGY). A volume of the atmosphere, usually extending over hundreds of thousands of square kilometres, within which the mean temperature of a vertical column of given height (some hundreds of metres) is not very different from place to place.

AIRSEC. See: *Aerodynamic time.*

AIRSHIP. A power-driven aircraft which is lighter than air.

AIR SHOWER. See: *Cosmic ray shower.*

AIRSPEED, EQUIVALENT. The indicated airspeed, corrected for the density of the air, i.e. for the altitude. It is equal to $V/\sqrt{\sigma}$, where V is the speed of the aircraft relative to the surrounding air and σ the factor by which the air density at the relevant altitude differs from the standard density at sea level.

AIRSPEED, INDICATED. The speed of an aircraft as displayed on an instrument actuated by the difference between two pressures dependent on the speed.

AIRSPEED, TRUE. The actual speed of an aircraft through the air, computed by correcting the indicated airspeed for altitude, temperature, position error (i.e. the error associated with the location of the pressure-measuring devices) and compressibility effect.

AIR STREAMLINE. A line along which the direction of motion of the air at any instant is tangential.

AIR TRAJECTORY. The path followed by an element of air, as given, for example, by the path followed by a small inert marker. For steady motion the air trajectories are identical with the streamlines in the field of flow.

AIR-WALL MATERIAL. See: *Air equivalent material.*

AIRY DISK: AIRY PATTERN. The Fraunhofer diffraction pattern due to a circular aperture uniformly illuminated; thus, the form of the image of a point source of light which would be produced by an aberration-free optical system. The variation of light intensity across the disk is given by

$$\left(\frac{2J_1(z)}{z}\right)^2,$$

where $z = \dfrac{2\pi r}{\lambda} \sin \alpha$, r is the distance from the centre of the image, α the semi-angle of the cone of rays forming the image and λ the wavelength.

AIRY POINTS. In metrology: two points on a standard bar which, if the bar is supported at them, result in the least effect on the horizontally projected length of the bar arising from the varying flexure of the bar due to changes in gravity or small errors in the exact point of support. Each point is distant from the centre point by $1/(2\sqrt{3})$ of the length of the bar. More generally, for n points of support the distance apart of the points is $l/\sqrt{(n^2 - 1)}$, where l is the length of the bar.

AIRY SPIRALS. Spiral interference patterns produced when quartz cut perpendicularly to the axis is examined in convergent circularly polarized light.

ALBEDO. (1) For radiant energy: the coefficient of diffuse reflection, i.e. the ratio of the total scattered energy to the incident energy. (2) For light: the ratio of the amount of light reflected in all directions by an element of a diffusely reflecting surface to the amount of light incident on that element. It is not necessarily a constant for any particular surface. In astrophysics the *Bond albedo* is used for determining the mean albedo. This is the ratio of the total light reflected in all directions from a sphere illuminated by parallel rays, to the light incident upon that sphere. (3) For cosmic rays: that fraction of the radiation outside the atmosphere which has its origin in interactions within the atmosphere. The *splash albedo* refers to all the outgoing radiation. The *geomagnetic albedo* is the fraction of the splash albedo which is returned to the atmosphere by the Earth's magnetic field. (4) For neutrons: the probability under specified conditions that a neutron entering into a region through a surface will return through that surface.

ALCOHOL. The hydroxyl derivative of a hydrocarbon. Of primary industrial importance are the primary alcohols, general formula $R \cdot OH$, where R represents the hydrocarbon group.

ALFVÉN WAVE. A magnetohydrodynamic wave which is propagated in the direction of the magnetic field. The mass displacement is perpendicular to the direction of propagation.

ALGEBRAIC EQUATION. In the widest sense, an equation $F(z) = 0$ not involving the operations of differentiation or integration. Its solution is a number or set of numbers. A more restrictive definition is to a polynomial equation

$$P(z) \equiv a_0 z^n + a_1 z^{n-1} + \dots + a_{n-1} z + a_n = 0,$$

the coefficients $a_0 \dots a_n$ normally being real.

ALGEBRAIC NUMBER. Any number that can be expressed as the root of an algebraic equation. Algebraic numbers include all integers and rational fractions, and also those irrational numbers that satisfy the above definition. See also: *Transcendental number.*

ALGORITHM. A specific sequence of operations which, in principle at least, can give the solution to a given problem. The sequence can be of either finite or infinite length.

ALIDADE. An accessory instrument used in plane table surveying consisting in its simplest form of a rule, fitted with sights, which gives the bearing of objects from the plane table station. Also known as a *sight rule*.

ALIGNMENT. In metrology: refers to the collinearity or coplanarity of a feature such as an axis or surface with a datum which is itself an axis or surface. Errors in alignment are the departures of such features from their proper positions relative to the datum and may be expressed either as a linear or angular displacement or as a combination of both.

ALIPHATIC COMPOUND. An open-chain, noncyclic, organic compound.

ALKALI. An inorganic base which is soluble in water.

ALKALI METAL. The metals occupying group 1 A of the Periodic Table, comprising Li, Na, K, Rb, Cs and Fr. They have one s electron in the outer shell, are thus monovalent, and are strongly electropositive.

ALKALINE EARTH METAL. The metals in group 2 A of the Periodic Table, comprising Be, Mg, Ca, Sr, Ba and Ra. The term is sometimes restricted to Ca, Sr and Ba. The predominant valency in this group is two, and the metals are all electropositive, although not so strongly as are the alkali metals.

ALLOBARS. Forms of an element having different isotopic compositions, i.e. different atomic weights. The term is often restricted to compositions which do not occur naturally.

ALLOMETRY. A relation between two variables x and y which can be put into the form $y = ax^n$. This is usually known as a simple allometry. It is characterized by the ordinary differential equation
$$\frac{1}{y}\frac{dy}{dx} = \frac{n}{x}.$$

ALLOMORPHISM. See: *Polymorphism*.

ALLOTROPY. The existence of different structural modifications or crystalline forms of the same element. The term was formerly restricted to the solid state, but has now been extended to include the existence of an element in two or more liquid or gaseous forms, i.e. the molecular structure may be involved as well as the crystal structure. See also: *Enantiotropy. Monotropy*.

ALLOTROPY, DYNAMIC. Allotropy in which the transition from one form to another is reversible, but the transition takes place over a range of temperatures rather than at a definite temperature. See also: *Enantiotropy. Monotropy*.

ALLOY. Any metal other than a metallic element. It may be a homogeneous mixture of metals, a solution of one or more metals in another, or an intermetallic compound. *Alloy structures* may be classified according to the more important factors controlling their formation, i.e. the atomic size, the electron concentration, and the electrochemical factor. Thus, we may speak of *size-factor compounds*, which owe their stability to the relative sizes of the component atoms permitting geometrical stacking; or *electron compounds*, in which, when size factor and electrochemical factor are favourable, the structure is stabilized at particular electron–atom ratios.

ALLOY THEORY. A theory, or group of theories, which attempts to explain the form taken by alloy phase diagrams, and the nature of the phases appearing in them, in terms of the atomic properties of the component elements.

ALMUCANTAR. (1) A small circle on the celestial sphere parallel to the observer's horizon and connecting points with the same altitude. (2) An instrument for measuring star positions, used as a substitute for a transit instrument.

ALPHA DECAY. Radioactive decay in which an α-particle is emitted. This lowers the atomic number of the nucleus concerned by two and its mass number by four.

ALPHA-DISPERSION. Of body tissues. See: *Body tissues, dielectric dispersion in*.

ALPHA-PARTICLE. A ^{4}He nucleus emitted during a nuclear transformation. By extension, any ^{4}He nucleus. An α-particle has zero spin.

ALPHA-RAYS. A stream of α-particles.

ALPHA-RAY SPECTROMETRY. (1) The measurement of the distribution of either the kinetic energy or the momentum of α-particles from a given radioactive nuclide. (2) The study of α-ray spectra and their implications.

ALPHA-RAY SPECTRUM. Describes the series of monoenergetic groups ("lines") of α-particles emitted from a given radioactive nuclide. Usually one group predominates, with an energy between 4 and 10 MeV. Other groups, of much lower intensity, constitute the fine structure of the spectrum, and are usually separated from the main group by a few kilo electron volts.

ALPHATRON. An ionization gauge in which the

source of the ionization is a small α-emitter.

ALTAZIMUTH MOUNTING. A mounting for an astronomical telescope which permits independent movements in altitude and azimuth.

ALTERNATING CURRENT. An electric current which alternately reverses its direction in a circuit in a periodic manner. The alternating current passes through its cycle of values once in every period, over which the average value is zero.

ALTERNATING CURRENT BRIDGE. A Wheatstone bridge network using a.c. of audio or radio frequency, for the measurement of resistance, impedance, capacitance or inductance.

ALTERNATION LAW OF MULTIPLICITIES. For atomic line spectra: states that the spectral terms observed for the neutral atoms of successive elements in the Periodic Table alternate between even and odd multiplicities. See also: *Multiplicity.*

ALTIMETER. An instrument for measuring height above the ground or above sea level. It may depend on the variation of pressure with height, or on the reflection of radio waves or light. See also: *Radio altimeter.*

ALTITUDE. (1) The vertical distance above ground or sea level. (2) The angular elevation of a heavenly body above the plane of the horizon, measured on the great circle passing perpendicular to that plane through the body in question and the zenith.

ALUMINOTHERMIC PROCESS. A process involving the reduction of a metallic oxide by the ignition of a mixture of the oxide and finely divided aluminium powder, with the production of a very high temperature. It is especially useful in the reduction of refractory oxides such as those of titanium or molybdenum.

AMALGAM. An alloy of mercury with another metal.

AMBIPOLAR DIFFUSION. In a uniformly ionized gas or plasma: the steady state diffusion of electrons and positive ions under the influence of a space-charge field. The *ambipolar diffusion coefficient* is the mean diffusion coefficient for the two types of charge carrier weighted inversely as their mobilities.

AMICI PRISM. See: *Spectroscope, direct vision.*

AMINO ACID. A carboxylic acid in which one hydrogen atom attached to the carbon chain has been replaced by an amino (NH_2) grouping. The simplest of these is amino-acetic acid, NH_2CH_2-COOH, in which the carboxylic acid concerned is acetic acid CH_3-COOH. The occurrence of amino acids in proteins renders them of biological importance. See also: *Fatty acids. Zwitter ion.*

AMMETER. An instrument for measuring electric current.

AMMONOLYSIS. A chemical reaction, comparable with hydrolysis, in which ammonia rather than water is the reactant from which the reaction gets its name. See also: *Hydrolysis.*

AMORPHOUS. Of a solid: having a complete absence of order in its atomic arrangement at large distances. Order at small distances (i.e. distances involving the first few neighbours) may, however, occur. The atomic arrangement may be described statistically by a distribution function. See: *Radial density function.*

AMOUNT OF SUBSTANCE. A basic quantity, the *mole,* in the SI system of units. It is defined as the amount of substance which contains the same number of molecules (or ions, or atoms, or electrons, as the case may be) as there are atoms in exactly 12 g of the pure carbon nuclide ^{12}C. See also: *Atomic mass unit, unified. Appendix II.*

AMPERE. The MKSA and SI unit of electric current. It is defined as "the constant current which, if maintained in two straight parallel conductors of infinite length, of negligible circular section, and placed one metre apart in a vacuum, would produce between these conductors a force equal to 2×10^{-7} newtons per metre of length". See: *SI Units. Appendix II.*

AMPERE, ABSOLUTE DETERMINATION OF. This is carried out by observing the force between coils which, because they can be wound with many turns, yield a larger force more suitable to deal with than the force between two infinitely long parallel wires specified in the definition of the ampere. The instrument used, which involves the balancing of the force between the coils, is known as an *ampere balance.*

AMPERE BALANCE. See: *Ampere, absolute determination of.*

AMPERE, INTERNATIONAL. The former standard of electric current, defined as the unvarying electric current which, when passed through a specified solution of nitrate of silver in water, will deposit silver at the rate of 0·001 1800 g/s. One international ampere is equal to 0·999 835 absolute amperes (abamperes). See also: *Abampere.*

AMPÈRE LAW. Relates the magnetic field H to the current which set up the field, by the equation $\int \mathbf{H} \cdot d\mathbf{l} = I$, where the line integral is taken round any closed path and I is the current enclosed. If the current is flowing in a conducting medium $I = \int \mathbf{J} \cdot d\mathbf{s}$ where $\mathbf{J}$ is the current density and $d\mathbf{s}$ is a vectorial area element. This reduces to curl $\mathbf{H} = \mathbf{J}$, which, if there is a time variation in charge density (i.e. if the current—and all other field quantities—are not constant) must be modified (as was done originally by Maxwell) to read curl $\mathbf{H} = \mathbf{J} + \dot{\mathbf{D}}$, where $\dot{\mathbf{D}}$ is the *displacement current density* and the sum $\mathbf{J} + \dot{\mathbf{D}}$ is known as the *total current density.* This is the first of Maxwell's equations.

AMPÈRE RULE. States, that the magnetomotive force generated by a current in a wire appears clockwise to an observer looking in the direction of the current.

AMPERE-TURN. The MKSA and SI unit of magnetomotive force (m.m.f.), or force causing a magnetic field. The m.m.f. produced by a current in a coil is defined as the product of the number of turns and the current in amperes. One ampere-turn is equal to $4\pi \times 10^{-1}$ CGS units or gilberts.

AMPHOLYTE. An amphoteric electrolyte.

AMPHOTERIC. Capable of functioning either as an acid or a base.

AMPLIFICATION FACTOR. Of a vacuum tube or thermionic valve: a measure of the effectiveness of the grid and anode voltage in controlling the anode current. It is given by $-\dfrac{\partial V_a}{\partial V_g}$ where ∂V_a is the change in anode voltage necessary to counteract a change of ∂V_g in grid voltage in order to keep the anode current constant.

AMPLIFIER. An apparatus for increasing the amplitude or power level of an electrical signal by using devices such as valves or transistors deriving energy from a power supply. It may be classified as a video frequency, audio frequency (a.f.), intermediate frequency (i.f.), radio frequency (r.f.) or ultra-high frequency (u.h.f.) amplifier, according to the frequency range in which it is designed to work.

AMPLIFIER, D.C. An amplifier whose bandwidth extends down to zero frequency. In spite of their name, d.c. amplifiers do not deal exclusively with undirectional current but may cope with fluctuations or even changes in sign.

AMPLITUDE. (1) Of an alternating quantity: the peak positive or negative value. (2) Of a celestial body: the angular distance of a star, etc., from the east or west point, measured on the horizon.

AMPLITUDE MODULATION. The process of imparting information to a carrier wave by causing the amplitude of the carrier to vary in accordance with an input signal. See also: *Modulation.*

AMPLITUDE SPLITTING. See: *Interferometer, classification of.*

AMSLER VIBROPHORE. A high-speed, direct-stress, fatigue-testing machine operating at resonant frequency.

AMYRIOTIC FIELD. A quantized field for which there are creation and annihilation operators satisfying specific commutation rules, and for which there is a vacuum state. See also: *Myriotic field.*

ANABATIC WIND. An upslope wind arising from the heating of a sloping surface by the Sun. An upslope motion of air or "valley breeze" due solely to the existence of the slope as such is not referred to as an anabatic wind.

ANAGLYPH. A pair of stereoscopic pictures, one in red and the other in blue or green, superimposed on the same surface and designed to be viewed through spectacles with one red and one green filter.

ANALEMMA. See: *Time, equation of.*

ANALYSER (POLARIZED LIGHT). A device which passes only that component of light which is polarized in a particular plane. It may be a Nicol prism or a polaroid sheet. See also: *Polarizing prism. Polaroid.*

ANALYTIC FUNCTION: HOLOMORPHIC FUNCTION: REGULAR FUNCTION. A function $f(z)$ of the complex variable $z = x + iy$ is analytic, holomorphic, or regular at a point on the z-plane if the function and its first derivative are finite and single-valued there. If this property applies to all points within a given region of the complex plane, $f(z)$ is an analytic (holomorphic or regular) function throughout the region. Points at which there is no derivative are known as *singular points* or *singularities*.

ANASTIGMATIC LENS: ANASTIGMAT. A lens or lens combination which is designed to minimize astigmatism even when used at a wide aperture over a wide field. It is usually also corrected for spherical and chromatic aberration, and coma.

ANCHOR RING. A surface generated by a circle rotating about an axis in its own plane, which does not pass through the centre. It is the shape of a doughnut with a hole in it. If r is the radius of the circle, and k the distance from the centre to the axis of revolution, the volume of the anchor ring is $2\pi^2 kr^2$ and its surface area is $4\pi^2 kr$. It is also known as a *torus*.

ANDRONIKASHVILI EXPERIMENT. The experimental confirmation by Andronikashvili, from the measurement of the period and decrement of a torsional pendulum immersed in liquid helium, of the theoretical prediction that liquid helium should show a Bose–Einstein condensation at sufficiently low temperatures.

ANELASTICITY. The departure from perfect elastic behaviour arising from the dependence of strain on variables other than stress, e.g. temperature. See also: *Elastic after-effect*.

ANEMOMETER. A device for measuring the speed of the wind but not its direction. The most usual types are the *cup anemometer*, employing a rotating cup; the *plate anemometer*, depending on the wind pressure on a plate; the *pressure tube anemometer*, employing the pitot principle; the *hot wire anemometer*, depending on the rate of heat loss; the *ionization anemometer*, depending on the measurement of the speed of positive ions originating at a corona discharge; and the *thermocouple anemometer*, based on the measurement of the temperature produced in a heat-insulated body by adiabatic compression as the flow is brought to rest.

ANGLE OF ATTACK. See: *Angle of incidence*.

ANGLE OF INCIDENCE. (1) For radiation incident upon a surface: the angle between the direction of incidence and the normal to the surface at the point of incidence. See also: *Reflection*. (2) In aerodynamics: the angle between the chord of an aerofoil and the direction of undisturbed flow in the absence of sideslip. Also called *angle of attack*.

ANGLE OF REFLECTION. For radiation incident upon a surface: the angle between the direction of reflection and the normal to the surface at the point of incidence. See also: *Reflection*.

ANGLE OF REFRACTION. For radiation passing from one medium to another: the angle between the direction of the refracted radiation and the normal to the surface separating the two media at the point of incidence. See also: *Refraction*.

ANGLE OF SITUATION. See: *Parallactic angle*.

ANGLE STANDARDS. Reference standards for engineering or metrological use. They fall into three main categories: *angle gauge blocks*, made of hardened steel, stabilized to ensure dimensional stability, which can be combined by wringing to produce various reference angles; *precision polygons*, i.e. blocks of regular polygonal form (of steel, glass or fused quartz), having reflecting peripheral faces; and *circular scales*, mainly for incorporation in angular measuring instruments. Angle gauge blocks are generally used with an autocollimator or precision level, precision polygons almost exclusively with an autocollimator.

ÅNGSTRÖM UNIT. A unit of length used in spectroscopy, X-ray diffraction, etc. The international Ångström was defined by assigning the value 6438.4696 Å to the wavelength of the cadmium red line in dry normal air at 15°C and standard atmospheric pressure. It was 10^{-8} cm to within 2 parts in 10^7, but is now defined as 10^{-10} m exactly. See also: *X-unit. Appendix II*.

ANGULAR CORRELATION FOR SUCCESSIVE RADIATIONS. For a disintegration process that results in the emission of two successive radiations (e.g. $\alpha-\gamma$, $\beta-\gamma$, $\gamma-\gamma$): the correlation between the directions in which individual radiations of each pair are emitted.

ANGULAR DISTRIBUTION. Of particles or photons, for collision processes: the variation of the differential cross-section for the emission of a given radiation in a nuclear reaction, with the angle of emission, usually given in the "centre-of-mass" system.

ANGULAR MOMENTUM. The product of the moment of inertia of a rotating body or system of bodies, as measured about the axis of rotation, and the angular velocity about that axis. Also called *moment of momentum*. It is a vector quantity, having the direction of the axis of rotation and a sense such that the vector points towards the observer if the rotation is clockwise as seen by him. The angular momentum remains constant in any isolated system.

ANGULAR MOMENTUM IN ATOMS AND NUCLEI. The concept of angular momentum is applied (1) to a particle or system of particles that spins about an axis (or behaves as though it did) as well as (2) a particle or system of particles that revolves in an orbit (or behaves as though it did). Both are quantized, the quantum of (1), the *spin angular momentum*, and (2), the *orbital angular momentum*, both being in units of $\hbar$, where $\hbar = h/2\pi$, h being Planck's constant.

ANGULAR MOMENTUM, INTRINSIC. The angular momentum excluding orbital angular momentum. The term is sometimes applied, however, to any part of the angular momentum which is not required to be analyzed further, e.g. that of an atomic nucleus in a nuclear reaction.

ANGULAR MOMENTUM OF LIGHT. The angular momentum h about the direction of propagation, which is assigned to a photon of circularly polarized radiation. The value is positive for left- and negative for right-circularly polarized radiation.

ANGULAR MOMENTUM OPERATOR. A linear vector operator which, acting on a given function of space and spin coordinates, generates the change produced by an infinitesimal rotation of the co-ordinate axes.

ANGULAR MOMENTUM, ORBITAL. The angular momentum excluding spin angular momentum.

ANHARMONIC OSCILLATOR. An oscillator in which the restoring force varies other than linearly with the displacement of the system from its equilibrium position.

ANHARMONIC OSCILLATOR SPECTRUM. A spectrum, e.g. a vibration spectrum or a Raman spectrum, for which a full explanation of the details requires the assumption of *anharmonicity* in the vibrations, i.e. of the existence of anharmonic oscillators.

ANION. A negatively charged ion, i.e. one which travels towards the anode.

ANISOTROPY. The exhibition by a medium of a variation in physical properties with the direction of measurement. Such a medium may be anisotropic in respect of some physical properties but isotropic in respect of others. See also: *Aelotropic. Isotropy.*

ANISOTROPY, PLASTIC. The directional variation of the mechanical properties of a single crystal, defined by stress–strain relationships, when deformation extends into the plastic region.

ANNEALING. The process of heat-treating a solid material (e.g. metal, alloy, glass) in order to achieve a required ductility, hardness, freedom from stress and strain, homogeneity, grain size, etc.

ANNIHILATION. The disappearance of a particle and its anti-particle resulting from their mutual collision, their energy being converted into *annihilation radiation* or other particles.

ANNIHILATION RADIATION. The electromagnetic radiation resulting from annihilation. In the case of a collision between a positive and a negative electron the annihilation radiation usually consists of two photons, each of energy about 0·511 MeV, emitted in opposite directions. It may, however, consist of three coplanar photons having the same total energy, i.e. $2m_0c^2$, where m_0 is the mass of each particle and c the speed of light. See also: *Radiationless annihilation.*

ANNUAL ABERRATION. The aberration of light associated with the motion of the Earth about the Sun.

ANNUAL VARIATION. A variation of a geophysical or astrophysical quantity which arises from annually recurring factors or which takes place at a sensibly constant annual rate. See also: *Geomagnetic field, annual variation of. Parallax, solar. Parallax, stellar. Secular variation. Star position, apparent, annual variation of.*

ANODE. (1) Any thermionic tube electrode operated at an appreciably positive potential. (2) The positive electrode in an electric arc or electrolytic cell. (3) A positive accelerating or focusing electrode in an electron-beam device such as a cathode-ray tube. (4) In general, a positive electrode.

ANODE CHARACTERISTIC. The relation between anode current and anode voltage in a thermionic valve or tube.

ANODE CURRENT. The current flowing to the anode in a thermionic tube or valve.

ANODE DARK SPACE. In a glow discharge tube: a thin dark region next to, and on the cathode side of, the anode glow. See also: *Discharge, glow.*

ANODE DROP. See: *Anode fall.*

ANODE FALL. The voltage drop between the anode and the positive column in a discharge tube or arc discharge. Also known as *anode drop.*

ANODE GLOW. In a glow discharge tube: a thin luminous layer occurring on the surface of the anode. See also: *Discharge, glow.*

ANODE RAYS. Positive ions emitted from the anode in a glow discharge.

ANODE RESISTANCE. The anode potential divided by the anode current in a thermionic tube

15

or valve.

ANODIZING. The formation of an adherent oxide film on a metal surface when the metal forms the anode in a suitable electrolyte.

ANOMALISTIC MONTH. The interval between successive passages of the Moon through perigee (i.e. through the point of its orbit which is nearest to the centre of the Earth). Its length is just over 27·5 mean solar days. See also: *Month*.

ANOMALISTIC YEAR. The interval between successive passages of the Sun through perigee, i.e. between successive passages of the Earth through perihelion. It is 365·2596 mean solar days. See also: *Year*.

ANOMALOSCOPE. An instrument used for the detection and classification of defective colour vision. It allows two colours to be mixed by the subject and the result to be matched with a third (standard) colour, e.g. a mixture of green and red may be matched against yellow.

ANOMALOUS DISPERSION. An anomaly in the regular variation of refractive index with wavelength observed in the region of an absorption band. The phenomenon occurs for light, neutrons, radio waves and X-rays. See also: *Dispersion. Gases, anomalous dispersion in. Neutron dispersion, anomalous. Radiowaves, dispersion of. Sellmeier dispersion formula. X-ray dispersion, anomalous.*

ANOMALOUS SERIES. Spectral series for which the quantum defect for the different members of the series does not vary in a smooth fashion with the total quantum number. The terms of such a series correspond to states of the atom in which two electrons are excited. Also known as *abnormal series*.

ANOMALOUS VISCOSITY. See: *Non-Newtonian liquids*.

ANORTHIC SYSTEM. The crystal system exhibiting the least symmetry. Crystals in this system possess no axis or plane of symmetry but may possess a centre of symmetry. They have no faces at right angles and are referrable only to three unequal axes, none of which is at right angles to another. Also called *triclinic system*.

ANSAE. The extremities of Saturn's ring system which, when viewed foreshortened from the Earth, appear as "handles" to the main sphere of Saturn.

ANTENNA. See entries under *Aerial*.

ANTICLASTIC SURFACE. A surface for which the centres of curvature in the two principal plane sections (i.e. normal sections of maximum or minimum radius of curvature) through the surface are on opposite sides of the surface, e.g. a saddle.

ANTI-COINCIDENCE CIRCUIT. An electronic circuit that produces a pulse only when *anti-coincidence* occurs, i.e. when a count occurs in a specified detector unaccompanied by a count in one or more other specified detectors within a given short interval of time.

ANTICYCLONE. A region (perhaps of some 10^6 km² in area) in which the atmospheric pressure is high and the isobars take the form of closed curves around the point of highest pressure. The winds associated with such a region circulate in a clockwise direction in the northern hemisphere and in an anti-clockwise direction in the southern.

ANTIFERROELECTRIC. A dielectric of high permittivity in which no net spontaneous polarization occurs. This lack of spontaneous polarization is believed to be associated with the existence of opposing directions of polarization in adjacent lines of ions in the crystal. The antiferroelectric properties disappear above the *antiferroelectric Curie point* at which the antiferroelectric also undergoes a change in crystal structure. See also: *Ferroelectricity*.

ANTIFERROMAGNETISM. The existence of a state analogous to the ferromagnetic state but with neighbouring spins anti-parallel instead of parallel. Anti-ferromagnetic substances exhibit paramagnetism that varies with temperature in a manner similar to that for ferromagnetic substances. The ordered arrangement of magnetic moments, and hence the antiferromagnetism, is destroyed above the *Néel temperature*. Conversely, the complete anti-parallel arrangement of spin vectors is attainable only at absolute zero. See also: *Diamagnetism. Ferromagnetism. Néel temperature: Néel point. Paramagnetism.*

ANTI-ISOMORPHISM. Apparent isomorphism in which two bodies, outwardly of the same crystalline form, have markedly different physical properties, particularly optical.

ANTI-NODE. See: *Node*.

ANTI-PARTICLE. A particle having the same mass as a given elementary particle, but whose other properties, while having the same magnitude, may some of them be of opposite sign, e.g. electric charge in the case of the electron and positron, magnetic moment in the case of the neutron and antineutron. On collision a particle and its antiparticle are mutually annihilated, with the emission of radiation or other particles. See also: *Annihilation radiation*.

ANTI-REFLECTION COATING. The coating applied to a glass surface to reduce the amount of light reflected at the surface. The application of such a surface is known as *blooming*.

ANTI-STOKES EMISSION. The emission of fluorescent, phosphorescent, or luminescent radiation of shorter wavelength than that of the incident radiation. It may arise, for example, by double excitation to an excited state of higher energy, two photons being absorbed and only one emitted. The corresponding spectral lines are known as *anti-Stokes lines*. See also: *Raman effect*.

ANTI-SYMMETRIC WAVE FUNCTION. See: *Fermi–Dirac statistics*.

ANTITRADES. Winds, occurring at varying heights, whose direction is opposite to that of the trade winds below. The height of transition from trades to antitrades increases from practically zero at latitudes of about $35°$ to several miles at lower latitudes. See also: *Doldrums. Trade winds*.

ANTONOFF RULE. States that, when two partially miscible liquids are mutually saturated, the interfacial tension between is equal to the difference between the separate tensions of the saturated phases. The rule fails if the initial spreading coefficient of the liquid pair is negative.

ANTOXIDANT. A substance added to prevent or reduce the oxidation which results in the deterioration of such materials as rubber, paints, etc.

APASTRON. Of the orbit of one star about another in a binary system: the point in that orbit which is farthest from the other star. See also: *Periastron*.

APERIODIC. Refers to the condition of a potentially oscillatory system in which oscillation does not occur following an initial disturbance, owing to the degree of damping present.

APERTOMETER. A device for measuring the numerical aperture of a microscope objective.

APERTURE. Of an optical system: the diameter of the exit pupil when the system is adjusted to infinity. The optical performance is then expressed by the *aperture ratio*, i.e. the diameter of the entrance pupil divided by the focal length of the system.

APERTURE, NUMERICAL. Of an optical system: the quantity $n \sin U$, where n is the refractive index of the medium in the *object space* (the space through which the light rays pass before passing through the refracting system) and U is the semi-angle of convergence of the cone of rays entering the entrance pupil of the objective from the object point.

APERTURE SPLITTING. See: *Interferometer, classification of*.

APHELION. Of the orbit of a body about the Sun: the point in that orbit which is farthest from the Sun. See also: *Perihelion*.

APLANATIC LENS. A lens in whose construction use is made of aplanatic points, whereby it possesses the property of giving a sharp image for rays making large angles with the axis.

APLANATIC POINTS. Two points on the axis of an optical system which have the property that rays proceeding from one of them shall all converge to, or appear to diverge from, the other.

APLANATIC REFRACTION (REFLECTION). Refraction (reflection) in which neither the spherical aberration nor the coma of the incident pencil is changed.

APLANATIC SURFACE. A surface at which aplanatic refraction or reflection occurs.

APOCHROMATIC LENS. A compound lens that is sensibly free from chromatic errors, from spherical aberration for two wavelengths, and from central coma for one wavelength.

APOGEE. Of the orbit of a body about the Earth: the point in that orbit which is farthest from the centre of the Earth. The term is sometimes used when speaking of the Sun's apparent orbit about the Earth. Similar terms (e.g. *Apojove, Aposaturnium*) are used to refer to orbits about other planets. See also: *Perigee*.

APOSTILB. A unit of surface luminance (brightness) applied to diffusing surfaces. It is the luminance of an ideal diffuser emitting one lumen per square metre.

APPEARANCE POTENTIAL. A measure of the minimum energy needed to ionize an atom or to excite an energy level in an atom, usually by electron impact. It is usually defined as the voltage through which an electron must fall to acquire the necessary energy.

APPLEGATE DIAGRAM. A diagram used to demonstrate the principles of velocity modulation (e.g. in klystron amplifiers). In such a diagram the axial positions of the electrons in a beam are plotted as functions of time and grouped according to their time of entry into the system.

APPLETON LAYER: F-LAYER. The higher of the two main strata or "layers" in the ionized atmosphere, the other being the Kennelly–Heaviside layer or E-layer. It occurs at above about 150 km. See also: *Ionosphere*.

APPROACH LIGHTS. A pattern of high-intensity lights so arranged as to indicate the correct approach to the runway for an aircraft about to land.

APPULSE. The close approach of two heavenly bodies as seen projected on the celestial sphere.

APSIS: APSE. The points in an elliptical orbit of a secondary body about a primary where the distance between the two bodies is a maximum or minimum (e.g. the perihelion and aphelion of the Earth's orbit). The line joining these two points is known as the line of *apsides* or *apse line*.

AQUADAG. A proprietary colloidal dispersion of graphite in water used for lubrication, as a coating in cathode ray tubes, as a sink for secondary electrons or as a post-deflection acceleration anode.

AQUEOUS HUMOUR. A fluid secreted in the eye, just behind the cornea.

ARAGO POINT. A bright spot which, owing to diffraction, appears at the centre of the shadow of a circular disk.

ARC. An electrical discharge in a gas, generally operated at a relatively low voltage and high current. Its uses, among others, are as a source of light or heat, to excite spectral lines or to produce ions. Some special types are described below.

ARC, CARBON. An arc between two carbon rods which are first touched together to start the arc and then separated by a few millimetres. It emits a brilliant light, and the anode crater of a high current carbon arc reaches a temperature of over 5000°C.

ARCHAEOMAGNETISM. The use of thermo-remanent magnetization in dating, by observing the difference between the direction of magnetization in baked clays and volcanic rocks, for example, compared with that at the same place today. The method involves a knowledge of the direction of the Earth's magnetic field at all relevant times and is complicated by reversals in that direction which have occurred at intervals.

ARCHIMEDES PRINCIPLE. States that the vertical hydrostatic force acting on an immersed body is equal to the weight of the fluid displaced by the body.

ARC-IMAGE FURNACE. See: *Furnace, arc.*

ARC, OPEN-FLAME. An arc in which the anode is vaporized and is ejected as a flame.

ARC, SINGING. An arc which, suitably connected to an inductance and capacity, oscillates at audio frequency. At higher frequencies radiowaves are emitted.

ARC, THERMIONIC. An arc in which the electron current from the cathode is provided predominantly by thermionic emission. The self-maintained thermionic arc is restricted to operation on refractory electrodes such as tungsten, carbon and molybdenum.

ARCTIC AIR. An air mass whose low temperature and humidity and stable thermal stratification are occasioned by its source in the anticyclonic areas over the arctic snow and ice and often, in winter, over the large continents.

ARC, VACUUM. An arc that is initiated either by high voltage breakdown or by the very close proximity of the electrodes. A complete explanation of the vacuum arc is still lacking.

ARDOMETER. A type of total radiation pyrometer in which the radiation is concentrated on a small (2–3 mm diameter) blackened platinum disk in an evacuated glass bulb.

AREA RULE. A term used in high-speed aerodynamics to describe either (1) certain theorems and formulae relating the wave drag and other properties of the flow round bodies at supersonic speeds to their cross-sectional area distributions, or (2) methods of reducing the wave drag of a wing–fuselage combination by suitable shaping of the fuselage.

AREOMETER. A device for determining the density of a liquid by measuring the loss of weight of a solid of known mass and volume suspended in the liquid.

ARGAND DIAGRAM. A graphical representation of a complex number by a vector. Rectangular axes are used, the real part of the number being measured along the abscissa and the imaginary part along the ordinate. The diagram permits, for example, the rapid compounding of vectors or the representation of a function of a complex variable.

ARIES, FIRST POINT OF. See: *Equinox, vernal. Right ascension.*

ARITHMETIC MEAN. See: *Mean: Average.*

ARMATURE. Material, usually ferromagnetic, which is mounted so as to be capable of rotation in the field of a magnet, as in a dynamo or electric motor.

ARMILLARY SPHERE. A medieval mechanical

device representing various great circles of the celestial sphere, and used to solve certain spherical triangles. In particular it was used to find, from the observed altitude and azimuth of a given celestial object, its right ascension and declination, i.e. its celestial longitude and latitude.

AROMATIC COMPOUND. A derivative of benezene.

ARRAY, SCANNING OF. For acoustic or radio arrays: the deflection of the directional beam pattern of the acoustic or aerial array without any physical movement of the array itself.

ARRHENIUS EQUATION. For the relation between the velocity constant k and the temperature T in a reversible reaction: states that $\left(\dfrac{d}{dT}\right) \ln k = \dfrac{A}{(RT^2)}$ where R is the gas constant and A may be considered as an activation energy.

ARRHENIUS–GUZMAN EQUATION. For the viscosity of a liquid η: states that $\dfrac{d \ln (\eta v^{1/3})}{dT}$ $= C_1 T^2$ where v is the specific volume, T is the temperature and C_1 is a constant.

ARTIFICIAL EAR. A device employed to calibrate earphones objectively for measurements on the acuity of hearing.

ASCENSIONAL DIFFERENCE. Of a celestial body: the difference between the right ascension and the oblique ascension of the body.

A-SCOPE. A form of cathode-ray tube display in which the horizontal displacement corresponds to distance and the vertical displacement to signal strength.

ASDIC. See: *Echo sounding. Sonar.*

ASPECT RATIO. Of a lifting wing surface: the ratio of the square of the span to the plan area of the wing.

ASSOCIATION, CHEMICAL. The reverse of dissociation. It is a reversible process which may be treated by the law of mass action and may involve any state of matter and ionic, covalent or hydrogen bonding. See also: *Dissociation, chemical.*

ASSOCIATION FACTOR. For a liquid: the ratio of the number of simple molecules which have formed groups by chemical association, to the number of such molecules that would be present in the absence of association.

A-STAR. See: *Stars, spectral classes of.*

ASTERISM. In a Laue X-ray diffraction pattern: the elongation of Laue spots into radial streaks. It is attributed to lattice distortion.

ASTEROID. One of a large number of small planets, all of diameter less than 770 km, most of which have near-circular orbits mostly lying between those of Mars and Jupiter, at radial distances between 240 and 700×10^6 km. The largest are *Ceres* (diameter 770 km), *Pallas* (560 km), *Vesta* (500 km) and *Juno* (250 km). Some small asteroids, however, have highly elliptic orbits and approach much nearer to the Sun. Nearly 1700 orbits have been computed but it is estimated that some 30 000 asteroids may exist. See also: *Solar system. Trojans.*

ASTIGMATIC FOCI. The two focal lines produced by an astigmatic system. The line at the greater image distance is known as the *primary*, *meridional* or *tangential focus*. The second line is called the *secondary*, or *sagittal focus*.

ASTIGMATIC MOUNTING. For a concave diffraction grating: a mounting designed to minimize astigmatism.

ASTIGMATISM. Of an optical system: the production, from a point source, of two short line images (*focal lines*) at right angles to each other, and at two different distances from the system. If a screen is placed at a suitable position between the two focal lines a circular image will be formed. This position is known as the *position of least confusion* and sometimes as the *circle of least confusion*. See also: *Lens, curvature of field of.*

ASTON DARK SPACE. In a glow discharge tube: a thin dark region immediately adjacent to the cathode. See also: *Discharge, glow.*

ASTROARCHAEOLOGY. The study of ancient structures which have, or may have, astronomical significance, as at Stonehenge.

ASTROBIOLOGY. See: *Astronomy.*

ASTROCHEMISTRY. See: *Astronomy.*

ASTRODYNAMICS. The application to astronautical problems of information gained from celestial mechanics, high-altitude aerodynamics, geophysics, navigation, propulsion theory, statistics, and so on. One such problem is the accurate computation of satellite orbits and spacecraft paths.

ASTROGRAPHIC CHART. See: *Carte du ciel: Astrographic chart.*

ASTROGRAPHIC TELESCOPE. A telescope specially designed for the taking of photographs of star fields for the accurate measurement of relative star positions. Sometimes referred to as an *Astrograph*. See also: *Galaxy machine*.

ASTROLABE. (1) A medieval instrument for measuring the altitude of a heavenly body, incorporating horizontal and vertical graduated circles. (2) A term sometimes applied to the armillary sphere. (3) A modern optical instrument (the *prismatic astrolabe*) designed to measure the precise instant at which a given star has a given altitude (usually 45° or 60°). The measurement depends on observing the conjunction of two images, one arising from a single reflection in a prism and the other suffering an additional reflection at a horizontal (mercury) surface. This conjunction takes place when the altitude of the star is equal to the appropriate angle of the prism. A development of this is the *impersonal astrolabe* in which a birefringent prism is inserted in the beam so as to produce parallel emergent beams which eliminate errors arising from the position of the eyepiece. (4) Any instrument for the measurement of the altitude of a celestial object.

ASTROMETRY. The precise measurement of stellar positions, either absolute or differential, usually from photographic plates. See also: *Galaxy machine*.

ASTRONAUTICS. The science of locomotion and travel outside the Earth's atmosphere, involving the problems of artificial satellites and of interplanetary journeys.

ASTRONOMICAL TIME. See: *Time*.

ASTRONOMICAL TRIANGLE. A spherical triangle on the celestial sphere.

ASTRONOMICAL UNIT. A unit of distance used to express distances in the solar system. It is defined as the geometrical mean distance of the Earth from the Sun and may be taken as $1.496 \pm 0.001 \times 10^8$km (about 93 million miles). It may be noted that the relative distances in the solar system are known to a much greater accuracy than is the astronomical unit. See: *Appendix II*.

ASTRONOMY. The investigation of the universe outside the Earth, comprising traditional astronomy (the study of distances, distribution and motion of celestial bodies and systems), *Astrophysics* (the study of the physical nature and behaviour of celestial bodies), *Astrochemistry* (the study of the chemical nature and behaviour of celestial bodies), *Astrobiology* (the study of the possibility of life existing elsewhere than on the Earth), and *Cosmology* (the study of the structure of the universe as a whole, including *Cosmogony* — the study of

the origin of the solar system). See also: *Radio astronomy. X-ray astronomy.*

ASTRONOMY, GEODETIC. (1) The observation of latitude, longitude and azimuth at the origins of independent surveys, or for the demarcation of astronomically defined international boundaries. (2) The observation of azimuth or longitude for geodetic triangulation or traverse. (3) The measurement of the deviation of the vertical (the difference between the astronomical and geodetic coordinates of a point) for a variety of purposes.

ASTROPHYSICAL TEMPERATURES. The temperatures of stellar surfaces. Four temperatures are in common use, namely: the *black-body temperature*, i.e. the temperature of a black body having the same specific emission as the star at a given wavelength; the *radiation temperature*, i.e. the temperature of a black body having the same specific emission as the star in a given range of wavelengths; the *effective temperature*, i.e. the temperature of a black body having the same total emission as a star; and the *colour temperature*, i.e. the temperature of a black body having the same distribution of relative intensities ("colour") as a star.

ASTROPHYSICS. See: *Astronomy*.

ASYMMETRIC CARBON ATOM. A carbon atom to which four different radicals are attached, resulting in optical isomerism.

ASYMMETRIC SYNTHESIS. The synthesis of optically active compounds from racemic mixtures.

ASYMPTOTE. To a plane curve: a straight line which is related to the curve in such a way that the distance between them approaches zero as both approach infinity, i.e. an asymptote is the tangent of the curve at infinity.

ASYNCHRONOUS MOTOR: INDUCTION MOTOR. An a.c. motor in which currents in the primary winding induce currents in the secondary winding (usually the rotor) which interact with the flux set up by the primary winding to produce rotation.

ATMOSPHERE. (1) The gaseous environment of any solid or liquid. (2) The gaseous envelope surrounding the Earth and retained in the Earth's gravitational field. The various layers, starting from sea level, are known as the *Troposphere* (up to about 14 km), *Stratosphere* (up to about 55 km), *Mesosphere* (up to about 80 km) and *Ionosphere* (up to over 150 km), which are separately defined. (3) A unit of pressure, viz. the *standard atmosphere*. See: *Atmosphere, standard*. (4) A cluster of impurity atoms (*a Cottrell atmosphere*) formed around dislocations in a solid. See also: *Cottrell atmosphere.*

ATMOSPHERE, COMPOSITION OF. The amounts of various gases in the Earth's atmosphere, usually expressed in parts per million of dry air by volume. In percentages the approximate values are nitrogen, 78%; oxygen, 21%; argon, 0·9%; carbon dioxide, 0·03%; other rare gases, 0·02%; the remaining 0·05% being made up of small amounts of methane, nitrous oxide, hydrogen and ozone.

ATMOSPHERE, RELATIVE DENSITY OF. The ratio of the atmospheric density at a given altitude to that at sea level.

ATMOSPHERE, STANDARD. A unit of pressure referring to sea level and a temperature of 0°C, and defined as the pressure exerted by a column of mercury 760 mm high, having a density of 13·5951 g/cm^3, the local value of the gravitational acceleration being taken as 980·665 cm/sec^2. This pressure is equal to 1 013 250 $dyne/cm^2$, i.e. 1013·250 mbar or 101 325 N/m^2 precisely. See also: *Atmospheric pressure. Normal temperature and pressure. Pressure. Appendix II.*

ATMOSPHERIC CIRCULATION. The broad pattern of wind velocity over the Earth, as a function of latitude, longitude and height.

ATMOSPHERIC CONVECTION. Usually refers to relatively small scale, approximately vertical, motions of the atmosphere which transfer heat, water vapour and other atmospheric constituents from one place to another under the influence of gravity. It is a special form of atmospheric turbulence.

ATMOSPHERIC CONVERGENCE. Used in meteorology and hydrodynamics generally to denote the accumulation of the mass of air within a given volume.

ATMOSPHERIC DIFFUSION. The processes by which heat and matter (e.g. water vapour, dust, natural and artificial radioactive matter, foreign gases, etc.) are transferred from one part of the atmosphere to another.

ATMOSPHERIC DIVERGENCE. Used in meteorology and hydrodynamics generally to denote the depletion of the mass of air within a given volume.

ATMOSPHERIC ELECTRICITY. A general term referring to the electrical charges existing in the Earth's atmosphere. These arise from ionization by radioactive material (in the Earth or in the atmosphere) and by cosmic rays, and their behaviour is dependent on the varying potential gradient between the positively charged ionosphere and the negatively charged Earth.

ATMOSPHERIC OPTICS. The study of the phenomena associated with the scattering, reflection, refraction and diffraction of light by the atmosphere or by particles (e.g. dust, water droplets, ice crystals, etc.) suspended in it.

ATMOSPHERIC POLLUTION. The presence of any undesirable substance in the open air. The term is also applied to the substance itself. Such a substance may obscure vision, may have a deleterious effect on terrestrial objects (including plants and animals), or may as grit or dust contaminate surfaces without any reaction.

ATMOSPHERIC PRESSURE. The pressure per unit area exerted by the atmosphere on an infinitesimal element of area taken at a given point. The meteorological unit is the *millibar* (10^3 $dyne/cm^2$ or 100 Pa) which is more convenient than the *bar* (10^6 $dyne/cm^2$ or 10^5 Pa). One standard atmosphere is, by definition, 1013·25 mbar or 101 325 Pa precisely. See also: *Atmosphere, standard. Appendix II.*

ATMOSPHERICS, RADIO. Short wave trains of electromagnetic radiation which arise from natural electrical disturbances in the atmosphere and interfere with radio reception. Also known as *static.*

ATMOSPHERIC SUBSIDENCE. A large-scale descent of air which is caused by the horizontal divergence of flow in the lower layers of the atmosphere or by the horizontal convergence in the upper.

ATMOSPHERIC TIDES. Oscillations of the atmosphere which are generated by the gravitational action of the Moon and Sun, or by solar thermal action.

ATOM. A unit of matter consisting of a single nucleus associated with one or more electrons. In the neutral state the number of these electrons is such as to balance the electrical charge on the nucleus, and the addition or removal of electron(s) results in the atom assuming a negative or positive state respectively (i. e. results in the formation of a negative or positive ion). The nature of the atom is, however, specified by the mass and charge of its nucleus rather than by its electrical state. The atom is the smallest particle of an element that can enter into chemical combination. See also: *Electron cloud. Nucleus* (1).

ATOM CORES, POLARIZATION OF. See: *Polarization.*

ATOMIC ABSORPTION COEFFICIENT. See: *Absorption coefficient for radiation.*

ATOMIC ABSORPTION SPECTROSCOPY. The study of the absorption spectra from excited atoms. See also: *Spectrum, absorption. Spectrum, atomic.*

ATOMIC BEAM FREQUENCY STANDARDS. Standards, based on the observation of magnetic hyperfine transitions by atomic beam techniques, for the measurement of frequency. A common type is the caesium beam frequency standard. Atomic frequency standards are commonly applied to transmissions of standard frequency and time. See also: *Second.*

ATOMIC BEAM: MOLECULAR BEAM. A narrowly defined stream of neutral atoms (molecules) moving through a highly evacuated enclosure, the distance between the atoms (molecules) both of the stream and of the surrounding space being so large that collisions or interatomic (intermolecular) forces can be neglected. Measurements with atomic and molecular beams have been used, among other things, for studying the kinetic properties of gases; the magnetic properties of atoms, molecules and nuclei; the hyperfine structure of spectra; and the interactions between gases and solids.

ATOMIC BOMB. See: *Nuclear explosives.*

ATOMIC CORE. An atom stripped of its valency electrons, i.e. whose electrons are all located in closed shells about the nucleus.

ATOMIC ENERGY. The popular name for *nuclear energy.*

ATOMIC ENERGY LEVELS. (1) The set of discrete energy states in which an isolated atom may exist. (2) The energy values associated with these states.

ATOMIC F-CURVE. For X-ray scattering: the curve which shows the variation of the atomic scattering factor, f, with $\sin \theta/\lambda$, where θ is the Bragg angle and λ the incident wavelength.

ATOMIC FORM FACTOR. The ratio of the atomic scattering factor to the atomic number. It approaches unity as the Bragg angle approaches zero.

ATOMIC HEAT. Of an element: the product of the atomic weight and the specific heat. For many solid elements at room temperature it is, according to Dulong and Petit's law, approximately $3R$, where R is the molar gas constant. The value of the atomic heat is thus about 25 J (or 6 cal) per degree.

ATOMIC MASS UNIT, UNIFIED. Defined as one-twelfth of the mass of the ^{12}C nuclide. It replaces the *atomic mass unit* (*a.m.u.*) which was related to the ^{16}O nuclide and the *atomic weight unit* (*a.w.u.*) which was related to the naturally occurring mixture of oxygen isotopes. It is greater than the a.m.u. by a factor of 1·000318 and than the a.w.u. by a factor of 1·000043. The numerical values of many of the fundamental physical constants must therefore be changed when expressed on the unified atomic mass scale. See also: *Amount of substance.*

ATOMIC NUMBER. Of an element: the number of protons in the atomic nucleus, sometimes known as the *proton number.* The product of the atomic number and the charge on the electron is the *nuclear charge.*

ATOMIC ORBITAL. A one-electron wave function describing the spatial behaviour of an electron of a given energy level in a particular atom. It is either a characteristic solution (eigenfunction) of the Schrödinger equation or an approximation thereto. Since the wave function represents the probability of finding an electron at a given point, or the distribution of its equivalent charge cloud, it can be regarded in a loose way as describing its orbit. The atomic orbitals can thus be regarded as the quantum mechanical counterparts of the orbits in Bohr's theory of atomic structure, hence the word orbital. See also: *Molecular orbital. Schrödinger equation.*

ATOMIC PHYSICS. The study of the physical properties of atoms when the atom is regarded as a whole, i.e. as a nucleus associated with which is a number of electrons. It is thus distinct from nuclear physics, which is the study of the nucleus alone, although there is inevitably some overlap.

ATOMIC RADIUS. Of an atom in a molecule or crystal: an imprecise quantity which may be considered basically as one-half of the internuclear distance between neighbouring atoms of the same kind. Its value in any specific instance depends on the nature of the atoms concerned and the types of bonding involved. Four main types of radius may be distinguished: the *ionic radius,* where neighbouring atoms consist of positive and negative ions; the *covalent radius,* where they are covalently bound to each other; the *metallic radius,* where they occur in a metallic crystal; and the *van der Waals radius,* where the two atoms are covalently bound to other atoms but not to each other.

ATOMIC RATIO. The ratio of quantities of different constituents in a given sample in terms of the numbers of atoms present, as opposed to the ratio of masses or volumes.

ATOMIC SCATTERING FACTOR. (1) For X-rays or electrons: the ratio of the amplitude of the wave scattered by an atom (equal for X-rays to ZA in the forward direction, where Z is the atomic number and A the amplitude of the incident wave) to that scattered by a single electron under the same conditions. It depends on the wavelength and the angle of scattering. See also: *Atomic f-curve*. (2) The scattering may also be expressed in terms of the distance from the centre of the atom to a point at which the amplitude of the scattered wave would be equal to that of the incident wave. For X-rays and electrons the atomic scattering factor is then the ratio of this distance to the corresponding distance for scattering by one electron. For neutrons it is stated absolutely as a length—the *scattering length*.

ATOMIC STOPPING POWER. See: *Stopping power*.

ATOMIC SUSCEPTIBILITY. The product of the atomic weight and the magnetic susceptibility per unit mass.

ATOMIC UNITS. A system of units employed in theoretical physics such that each of the universal constants e (electronic charge), m (electronic mass) and h (Planck's constant) has the value unity. The numerical values of these constants in any other system of units are then the conversion factors from atomic units to the system in question.

ATOMIC WEIGHT. Of a given specimen of an element: the mean weight of its atoms expressed on the unified atomic mass scale. See also: *Atomic mass unit, unified*.

ATOMIZATION. The breaking up of a stream of liquid into small particles or droplets to increase the specific surface.

ATOMS IN MOLECULES METHOD. For the study of molecular structure: the description of a molecular structure in terms of the isolated states of its constituent atoms, the process of molecular formation being regarded as a perturbation of the isolated atomic states.

ATTACHING GAS. A gas in which electron attachment occurs.

ATTACHMENT COEFFICIENT. For a swarm of electrons drifting under the influence of a uniform electric field through a gas in which electron attachment occurs: the chance that an electron will form a negative ion in unit distance of drift. See also: *Electron attachment*.

ATTACHMENT PROBABILITY. See: *Electron attachment*.

ATTEMPT FRÈQUENCY. In α-decay theory: the frequency with which an α-particle presents itself at the Gamow barrier.

ATTENUATION COEFFICIENT: ATTENUATION CONSTANT. For an electrical transmission line: the real part of the propagation coefficient.

ATTENUATION COEFFICIENT FOR RADIATION. The quantity α, in the expression $\alpha \Delta x$ for the fraction of a particular type of radiation removed by attenuation in passing through a thin layer of matter of thickness Δx. It is the attenuation coefficient of that matter for that type of radiation. It is a function of the energy of the radiation and the nature of the matter, and, according as the thickness is expressed in terms of length, mass per unit area, moles per unit area, or atoms per unit area, the attenuation coefficient is termed the *linear*, *mass*, *molar*, or *atomic* attenuation coefficient respectively. The attenuation coefficient is greater than the absorption coefficient. See also: *Absorption coefficient for radiation*.

ATTENUATION FACTOR FOR IONIZING RADIATION. For a given attenuating body in a given configuration: the factor by which a radiation quantity at some point of interest is reduced owing to the interposition of the body between the source of radiation and the point of interest.

ATTENUATION OF IONIZING RADIATION. The reduction of a radiation quantity upon the passage of radiation through matter, resulting from all types of interaction with matter. The radiation quantity may be, for example, the particle flux density or the energy flux density. The effect does not include the geometric reduction of the radiation quantity with distance from the source (*geometric attenuation*).

ATTENUATOR. An energy-absorbent material placed in a transmission system to absorb energy flowing in the system. See also: *Electrical attenuation. Electrical attenuator*.

AUDIOMETER. An instrument for measuring the acuity of hearing. Measurements may be made with speech signals or with tone signals.

AUFBAUPRINZIP. A scheme, due to Bohr, whereby the electron configuration of a neutral atom of nuclear charge Z was obtained from that of a neutral atom of nuclear charge $Z - 1$ by adding an electron of appropriate quantum number.

AUGER EFFECT. The emission of an electron (*Auger electron*), rather than an X-ray photon, from an atom after the removal of an electron from one of the inner shells. The effect may be considered as the internal conversion of the X-ray photon that would otherwise have been emitted. See also: *Fluorescence yield*.

AUGER EMISSION SPECTROSCOPY. The analysis of the atomic composition of materials from the distribution of the Auger electrons emitted by them, when irradiated by electrons of appropriate energy. See also: *Auger effect. X-ray fluorescence analysis.*

AUGER SHOWER. See: *Cosmic-ray shower.*

AUREOLE. See: *Coronae (meteorology).*

AURORAE. Atmospheric phenomena consisting of the appearance of luminous arcs, rays, streamers, etc., of green, red or yellow colour. They are caused by the entry into the high atmosphere of a stream of charged particles originating in the Sun, and are usually limited to two zones girdling the Earth at about 23° from the North and South magnetic poles. Since the light emission may extend to heights of some hundreds of kilometres, the lights may be observed at considerable distances as a dawn-like glow along the horizon, hence the name *aurora borealis* (northern dawn) and *aurora australis* (southern dawn). The aurorae are also known as *polar lights.*

AURORAL LINE. A green line of atomic oxygen at 5577 Å, first observed in the aurorae and later reproduced in the laboratory. Many other lines appear in aurorae and the term is becoming obsolete. (*Note:* 1 Å is equal to 10^{-10} m.)

AURORAL TRANSITIONS. Transitions giving rise to strong features of the auroral spectrum.

AURORAL ZONES. The regions in each hemisphere, situated above about magnetic latitude 67°, within which aurorae are of most frequent occurrence.

AUSTENITE. Originally a solid solution of carbon in γ-iron. Now describes any solid solution based on γ-iron.

AUTOCATALYSIS. Catalysis of a reaction by one of the products of that reaction.

AUTOCOLLIMATOR. An instrument for measuring to a high degree of precision small changes in the inclination of a reflecting surface. Also known as an *autocollimating telescope.*

AUTOELECTRONIC EMISSION. The emission of electrons from the surface of a conductor by the application of a sufficiently high voltage gradient (about 10^7 V/cm) at ordinary temperatures. See also: *Cold emission.*

AUTOIONIZATION. An effect analogous to the Auger effect but concerned with the optical rather than the X-ray energy levels of an atom.

AUTOMATIC FRACTION COLLECTOR. A device for the automatic collection of the effluent in chromatography as a series of fractions of more or less equal volume. It may be time-actuated or weight-actuated, or may depend on drop counting or on the direct collection of constant volumes.

AUTORADIOGRAPH. A photographic record of the radiation from radioactive material in an object (which may be of scientific, technological, medical or biological interest), made by placing the object in close apposition to a photographic film. From an autoradiograph the location and amount of radioactive materials in the object of interest may be determined.

AVALANCHE. (1) In meteorology: a hurtling mass of snow which descends a mountain side, carrying with it rocks or ice. (2) Of a burst of ions: all the ions produced from a single primary ion through the process of cumulative ionization. The term is applied mainly to counters but is sometimes used to refer to a cascade of electrons or nucleons in cosmic rays. See also: *Cosmic-ray cascade. Ionization, cumulative.*

AVALANCHE WIND. A wind, usually very strong, experienced ahead of an avalanche. It is probably produced by the direct thrust of the avalanche on the air ahead. Some contribution may also be made by the weight of the ice crystals suspended in the air.

AVERAGE LIFE: MEAN LIFE. The average lifetime for an atomic or nuclear system in a specified state. For an exponentially decaying system, it is the average time for the number of atoms or nuclei in a specified state to decrease by a factor of e (2·718 ...). This is $1/\ln 2$ times the half-life and, for a radioactive nuclide it is the reciprocal of the disintegration constant.

AVERAGE: MEAN. One of a small number of specific parameters used in summarizing statistical data, the most useful measure of which is the *arithmetic mean*, defined, for a series of values $x_1, x_2, ..., x_n$, as $(x_1 + x_2 + \cdots + x_n)/n$. In the case of a frequency distribution with f_i members in a class interval, each class having a mean x_i ($i = 1, 2, ..., n$) then the arithmetic mean is given by $(f_1 x_1 + f_2 x_2 + \cdots + f_n x_n)/N$, where $N = f_1 + f_2 + \cdots + f_n$. This also defines the *weighted mean* of the quantities x_i, the effect of the x_i then being proportional to the weights f_i, which allow for their relative reliability or importance. See also: *Moving average.*

AVOGADRO CONSTANT. The number of molecules (or ions, atoms, electrons etc., as specified) in a mole of substance. It is a fundamental physical constant, of magnitude $6 \cdot 022045$ ($\pm\ 0 \cdot 000031$) $\times$ 10^{23} mol $^{-1}$. Also known as *Avogadro's number* and in Germany as *Loschmidt's number*. See also: *Loschmidt number*. *Mole*. See: *Appendix III*.

AVOGADRO LAW. States that equal volumes of different gases at the same temperature and pressure contain equal numbers of molecules. Also known as *Avogadro's hypothesis*.

AVOGADRO NUMBER. The number of molecules (or ions, atoms, electrons etc., as specified) in a mole of substance. It is a fundamental physical constant, of magnitude $6 \cdot 0225 \times 10^{23}$ molecules per mole. Also known as *Avogadro's constant* and in Germany as *Loschmidt's number*. See also: *Loschmidt number*. *Mole*.

AXIAL-FLUX MOTOR. See: *Linear motor*.

AXIAL LENGTHS. In crystallography: the lengths of the edges of the unit of pattern or unit cell from which, by repetition in three dimensions, the regular structure of a crystal is built up. Together with the interaxial angles they define the shape and size of the fundamental repeating unit.

AXIAL RATIOS. In crystallography: the ratios of the lengths of pairs of edges of a unit cell. Such ratios will usually be the same as those defined from a macroscopic examination of the crystal faces, where such examination is possible.

AXON. See: *Nerve*.

AYRTON EQUATION. For a carbon arc: connects the current and voltage of a normal arc operating in the "quiet" zone, i.e. when there is no hissing noise.

AZEOTROPES. Constant boiling point mixtures obtained in the distillation of binary liquid mixtures which deviate from Raoult's law. It is not possible to separate such mixtures into the pure components by distillation; one pure component and a *constant-boiling mixture* are obtained.

AZIMUTH. (1) Of a celestial body: the arc of the horizon intercepted between the north or south point of the horizon, according as the observation is made in the northern or southern hemisphere, and the point where the meridian plane passing through the body cuts the horizon. (2) In general: a bearing or similar angle measured in a horizontal plane.

b

BABINET COMPENSATOR. A microscope accessory used between crossed Nicol prisms for measurements of double refraction in crystal plates. It consists of a calibrated pair of thin quartz wedges, cut with their edges respectively parallel and perpendicular to the optic axis and adjustable to give a parallel slab of variable thickness. The latter is adjusted to annul the displacement of interference figures brought about by the insertion of a crystal section.

BABINET PRINCIPLE. States that: complementary diffracting screens (i.e. screens for which the opaque portions of one correspond to the transparent portions of the other, and vice versa) produce identical intensity distributions.

BACK–GOUDSMIT EFFECT. The breakdown of the coupling between the nuclear spin angular momentum and the total angular momenta of the atomic electrons under the influence of a relatively small magnetic field. The effect alters the hyperfine structure pattern in both distribution and number of line components. See also: *Spectrum, hyperfine structure in.*

BACKGROUND. (1) A general term for unwanted effects in physical measurements above which a phenomenon must show itself in order to be measured. (2) Naturally occurring effects as opposed to those arising from human activities, e.g. as in *background radiation*, when considering the radiation to which mankind is exposed.

BACKING (METEOROLOGY). The anti-clockwise rotation of the wind direction. The opposite of veering.

BACKING VACUUM. In a vacuum pumping system: the pressure at the backing pump.

BACKLASH. The slackness or lost motion between two elements of an engineering mechanism, i.e. the amount which the first has to move before communicating its motion to the second. It may arise from wear or inaccurate machining.

BACK POINTER. See: *Beam direction indicator* (*radiotherapy*).

BACK REFLECTION METHOD. In X-ray diffraction: a technique in which the diffraction pattern is observed at Bragg angles approaching 90°, i.e. the diffracted rays travel back close to the incident beam. The method takes advantage of the high angular resolution attained as the Bragg angle approaches 90°.

BACK SCATTER. (1) The scattering of a beam of particles or radiation through angles greater than 90°. (2) The actual scattered particles or radiation. See also: *Forward scatter.*

BACK STREAMING. In a vacuum system: the continuous transfer of pumping fluid vapour from the pump to the vacuum system. Its effect can be reduced by the introduction of a cooled baffle.

BACKWARD DIODE. A semiconductor *p–n* junction diode closely related to the tunnel diode. It may be regarded as a conventional rectifying diode with substantially zero reverse breakdown voltage, the breakdown mechanism being tunnelling. See also: *Tunnel diode.*

BAFFLES AND BAFFLE VALVES. Physical obstructions to gas flow which are introduced to aid condensation or for the collection of particulate matter from gas streams.

BAILEY BEADS. A phenomenon observed during total solar eclipses and caused by the irregularities of the Moon's limb. At the approach of totality these irregularities are seen as small separate specks which have the appearance of a string of beads. The last visible bead heralds the beginning of totality with the appearance of the flash spectrum and the corona.

BALANCE. (1) An instrument used in weighing. A modern balance used for scientific measurements consists essentially of an equal-arm lever or beam of metal (brass or an aluminium alloy almost always) which turns about a central rigidly attached horizontal knife edge (usually of agate). The loads to be compared are applied at parallel knife edges rigidly attached at the ends of the beam, usually by means of appropriate pans. (2) A condition of partial or complete equilibrium or adjustment.

BALANCE, CHAIN. A balance in which the unwinding of a chain suspended from the beam is used for final balancing, as distinct from the movement of a rider along the beam.

BALANCE, DAMPED. A balance equipped with means for damping the movement of the suspended system, of particular application to balances provided with knob-controlled weight manipulation and read by means of an optically projected graticule. Air damping is usually employed but magnetic, electromagnetic and oil damping have all been used.

BALANCED LOAD. (1) In supply systems with two or more interconnected voltage sources (e.g. three-wire d.c. and polyphase a.c.), a load such that the currents supplied by each of the individual voltage sources are equal. In a.c. systems the voltage–current phase relationships are also equal. (2) A set of equal impedances used in an interconnected system.

BALANCE, FILM. An instrument, incorporating a torsion balance, for measuring the two-dimensional pressure of surface films spread on water or other liquids.

BALANCE, HYDROSTATIC. A balance used for determining the density of liquids from the loss of weight resulting from the immersion of a body of known volume.

BALANCE, MAGNETIC. A balance for the measurement of the small forces involved in the determination of para- and diamagnetic susceptibility. The choice of balance is decided by the accuracy required, the order of magnitude of the susceptibility, the temperature range, and the nature and amount of the material concerned. See also: *Curie–Chéveneau balance. Faraday balance. Gouy balance. Kapitza balance. Quincke balance. Rankine balance. Sucksmith ring balance.*

BALANCE, PRESSURE. A balance used to measure the pressure of a fluid (liquid or gas). In one type the pressure exerted by the fluid is balanced by a mechanical force. In another the unknown fluid pressure is transmitted to another fluid whose pressure can be measured.

BALANCE, RIDER FOR. A small weight used on the arm of a chemical balance for final adjustment.

BALANCE, SINGLE-PAN. A balance in which the second pan is avoided by having a sliding weight on a beam.

BALANCE, SORPTION. A balance for the gravimetric estimation of the quantity of material adsorbed on solids from the vapour phase.

BALANCE, SPRING. A balance which measures the load by the deflection of a calibrated spring.

BALANCE, TORSION. A very sensitive balance in which a small force (e.g. gravitational or magnetic) is determined from the twisting force set up in a wire or fibre. The force is caused to act at one end of a small beam suspended, in the case of a vertical balance, from the wire or fibre. The twisting force measured is that required to restore the beam to its original position when the external force is applied. Such vertical torsion balances are used chiefly for gravitational forces. Horizontal wires are used for microchemical weighing and for the measurement of magnetic susceptibilities. In the *Eötvos torsion balance*, used extensively for differential measurements in geogravitational surveying (e.g. for prospecting), two masses attached to a beam are spaced apart vertically as well as horizontally, permitting the measurement of the local distortion of the Earth's gravitational field arising from the presence of geological features such as hidden mineral deposits.

BALANCE WHEEL. A spring-controlled wheel, the frequency of oscillation of which is used for the measurement of time intervals (as in a watch or clock). The time of vibration of such a wheel depends on the stiffness of the spring and the moment of inertia of the wheel. It is independent of orientation or of the local value of gravity, but is markedly dependent on temperature unless special precautions are taken.

BALLISTICS. The study of the motion of projectiles. *Internal ballistics* is concerned with the phenomena occurring before a shell leaves the muzzle. *External ballistics* is concerned with the motion of a projectile after leaving the gun or launching ramp. For a conventional gun this is the period of free flight, but for a rocket it will include the motion while the rocket motor is still burning.

BALLISTO-CARDIOGRAPH. An instrument for recording gross mechanical movement of the body as a whole resulting from movements of the heart and blood. The oscillations of the body, which lies horizontally on a freely suspended table, are recorded as a series of waves which can be correlated with events in the cardiac cycle. The record obtained is known as a *ballisto-cardiogram*.

BALL MILL. A grinding mill in which materials are pulverized by collision with pebbles or metal balls and the shell of the mill.

BALLOON. A lighter-than-air aircraft consisting essentially of an envelope containing a light gas. It may have provision for housing crew and passengers but has no power plant. It may float freely in the air (a *free balloon*) or may be captive (a *kite balloon*).

BALMER CONTINUUM. In the spectra of Sun, stars and gaseous nebulae: the region of continuous absorption beyond the Balmer limit. It corresponds to the photoelectric detachment of electrons from the second level of the hydrogen atom.

BALMER DISCONTINUITY. An intensity discontinuity at the Balmer limit, particularly noticeable for B-stars. It is of importance in distinguishing between dwarfs and supergiants in early-type stars, and also for other studies of luminosity effects.

BALMER LIMIT. In the spectra of the Sun, stars and gaseous nebulae: the limit towards which the lines of the Balmer series are seen to crowd and beyond which they merge into a continuum. It corresponds to a wavelength of about 365 nm (3650 Å).

BALMER SERIES. A series in the line spectrum of the hydrogen atom. See also: *Hydrogen atom, line spectrum of.*

BANACH SPACE. One of the abstract spaces studied in the theory of functional analysis. The Hilbert spaces, of importance in quantum mechanics, represent special types of Banach space.

BAND HEAD. In molecular spectra: that edge of a spectral band which is sharply defined. On one side the intensity is zero but on the other the intensity falls off gradually. According as the intensity decreases towards the shorter or longer wavelengths, the band is said to be *degraded* (or *shaded*) to the violet or red. The wavelength of the band head is often important as a means of identification.

BAND-PASS FILTER. An optical or electrical filter which transmits a single band of frequencies. Such a filter may be classified as broad or narrow according to the range of frequencies transmitted. The frequency range within which transmission takes place without appreciable attenuation is termed the *pass-band.* See also: *Stop band.*

BAND THEORY OF SOLIDS. A theory, associated with the collective electron theory of solids, which postulates the existence of zones or bands of allowed electron energy levels. Various properties of solids have been explained in terms of these bands. See also: *Bloch theory. Brillouin zones. Collective-electron theory. Conduction band. Electron energy band. Ferromagnetism, band theory of.*

BAND WIDTH. Of an electrical device or circuit: the range of frequency within which the performance with respect to a given characteristic (e.g. the gain of a receiver or amplifier) falls between specified limits.

BAR. A unit of pressure, defined as 10^6 dyne/cm² (10^5 N/m²). A more convenient unit is the *millibar.* One standard atmosphere is, by definition, 1013·25 mbar precisely. See also: *Atmosphere, standard. Appendix II.*

BARBA LAW. States that similar test pieces, when strained to fracture in a tensile test, deform in a similar manner. Thus, for cylindrical test bars the same value for the elongation is obtained for a given material if the ratio of length to diameter is the same.

BARKER INDEX. The basis of a method of identification for crystalline substances from measurements of interfacial angles. The system offers a practical set of rules which ensure that independent crystallographers measuring crystals of the same substance will choose the same set of angles for identification. Crystals are listed in the various systems of symmetry.

BARKHAUSEN EFFECT. The occurrence of a series of discontinuous steps (*Barkhausen discontinuities*) in the magnetization of a ferromagnetic material as the magnetizing field is changed continuously. The discontinuities are produced by irreversible changes in the domain structure of the material. Their size may be increased by subjecting the specimen to suitable stresses.

BARKHAUSEN–KURZ OSCILLATOR. A high-frequency oscillator in which the frequency of oscillation depends mainly upon the electron-transit time within the tube.

BARLOW LENS. A negative lens used to decrease the convergence of the beam from the object glass of a telescope, and so to increase the effective focal length.

BARN. A unit of area used in expressing a nuclear cross-section. It is equal to 10^{-24} cm² (10^{-28} m²). This value was chosen since the cross-sections for all processes for most nuclei lie within the range 10^{-19} to 10^{-27} cm². See: *Appendix II.*

BARNETT EFFECT: MAGNETIZATION BY ROTATION. The magnetization of an initially unmagnetized specimen by the rotation of the specimen in the absence of any external magnetic field. The discovery of the effect by Barnett, in 1914, was the first successful experiment in the field of gyromagnetic phenomena. The inverse effect, the slight rotation of a suspended iron cylinder when suddenly magnetized, was discovered at about the same time by Einstein and de Haas. See also: *Einstein–de Haas effect: Rotation by magnetization.*

BAROCLINIC (METEOROLOGY). Of the atmosphere: denotes a characteristic condition of the atmosphere whereby a variation in temperature (and hence density) exists in an isobaric surface, leading to a variation of wind with height. When the density in an isobaric surface shows no variation, the atmosphere is said to be *barotropic*.

BAROGRAPH. A recording barometer, usually of the aneroid type.

BAROMETER. An instrument for measuring the pressure of the atmosphere. It can either be of the absolute type, in which the pressure is balanced directly against a column of mercury, or of the indirect type, in which the pressure is measured by the deflection it produces on a suitable elastic structure.

BAROMETER, ANEROID. A barometer depending for its action on the response of an evacuated capsule of thin corrugated metal to changes in atmospheric pressure. The relative movement of the end faces of the capsule as a result of such changes is transmitted to a pointer moving over a suitably graduated dial or to the recording pen of a barograph.

BAROMETER CONVENTIONS. Agreed conventions relating to the value of temperature and gravitational acceleration at which mercury barometers are adjusted to read directly in accepted pressure units. The conditions adopted by the World Meteorological Organization are: a standard instrumental temperature of 0°C and a standard value of g of 980·665 cm/sec^2 (9·80665 m/sec^2).

BAROMETER, FORTIN. A mercury barometer in which the mercury level in the reservoir is always brought to the same height before a reading of the main column is taken, thus permitting the use of a fixed scale.

BAROMETER, KEW PATTERN. A mercury barometer in which the scale attached to the tube is graduated to allow for the small changes in the level of the mercury in the reservoir which accompany changes in the level of the mercury in the tube, and thus permits the atmospheric pressure to be recorded by one measurement.

BAROMETER, NEWMAN. A portable movable scale barometer in which the zero of the scale is adjusted to the level of the mercury in the reservoir, instead of vice versa as in the Fortin barometer. A similar instrument which is not portable is known as the *observatory barometer*.

BAROMETRIC REDUCTION. In meteorology: the correction of barometric readings to a standard datum level, usually mean sea level, so as to render possible the comparison of pressure readings at stations located at various heights. See also: *Temperature, virtual*.

BAROMETRIC TENDENCY. The rate of change of atmospheric pressure with time, defined for synoptic meteorological purposes as the change in pressure over the 3 h period preceding the time of a standard hour of observation at a given place. The form of the tendency, e.g. "rising then steady", is also classified.

BAROSTAT. A device by means of which the pressure in an enclosure can be kept at a constant value.

BAROTROPIC (METEOROLOGY). See: *Baroclinic*.

BARRETER. A device for ensuring that the current in the circuit to which it is connected remains substantially constant over a wide range of voltage. It consists essentially of a thin iron wire filament enclosed in a glass envelope which contains hydrogen at reduced pressure, the resistance increasing proportionally with voltage. The voltage and current ranges within which the instrument will prove effective are governed by the dimensions of the iron wire filament.

BARRIER LAYER. A localized variation in electrostatic potential between two adjoining regions in which the potential is relatively uniform, which may be thought of as an electrical double layer. The term usually refers to barriers in solids but such layers may also be found in solid–liquid or solid–gas–liquid systems. The effect of a barrier layer is to restrict the flow of current carriers, and such layers at semiconductor surfaces or within semiconductors are responsible for rectification (*barrier-layer rectification*). Since barrier layers involve the storage of opposite charges, they possess self-capacitance: the *barrier-layer capacitance*. See also: *Photovoltaic cell*.

BARRIER LAYER PHOTOCELL. See: *Photovoltaic cell*.

BARRIER, NUCLEAR: BARRIER, POTENTIAL: BARRIER, GAMOW. The region of high potential energy through which a charged particle must pass on leaving or entering an atomic nucleus. See also: *Coulomb barrier*.

BARRIER WIDTH. The width of the barrier layer formed at the boundary between two conductors.

BARTLETT FORCE. One of the possible types of exchange force between nucleons. The Bartlett force changes sign if the spin coordinates are interchanged.

BARYON. A collective term for fermions with mass greater than or equal to that of a nucleon and sometimes for their anti-particles the *anti-baryons*.

BARYON NUMBER. The number of baryons minus the number of anti-baryons in a system. It is believed to be conserved in any conceivable process, i.e. only a baryon–anti-baryon pair may be created or annihilated.

BASAL PLANE. The plane normal to the unique axis in trigonal, tetragonal and hexagonal crystals. In general, a plane normal to the axis of a crystal prism or forming the base of a pyramid.

BASALTIC LAYER. The lower of the two layers that may occur in the Earth's crust, the upper layer being the *granitic layer*. The basaltic layer underlies the deep ocean basins to a thickness of about 10 km, but the granitic layer is absent. See also: *Earth, crust of.*

BASE. A substance having a tendency to gain a proton in a chemical reaction. See also: *Acid.*

BASE LINES (GEODESY). Survey lines, of which the lengths are very accurately measured, used as a basis for subsequent triangulation. Until fairly recently tapes or wires made of invar were used in measuring out base lines, but light or radar equipment is now being increasingly employed. See also: *Comparator, Väisälä. Geodimeter. Mekometer. Tellurometer. Triangulation.*

BASE PRESSURE. The suction acting on the base of a moving projectile and giving rise to drag (*base drag*). The magnitude varies appreciably with Reynolds number and Mach number.

BATHOCHROMIC SHIFT. The shift of an absorption band to a region of longer wavelength, thus giving visible colour to a previously colourless substance. The phenomenon is associated with the introduction of unsaturated linkages or groupings (known as *chromophores*) into the molecules concerned. The opposite effect, i.e. a shift to a shorter wavelength, is a *hypsochromic* shift.

BATTERY. Two or more electrical cells connected and used as a single unit, as in an accumulator. When it is connected to a load of substantially constant voltage which is just sufficient to keep the battery in a fully charged condition, it is known as a *floating battery*.

BAUSCHINGER EFFECT. The phenomenon whereby the mechanical properties of a metal for tensile loading are different from those for compressive loading as a result of its having been plastically deformed in tension and unloaded at a temperature at which it shows appreciable work hardening and very little recovery of mechanical properties.

BEAM COUPLING COEFFICIENT. A factor used in evaluating the interaction between charged particles and an electric field. It is a measure of the reduction in energy transfer caused by the finite transit time of the particle and by the spatial variation, if any, of the field.

BEAM CURRENT. The current represented by a stream of electrons in a cathode-ray tube, travelling wave tube, etc. It is controlled by the variation of the negative bias on an electrode close to the cathode.

BEAM DIRECTION INDICATOR (RADIO-THERAPY). A device which indicates the direction of radiation in the body. Such a device is the *back pointer*, a mechanical or optical system which indicates the point at which the axis of the beam emerges from the patient (*emergent ray point* or *exit point*).

BEAM DIVERGENCE. The angular spread of a beam of radiation which is nominally collimated. The *beam edge* is conventionally taken as the locus of positions at which the intensity is 10% of that along the axis.

BEAM EXTRACTOR. A magnetic or electrostatic device the operation of which allows the particles in a cyclic accelerator to leave the vacuum chamber at the end of their path.

BEAM FACTOR. Of a searchlight beam: the solid angle subtended at the source by the (circular) region over which the light intensity has dropped to one-half its value at the centre of the beam. It is a measure of the performance of the searchlight.

BEAM-FOIL SPECTROSCOPY. A technique, based on the excitation and subsequent analysis of fast heavy ions in a thin foil, for the investigation of atomic spectra, the energy levels of atoms and ions, mean lifetimes of excited states, and atomic fine-structure effects.

BEAM POWER VALVE: BEAM POWER TUBE: BEAM TUBE. An electron beam tube in which use is made of directed electron beams to contribute substantially to its power-handling capability.

BEAM SWITCHING: LOBE SWITCHING. In radar: a method of determining the direction of a target by successive comparisons of the signals corresponding to two or more beam directions differing slightly from the target direction.

BEARING. (1) The horizontal angle between the direction from an observer of an object of interest and a given reference direction. (2) A member used to support, guide or restrain moving elements of a machine. (3) That part of a beam, girder or similar structure that rests on the support.

BEARING PLANE. Of a beam balance: the plane in which, if the sensitivity of the balance is to be independent of the load, all three knife edges of the balance (two pan supports at the end of the beam and the central knife edge) should lie. In practice such a plane can be achieved only over a limited range of loads.

BEATS. A series of alternate maxima and minima in the amplitude of vibration produced by the interference of two wave trains having slightly different frequencies. The *beat frequency* is the difference between the two wave frequencies. Two ultrasonic "tones" may produce an audible *beat tone*. An analogous phenomenon at radio frequencies is responsible for the heterodyne effect.

BEAUFORT NOTATION. A notation of letters, with alternative symbols, used primarily in synoptic meteorology to represent the phenomena of weather.

BEAUFORT SCALE OF WIND FORCE. A nonlinear scale of wind speed used in synoptic meteorology and for nautical purposes. It was defined originally in terms of the amount of canvas that a man-of-war could carry, and later in terms of the effect of wind on, for example, waves and trees. If W is the wind speed in knots and B is the corresponding number on the Beaufort scale, then $W = 1.65 \sqrt{B^3}$ approximately.

BECKE LINE. The basis of a method, due to Becke, for determining the relative refractive indices of two minerals (or a mineral and its mounting) in contact in a microscope section. A narrow line of light appears at the junction (on account of refraction and total internal reflection) and, as the microscope objective is raised, this line travels into the medium of higher refractive index.

BECQUEREL. The SI unit of radioactivity, which is to replace the curie. It is equal to one disintegration per second, hence 1 curie is equal to 3.7×10^{10} becquerel. See: *Appendix II*.

BECQUEREL EFFECT. The production of a potential difference by the illumination of one of two similar electrodes (of metal, metallic oxide, halide or sulphide) immersed in a suitable electrolyte, or of metal electrodes with materials such as selenium sandwiched between them. The effect is a special case of the photovoltaic effect. See also: *Photovoltaic cell*.

BEER–BOUGUER LAW OF ABSORPTION. See: *Bouguer–Lambert law of absorption: Bouguer–Beer law*.

BEER LAW. States that the absorption of light in a solution may be expressed as $I = I_0 e^{-kct}$, where I_0 is the incident and I the transmitted intensity, k is the absorption coefficient for unit concentration, c is the concentration and t is the thickness of solution through which the light is transmitted.

BEEVERS–LIPSON STRIPS. Tables of cosine and sine functions set out on strips of card and used in the evaluation of multidimensional Fourier series for crystal structure analysis. Their use reduces the evaluation of such a series to that of a succession of one-dimensional series.

BEILBY LAYER. A microcrystalline or amorphous layer formed on the surface of metals by polishing, and having physical properties different from those of the bulk metal. The depth of the layer is between 50 and 200 Å, but there is no sharp dividing line between the layer and the unaffected bulk metal. (*Note:* $1 \text{ Å} = 10^{-10}$ m.)

BEL. A logarithmic unit used for expressing power ratios. The number of bels corresponding to a power ratio W is $\log_{10} W$. It is usually more convenient to use one-tenth of this unit, the *decibel*. The power ratio W will then correspond to $10 \log_{10} W$ decibels. The use of bels and decibels has also been extended to cover intensity ratios.

BELL. A sound-emitting metal device, operated by striking. It can be considered theoretically as a specialized form either of a tube closed at one end or of a suitably bent metal plate. The sound emitted is characterized by overtones which vary slightly during the decay of the sound and are not exact multiples of the lowest or *hum* tone.

BÉNARD CELLS. See: *Convection, cellular*.

BENCH MARKS (SURVEYING). Marks inscribed on a building, pavement or special structure, to indicate points whose heights above some assumed datum have been accurately determined.

BENDING MOMENT. Of an elastic beam, at any imaginary transverse section: the algebraic sum of the moments of all the forces to either side of the section, i.e. the resultant couple of the forces involved. The bending moment, M, is related to the radius of curvature, R, of the central line by the *Bernouilli–Euler* law: $M = EI/R$, where E is Young's modulus and I is the moment of inertia of the cross-section of the beam about an axis which is normal to the plane of bending and passes through the central line.

BENDING, PURE. Of an elastic beam: bending

arising from the application of a pair of equal and opposite couples at each end of the beam.

BEND PLANE. See: *Polygonization.*

BENZENOID. Descriptive of all compounds containing aromatic six-membered rings, and thus related to benzene.

BERNAL CHART. A chart used for indexing X-ray diffraction photographs from single crystals. From such a chart may be read the axial and radial cylindrical coordinates of that point in reciprocal space which corresponds to any particular X-ray reflection.

BERNOUILLI EQUATION. (1) For an inviscid fluid in steady flow, where the only body forces are gravitational: states that, along a streamline,
$\dfrac{p}{\varrho} + \dfrac{v^2}{2} + gh$ = constant where p, ϱ and v are the pressure, density and velocity respectively, g is the gravitational acceleration and h is the height above a fixed datum. Where the body forces are due partly to effects other than gravity (e.g. electromagnetic forces) the term gh should be replaced by the potential energy per unit mass. (2) A first order non-linear differential equation of the type
$\dfrac{dy}{dx} + f(x)\, y = r(x)\, y^n$. It may be rendered linear by the substitution $t = y^{1-n}$, giving
$\dfrac{dt}{dx} + (1-n) ft = (1-n) r$.

BERNOUILLI–EULER LAW. See: *Bending moment.*

BERTHELOT METHOD. For measuring the latent heat of vaporization of a liquid: a method depending on the determination of the rise in temperature of a water bath inside which a given amount of vapour is condensed.

BERTHELOT PRINCIPLE. States that a chemical reaction can only occur spontaneously if accompanied by an evolution of heat. There are numerous exceptions.

BERTRAND LENS. An auxiliary lens often present in a polarizing microscope which, when in use, permits the lens system to be focused at infinity.

BESSEL EQUATION. The name given to the second-order linear ordinary differential equation

$$x^2 \frac{d^2 y}{dx^2} + x \frac{dy}{dx} + (x^2 - v^2)\, y = 0$$

where v is a constant. Both x and v may be complex. The solutions of this equation are known as *Bessel*

functions. Any multiple of the function $J_v(x)$ defined by the series

$$J_v(x) = \sum_{m=0}^{\infty} \frac{(-1)^m\, (z/2)^{v+2m}}{m!\, \Gamma(v + m + 1)}$$

is a solution of Bessel's equation. The function $J_v(x)$ is called a *Bessel function of order v and argument x of the first kind.* Bessel functions of the second kind are known as *Neumann functions*, or *Weber's Bessel functions of the second kind*, and those of the third kind *Hankel functions.*

BESSEL FUNCTION, MODIFIED. A Bessel function with an imaginary argument.

BESSEL FUNCTION, SPHERICAL. A Bessel function of the first kind whose order is half of an odd integer (either positive or negative).

BESSELIAN YEAR. That tropical year which begins when the mean longitude of the Sun is exactly 280° (18 h 40 m), this instant being near to the beginning of the civil year. See also: *Year.*

BE STAR. An irregular variable star of spectral type B (but occasionally O or A) exhibiting hydrogen emission lines in its spectrum.

BETA-DECAY. Radioactive decay in which a β-particle is emitted or in which orbital electron capture occurs. This changes the atomic number of the nucleus concerned by plus or minus one, but does not change its mass number. The *Fermi theory of β-decay* involves the assumption of the creation of leptons in the decay process, the specification of appropriate selection rules, and the conservation of leptons. It gives the probability per unit time of the ejection of a β-particle whose energy lies between E and $E + dE$.

BETA-DISPERSION. Of body tissues. See: *Body tissues, dielectric dispersion in.*

BETA FILTER: K-BETA FILTER. A filter used in X-ray diffraction studies to remove essentially all the $K\beta$ radiation from a beam of characteristic K X-rays. It usually takes the form of a foil of an element which has an absorption edge between the $K\alpha$ and $K\beta$ wavelengths.

BETA-FUNCTION. This is defined as $B(m, n)$ $= \int_0^1 x^{m-1}(1 - x)^{n-1}\, dx$ where m and n are real numbers greater than zero. It can be expressed in terms of the gamma-function by the equation $B(p, q)$ $= \dfrac{\Gamma(b)\,\Gamma(q)}{\Gamma(p + q)}$. The beta-function is also called an *Eulerian integral of the first kind.* See also: *Gamma-function.*

signifies radiation composed of β-particles. See also: *Electron*.

BETA-PARTICLE. An electron, of either positive charge (β^+) or negative charge (β^-), which has been emitted by an atomic nucleus or neutron in the process of a transformation. The term *beta-rays*

BETA-RAY SPECTROMETER. An instrument designed for measuring the distribution of energy and momentum in β-ray spectra, usually by counting the β-particles after deflection and focusing by a magnetic or electric field. The particles are focused into a momentum spectrum on to a suitably placed photographic plate, or may be detected by proportional or scintillation counters.

BETA-RAY SPECTROMETER, TRANSMISSION FACTOR OF. For a given β-ray spectrometer and β-particles of a given energy: the factor by which a spectrometer fails to measure the number of β-particles of the energy in question.

BETA-RAY SPECTRUM. The distribution in energy or momentum of the β-particles emitted in a β-decay process. The spectrum is continuous from zero to a maximum equal to the nuclear energy change in the transition. Sharp lines, superimposed on the continuous spectrum, arise from internal conversion electrons and do not constitute a part of the β-ray spectrum proper.

BETATRON. A particle accelerator in which electrons are accelerated by magnetic induction. The electrons travel in approximately circular orbits and are accelerated by the electric field produced when the magnetic flux within the orbit is increased. The magnetic field is arranged to increase as the energy of the electron increases, in such a way as to keep the electrons on a fixed orbit. The practical upper limit of energy attainable by a betatron, which is set by the radiation losses of the electrons, is about 500 MeV. Betatrons are used for nuclear physics research, radiography and radiotherapy. See also: *Particle accelerator*.

BETATRON OSCILLATIONS. Oscillations of electrons about the "fixed orbit" in a betatron. Similar oscillations will occur in any particle accelerator which has an equilibrium orbit, e.g. in a cyclotron with slight orbit spiralling.

BETHE–BLOCH FORMULA. For the derivation of the linear stopping power, S_l, of a material for a fast charged particle: states that $S_l = \dfrac{4\pi e^4 z^2}{m_0 V^2}\, nB$, where ze is the charge on the particle, m_0 is the mass of the electron, $V(=\beta c)$ is the velocity of the particle, n is the number of atoms per cm^3 of sub-

stance, c is the speed of light, and B, the *stopping number*, is given by
$$Z\left[\ln\frac{2m_0 V^2}{I} - \ln(1-\beta^2) - \beta^2 - \frac{C}{Z}\right],$$
where Z is the atomic number of the stopping material, I is the average excitation potential of the material and C is a correction term for the ineffectiveness of the K electrons due to polarization (the *density effect*). See also: *Stopping power*.

BETHE–HEITLER THEORY. For the energy loss of charged particles: the theory uses Dirac's equation for the electron and Born's approximation to obtain the interaction of the electron with the field of the nucleus and leads to expressions for the radiative scattering cross-sections, and hence the rate of energy loss with distance. According to this theory this rate depends on the extent to which the nucleus is screened by the atomic electrons and on the atomic number of the scattering medium, and is proportional to the electron energy.

BETHE–SLATER CURVE. A plot of the exchange energy against the ratio of the interatomic distance to the calculated radius of the $3d$ shell for the transition elements, this ratio being a measure of the degree of overlap of the $3d$ wave functions of the separate atoms. A large positive exchange interaction implies the existence of ferromagnetism.

BETTI RECIPROCAL THEOREM. States that, if an elastic body is subjected to two systems of body and surface forces, then the work that would be done by the first system of forces in acting through the displacements due to the second system is equal to the work that would be done by the second system acting through the displacements due to the first. The theorem finds numerous applications in the theory of elasticity.

BEVATRON. The name given to the 6·2 GeV proton synchrotron at the Lawrence Radiation Laboratory in California—from BeV (billion electron volt). [*Note:* the U.S. billion is equal to 10^9, for which the internationally agreed prefix is now giga (symbol G).]

BH CURVE. A curve showing the relation between the induction (**B**) in a magnetic specimen and the magnetic field (**H**) in which the specimen is placed. See also: *Magnetization curves*.

BIAS VOLTAGE. Voltage applied to an electrode whereby its influence on the flow of current between two other electrodes is significantly affected. The most familiar case is that of the grid bias in a thermionic tube. It is also used of the voltage applied to a control electrode in a solid state device, where current flow is controlled by carrier injection.

BIAXIAL CRYSTAL. A crystal for which there are two different directions, the two optic axes, along which light travels with a single velocity, i.e. with no double refraction. Crystals of orthorhombic, monoclinic and triclinic symmetry are of this type, and such crystals possess three principal refractive indices.

BIFILAR MICROMETER. An instrument attached to the eye end of a telescope for measuring the angular separation and orientation (position angle) of a visual double star. It comprises two parallel wires, each movable by micrometer head screws (for the separation measurement), and a third fixed wire (for aligning the two stars) perpendicular to the other two, the whole being rotatable (for the orientation measurement).

BIFILAR SUSPENSION. The suspension of a body by two parallel vertical wires or threads, which give a considerable controlling torque. If the wires are of length x and are distance d apart, the period of torsional vibration of a suspended body of moment of inertia I and mass m is $T = 4\pi \sqrt{\dfrac{Ix}{mgd^2}}$, where g is the gravitational acceleration.

BIFILAR WINDING. A type of winding used extensively in wire-wound low inductance resistors, and sometimes in high-frequency electronic circuits, to isolate the capacitance to primary and earth of filament transformers from directly heated cathodes. It consists of two identical wires wound closely together side by side on a former made of insulating material, the currents in the two wires flowing in opposite directions.

BIHARMONIC EQUATION. An equation arising in the study of hydrodynamics and elasticity. It may be stated as

$$\nabla^4 \phi \equiv \left(\frac{\partial^2}{\partial x^2} + \frac{\partial^2}{\partial y^2} + \frac{\partial^2}{\partial z^2} \right)^2 \phi = 0.$$

BILATERAL SLIT. A type of precision slit used in spectrometers and spectrographs in which two metal strips are so mounted as to be movable symmetrically to or from each other by the rotation of a fine precision screw, the distance between them capable of being read with great accuracy (0·001 mm).

BILBY TOWER. A steel tower on which a theodolite or triangulation beacon can be raised 30 m or more above the ground. It can be erected and dismantled in a few hours and is easily transportable.

BILINEAR FORM. A polynomial in two sets of variables, which is linear and homogeneous in each.

BILINEAR TRANSFORMATION. A transformation from one real or complex variable to another such that the equation relating the two variables is linear in each variable.

BIMETALLIC STRIP. A composite strip consisting of two metals or alloys with different coefficients of thermal expansion, which are usually roll-welded together. Such strips bend when the temperature changes and are used to indicate temperature, to operate thermostats and, in conjunction with contactors, to protect electric motors against overloading.

BIMIRROR: FRESNEL MIRROR. A pair of plane mirrors slightly inclined to one another, used for producing two coherent images in interference experiments.

BIMOLECULAR REACTION. An elementary reaction between two molecules or atoms. If, in the whole of a chemical reaction of which the elementary reaction forms a step, the overall rate is determined by that step, the entire reaction is also termed bimolecular.

BINARY ENGINE. A heat engine using two different working substances for the high- and low-temperature ranges of the cycle respectively.

BINAURAL LOCATION OF SOUND. The location of sound by the use of two ears. At low frequencies this is believed to be done through the difference between the times of arrival at the two ears, and at high frequencies through the difference of intensity; but a complete theory is still lacking.

BINAURAL PRESENTATION: BINAURAL HEARING. The simultaneous reception of a message or signal by both ears. See also: *Dichotic presentation: Dichotic listening.*

BINDING ENERGY. (1) For a particle in a system (e.g. a particle in a nucleus or an electron in an atom): the net energy required to remove it from the system. Sometimes called *separation energy.* (2) For a system (e.g. a nucleus): the net energy required to decompose it into its constituent particles.

BINDING FRACTION. For a given nucleus: the average binding energy per nucleon, i.e. the total binding energy divided by the mass number.

BINOCULARS. A pair of telescopes for use with both eyes simultaneously, usually with focusing tubes controlled by a common screw adjustment. In· *prism binoculars* each telescope consists of an objective, an eyepiece and some form of prism system to invert and reverse the image, and to shorten the tube length as compared with an ordinary telescope of equal power. In the *opera glass* two Galilean telescopes are employed, with correspondingly small magnifications. No real image is formed so that the image is not inverted and no prism system is required to produce an erect image.

BINOMIAL COEFFICIENTS. The coefficients which occur in the statement of the binomial theorem. They are usually denoted by the symbol $\binom{n}{r}$. If n and r are positive integers and $n > r$, then $\binom{n}{r} = \dfrac{n!}{r!\,(n-r)!}$, but if n is not a positive integer

$$\binom{n}{r} = \frac{n(n-1)\cdots(n-r+1)}{r!}.$$

BINOMIAL DISTRIBUTION. The probability distribution that arises when we have a series of n trials and the probability p of success is the same for each trial. The probabilities of $0, 1, 2, \cdots, n$ successes, in a series of n trials, are then the respective terms of the binomial expansion $(q + p)^n = q^n + nq^{n-1}p + \cdots + nqp^{n-1} + p^n$, where $q(=1-p)$ is the probability of failure.

BINOMIAL THEOREM. This states that, if n is a positive integer, $(a + b)^n = \sum\limits_{r=0}^{r} \dfrac{n!}{r!\,(n-r)!}\,a^{n-r}b^r$, and that, if n is not a positive integer, the series $\sum\limits_{r=0}^{\infty} \dfrac{n(n-1)\cdots(n-r+1)}{r!}\,x^r$ converges to the sum $(1 + x)$ provided $-1 < x < 1$.

BIOCHEMISTRY. The study of chemical reactions that can occur in living matter. It may conveniently be divided into *physiological biochemistry* (concerning normal organisms), *pathological biochemistry* (concerning diseased organisms) and *pharmacological biochemistry* (concerning organisms into which an abnormal substance or an excess of a normal substance has been introduced).

BIOCOLLOIDS. All naturally occurring organic macromolecular substances and certain micellar or low-molecular-weight colloids.

BIOELECTRICITY. Electricity of biological origin, from nerves, muscles, etc. A number of vital activities at cellular and higher levels are largely controlled by the electrical properties of living material, the sources of electrical potential in such material being mainly liquid junctions between different solutions; phase boundaries; different solutions separated by permeable or semi-permeable membranes; oxidation-reduction systems; and systems involving the relative movement of solid and liquid phases. Structural potential gradients, as from films of oriented electric dipoles, also occur.

BIOELECTRIC POTENTIALS. Electrical potentials which are detectable in the body. A wide variety of sites are the seat of such potentials, e.g. the heart, the muscles, the brain and the eye. Diagnostic tests known as electrocardiography, electromyography, electroencephalography and electroretinography respectively, have been developed in these instances.

BIOLOGICAL ENGINEERING. The collaborative application of engineering, physics and mathematics in medicine and biology, as in the development of the cardiac pacemaker, heart–lung machines and radio pills.

BIOLOGICAL STIMULATION. The production of the characteristic activity of a biological tissue by a sufficiently intense change of environment. The change may be mechanical, chemical, thermal or electrical.

BIOLUMINESCENCE. The production of light by living organisms, such as glow-worms, some bacteria and some fungi. Compounds having similar properties to bioluminescent compounds can now be produced synthetically.

BIOMECHANICS. The study of those properties of living systems which enable them to exhibit coordinated movements.

BIOMETRY. The application of statistical methods to biological measurements.

BIOPHYSICS. The application of the methods and principles of physics to the investigation of biological systems. It may be concerned with the physics of biological systems themselves (e.g. the physical basis of colour vision or the physics of the nervous system), with the biological effects of physical agents (e.g. ionizing radiation), or with the use of physical methods in the study of biological problems. See also: *Health physics. Hospital physics. Medical physics.*

BIORHEOLOGY. The rheological study of body fluids (e.g. blood, synovial fluid, mucus, cerebrospinal fluid, intra-ocular fluid), muscle, bone, etc.

BIOT. The constant current which, if maintained in two straight parallel conductors of infinite length, of negligible circular section, and placed 1 cm apart *in vacuo*, would produce between these conductors a force of 2 dyne/cm length. It is the unit of current in the CGS electromagnetic system of units, and is equal to 10 A, i.e. to 1 *abampere*.

BIOT LAWS. State that the rotation produced by an optically active medium is proportional to the length of the light path in the medium, to the concentration (for solutions), and to the inverse square of the wavelength of the light. The last is only approximately true.

BIOT–SAVART LAW. States that the magnetic field due to a current flowing in a long straight conductor is directly proportional to the current and inversely proportional to the distance of the point of observation from the conductor.

BIPLATE. (1) Two plane parallel pieces of glass cemented together with a slight angle between them, and used for producing a double image of a slit for interference experiments. (2) Two half-wave plates of birefringent material cemented together with an angle of 90° between their axes and used for detecting traces of optical polarization. See also: *Bravais biplate*.

BIPRISM. (1) *Fresnel biprism:* a single prism of very obtuse angle that can be considered as two prisms of very acute angle placed base to base. It is used for producing two coherent images from the same source. (2) *Hüfner rhomb:* a single rhomb of glass that can be considered as two prisms of large angle placed base to base. It is used for bringing two separated beams of light into juxtaposition for photometric comparison.

BIQUARTZ. A sensitive analyser for saccharimeters formed from two pieces of quartz, one laevorotatory and the other dextrorotatory, cemented side by side and used in conjunction with a Nicol prism. The thickness of the quartz piece in each case is such that it will rotate the plane of polarization of yellow light through 90°.

BIRADICAL: DIRADICAL. A molecule with an even number of electrons, of which two are unpaired.

BIREFRINGENCE: DOUBLE-REFRACTION. The property, exhibited by anisotropic crystals, of possessing two refractive indices, i.e. of allowing light to pass through them with two different velocities. A ray of light, passing into such a crystal will, unless it is parallel to certain directions (*optic axes*), be split into two rays which pursue different paths and are polarized at right angles. Only one of these, the *ordinary ray*, obeys Snell's law. The other is known as the *extraordinary ray*. The corresponding *refractive indices* are also designated as *ordinary* and *extraordinary* respectively. See also: *Form birefringence*.

BIRGE–MECKE RULE. A semiquantitative relationship between certain molecular constants of the various electronic states of a diatomic molecule. It states that ω_e/I_e or $\omega_e r_e^2$ is constant, where ω_e is the equilibrium vibrational frequency, I_e is the moment of inertia and r_e is the equilibrium internuclear distance. See also: *Clark rule. Morse rule*.

BIRGE–SPONER EXTRAPOLATION. A method of determining dissociation limits for diatomic molecules when the convergence limit of vibrational bands cannot be directly observed. It is based on the assumption that the vibrational energy levels converge to a limit at a finite value of the vibrational quantum number, which is justified only when the force between the nuclei at a large internuclear distance varies as a power of the reciprocal of that distance greater than the cube. See also: *Dissociation energy. Dissociation limit*.

BISECTRIX. Of a biaxial crystal: the bisector of the angle between the optic axes. The bisector of the acute angle is the *acute bisectrix* and that of the obtuse angle the *obtuse bisectrix*.

BIT. In computer terminology: an abbreviation for *binary digit*, the smallest unit of information, i.e. 1 or 0.

BITTER FIGURE. A pattern showing magnetic domains, obtained by spreading a film of a colloidal suspension of magnetite particles over a strain-free surface of a ferromagnetic material. The magnetite particles migrate to points where the magnetic field has a local maximum, and thus reveal the underlying domain structure. The pattern may be observed under a microscope. It is also known as a *magnetic powder pattern*.

BLACK BODY. One which absorbs completely any heat or light radiation reaching it and reflects none. It remains in equilibrium with the radiation reaching and leaving it, and at a given steady temperature emits radiation (*black-body radiation*) with a flux density and spectral energy distribution which are characteristic of that temperature and is described by Planck's radiation formula. In practice no substance is an ideal black body and its nearest approach is a uniform temperature enclosure having a small hole or slit. See also: *Radiation, laws of*.

BLACK-BODY TEMPERATURE. The temperature of a body as measured by a radiation pyrometer. It is usually appreciably less than the true temperature of the body. See also: *Brightness temperature: Luminance temperature*.

BLACK HOLE. In space: the result of the complete gravitational collapse of a celestial body. Within such a hole space-time is so strongly curved that no signal of any kind can escape. There is evidence that the secondary component of the binary star system ε Aurigae is a black hole. See also: *Neutron star. Schwartzchild radius. Stellar evolution. White hole*.

BLASIUS SOLUTION. The solution of the differential equation $F'''(\eta) + F(\eta) F''(\eta) = 0$ for which

$F(0) = F'(0) = 0$, $F'(\infty) = 1$. It gives the velocity distribution in the laminar boundary layer of a fixed semi-infinite flat plate in an otherwise uniform stream.

BLAZE. The flat side of a groove in a diffraction grating.

BLOCH FUNCTIONS. Solutions of the Schrödinger equation with a periodic potential (e.g. for an electron moving in a crystal lattice). They are of the form $\psi_k = u_k(\mathbf{r})\, e^{i\mathbf{k}\cdot\mathbf{r}}$, where $\mathbf{r}$ is the position vector, with components x, y, z; $u_k(\mathbf{r})$ is a function which is periodic in x, y, z with the periodicity of the potential, i.e. with the period of the lattice; and $\mathbf{k}$ is a wave number vector which may be defined in terms of the reciprocal lattice of the crystal.

BLOCH–NORDSIECK METHOD. A quantum method for removing the theoretical infinity in the emission of low-frequency radiation by a charged particle, which appears as a consequence of classical electrodynamics and is known as the *infrared catastrophe.*

BLOCH THEORY. A theory allowing for the effect on the motion of free electrons of the periodic field of a crystal lattice and thus permitting a distinction between metal and non-metal in the electron gas concept. It is based on the use of a wave function which introduces the periodicity of the lattice by modulating the original plane wave function. See also: *Bloch functions.*

BLOCH WALL. The transition layer, which may be as great as $1 \cdot 2\ \mu m$, which separates two domains magnetized in different directions and in which the change in spin direction between the two domains is assumed to take place gradually. Also known as the *domain boundary.*

BLOCKING. Of a beam of protons or other ions: the prevention of the beam from leaving a crystal in the direction of a close-packed row or along a densely populated plane of atoms, owing to the obstacle presented by such a row or plane. See also: *Proton scattering microscopy.*

BLONDEL–REY LAW. Gives an expression for the apparent point brilliance, B, of a flashing light in terms of the point brilliance, B_0, during a flash of duration t seconds. It states that $B = B_0[t/(a + t)]$ where a is a constant lying between $0 \cdot 1$ and $0 \cdot 3$ s, and equal to about $0 \cdot 2$ s when B is near the threshold for white light. If t is very small the expression approximates to $B = B_0 t/a$. The law applies up to about 5 flashes per second, with flashes of approximately $0 \cdot 1$ s duration or less. See also: *Point brilliance.*

BLOOD RAIN. Rain which is red in colour due to the capture by the raindrops of red dust carried in the air in which the raindrops form or in the air below this. The phenomenon is most likely to occur near desert or arid regions.

BLOOMING. Of a glass surface: the application of an *anti-reflection coating* to reduce the amount of light reflected at the surface.

BLUFF BODY. A body of such a shape that, when it is placed in a uniform stream, separation of fluid flow occurs well before the downstream end of the body.

BODE LAW. An empirical relation giving the mean distances of the planets from the Sun, being based on the series 0, 3, 6, 12, 24 . . . If 4 is added to each number in this series the resulting figures, divided by 10, give quite closely the mean distances in Astronomical Units up to Uranus. The law is also known as the *Titius-Bode law.*

BODY-CENTRED CUBIC STRUCTURE. A type of crystal structure in which the atoms are located at the corners and centre of a cubic unit cell.

BODY-CENTRED LATTICE. See: *Space-lattice: Bravais lattice.*

BODY-TISSUES, DIELECTRIC DISPERSION IN. The variation of dielectric constant with frequency. *Alpha-dispersion* occurs in cellular membranes on account of a Maxwell–Wagner type of relaxation mechanism, with a relaxation frequency less than 1 kHz. *Beta-dispersion* occurs in the megacycle region and is attributed to the breakdown of the cellular units and to the relaxation of the protein molecules. *Gamma-dispersion* is associated with the water in the tissues and follows closely the behaviour of pure water. It occurs at very high frequencies, with a relaxation wavelength of about $1 \cdot 7$ cm. See also: *Maxwell–Wagner effect.*

BOHNENBERGER EYEPIECE. An eyepiece, containing an unsilvered 45° flat glass plate, used in the measurement of the level error of a meridian circle.

BOHR FREQUENCY CONDITION. Gives the frequency, ν, associated with the quantum of radiation emitted by an atom making a transition from energy E to a lower energy E_0 as $E - E_0 = h\nu$, where h is Planck's constant. The reverse condition holds for absorption.

BOHR MAGNETON. The unit of atomic magnetic moment, denoted by β or μ_B. It is equal to $eh/4\pi mc$, where e is the electronic charge, h is Planck's constant, m is the electron rest mass and c is the speed of light. It is the moment of a single electron spin and its value is about $9 \cdot 27 \times 10^{-21}$ erg/gauss or $9 \cdot 27 \times 10^{-24}$ JT^{-1}. See also: *Nuclear magneton. Appendix III.*

BOHR MAGNETON NUMBER. The magnetic moment per atom expressed in Bohr magnetons, usually denoted by n_0.

BOHR ORBIT. One of the electron orbits permitted by Bohr's theory of the atom, in which the electron has an orbital angular momentum which is an integral multiple of $h/2\pi$, where h is Planck's constant.

BOHR RADIUS. The radius of the smallest electron orbit in Bohr's model of the hydrogen atom. It is equal to $h^2/4\pi^2 me^2$, where h is Planck's constant, m is the rest mass of the electron and e the electronic charge. It has a value of 0.53×10^{-8} cm.

BOHR THEORY. A combination of Rutherford's concept of the atom as a central, positively charged nucleus surrounded by planetary electrons, with the ideas of the quantum theory. The possible electron orbits are restricted to those whose orbital angular momentum is an integral multiple of $h/2\pi$, where h is Planck's constant. The jump of an electron from its orbit to one of a smaller radius is accompanied by the emission of electromagnetic radiation, one quantum of which (i.e. $h\nu$, where ν is the frequency) is equal to the difference in energy between the two orbits concerned.

BOILING. The conversion of liquid to vapour, accompanied by the formation of bubbles. The term is usually restricted to instances in which the vapour pressure of the liquid is equal to, or greater than, the atmospheric pressure above the liquid.

BOILING POINT. The temperature at which a liquid boils in an open vessel, i.e. at which the saturation vapour pressure equals the pressure of the atmosphere. See also: *Ebullioscopic constant*.

BOLIDE. Another name for a fireball, i.e. a very bright meteor. It is not to be confused with ball lightning, which is purely electrical in origin.

BOLOMETER. A very sensitive instrument for the measurement of small amounts of radiant heat. It depends on the measurement of the change of resistance of a metal foil arising from the absorption of radiation. A particularly sensitive type, the *superconducting bolometer*, makes use of the rapid change in resistance which takes place in a superconducting material under the influence of radiant energy.

BOLOMETRIC SCALE. A scale of stellar magnitude based on the measurement of radiant energy. The *bolometric magnitude* may not be the same as the visual magnitude and the difference is known as the *bolometric correction*. The bolometric correction cannot be measured, but only estimated, since the absorption by the Earth's atmosphere is unknown.

BOLTZMANN CONSTANT. The gas constant per atom (or per molecule for a molecular gas). It is equal to the gas constant per mole divided by Avogadro's constant, and may also be considered as the ratio of the mean total energy per atom (or molecule) in a gas to the absolute temperature. Its value is 1.380×10^{-16} erg/deg, or 1.380×10^{-23} J/deg. See also: *Electron temperature. Kinetic energy. Appendix III*.

BOLTZMANN DISTRIBUTION LAW. For a physical system consisting of a large number of independent particles in statistical equilibrium: a law giving the average number of particles within the system having positions and speeds within well-defined limits. The distribution function, f, i.e. the number of particles which at time t have such specified positions and speeds, is given by $f = A \exp(-E/kT)$, where A is a normalization constant, E is the energy of the particle as a function of position and speed, k is Boltzmann's constant and T is the absolute temperature.

BOLTZMANN EQUATION: TRANSPORT EQUATION. Gives the distribution function, f, for a system of particles which are not in equilibrium. For monatomic particles it is of the form $\dfrac{\partial f}{\partial t} + \mathbf{V} \cdot \operatorname{grad} f + \mathbf{F} \cdot \operatorname{grad}_v f = \left(\dfrac{\partial f}{\partial t}\right)_c$ where $\mathbf{V}$ is the velocity, $\mathbf{F}$ is the acceleration, $\operatorname{grad}_v f$ is the gradient of f in velocity space and $\left(\dfrac{\partial f}{\partial t}\right)_c$ is the rate of change of f due to collisions. See also: *Boltzmann distribution law*.

BOLTZMANN FACTOR. A name sometimes given to the Stefan–Boltzmann constant.

BOLTZMANN RATIO. In Boltzmann statistics: the ratio of the mean occupation numbers of two quantum states.

BOLTZMANN STATISTICS. The statistics of weakly interacting, distinguishable, particles in equilibrium. The mean number of particles, n_i, in the ith quantum state is given by $n_i = w_i \exp(\mu - \varepsilon_i/kT)$ where w_i is the degeneracy of the state, μ is the chemical potential, ε_i the energy of the state, k is Boltzmann's constant and T the absolute temperature. Boltzmann statistics are valid only if n_i is very much less than one, otherwise more specialized statistics apply.

BOLTZMANN SUPERPOSITION PRINCIPLE. For a large number of small particles undergoing thermal agitation while subjected to the action of a superposed field (electric, magnetic, gravitational or inertial): states that, when statistical equilibrium is reached, the number of particles per unit volume in a region of the field where the potential energy of a particle is E, is equal to $N_0 \exp(-E/kT)$ where N_0 is the number of particles per unit volume in the region of the field where E is zero, k is the Boltzmann constant and T the absolute temperature of the system of particles.

BOLUS (RADIATION THERAPY). Material whose absorbing and scattering properties for a given radiation are similar to those of the human body. The material is used to fill air gaps adjacent to the treatment zone.

BOND. A union between atoms arising from their tendency to acquire or adopt electronic configurations in which at least the outer shell of electrons is closed. See also: *Hydrogen bond. Van der Waals forces.*

BOND ALBEDO. See: *Albedo.*

BOND, COORDINATE. The type of bond operating between groups in a complex (i.e. coordination) compound. See also: *Complex compound.*

BOND, COVALENT. A bond in which electrons are equally shared between two or more atoms. See also: *Bond, valency.*

BOND, DATIVE. A covalent bond in which both electrons forming the bonding pair are supplied by one atom. Such bonds exhibit some degree of polarity and are sometimes called *semipolar bonds.*

BOND, DOUBLE. See: *Bond, multiple.*

BOND, DOUBLE, PROTONATED. A four-electron bond formed by electron-deficient atoms and hydrogen atoms.

BOND, ELECTROVALENT. See: *Bond, ionic.*

BOND ENERGY, MEAN. The average energy required to break a bond in a binary compound containing only one kind of bond. See also: *Bond strength.*

BOND, HETEROPOLAR. A covalent bond characterized by an unequal distribution of charge, i.e. by a difference in polarity between the bonded atoms.

BOND, HOMOPOLAR. The common type of covalent bond for which there is no difference in polarity between the bonded atoms.

BOND, INTERMOLECULAR. A bond between atoms in different molecules.

BOND, INTRAMOLECULAR. A bond between atoms in the same molecule.

BOND, IONIC. A bond between two atoms, the first of which acquires electron(s) from the other to complete its valence shell, the second achieving a closed inner shell by losing electron(s) to the first. Also known as an *electrovalent bond* or *polar bond.*

BOND LENGTH. The internuclear distance between a pair of atoms between which a bond exists.

BOND, METALLIC. A type of bond in which the valency electrons of the constituent atoms are free to move in the periodic lattice of a metal. The atoms of the metals are sometimes said to exist in a sea of electrons. This type of bonding is the basis of the observed metallic properties.

BOND, MULTIPLE. A bond arising when two atoms are joined by more than one covalent bond. When two covalent bonds are involved the bond is said to be a *double bond.* With three covalent bonds it is a *triple bond.*

BOND ORDER. A measure of the number of electrons contributing to a bond. For a bond between two atoms it is defined as one-half of the total number of electrons taking part in the bond.

BOND, POLAR. See: *Bond, ionic.*

BONDS, DELOCALIZATION OF. See: *Molecular orbital.*

BOND, SEMIPOLAR. See: *Bond, dative.*

BONDS, EQUIVALENT. Two bonds which possess identical physical properties such as bond length, spatial orientation, etc.

BOND STRENGTH. (1) The mean bond energy. (2) More usually, the energy required to break a specified bond in a molecule. See also: *Bond energy, mean.*

BOND, TRIPLE. See: *Bond, multiple.*

BOND, VALENCY. A covalent bond. It may be described by the Heitler–London theory or, in some cases, by the "mixing" of or resonance between alternative structures. See also: *Heitler–London theory. Resonance of chemical bonds.*

BOOLEAN ALGEBRA. The algebra of classes. The classes range from the empty class (symbol 0) to the universal class (symbol 1), and, in addition to the operations of addition and multiplication (which in the present context are without arithmetical implications), Boolean algebra employs a symbol ($\subset$) for inclusion (A is said to be contained or included in B if every member of A is also a member of B). Boolean algebra is used in computer theory and in switching theory.

BORDONI PEAK. A maximum in the internal friction of plastically deformed metals which is observed when the friction is measured as a function of temperature at a constant frequency. It is especially noticeable in face-centred cubic metals and, in the frequency range 10^3 to 10^6 Hz, it appears at about one-third of the Debye temperature.

BORE. See: *Tidal wave.*

BORN APPROXIMATION. In quantum mechanics: an application of perturbation theory to scattering problems, for the calculation of wave functions and cross-sections of collision processes. The Born approximation assumes that the interaction energy between the colliding particles is small compared with their kinetic energy.

BORN–HABER CYCLE. A cycle of chemical and physical measurements made at various stages in the decomposition of a compound, from which experimental values of lattice energies or electron affinities may be calculated.

BORN–MAYER EQUATION. Represents the binding energy of an ionic crystal as the sum of the Coulomb energy of the lattice, the repulsive energy between the ions and the attractive (van der Waals) energy between them.

BORN–OPPENHEIMER APPROXIMATION. An approximation used in considering the electronic behaviour of molecules or solids, according to which, under certain conditions, nuclear motion may be neglected, the interacting particles being treated as point charges and masses which interact electrostatically.

BORN–OPPENHEIMER METHOD. A method for calculating the force constants between atoms in a molecule or solid, on the assumption that the electrons follow the motions of the nuclei adiabatically.

BORN–VON KÁRMÁN BOUNDARY CONDITIONS. Conditions imposed in the calculation of specific heats to allow for the fact that the longitudinal and transverse wave velocities in a solid are not constant over the whole spectrum of frequencies, as had been assumed by Debye in his theory of specific heats, in which no detailed account was taken of the structure of the solid or of the interatomic forces involved.

BORRMANN EFFECT. The production of light or dark diffraction lines when characteristic X-rays are generated in a single crystal by X-rays. The lines are similar to the Kossel lines occurring when characteristic X-rays are generated in a single crystal by an electron beam, and to the Kikuchi lines which are observed in electron diffraction.

BOSE–EINSTEIN CONDENSATION. For a vapour to the molecules of which Bose–Einstein statistics apply: the condensation of the vapour to a state in which some of the molecules have a momentum of nearly zero instead of having their momenta spread over a large range of values. This is analogous to a liquid whose molecules are in contact, instead of ranging over a large volume. The process, which was formerly regarded as hypothetical, is now believed to be related to the transition that occurs between the two forms of liquid helium at about 2·2 K.

BOSE–EINSTEIN LIQUID. The name given to the quasi-liquid formed by the separation of molecules of lowest transitional energy in the Bose–Einstein condensation.

BOSE–EINSTEIN STATISTICS. The form of quantum statistics, having to do with the distribution of particles among various allowed energy values, applicable to an assembly of particles in which many particles are allowed in a given state. Their wave functions have symmetry properties such that the wave function remains unchanged if any two particles are interchanged. Such wave functions are said to be *symmetric*. At sufficiently high temperatures, where a large number of energy levels are excited, Bose–Einstein statistics, like Fermi–Dirac statistics, reduce to the classical Maxwell–Boltzmann statistics. See also: *Fermi–Dirac statistics.*

BOSON. A particle to which Bose-Einstein statistics apply. Such particles do not follow the Pauli exclusion principle. Photons, pions, α-particles, and all nuclei of even mass number are bosons. Their spin, in common with that of all bosons, is zero or integral. See also: *Fermion.*

BOSON, INTERMEDIATE. A particle postulated on the hypothesis that all weak interactions are second-order processes generated through the emission and absorption of an intermediate boson field. The role of the intermediate bosons (or *W mesons*) would be analogous to that of photons in electromagnetic interactions or pions and other mesons in strong interactions.

BOUGUER HALO. See: *Brocken, spectre of.*

BOUGUER–LAMBERT LAW OF ABSORPTION: BOUGUER–BEER LAW. A statement of the exponential nature of the absorption of light in a homogeneous transparent medium which, as given in the general definition "Absorption coefficient for radiation", may be written $I = I_0 e^{-kx}$ where I is the intensity of light transmitted, I_0 is the incident intensity, x is the thickness of the transmitting layer, and k is the absorption coefficient. Where use is made of logarithms to the base 10, instead of natural logarithms, so that $I = I_0 \times 10^{-ax}$ the absorption coefficient, α, is known as the *optical extinction coefficient*. See also: *Absorption coefficient for radiation*.

BOUGUER PHOTOMETER. A photometer involving the use of two translucent sources which are illuminated by two light sources that are to be compared.

BOUNDARY CONDITIONS. See: *Boundary-value problems*.

BOUNDARY, LARGE-ANGLE. See: *Grain boundary, large-angle*.

BOUNDARY LAYER. For an incompressible fluid flowing past a stationary object: the thin layer (usually about 50–100 μm) immediately next to the stationary object within which viscous stresses cannot be neglected. In a *laminar* boundary layer the streamlines are smooth; in a *turbulent* boundary layer they are irregular and unsteady, with large eddy motions. The behaviour of the boundary layer is important in considering such problems as lift and drag.

BOUNDARY-LAYER CONTROL. Of an aerofoil: the prevention of the transition of the boundary layer from a laminar to a turbulent state, or of boundary layer separation, by the incorporation of suction to remove the boundary layer adjacent to the surface, or by the injection of higher-energy air into the layer, e.g. by air blown through a narrow slot.

BOUNDARY-LAYER SEPARATION. For a fluid flowing past a solid surface: the separation of the boundary layer from the surface by a stagnant region of slow-moving fluid in which eddies are formed with the consequent breakdown of streamline flow. See also: *Boundary-layer control*.

BOUNDARY SCATTERING OF PHONONS. The scattering of phonons at the boundary surfaces of a solid. When the mean free path of the phonon exceeds the dimensions of the specimen, i.e. at low temperatures and for sufficiently small samples, this results in a reduction in thermal conductivity.

BOUNDARY-VALUE PROBLEMS. Problems arising in connection with the solution of differential equations when, for certain values of the independent variables, the solution has to meet specified requirements, usually imposed as a result of physical considerations. These requirements are known as *boundary conditions* or *boundary values*.

BOUNDARY WAVE. A concept due to Young according to which the light diffracted by a hole in an opaque screen is expressed as the sum of the geometrical wave (i.e. one transmitted according to geometrical optics) and an *edge* or *boundary wave*, which originates at the edge of the hole and is propagated as a cylindrical wave with non-uniform amplitude.

BOUNDARY WAVELENGTH: QUANTUM LIMIT. In a continuous X-ray spectrum: the shortest wavelength present. The term *quantum limit* is also used to denote the energy appropriate to that wavelength. See also: *X-rays*.

BOUND ELECTRON. An electron which is not free to move under the influence of an external field, e.g. an atomic electron in a stationary state, or an electron bound to an impurity in a solid. See also: *Conduction electron. Free electron*.

BOUND ENERGY. A term sometimes used to denote the difference between the internal and free energy of a thermodynamic system. It is given by TS, where T is the absolute temperature and S the entropy.

BOUND STATE. Of an atomic nucleus: the state in which the nucleus is stable against the spontaneous emission of any nucleon. When a nucleon can be spontaneously emitted the nucleus is said to be in a *virtual state*. See also: *Virtual state*.

BOURDON TUBE. A spiral of hollow metal tubing which responds to changes in its internal pressure by coiling or uncoiling as the case may be. It may be used to indicate changes in temperature or pressure. See also: *Pressure gauge, Bourdon. Thermometer, liquid-filled*.

BOW WAVE. The surface wave stemming from the bow of a vessel in motion. Often used for the shock-wave system from the front of a body in supersonic motion, which should properly be described as the *bow shock wave*.

BOYLE LAW. States that, for a given mass of gas at constant temperature, the product of pressure and volume is constant, i.e. the volume of the gas varies inversely as the pressure. The law holds only for ideal or perfect gases and fails for all real gases to a greater or lesser extent.

BOYLE TEMPERATURE. The temperature at or near which Boyle's law provides a good approximation to the true equation of state for a given gas. See also: *Virial coefficients*.

BRACE POLARIZER. A thin plate of calcite immersed in a liquid of high refractive index, the plate being inclined to the incident beam so that the extraordinary ray suffers total internal reflection.

BRACE SPECTROPHOTOMETER. A visual instrument in which the comparison field is produced in a dispersing prism by means of a second collimator.

BRACHISTOCHRONE. A curve joining two points, along which a particle moves under the action of an assigned conservative force system (commonly a constant gravitational force) in the least possible time. The motion experienced by the particle is known as *brachistochronic motion*.

BRACKETT SERIES. A series in the line spectrum of the hydrogen atom. See also: *Hydrogen atom, line spectrum of.*

BRAGG CURVE. A curve in which the specific ionization along the path of a beam of α-particles in a gas (usually air) is plotted against the distance from the α-particle source, and from which the range of the α-particles may be determined. Similar curves may also be plotted for other ionizing particles. See also: *Range of ionizing particles.*

BRAGG CUT-OFF WAVELENGTH. The wavelength above which Bragg reflection cannot occur. For any particular crystal this will be given by twice the highest crystal spacing capable of reflection, from the Bragg equation $\lambda = 2d \sin \theta$, where λ is the wavelength, d is the spacing and θ the glancing angle of reflection.

BRAGG EQUATION: BRAGG LAW. The equation setting out the condition for the diffraction ("reflection") of a parallel beam of monochromatic (monoenergetic) X-rays from a crystal, which may be stated: $n\lambda = 2d \sin \theta$, where n (an integer) is the order of diffraction ("reflection"), λ is the wavelength of the X-rays, and d is the distance between parallel atomic planes in the crystal which give diffraction maxima when the incident beam is inclined at an angle θ to these planes. The diffracted ("reflected") beam is also inclined at the same angle. The Bragg equation also holds for the diffraction of electrons, neutrons, etc.

BRAGG REFLECTION. The diffracted beam produced by reinforcement from successive members of a set of crystal planes when the Bragg equation is satisfied.

BRAGG SPECTROMETER. An instrument for the X-ray analysis of crystal structure in which a monochromatic beam of X-rays was diffracted by a crystal and the diffracted beam detected and measured by an ionization chamber. Its construction resembled that of an optical spectrometer, collimation being effected by a sequence of slits or pinholes. Instruments developed from this are now known as *diffractometers*. See also: *Diffractometer.*

BRANCHED-CHAIN COMPOUND. An aliphatic organic compound in which the chain of carbon atoms contains one or more branches.

BRANCHES. In spectroscopy. See: *Spectra, molecular, branches in.*

BRANCHING DECAY. Radioactive decay which can proceed in two or more different ways.

BRANCHING FRACTION. In branching decay: the fraction of nuclei which disintegrate in a specified way. It is usually expressed as a percentage..

BRANCHING RATIO. In branching decay: the ratio of the branching fractions for two specified modes of disintegration.

BRANCHING RULE. For spectra: a rule used in the construction of atomic term diagrams to obtain the multiplicities of terms. It is concerned with the effects of the addition of extra electrons to given states.

BRASS. A generic name for a broad class of copper–zinc alloys. Apart from the widespread use of various types of brass in industry the structure and properties of the brasses are of interest in the study of metallic behaviour, for example in studying the relationship between the preferred structure of an alloy and its electron/atom ratio.

BRAVAIS BIPLATE. A biplate used as a sensitive detector of traces of optical polarization. It consists of two half-wave plates of birefringent material each cut parallel to its axis, cemented together with an angle of 90° between the axes. The action of the biplate is based on the fact that a half-wave plate of the type used shows, between crossed polarizers, the *sensitive tint* (i.e. a tint sensitive to a slight change in polarization), otherwise known as the *tint of passage*.

BRAVAIS LATTICE: SPACE LATTICE. An infinite three-dimensional array of points in space, such that each point has the same environment. Only fourteen distinct arrays or lattices of this kind are possible and these may be referred to one or other of the seven crystal systems. A *primitive lattice* (symbol P) has a lattice point at each corner of the appropriate unit cell. A *body-centred lattice* (symbol I) has, in addition, a point at the "centre" of the unit cell. The *face-centred lattice* (symbol F) has a point at the centre of each face as well as one at each corner, and the remaining type of lattice (symbol C) has, instead, points at the centres of only one pair of faces. Although a lattice is a geometrical concept and not a crystal, and therefore cannot scatter radiation, the structural motif (some particular atomic or molecular group) the repetition of which in three dimensions constitutes a crystal, may always be associated with a lattice point at each repetition, i.e. the motif may be replaced by a representative point situated at a point of the lattice.

BREAKAWAY. Of fluid flow: boundary layer separation without subsequent reattachment to the surface. See also: *Boundary layer. Boundary-layer separation.*

BREAKDOWN VOLTAGE. The voltage at which a given insulator or insulating material fails to withstand the voltage and becomes conducting. The term is often applied to the passage of electric current through a gas.

BREAKING STRENGTH. See: *Tensile strength: Ultimate tensile stress.*

BREEDING. Of fissile material: the transformation of a fertile substance into a fissile substance in a nuclear reactor when the number of fissile nuclei produced is greater than the number of fissile nuclei consumed, i.e. when the conversion ratio is greater than unity. See also: *Nuclear reactor, conversion in.*

BREEDING GAIN. Of a nuclear reactor: the ratio of the excess of fissile nuclei produced over fissile nuclei consumed to the fissile nuclei consumed, i.e. the breeding ratio minus one.

BREEDING RATIO. Of a nuclear reactor: the ratio of the number of fissile nuclei produced to the number of fissile nuclei consumed when it is greater than unity, i.e. the conversion ratio when it is greater than unity.

BREIT–WIGNER FORMULA. Expresses the capture or scattering cross-section of an atomic nucleus in terms of the energy of the bombarding particle when this energy is near to the resonance energy of the compound nucleus. For example, the absorption cross-section of a nucleus for a neutron of energy E, near a resonance of energy E_0, is given by

$$\frac{\sigma_0}{1 + \dfrac{(E - E_0)^2}{(\Gamma/2)^2}}$$

where σ_0 is the resonance cross-section and Γ is the resonance width.

BREMSSTRAHLUNG. The electromagnetic radiation associated with the deceleration of charged particles. It means, in German, "braking radiation", but is also applied to the radiation associated with the acceleration of charged particles. The best known example is the continuous X-ray spectrum emitted by an X-ray tube. See also: *Inner bremsstrahlung: Internal bremsstrahlung. X-rays.*

BREWSTER ANGLE. See: *Brewster law.*

BREWSTER BANDS: BREWSTER FRINGES. Interference fringes which are seen when white light is viewed through two plane parallel plates whose thicknesses are nearly equal.

BREWSTER LAW. States that, when light is reflected by a refracting medium, the tangent of the polarizing angle is equal to the refractive index; or, in other words, when light is reflected at the polarizing angle the reflected and refracted rays are 90° apart. The *polarizing angle*, namely the angle at which the reflected ray is completely plane, polarized, is also known as the *Brewster angle*. The concepts involved are also applicable to electromagnetic waves in general, particularly radio waves. For reflection at the boundary between two media of different refractive indices the tangent of the polarizing angle is equal to the ratio of the refractive indices.

BREWSTER STEREOSCOPE. A prismatic device for viewing a pair of stereoscopic pictures of which the separation is greater than the interocular distance.

BRIDGE, ELECTRICAL. A term referring to any one of a variety of electrical networks employing a bridge circuit, which is essentially a four-terminal network to the input terminals of which is applied an e.m.f., the output terminals being "bridged" by a detector. By adjusting one or more components in the network the voltage across the "bridge" can be brought to zero and, in this unique condition, specific relations exist between the values of the components in the system, which may then be used for measuring one component in terms of another. Some of the best known examples are the *Wheatstone bridge* and *Carey–Foster bridge* for measuring resistances, the *Kelvin double bridge* (with eight arms instead of four) for comparing low resistance standards, the *Wien bridge* for a.c. capacitance, the *Nernst high-frequency capacitance bridge*, the *Heaviside mutual inductance bridge*, and the *Maxwell bridge*, for the comparison of inductance with capacitance.

BRIGHTNESS. (1) The amount of light that appears to be emitted by a body, considered subjectively, sometimes described as *subjective brightness.* (2) A quantitative term more correctly referred to as luminance or luminous intensity.

BRIGHTNESS TEMPERATURE: LUMINANCE TEMPERATURE. Of a radiating body: the temperature of a black-body radiator which has the same photometric luminance as the radiating body at a specified wavelength. The brightness temperature is always lower than the true temperature.

BRILLOUIN FUNCTION. A mathematical function appearing in the quantum theory of para- and ferromagnetism.

BRILLOUIN ZONES. In the band theory of solids: zones of allowed electron energy levels in a crystal. They may be represented in reciprocal space as closed polyhedra, the relationships between which are related to the electrically conducting properties of the crystal. Gaps in the allowed energy levels occur for electrons whose energy and direction are such that they can suffer Bragg reflection. In some crystals the Brillouin zones are separated, i.e. the gaps for differently directed electrons coincide and leave a range of energy that is forbidden to electrons whatever their energy and direction. In other crystals the Brillouin zones overlap, i.e. the gaps for one direction are no longer gaps for other directions and no resultant gap exists.

BRITISH THERMAL UNIT (B.T.U.). The energy required to raise the temperature of 1 lb of pure water by 1 deg F at normal atmospheric pressure. It may be referred to a particular temperature (e.g. the 60°F B.t.u. which refers to a rise of temperature from 60°F to 61°F), or to the *mean B.t.u.,* which is 180th part of the energy required to raise the temperature from 32°F to 212°F. The most recent definition of the B.t.u. is in terms of the joule and is independent of temperature. One international table B.t.u. is defined as equal to 1·05506 k J. With this definition the specific heats in the British system, with the B.t.u., are numerically identical with those in the CGS system, with the calorie. (*Note:* 1 therm = 10^5 B.t.u.). See: *Appendix II.*

BRITTLE–DUCTILE FRACTURE. See: *Fracture. Transition, brittle–ductile.*

BROADBAND ELECTROMAGNETIC TESTING. An extension of the sinusoidal eddy-current methods of non-destructive testing to the use of currents varying in a more complex fashion with time, i.e. covering a broad band of frequencies.

BROAD BEAM. In beam attenuation measurements for ionizing radiation: a beam in which the unscattered and much of the scattered radiation reaches the detector. See also: *Narrow beam.*

BROCKEN, SPECTRE OF. The shadow of an observer, sometimes appearing to be of gigantic proportions, cast by the Sun on a bank of fog, mist or cloud. It is often seen from a mountain top or aeroplane. Sometimes the shadow appears to be surrounded by rainbow coloured rings (a *glory*) together with a white fog bow. This phenomenon is known as a *Bouguer halo.*

BRONZE. A term for copper alloys in which the principal addition is tin. It has, however, been extended loosely to cover a wide range of alloys not containing tin.

BROWNIAN MOTION. The irregular, random movements of small particles suspended in a fluid, arising from the thermal motion of the molecules of the fluid. The distribution of particle velocity is similar to that of a gas molecule of the same weight as the particle. The theory of the Brownian motion may also be applied to the "noise" in electrical networks (*Nyquist's formula*) due to the random movement of electrons, and to other problems involving random movement.

BRÜCKNER CYCLE. A supposed cycle of about 35 years in weather elements such as temperature and rainfall. Its validity has not been established.

B-SCOPE. A form of cathode-ray tube display in which the horizontal displacement corresponds to distance and the vertical displacement to azimuthal angle.

B-STAR. See: *Stars, spectral classes of.*

BUBBLE CHAMBER. An instrument for rendering visible the tracks of ionizing particles. Essentially it is a vessel filled with a highly superheated transparent liquid (commonly hydrogen or helium), the passage of an ionizing particle through which is marked by the appearance of a series of bubbles along the particle's path. The superheating is achieved by suddenly relieving the pressure to which the liquid is subjected. Because of the relatively high density of liquids, bubble chambers are usually superior to cloud chambers for the study of the interactions of ionizing particles with matter. See also: *Cloud chamber. Spark chamber. Streamer chamber.*

BUCKLING. (1) Of a structural system: the deformation arising when the loading conditions are such that the original configuration of the structure becomes unstable. The buckling is said to be *elastic* if the original configuration is regained when the loads are removed. (2) In nuclear reactor theory: a measure of the curvature (or "buckling") of the neutron flux density distribution. For a bare reactor the *geometric buckling*, B_g^2, is the first (i.e. lowest) eigenvalue of the equation $\nabla^2\phi(r) + B^2\phi(r) = 0$, where ∇^2 is the Laplacian operator and $\phi(r)$ is the neutron flux density at the point r, with the condition that this density is zero at the extrapolated boundary of the assembly. The geometric buckling, which is sometimes defined as the negative of the *Laplacian*, is dependent only on the shape and dimensions of the assembly. The *material buckling*, B_m^2, provides a measure of the multiplying properties of an assembly as a function of the materials and their disposition. It is the value of B^2 satisfying the equation $k_\infty \exp(-B^2\tau) = 1 + L^2B^2$, where k_∞ is the infinite multiplication factor, τ the age, and L the diffusion length of the neutrons. For a critical reactor the geometric and material buckling have the same value, known simply as the *buckling*.

BUDDE EFFECT. The increase in volume of certain gaseous systems which occurs on exposure to light if the molecules are photochemically dissociated into atoms which then recombine with the liberation of the binding energy as heat. Chlorine gas offers a typical example of this effect. See also: *Draper effect.*

BUFFER SOLUTION. A solution whose hydrogen ion concentration remains nearly constant even after appreciable additions of acid or base.

BULK MODULUS: COMPRESSION MODULUS. (1) Of an elastic isotropic material: the change in pressure which produces unit fractional change in volume. (2) Of a liquid: the change in pressure which produces unit fractional change in density. See also: *Elastic modulus.*

BUNCHING. Of a beam of particles: any variation in the density of particles along the beam. The extreme limit would arise from the emission of pulses of monoenergetic particles from a source, but the term usually refers to some form of axial focusing.

BUNN CHART. A chart used in indexing X-ray diffraction powder photographs of substances of tetragonal or hexagonal symmetry.

BUNSEN PHOTOMETER. A device for the comparison of the luminous intensities of light sources. It consists of an opaque white screen (the Bunsen disk) having a translucent spot in the middle, and the required comparison is effected by finding a position, with the screen situated between the sources, where both sides of the screen appear identical. Application of the inverse square law then gives the required result. The photometer is sometimes known as the *Bunsen grease-spot photometer*. Mirrors are used to permit simultaneous viewing of both sides of the screen. These are known as *Rudorff mirrors*.

BUOYANCY. The upward pressure or force on a body which is wholly or partially immersed in a fluid. The force is equal to the weight of the fluid displaced by the immersed part of the body and acts through the centre of gravity of the displaced fluid. This point is termed the *centre of buoyancy*. For stable equilibrium the centre of gravity of the body must be below the centre of buoyancy.

BURETTE. A glass tube, usually calibrated in millilitres, open at the top and having a ground glass tap at the bottom. It is mainly used in chemical analysis. A *gas burette* is a calibrated U-shaped tube used for measuring the volume of a gas and passing the gas into a chemical reagent and back again.

BURGERS VECTOR. Of a dislocation: the translation vector associated with a particular dislocation in a crystal. Its magnitude measures the *strength* of the dislocation. See also: *Dislocation.*

BURNING VELOCITY. In combustion: the velocity of the reaction zone front with respect to the unburned gas.

BUSBAR. A rigid conductor (or conductors) to which electricity supplies are fed and from which loads are supplied.

BUSHING. An insulator whose distinctive function is to insulate an electrical conductor where it passes through a partition such as the tank of a transformer or the wall of a sub-station.

C

CABLE. An electrical conductor surrounded by a dielectric and a sheath, or an assembly of conductors so surrounded, used for the transmission of electric power or for communication. See also: terms beginning *Electrical cable.*

CADMIUM CUT-OFF. The neutron energy value above which cadmium is "transparent" to neutrons and below which it is "opaque". It is about 0·4–0·5 eV, but varies with the experimental configuration. The value of the *effective cadmium cut-off* is determined, for a particular configuration, by the condition that, if a cadmium cover surrounding a detector were replaced by a fictitious cover opaque to neutrons with energy below this value and transparent to neutrons with energy above this value, the observed detector response would be unchanged. (*Note:* 1 eV = 1·6022 × 10^{-19} J approx.)

CADMIUM DIFFERENCE. The difference in the effect observed with an uncovered neutron detector and that with the detector covered with a thin layer of cadmium, when the detector is exposed to a neutron beam. This difference gives the effect caused by epithermal neutrons.

CADMIUM NEUTRONS. Neutrons whose energy lies below that of the cadmium cut-off.

CADMIUM RATIO. The ratio of the response of a bare neutron detector to that under the same conditions when the detector is covered by cadmium of a specified thickness.

CAILLETET PROCESS. A method used for the liquefaction of oxygen, based on the cooling which takes place when a gas expands freely from a higher to a lower pressure, doing both external and internal work.

CALCULATING MACHINE. A machine which can perform one or more of the four fundamental arithmetical operations when the figures with which the machine is to operate are suitably designated, e.g. by a keyboard or by setting levers. It may be hand operated or electrically driven. See also: *Computer.*

CALCULUS OF RESIDUES. The study of the Cauchy integral and theorem. The key theorem in this study is *Cauchy's residue theorem* which states that, if $f(z)$ is an analytic function, except for a finite number of singularities, in a region which contains a simple closed contour C and its interior D, $\int_C f(z)\,dz = 2\pi i \Sigma R$, where ΣR denotes the sum of the residues of $f(z)$ at those of its singularities which lie in D, it being understood that no singularities lie on C. See also: *Analytic function: Holomorphic function. Cauchy integral. Cauchy theorem.*

CALCULUS OF VARIATIONS. The study of the maximum and minimum properties of definite integrals. Examples are: the brachistrone problem, the Euler equation, and the isoperimetric problem. See also: *Brachistrone. Euler equation. Isoperimetric problem.*

CALLIER COEFFICIENT. Refers to the variation of the measured density of a photographic negative with the direction of illumination. It is defined as the ratio of the density measured with parallel light to that measured with totally diffuse light, and has an average value of about 1·5.

CALM DAYS: QUIET DAYS. Five days selected each month as international magnetic "quiet" or "calm" days, on the basis of many magnetic observations. See also: *Disturbed days: Active days.*

CALORIE. A unit of heat originally defined as the quantity of heat required to raise the temperature of 1 g of water by 1°C at normal atmospheric pressure. It may be referred to a particular temperature (e.g. the 15°C calorie, which refers to the rise in temperature from 14·5° to 15·5°C, or to the *mean calorie,* which is 100th part of the heat required to raise the temperature from 0°C to 100°C). The calorie is now internationally defined as 4·1868 J exactly and is known as the *international table calorie* or cal$_{IT}$. The *defined thermochemical calorie,* used mainly in the U.S.A., is defined as 1 cal$_{thermochem}$ = 4·184 J exactly and the 15° *calorie* as 4·1855 J exactly. The *Cal* or large calorie = 1 kcal = 1000 cal. See: *Appendix II.*

CALORIFIC VALUE. Of a material: the heat released by combustion of unit quantity of the material under specified conditions. It can be defined only in terms of these conditions and is not necessarily the same as the heat of combustion, conversion to which may be a complex procedure. However, the calorific value obtained by using, for example, a bomb calorimeter is negligibly different from the heat of combustion.

CALORIFIC VALUE, GROSS. The calorific value measured when the water formed during combustion (and any water originally present) is in the liquid state. It may be referred either to constant volume or constant pressure. The gross calorific value is that obtained by conventional experimental methods.

CALORIFIC VALUE, NET. The calorific value measured when the water formed during combustion (and any water originally present) is in the gaseous state. It may be referred either to constant volume or constant pressure.

CALORIMETER. An instrument used for measuring heat content. The term may refer to the complete apparatus or be restricted to the sample container and its contents. If the latter remains at a constant temperature, the calorimeter is said to be *isothermal*. If it does not, as is usually the case, the calorimeter is said to be *non-isothermal*.

CALORIMETER, ADIABATIC. A calorimeter in which heat losses to its surroundings are virtually eliminated by the use of efficient insulation.

CALORIMETER, ANEROID. A bomb calorimeter in which the bomb is enclosed in a vacuum chamber instead of being immersed in water.

CALORIMETER, BOMB. A calorimeter in which the sample is burnt in oxygen under pressure in a "bomb" immersed in water, the temperature of which is measured. Its most important uses are for measuring the heats of formation of pure chemicals and heats of combustion of fuels and foods, and for judging the quality of explosive materials or propellants.

CALORIMETER, CONTINUOUS FLOW. A calorimeter, used for determining the heat capacity of liquids and gases, in which heat is added to the flowing fluid at a constant rate.

CALORIMETER, DEWAR. A calorimeter in which the sample container is surrounded by a vacuum jacket to minimize heat losses.

CALORIMETER, DIFFERENTIAL. One of a pair of identical calorimeters (*twin calorimeters*) which, when used together, eliminate errors due to external heat transfer. The unknown amount of heat is introduced into one calorimeter and the difference in temperature between the two calorimeters (which have essentially the same external heat transfer) is observed. Such calorimeters are often used in microcalorimetry.

CALORIMETER, GAS. A calorimeter used in the analysis of combustible gases to determine calorific value and moisture content. It may be of the still-water type, used for small quantities of gas, or, for large volumes, may be of the flow type, either recording or non-recording.

CALORIMETER, ICE. An isothermal calorimeter in which the amount of ice melted is used as a measure of the amount of heat added. In the Bunsen ice calorimeter the amount of ice melted is determined indirectly from its change in volume on melting.

CALORIMETER, NERNST. A calorimeter, used extensively for measurements at low temperature, in which the calorimeter jacket temperature is kept constant.

CALORIMETER, TWIN. See: *Calorimeter, differential*.

CALORIMETRY. The measurement of heat content. See also: *Calorimeter* and terms following it.

CALORIMETRY, INTERNAL. The measurement of heat absorption or evolution associated with a physical change of state.

CALORIMETRY, MICRO-. The measurement of small amounts of heat. Differential twin calorimeters are often used but another type is the *radiation balance* (or *radio balance*) in which Peltier cooling is used to balance the small amount of heat to be measured. The Bunsen ice calorimeter has also been used.

CALORIMETRY, X-RAY. The measurement of the intensity of an X-ray beam by the transformation of its energy to heat.

CALUTRON. An American code name for an electromagnetic isotope separator for ^{235}U, based on a focusing mass spectrometer. The name was derived from the location of the instrument at the University of California.

CAMERA LUCIDA. A device used in making sketches of microscopic objects in which an image of the drawing board is superimposed on the microscope image by the use of a beam-splitting prism mounted over the eyepiece.

CAMERA OBSCURA. A box or chamber within which an image of an external scene is formed by means of a small aperture or lens. It was used in early studies of solar phenomena (e.g. sunspots) and, more recently, has been used for checking the accuracy of radio-navigation devices.

CAMERA, PHOTOGRAPHIC. An apparatus in which sensitive emulsions are exposed to light under controlled conditions, to record an image of a scene or object, using an image-forming means such as a lens or pinhole. The term is often extended to include apparatus for the recording of images or other data obtained from radiation other than light, e.g. X-rays, electrons, atomic particles, etc.; and such terms as X-ray camera are therefore in common use.

CAMERA TUBE. An electron beam tube, incorporating a photosensitive electrode, for deriving from an optical image a corresponding electrical signal. The electron beam of this tube usually scans the surface of the photosensitive electrode, on which the optical image is projected. This electrode may be photoemissive, as in the *iconoscope* and *orthicon*; or may be photoconductive, as in the *vidicon*. See also: *Iconoscope. Orthicon. Vidicon.*

CAMOUFLAGE. The disguise of an object to prevent or delay recognition of its true nature. It is associated almost exclusively with visual deception, either by concealment (as in defence) or mimicry (as in attack).

CAN. For nuclear reactor fuel. See: *Nuclear reactor fuel.*

CANAL RAYS: POSITIVE RAYS. Positively charged particles, consisting of atoms or molecules of gaseous matter, which are produced in an electric discharge at low pressure. They are impelled towards the cathode and will pass through it, if it is perforated, in the form of a beam. The separation of the positive ions in such a beam forms the basis of the mass spectrograph and mass spectrometer. See also: *Mass spectrography: Mass spectrometry.*

CANCER, FIRST POINT OF. The point at which the ecliptic touches the tropic of Cancer. See also: *Solstice, summer.*

CANDELA. The unit of luminous intensity. It is defined as "the luminous intensity, in the perpendicular direction, of a surface of 1/600000 square metres of a black body at the temperature of freezing platinum under a pressure of 101 325 newtons per square metre". The candela is nearly 2% less than the candle, which it has replaced. See also: *Light, primary standard of. SI units. Appendix II.*

CANDLE. The former unit of luminous intensity, now replaced by the candela. One candle is about 1·02 candela.

CANDLE POWER. Of a luminous source: the light-radiating capacity in terms of the luminous intensity expressed in candelas. See also: *Illumination. Luminance.*

CANDLE POWER, MEAN SPHERICAL. The average value of the candle power of a luminous source in all directions. See also: *Illumination. Luminance.*

CANDLE, STANDARD. Refers to a number of "standard" sources of light which have now been superseded. See also: *Candela. Lamp, standard. Light, primary standard of.*

CANONICAL. Describes a standard form of function, rule, or equation, especially when the form is simple.

CANONICAL ASSEMBLY. A statistical assembly of identical systems with possible energy levels $E_1, E_2, ..., E_i, ...$, such that the relative probability of finding any given system in energy state E_i equals $\exp(-\alpha E_i)$, where α is a constant, the same for all systems. Such an assembly, or ensemble, is obtained in the limit from a *microcanonical assembly* of N identical systems having their tota energy between E and $E + dE$, such that as E and N tend to infinity E/N remains constant and equal to the average energy of an individual system.

CANONICAL DISTRIBUTION. A distribution giving the proportion of systems in an assemblage (e.g. molecules in a gas) whose momenta lie between $p_1 ... p_n$ and $p_1 + dp_1 + ... + p_n + dp_n$, and whose associated coordinates lie between $q_1 ... q_n$ and $q_1 + dq_1 + ... + q_n + dq_n$. It is given by $A \exp(-\text{energy}/\Theta)$, where A is a constant and $\Theta = kT$, k being Boltzmann's constant and T the absolute temperature.

CANONICAL EQUATIONS OF MOTION. For the nth set of generalized coordinates, p_n, and momenta, q_n, in a conservative dynamical system: the equations $\dfrac{\partial H}{\partial p_n} = \dot{q}_n$ and $\dfrac{\partial H}{\partial q_n} = -\dot{p}_n$, where H is the Hamiltonian function.

CANONICAL TRANSFORMATION. A transformation from one set of generalized coordinates and momenta to another, so that the form of the canonical equations of motion is preserved.

CANYON. A term formerly used to denote a long, narrow, heavily shielded building, or major part of a building, concerned with the processing of highly radioactive materials.

CAPACITANCE. (1) The ratio of the electric charge acquired by a body to the resultant change of potential. Usually expressed in terms of coulombs per volt (i.e. farads), or its submultiples. Formerly known as *capacity*. (2) Acoustic capacitance is defined as the negative imaginary part of acoustic impedance.

CAPACITANCE COEFFICIENTS. The coefficients which, together with the coefficients of induction, specify the charges on a system of conductors in terms of their potentials.

CAPACITANCE, DISTRIBUTED. The capacitance inherent in any coil because of the adjacent turns, layers, windings, etc., which are separated by dielectric material, and in which potential differences occur.

CAPACITIVE LOAD. An electrical load whose reactance component is negative, i.e. a load that acts like a combination of resistance and capacitance.

CAPACITOR. A system of conductors (usually in the form of plates) separated by a dielectric, used to secure an increased capacitance or to achieve a capacitance of specified value. Formerly known as a *condenser*.

CAPACITY. (1) The former name for *capacitance*. (2) In electrochemistry: the quantity of useful current which can be delivered by a cell or battery under specified working conditions. Often expressed in ampere-hours.

CAPILLARITY. Refers to those phenomena associated with the surface tension of liquids, originally manifested in the rise of liquids in capillary tubes.

CAPILLARY CONSTANT. The product of the height (h) to which a liquid rises in a capillary tube, and the radius (r) of the tube. It is related to the surface tension γ by $rh = 2\gamma/gd$, where g is the acceleration due to gravity and d is the difference between the densities of liquid and vapour. The capillary constant is often denoted by a^2, having the dimensions of area.

CAPILLARY DEPRESSION. The depression of the meniscus of a mercury surface.

CAPILLARY FORCES. Intermolecular forces which arise particularly in capillary tubes, where liquid, solid and gas interfaces meet. They depend on the nature of the surface forces as well as the temperature.

CAPILLARY WAVES. Waves on the surface of a liquid, whose wavelength is so small that the effect of gravity on the speed of propagation can be neglected. This speed is then given by $(2\pi\sigma/\varrho\lambda)^{1/2}$, where σ is the capillary constant, ϱ the density, and λ the wavelength. Measurement of the speed then serves to determine the surface tension.

CAPRICORN, FIRST POINT OF. The point at which the ecliptic touches the tropic of Capricorn. See also: *Solstice, winter*.

CAPTURE. Of atomic particles, etc.: any process by which an atomic or nuclear system acquires an additional particle.

CAPTURE γ-RADIATION. The γ-radiation emitted in radiative capture.

CARATHÉODORY PRINCIPLE. Refers to a series of axioms from which the principles of classical thermodynamics may be developed without consideration of the properties of heat engine cycles. The treatment circumvents many objections to the more usual procedure (e.g. that this procedure tends to obscure the universal validity of thermodynamics when applied to macroscopic physical processes), and facilitates the application of thermodynamics to complex systems.

CARBON CYCLE. A series of consecutive thermonuclear reactions resulting in the formation of a helium nucleus from a carbon nucleus and four protons. Nitrogen and oxygen nuclei are produced in the intermediate stages of the process, and the carbon nucleus is regenerated at the end. It may thus be said to act as a catalyst in the transformation of hydrogen into helium. The energy liberated is thought to be the main source of energy in a large class of stars, particularly the hotter stars. See also: *Proton–proton chain. Solar energy*.

CARBON FIBRE. A generic term for organic fibre that has been heated to a temperature well in excess of its decomposition temperature. It may be in the form of amorphous carbon or of graphite, depending on the chemical composition and processing conditions. Commercially available products are designated carbon fibre when they have been heated to about 1000°C and graphite fibre when they have been heated at temperatures of 2500°C or more. They are often cemented together in uniaxial array with epoxy or other thermosetting resins. One well-known use of these fibres is in space and underwater exploration on account of their temperature resistant properties and their high strength, high stiffness and low density. They are also used for thermal insulation in various forms of electric heating, in ablative materials and in gear wheels and bearings, to name a few applications.

CARBON RESISTOR. A resistor composed of granules of carbon pressed together, usually in rod form. Such resistors are widely used in electrical circuits, may also serve as heating elements, and have proved useful as resistance thermometers at very low temperatures.

CARBON RHEOSTAT. A variable carbon resistor.

CARDIAC PACEMAKER. An electronic device which delivers a series of controlled electrical impulses to the heart, causing the muscles to contract regularly, thus being able to produce a steady beat at the normal rate. It may be inserted into the body or may be external to the body.

CARDIOID CONDENSER. A condenser used for dark-ground illumination in the microscope. It employs two reflections, the second of which is at a surface whose theoretical shape is a surface of revolution having a cardioid as the meridian section.

CAREY–FOSTER BRIDGE. See: *Bridge, electrical*.

CARLSBAD TWIN. Refers to a characteristic form of twinning in the mineral orthoclase, in which the two halves of the twin are related by a rotation of 180° about the vertical crystallographic axis.

CARNOT CYCLE. A reversible sequence of operations forming the working cycle of an ideal heat engine of maximum thermal efficiency. It consists of isothermal expansion, adiabatic expansion, isothermal compression, and adiabatic compression to the initial state. The operation of the cycle, which is usually given on a pressure–volume diagram, illustrates the second law of thermodynamics.

CARNOT THEOREM: CARNOT LAW. States that no engine operating between two given temperatures is more efficient than a perfectly reversible engine (i.e. a "Carnot" engine) operating between the same two temperatures.

CARRIER. (1) An entity capable of carrying electric charge through a solid. In differing circumstances electrons, holes or ions may serve as carriers. (2) A wave train suitable, after modulation, for transmitting information. (3) An element associated with traces of radio-isotopes of the same element, or of an analogous element, giving a quantity of material sufficiently large to ensure that it follows chemical or physical processes in the manner characteristic of bulk matter.

CARRIER CONCENTRATION. The number of electric current carriers per unit volume.

CARRIER FREQUENCY. The frequency of a wave train on which modulation is imposed for the purpose of transmitting information.

CARTE DU CIEL: ASTROGRAPHIC CHART. A series of photographic charts of the whole sky, organized by the International Astronomical Union, to help in cataloguing star positions. A network of lines is superimposed on the photographs and the positions of the stars are given by comparison with the positions of a limited number of *reference stars*.

CARTESIAN OVAL. The locus of a point P which moves so that its distances from two fixed points A and B satisfy the relation $aPA + bPB = k$, where a, b and k are constants. If $a = b = 1$, the oval is an ellipse with foci A and B.

CASCADE CONNECTION. Of a number of electrical networks: a connection such that the output terminals of the first network are connected to the input terminals of the second, the output terminals of the second to the input terminals of the third, and so on.

CASCADE PARTICLE. See: *Hyperon*.

CASCADE SHOWER. The successive production of electrons and high-energy photons by bremsstrahlung and pair production.

CASCADE UNIT. In the theory of electron–photon cascade showers: the average distance in which the energy of an electron is reduced by bremsstrahlung to $1/e$ of the initial energy, and the almost identical mean range for the absorption of photons by pair production.

CASCADING. In high-voltage insulators: the flashing over under excess voltage, where the arc clings to the surface of a string of insulators.

CASE HARDENING. The production of a hard surface layer on steel by heating it in a suitable medium containing carbon (e.g. molten sodium cyanide) and subsequent quenching.

CASSINIAN OVAL. The locus of a point P which moves so that the product of its distances from two fixed points is a constant.

CASTING. (1) Any process for forming metal objects of predetermined size and shape by transferring molten metal of suitable composition into a mould of the required form, and allowing it to solidify. (2) The product of such a process.

CAT. See: *Clear air turbulence* (*CAT*). *Windtunnel, compressed air* (*CAT*).

CATADIOPTRIC. Involving both the reflection and refraction of light. See also: *Catoptric*; *Dioptric*.

CATALYSIS. The phenomenon of the acceleration (*positive catalysis*) or retardation (*negative catalysis*) of the speed of a chemical reaction by a substance which does not itself appear in the final product and undergoes no permanent change. Such a substance is known as a *catalyst*.

CATALYST. See: *Catalysis*.

CATALYST, HETEROGENEOUS. A catalytic reaction occurring at phase interphases. Also known as *contact* or *surface* catalysis.

CATALYST, HOMOGENEOUS. A catalytic reaction taking place in a single phase.

CATALYTIC POISON. A substance which inhibits the activity of a catalyst.

CATAPHORESIS: ELECTROPHORESIS. The migration of charged colloid particles in an electric field. It is the converse of the Dorn effect. One of its principal uses is as an aid to the purification of proteins of various types (enzymes, hormones, viruses, etc.).

CATASTROPHE THEORY. A mathematical theory which, unlike the calculus of Leibnitz and Newton, deals with systems which do not exhibit "smooth" behaviour but involve sudden discontinuities ("catastrophes"). Examples include shock waves, turbulent flow, the stability of engineering structures and many biological processes. Only seven basically different types of catastrophe are said to occur.

CATHETOMETER. An instrument used for measuring small vertical distances, e.g. the difference in height between two points, by noting the successive positions of a telescope which slides along a vertical scale and is focused in turn on the two points whose vertical separation is required.

CATHODE. The name given to the electrode, usually metallic, which is at a negative potential in any system carrying electric current.

CATHODE CRATER. A crater formed in a cathode surface by positive ion bombardment.

CATHODE DARK SPACE. In a glow discharge tube: the entire region between the cathode and the negative glow. It includes the Aston dark space and the cathode layers. Also known as the *cathode fall region*, *Crookes dark space* or the *Hittorf dark space*. See also: *Cathode fall. Discharge, glow*.

CATHODE FALL. In a glow discharge tube: the drop in potential which occurs in the small distance between the cathode and the plasma boundary. It extends from the cathode to the negative glow and the region involved is known as the *cathode fall region*, the *cathode dark space*, the *Crookes dark space*, or the *Hittorf dark space*. See also: *Cathode dark space. Discharge, glow*.

CATHODE FOLLOWER. A valve amplifier circuit with input connections between anode and grid, and output connections between cathode and anode. The output voltage "follows", i.e. is in phase with, the input voltage.

CATHODE GLOW. In a glow discharge tube: the glow between the Aston dark space and the cathode dark space. See also: *Discharge, glow*.

CATHODE LAYERS. In a glow discharge tube: one or more faint layers adjacent to, and on the anode side of, the Aston dark space. See also: *Discharge, glow*.

CATHODE-RAY OSCILLOSCOPE. A cathode-ray tube used as an oscilloscope. See also: *Cathode-ray tube*.

CATHODE RAYS. Streams of electrons emitted from the cathode in low pressure discharge tubes or similar devices.

CATHODE-RAY TUBE. A vacuum device in which a beam of electrons emitted from the cathode is accelerated to a relatively high-voltage anode. The electrons are concentrated into a narrow beam by electrostatic or electromagnetic fields and strike a fluorescent screen causing a bright spot. The beam may be deflected by internal plates or external coils by the application of appropriate voltages, and the position of the spot on the fluorescent screen deflected accordingly. The cathode-ray tube may be used as an oscilloscope for studying current or voltage wave-forms, as the picture tube of a television receiver, etc.

CATHODE SPOT. The luminous region surrounding the carbon cathode in a carbon arc, operating in air, when the cathode has worn away to a point.

CATHODE SPUTTERING. A method of depositing extremely thin layers of metal on a surface, by bombarding a metal cathode with positive ions in a moderate vacuum and allowing the metallic atoms released by the bombardment to fall uniformly upon the object to be plated.

CATION. A positively charged ion, i.e. one which travels towards the cathode.

CATOPTRIC. Involving the reflection of light. See also: *Catadioptric. Dioptric.*

CAUCHY DISPERSION FORMULA. Refers to the curve of normal dispersion. It is

$$n = A + \frac{B}{\lambda^2} + \frac{C}{\lambda^4}$$

where n is the refractive index, λ the wavelength and A, B, C are constants characteristic of the medium. See also: *Dispersion.*

CAUCHY INTEGRAL. If C is a simple closed contour and $f(z)$ is a holomorphic function in a region containing C and its interior D, then, if a is any point of D, the function $f(a)$ may be represented by the Cauchy integral

$$f(a) = \frac{1}{2\pi i} \int_C \frac{f(z)\, dz}{z - a}.$$

CAUCHY RELATIONS. A series of relationships between the elastic constants of a crystal based on the assumption that all the forces between the atoms are central forces. This assumption is not justified.

CAUCHY RESIDUE THEOREM. See: *Calculus of residues.*

CAUCHY–RIEMANN EQUATIONS. Equations arising in the theory of functions of a complex variable. If $\dfrac{\partial u}{\partial x} = \dfrac{\partial v}{\partial y}$ and $\dfrac{\partial u}{\partial y} = -\dfrac{\partial v}{\partial x}$, where u and v are both functions of x and y, these equations, the Cauchy–Riemann equations, will be satisfied for an analytic function $(u + iv)$ of the complex variable $z = (x + iy)$. The equations are a necessary but not a sufficient condition for the function $(u + iv)$ to be analytic.

CAUCHY THEOREM. States that, if C is a closed contour and $f(z)$ is a holomorphic function in a region containing C and its interior, then $\int_C f(z)\, dz = 0$.

CAUSTIC. Neighbouring rays emanating from the same object point and passing through an optical system will not pass through the same image point but, owing to aberration, will intersect at different points. The totality of these point images lies on a surface known as a caustic.

CAVENDISH EXPERIMENT. A direct determination of the gravitational constant by the measurement, using a torsion balance, of the mutual attraction between two bodies of small size and known mass. The experiment was performed by Cavendish in 1798.

CAVITATION. The formation of local cavities in a liquid as a result of the reduction of total pressure. Collapse of these cavities produces large impulsive pressure which may cause considerable damage to neighbouring solid surfaces. Cavitation may be induced by ultrasonic waves and used, among other things, for drilling and cutting; or it may occur in the operation of hydraulic pumps and turbines, or behind the blades of a ship's propeller, with destructive erosion.

CAVITATION NUMBER: CAVITATION PARAMETER. A non-dimensional parameter providing a measure of the risk of cavitation occurring in a particular flow system. It is given by $(P - p_c)/\frac{1}{2}(\varrho U^2)$, where P is the static pressure at a reference position in the flow, p_c is the vapour or cavity pressure, ϱ is the mass density of the liquid, and U is the velocity of the stream past the reference position.

CAVITATION TUNNEL: WATER TUNNEL. Experimental equipment used for testing and research, in which cavitation phenomena are reproduced.

CAVITY ABSORBENT. A cavity, essentially a Helmholtz resonator, consisting of an enclosed volume of air connected by a short narrow tube or neck with a region containing a field of sound, and used in buildings for sound attenuation. The maximum amount of sound energy is absorbed at the natural fundamental frequency of the cavity $\dfrac{c}{2\pi}\sqrt{\dfrac{a}{lV}}$, where c is the speed of sound, a the cross-sectional area of the neck, l the length of the neck and V the volume of the cavity. For high frequencies (say above 1000 Hz) coupled resonators are often used, consisting, for example, of a large number of equally spaced holes drilled in an air-tight sheet located at a fixed distance from a solid boundary.

CAVITY FIELD. In the calculation of the field acting on a molecule or group of molecules in a dielectric: the field, due to the external applied field, which would exist in a cavity in a continuum having the electrical properties of the bulk material, if the molecule or group of molecules under consideration, assumed to be present in the cavity, were in fact absent.

CAVITY FLOWS. Liquid flows involving coherent gas-filled spaces, often, but not always, produced by cavitation.

CAVITY RADIATION. Synonym for black-body radiation. See also: *Black body.*

CAVITY RESONATOR. An electrical circuit, resonant to radio frequencies, in the form of a closed box. It will resonate at a number of frequencies, depending on the dimensions of the cavity, but in practice only the lower modes of oscillation are important. Cavity resonators are used, for example, as resonant circuits in microwave valves (e.g. the klystron and magnetron), as tuning elements, in wave meters and filters, and in frequency measurements.

CAVITY RESONATOR, FORM FACTOR OR SHAPE FACTOR OF. The quantity Qd/λ where Q is the selectivity of the resonator, d is the skin depth (i.e. the depth of penetration of the current in the cavity walls) and λ is the wavelength.

CAVITY RESONATOR, SELECTIVITY OF. This is defined as

$$2\pi \left(\frac{\text{energy stored}}{\text{energy lost/cycle}} \right).$$

It is approximately equal to v/da where v is the volume of the cavity, a is the area of the inner surface, and d is the skin depth (i.e. the depth of penetration of the current in the cavity walls).

C-CENTRE. See: *Colour centres.*

CEILING TEMPERATURE. In polymerization: the temperature above which the formation of a long-chain polymer from monomer at a given concentration is impossible.

CELESTIAL EQUATOR. The great circle in which the plane of the Earth's equator cuts the celestial sphere.

CELESTIAL SPHERE. A sphere of infinite radius on which the stars are envisaged as lying. Any point on the Earth, at any position in its orbit round the Sun, may be considered to be the centre of this sphere (except that a very few stars have an annual parallax of up to about 1 sec). The *poles* of the sphere are the points at which it is cut by the prolongation of the Earth's axis of rotation; and the sphere may be imagined to rotate about this prolonged axis once in 24 sidereal hours, although the exact situation is confused by the effects of precession of the equinoxes, nutation and other slight irregularities of rotation.

CELL, NOTCHED-ÉCHELON. A device that enables the complete absorption curve of a liquid to be obtained from a single exposure through a spectrograph. It involves the use of a series of steps giving ten thicknesses of liquid in geometrical progression and the provision of a notch which allows a reference beam to bypass the sample.

CEMENTATION. The process whereby elements are introduced into the outer layers of metal objects by diffusion at elevated temperatures.

CEMENTITE. The brittle iron carbide constituent (Fe_3C) of steel and cast iron.

CENT (NUCLEAR REACTOR TECHNOLOGY). A unit of reactivity equal to one-hundredth of a dollar. See also: *Dollar (nuclear reactor technology).*

CENTIGRADE HEAT UNIT (C.H.U.). A little used unit of heat representing the quantity of heat required to raise the temperature of 1 lb of pure water by 1°C at normal atmospheric pressure and a specified temperature. Also known as *centigrade thermal unit (C.T.U.)* or *pound-calorie.*

CENTRAL FORCES. Forces which are directed towards or away from some fixed centre or origin and whose magnitudes are functions of position with respect to that origin. Examples are found in the coulomb interaction effects and in planetary motion. *Central nuclear forces* are those resulting from a spherically symmetrical potential distribution. Although they are convenient to handle theoretically, the forces between nucleons are not always of the central type.

CENTRE OF COMPRESSION AND TWIST. That point, in each cross section of a twisted thick beam, at which the elastic strain is zero. Also known as the *neutral point.*

CENTRE-OF-MASS SYSTEM. See: *Coordinates, laboratory system of.*

CENTRE OF OSCILLATION. Of a compound pendulum: a point which, when the pendulum is at rest, is vertically below the point of suspension at a distance equal to the length of the equivalent simple pendulum.

CENTRE OF PRESSURE. (1) Of an aerofoil: that point at which the resultant of the aerodynamic forces on the aerofoil cuts the chord line or some similar reference line. Its distance behind the leading edge is usually given as a fraction of the chord length. (2) Of a surface immersed in a fluid: that point at which the resultant pressure on the surface may be taken to act.

CENTRE OF SYMMETRY. In a crystal: the point about which like faces or edges are arranged in pairs in corresponding positions on opposite sides. Similarly, the point within the unit cell about which the atomic positions are related in the same fashion.

CENTRIFUGAL FORCE. An apparent force experienced by a body in a reference frame which is rotating with respect to a stationary frame. For a body of mass m, moving in a circle of radius r with angular velocity ω, this force is equal to $m\omega^2 r$ and its direction is outwards along the radius. It is, mathematically, an imaginary inertial force introduced to make valid the use of Newton's third law of motion in a moving frame of reference, and may be regarded as the reaction to the centripetal force.

CENTRIFUGAL SEPARATOR. A device for the separation of solid, liquid and gaseous phases, or of different materials in the same phase. It is based upon the use of centrifugal acceleration, which is imparted by the rotation of the containing vessel. See also: *Cyclone separator*.

CENTRIFUGE. A machine used for the separation of solids from liquids or liquids from other liquids by high-speed rotation. Ordinary centrifuges are characterized by speeds of about 5000 rpm. *Ultracentrifuges* may have speeds of 60000 rpm or more.

CENTRIPETAL FORCE. The force which acts on a body which is moving about a fixed point so as to deflect that body from a straight path. For a body of mass m, moving in a circle of radius r with angular velocity ω, this force is equal to $m\omega^2 r$ and its direction is inwards along the radius; and the corresponding *centripetal acceleration* is $\omega^2 r$.

CEPHEID VARIABLES. A group of variable stars of short period whose brightness varies in a specified way, and of which the star δ-Cephei is the prototype. A similar group of stars is represented by the *RR Lyrae* or "*cluster variables*" which have a shorter period and appear to be distinct in physical characteristics. The existence of other short period groups is suspected, each of which is characterized by its own period–luminosity relation. See also: *Stars, variable*.

CERAMICS. Traditionally refers to pottery, porcelain, and refractory or high-temperature oxides. More recently the term has been used to cover any refractory material, for example carbides, nitrides and intermetallic compounds, and a whole series of new uses has been found for such materials, e.g. in turbine blades and as fuel for nuclear reactors.

ČERENKOV DETECTOR. A detector for charged particles which consists essentially of a transparent medium which emits Čerenkov radiation and is viewed by a photomultiplier tube.

ČERENKOV RADIATION. Bluish light emitted when a charged particle moves in a transparent medium with a speed greater than that of light in the same medium. The phenomenon involved (the *Čerenkov effect*) is that of an electromagnetic shock wave; and is the optical analogue of the sonic boom or bang.

CERES. See: *Asteroid*.

CERMET. A metal–ceramic mixture, usually fabricated into a suitable shape, in which are combined the desirable properties of both types of material, e.g. the high thermal conductivity of a metal and the heat-resisting properties of a ceramic.

CGS UNITS. See: *Units, CGS*.

CHAIN REACTION. A series of reactions in which one of the agents necessary to the series is itself produced by the reactions so as to cause similar reactions. Examples are: *chemical chain reactions* and *nuclear chain reactions*.

CHAIN REACTION, NEUTRON. A nuclear chain reaction in which the agent is a neutron which has been produced in the fission process. According as the resulting neutron population decreases with time, remains constant, or increases, the reaction is said to be *convergent (sub-critical)*, *self-sustained (critical)*, or *divergent (supercritical)*, respectively. See also: *Chain reaction. Multiplication factor: Multiplication constant. Neutron multiplication*.

CHANDLER WOBBLE. The wandering of the celestial poles relative to the stars. It has a periodicity of about 14 months and causes minute variations in the meridian.

CHANDRASEKHAR ABSORPTION CONSTANTS. Constants expressing the absorption of stellar radiation by interstellar clouds. The clouds are pictured, not as separate, but as a continuous cloud of fluctuating density.

CHANDRASEKHAR METHOD OF ENVELOPES. A method used in the study of the behaviour and eventual breakdown of certain models of stars of large masses, which takes full account of the radiation pressure, neglected in the usual mass–luminosity treatments.

CHANNELLING: STREAMING. The increased transmission of electromagnetic or particulate radiation through a medium resulting from the presence of extended voids or other regions of low attenuation.

CHANNEL WIDTH IN NUCLEAR REACTION THEORY. That part of the total energy width of a nuclear energy level which corresponds to decay through a particular mode or channel.

CHAPMAN–JOUGET STATE. A point on the Hugoniot curve of pressure versus temperature in a detonating gas at which the detonation velocity equals the sum of the particle velocity and the sound velocity in the burned gas.

CHARACTERISTIC CURVE. (1) In photography: a curve showing, for a photographic emulsion, the variation in photographic density with the logarithm of exposure under specified conditions of processing. (2) Of a discharge: the curve showing the variation in resultant current with applied voltage.

CHARACTERISTIC EQUATION. See: *Equation of state*.

CHARACTERISTIC IMPEDANCE. (1) Of a transmission system: the square root of the ratio of the series impedance to the shunt admittance. (2) In an electrical cable: the impedance which would be measured at the input terminals of an infinitely long line having the same characteristics per unit length as the cable in question. See also: *Acoustic impedance, characteristic*.

CHARACTERISTIC OF A LOGARITHM. See: *Logarithm*.

CHARACTERISTIC RADIATION. X-rays consisting of a series of groups of discrete wavelengths which are characteristic of the emitting element and independent of its state of chemical combination. Characteristic radiation arising from the absorption of X- or γ-rays is sometimes called *fluorescence X-rays*. See also: *X-rays*.

CHARACTERISTICS (IN THE SOLUTION OF HYPERBOLIC DIFFERENTIAL EQUATIONS). Refers to certain curves from which the calculation of the higher derivatives from the partial differential equation is not possible.

CHARACTERISTIC VELOCITY OF AN ELECTROMAGNETIC WAVE FRONT. The velocity, in a homogeneous medium, of the boundary between a finite moving field and a field-free space. It is the same as the *phase velocity*, and is given by $1/\sqrt{\mu\varepsilon}$ where μ is the permeability and ε is the dielectric constant of the medium.

CHARACTERISTIC WAVELENGTH OF AN ELECTROMAGNETIC WAVE FRONT. The characteristic velocity divided by the frequency.

CHARCOAL. The residue from the destructive distillation of wood or animal matter. The open structure of charcoals confers high adsorptive capacity which may become apparent, however,

only when the charcoal is *activated*, i.e. when organic products are removed.

CHARGE CHANGING AND CHARGE NON-CHANGING CURRENTS. See: *Neutral currents*.

CHARGE CONJUGATION. The operation of changing the signs of all electric charges and electromagnetic fields in a system. A reaction and its conjugate reaction (in which each particle is replaced by its anti-particle) are identical.

CHARGE DENSITY, ELECTRIC. The charge per unit volume. It is sometimes used to mean the *surface charge density*, which is the charge per unit area of a conductor, for example, where the charge must reside on the surface.

CHARGED PARTICLE TRAPPING. The containment of charged particles in a magnetic field, as in the cyclotron (*absolute trapping*) or certain machines used in experiments on controlled thermonuclear reactions (*adiabatic trapping*).

CHARGE, ELECTRIC. A quantity of electricity, resident, for example, on a body, or transmitted through a medium. The unit of charge in the International System (SI) is the coulomb. The "natural" unit is the charge on the electron, which is $1·6022 \times 10^{-19}$ C. See also: entries under *Units*. See: *Appendix III*.

CHARGE EXCHANGE. The transfer of electrical charge between two colliding particles, e.g. between an ion and a neutral atom in a plasma.

CHARGE INDEPENDENCE. A hypothesis which postulates that the nuclear forces between any two nucleons are the same, provided they have the same angular momentum and spin.

CHARGE INVARIANCE. A hypothesis which postulates that the nuclear forces between any two nucleons are *invariant* under rotation in isotopic spin space. This implies that the electric charges are independent of the nuclear forces and vice versa.

CHARGE-TRANSFER COMPLEX. A complex between molecules, radicals, ions or atoms, having an excited state characteristic of the complex as a whole, in which, relative to the ground state, essentially one electron has been transferred from one part of the complex (the "electron donor") to the other part (the "electron acceptor").

CHARLES LAW: CHARLES–GAY-LUSSAC LAW. For a perfect gas: states that the volume of a given mass of any gas at constant pressure rises with temperature by a constant fraction of the volume at 0°C, i.e. the coefficient of expansion is the same for all gases. The fraction is about $\frac{1}{273}$ or 0·003 66 per °C and is a fair approximation for many gases.

CHARM. See: *Quarks. Quantum number internal.*

CHARMONIUM. See: *J/Psi particle.*

CHARPIT EQUATIONS. Equations employed in a method developed by Lagrange and perfected by Charpit for finding a complete integral of the general first-order partial differential equation in two independent variables.

CHARPY TEST. See: *Impact tests.*

CHATTOCK GAUGE. A sensitive pressure gauge for measuring very small pressure differences. It consists essentially of a tilting manometer which balances the pressure to be measured against the pressure head caused by the tilt.

CHELATION. The formation of a cyclic chemical structure through hydrogen bonds or coordination linkages between electron donors and acceptors.

CHEMICAL CONSTANT. See: *Vapour pressure, general equation for.*

CHEMICAL EQUILIBRIUM. The condition of a system in which the material composition of any given macroscopic portion of matter, under specified conditions of temperature and pressure, remains unchanged with time, i.e. in which opposing reactions take place at equal rates. The existence of chemical equilibrium is a necessary, but not sufficient, condition for true thermodynamic equilibrium.

CHEMICAL EXCHANGE REACTION. See: *Isotopic exchange reaction.*

CHEMICAL POTENTIAL. A thermodynamic variable bearing a relation to the flow of matter similar to that which temperature bears to the flow of heat, i.e. matter flows in a chemical reaction from a region of higher chemical potential to one of lower.

CHEMICAL REACTION, COLLISION THEORY OF. A theory which relates the rate of a chemical reaction to the collision diameter of the participating molecules and to their kinetic energies. See also: *Collision diameter.*

CHEMICAL REACTION, HETEROGENEOUS. A reaction in which more than one phase is present at some stage during the reaction. It is usual to consider only those reactions which involve at least one solid phase.

CHEMICAL REACTION, HETEROLYTIC. A reaction in which a covalent bond is formed by coordination (both electrons supplied by one of the combining species) or is broken by the reverse process (*heterolysis*).

CHEMICAL REACTION, HOMOGENEOUS. A reaction in which only one phase is present.

CHEMICAL REACTION, HOMOLYTIC. A reaction in which a covalent bond is formed from electrons supplied one by each of the combining species (*colligation*) or is broken by the reverse process (*homolysis*).

CHEMICAL REACTION, ORDER OF. (1) With respect to a component of a chemical reaction: the order to which the concentration of that component is raised in the expression for the rate of the reaction. (2) With respect to a chemical reaction as a whole: the sum of the orders of the individual components.

CHEMICAL REACTION, REVERSIBLE. A reaction with appreciable reaction rates in both directions. It is therefore incomplete, a mixture of reactants and reaction products being obtained unless one of the products is removed as it is formed.

CHEMICAL REACTIONS, CONSECUTIVE. Reactions which proceed from reactants to final products through one or more intermediate reactions.

CHEMICAL REACTION, SPECIFIC REACTION RATE CONSTANT OF. A constant which, when multiplied by the product of the concentrations of the components of a chemical reaction, each raised to the appropriate power, gives the rate of the reaction. Also known as *velocity constant*. See also: *Chemical reaction, order of.*

CHEMICAL REACTION, SUBSTITUTION. A reaction in which one atom or group is replaced by another. The term was originally restricted to the particular class of organic reactions in which a hydrogen atom of a hydrocarbon was replaced by another atom or group.

CHEMICAL REACTION, TEMPERATURE CO-EFFICIENT OF. The way in which the rate of a chemical reaction varies with temperature, a good measure of which is the experimentally observed activation energy, E, which enters as $\exp(E/RT)$ in the expression for the reaction rate, where R is the gas constant per mole and T the absolute temperature.

CHEMICAL REACTION, UNIMOLECULAR. An elementary reaction that involves only one reactant molecule. Such a reaction can strictly only occur in the gas phase, but some non-ionic reactions in solution are often treated as if they were unimolecular.

CHEMICAL REACTIVITY. There is no agreed quantitative definition of this term, but, for the reactivity of a given substance for a reaction with a specified reagent, the reaction rate constant at some agreed temperature is usually employed as a quantitative measure of the reactivity. See also: *Chemical reaction, specific reaction rate constant of.*

CHEMICAL RECORDER. A device which records signals on impregnated paper through the chemical action produced by the flow of current in the chemical with which the paper is impregnated.

CHEMILUMINESCENCE. The emission of light during a chemical reaction at an intensity which cannot be attributed to temperature radiation. It arises from the fluorescence of electronically excited molecules or atoms.

CHEMISORPTION. See: *Absorption, chemical.*

CHEMISTRY. The study of the chemical properties of elements and compounds, and of the causes, courses and consequences of the reactions in which they take part.

CHÉVENARD MACHINE. A mechanical testing machine for very small specimens.

CHILD–LANGMUIR EQUATION. Relates the current density, I, which can be drawn under space-charge-limited conditions between infinite parallel plane electrodes to the potential, V, applied between the electrodes. It may be written as $I = AV^{3/2}$, where A is a constant, known as the *perveance*. The equation is a special case of *Child's equation* connecting anode current and anode voltage for a diode, which may be written as $C = KV^x$, where C is the anode current, V the anode voltage, and K, x are constants.

CHIP. See: *Microcircuit.*

CHI-SQUARED TEST. One of a large class of statistical significance tests, sometimes termed goodness-of-fit tests, based on a single test statistic whose distribution is of a particular type (the χ^2 type) appropriate to a series of data which are quantitative only in that it is known into which of a number of different categories they fall.

CHLADNI FIGURES. The figures, demonstrating the vibrations of a glass or metal plate, obtained when sand, sprinkled on the plate, collects along the nodal lines.

CHOKE. An arrangement (e.g. an inductance coil) whereby energy at high frequency is prevented from leaving a system at some point while energy at much lower frequencies may be extracted or fed in as required.

CHOKE COUPLING. An arrangement for the connection of two sections of a transmission system which provides a minimum loss of high-frequency energy without the necessity for conductive contact between the sections.

CHOKING OF NOZZLES. (1) A convergent nozzle or one of constant cross-sectional area is said to be choked when the velocity at the exit is sonic. (2) A subsonic wind tunnel is choked when the velocity is sonic across a section at the model, so that no further increase of speed is possible. (3) In a supersonic wind tunnel, choking occurs when the model is so large that supersonic flow past it cannot be established.

CHOLESTERIC PHASES. See: *Liquid crystals.*

CHONDRITES. Debris from broken planets (probably the asteroids), believed to be in or near the same state that the rocky matter of the inner planets initially assumed when the solar system was formed. About 85% of meteorites are chondrites and 50% or over of a chondrite is made up of *chondrules*. See also: *Chondrules.*

CHONDRULES. Small spheroidal structures, 0·5–3 mm across, which make up over 50% of the volume of some chondrites. Their composition suggests that they were once dispersed droplets of molten lava, and possibly represent the first material to condense when the solar system was formed. See also: *Chondrites.*

CHOPPER. (1) For neutrons: a mechanical shutter, opaque to neutrons apart from slits or openings which pass through it, rotating at a high speed in a well-collimated beam of neutrons so as to produce bursts of neutrons. Such a chopper is an essential component of a time-of-flight neutron spectrometer and may be designated as *fast* or *slow* according to the speed of the neutrons being "chopped". (2) For light: a device for periodically interrupting the light in an instrument, e.g. to permit the use of a.c. amplifiers in measuring apparatus, or to produce light pulses for use in opto-acoustic detection.

CHORD: CHORD LENGTH (AERODYNAMICS). The length of that part of the chord line intercepted by the extremities of the leading and trailing edges of an aerofoil section.

CHORD LINE (AERODYNAMICS). The straight line joining the centres of curvature of the leading and trailing edges of an aerofoil section.

CHRISTIANSEN FILTER. A selective light filter which may operate in the infrared, visible or ultra-violet region of the spectrum. Such a filter comprises small solid crystals suspended in a liquid having a different dispersion but with a refractive index equal to that of the crystals at the wavelength at which maximum transmission is required.

CHROMATIC ABERRATION. An aberration in an optical system arising from the variation of refractive index and focal length with wavelength.

CHROMATIC COLOUR. A colour which is not a grey.

CHROMATICITY. The colour quality of light, definable by its chromaticity coordinates. See also: *Chromaticity coordinate.*

CHROMATICITY CHART: CHROMATICITY DIAGRAM. A plane diagram formed by plotting one of the three chromaticity coordinates against either of the others.

CHROMATICITY COORDINATE. The ratio of any one of the tristimulus values of a sample to the sum of the three tristimulus values. Since the sum of the three chromaticity coordinates is unity, any two of them determine the *chromaticity.*

CHROMATIC PARALLAX. In an optical instrument: the parallax between a line image and a graticule when the wavelength of the incident light is varied.

CHROMATIC SENSATION. One which gives a sensation of colour, i.e. not a grey.

CHROMATOGRAPHY. A separation process by which mixtures of chemical compounds may be resolved. Typically, the mixture, usually dissolved in a small volume of suitable liquid, is placed on a retaining medium (the *sorbent*), perhaps in the form of a packed column, through which a solvent (liquid or gas) is allowed to percolate, the components of the mixture being thereby moved differentially in zones along or through the medium. In *adsorption chromatography* the components of the mixture are retained on the sorbent by surface forces. In *partition chromatography* the sorbent is a relatively inert solid (the most important being paper, in *paper chromatography*) and the separation of the components depends on the partition of each between the mobile and stationary phases. In *gas-phase chromatography* the mobile phase is a gas and the sorbent usually consists of an inert solid containing a sorbed organic liquid. In *ion-exchange chromatography* use is made of sorbents containing ionizable or functional groups capable of undergoing chemical reaction with components of the mixture to be separated. This type of chromatography is applicable to the separation of different ionic species. Originally the various components were recognized by their colours, but other properties (e.g. fluorescence in ultra-violet light) are also used.

CHROMIZING. The process of producing a surface layer of high chromium content on iron and steel, by heating to a high temperature in a packing material containing chromium salts or an atmosphere containing $CrCl_2$.

CHROMO-OPTOMETER. An optometer in which the chromatic aberration of the eye is used to determine its refraction.

CHROMOPHORE. See: *Bathochromic shift.*

CHROMOSOME. A unit of structure in the nucleus of a biological cell, which contains the genes which control heredity. The number of chromosomes present in the cell is constant for a given species.

CHROMOSPHERE. That layer of the Sun's atmosphere that can be seen for a few seconds as a red coloured crescent at certain periods of a solar eclipse. It occupies the region from about 0 to 8000 km above the level represented by the sharp limb of the visible Sun. It is above the photosphere and reversing layer. See also: *Solar atmosphere.*

CHRONOGRAPH. An instrument for recording the time at which a given event takes place. It may involve pen recording on a rotating drum or photographic recording of an oscillograph.

CHRONOMETER. A balance wheel type of clock fitted with a very accurate escapement.

CHRONON. An assumed quantum of time. See also: *Time.*

CINE-RADIOGRAPHY. See: *Flash radiography.*

CIRCLE DIAGRAM. An Argand diagram used in electrical engineering to represent, for example, the magnitude and phase of a vector impedance. See also: *Argand diagram.*

CIRCLE OF LEAST CONFUSION. See: *Astigmatism.*

CIRCUIT. The closed path along which electric currents can flow. See also: entries under *Electrical circuit.*

CIRCUIT BREAKER. A switch designed to interrupt a circuit automatically when an excessive current flows.

CIRCUIT ELEMENT, BILATERAL. A circuit element through which current may flow in either direction.

CIRCUITS, TOPOLOGY OF. The study of networks of physical components in terms of their connective properties only, so that the laws of topology can be applied.

CIRCULAR ELECTRIC WAVE. A wave in which the lines of electric force are circles. It can be generated by a circular filament carrying a current which is uniform along the filament.

CIRCULAR FUNCTION. (1) For a real argument: a trigonometric function. (2) For a complex argument: a hyperbolic function.

CIRCULAR MAGNETIC WAVE. A wave in which the lines of magnetic force are circles, e.g. in the magnetic field in a plane at right angles to a radiating dipole.

CIRCULATING ELECTROMAGNETIC WAVE. A wave in which the equiphase surfaces are semi-infinite planes originating from an axis—the *axis of circulation*.

CIRCULATION. The line integral of a vector around a closed contour in a moving fluid. The vector usually considered is the flow velocity, $\mathbf{v}$, and the circulation is then $\oint_c \mathbf{v} \cdot d\mathbf{s}$, where $d\mathbf{s}$ is a vector element of the path.

CISSOID CURVE. A curve which is defined by $y^2 = x^3/2(a - x)$. It has a cusp at the origin and $x = 2a$ as asymptote.

CIS–TRANS ISOMERISM. A type of geometrical isomerism in which differences in the spatial arrangement of atoms or groups lead to differences in properties other than optical activity.

CIVIL TIME. See: *Time*.

CIVIL TWILIGHT. See: *Twilight*.

CLADDING. Of nuclear reactor fuel: See: *Nuclear reactor fuel*.

CLAIRAUT AND STOKES THEOREMS. Theorems dealing with the form of an equipotential surface which encloses all the matter of a rotating body; of importance in geophysics.

CLAIRAUT EQUATION. A first-order differential equation of the type $y = xp + f(p')$. Its general solution is $y = cx + f(c)$ where c is an arbitrary constant.

CLAPEYRON EQUATION. See: *Clausius–Clapeyron equation: Clapeyron equation*.

CLARK RULE. A modification of the Morse rule relating r, the equilibrium internuclear distance of a diatomic molecule, and ω, the equilibrium vibrational frequency. It states that $\omega r^3 \sqrt{n} = k - k'$ where n is the group number, assigned according to the number of shared electrons, k is a constant characteristic of the period, and k' a correction for singly ionized molecules of that period (zero for neutral molecules). See also: *Birge–Mecke rule. Morse rule*.

CLASSICAL PHYSICS. That part of physics in which matter or energy is considered to be a continuum, e.g. classical mechanics as opposed to quantum mechanics, or classical thermodynamics as opposed to quantum statistics.

CLATHRATE. A type of substance having a crystal structure in which certain of the atoms are arranged in the form of a cage, which permits the imprisonment (*clathration*) of small unrelated atoms or molecules without ordinary chemical union between the two species.

CLAUDE–HEYLANDT LIQUEFACTION PROCESS. A modification of the Claude process in which some of the precooled air is diverted from the compressor to a nozzle, where adiabatic expansion produces liquefaction, the unliquefied air being recirculated.

CLAUDE LIQUEFACTION PROCESS. A method of liquefaction in which compressed air which has been cooled by its adiabatic expansion in a cylinder is used to precool the incoming air before it, in turn, is expanded.

CLAUSIUS–CLAPEYRON EQUATION: CLAPEYRON EQUATION. An equation which gives the relation between the vapour pressure of a liquid and its temperature. It may be written as $\dfrac{dp}{dT} = \dfrac{\Delta H}{T\Delta V}$, where p is the pressure, T the temperature, ΔH the change in heat content and ΔV the change in volume.

CLAUSIUS–MOSOTTI EQUATION. An equation which gives the relation between the total polarization, P, of a dielectric and, k, the dielectric constant. It may be written as $P = \dfrac{(k - 1)}{(k + 2)} \cdot \dfrac{M}{\varrho}$, where M is the molecular weight and ϱ the density.

CLAUSIUS THEOREM. States that, for a system undergoing any reversible cycle of changes in which it returns finally to its initial state, $\oint \dfrac{\Delta Q}{T} = 0$, where ΔQ represents the infinitesimal quantity of heat absorbed by the system at $T°$K.

CLEAR AIR TURBULENCE (CAT). Strong atmospheric turbulence occasionally encountered at great heights above the Earth's surface, notably near a jet stream. This turbulence is invisible and presents a hazard to aircraft.

CLEAVAGE. Of a crystal: the property of splitting much more readily along a certain few planes than along any others.

CLIMATE. Of a given locality: the average weather condition in that locality. The weather statistics involved include mean values, distributions, standard deviations and extremes, and refer in general to monthly, seasonal or annual values obtained from a sample of several decades. The study of these statistics is known as *climatology*. See also: *Weather*.

CLINOMETER. An instrument used to determine the angular relationship between a sloping surface (e.g. the deck of a ship) and the horizontal plane.

CLOCK. An instrument for measuring time. Clocks can be broadly divided into pendulum clocks, balance wheel clocks, quartz clocks and atomic clocks, depending on the regulating device employed. See also: *Second*.

CLOCK, ATOMIC. A clock in which the regulating device is the characteristic vibration frequency of a given atom or molecule in particular electric or magnetic states. See also: *Second*.

CLOCK, LE ROY. The first clock (1752) to incorporate a type of chronometer escapement.

CLOCK, OBSERVATORY. A clock, of the pendulum type, to which the impulses necessary to keep the pendulum in motion are transmitted electrically from a master pendulum.

CLOCK OR WATCH, BAROMETRIC COEFFICIENT OF. The variation in the rate arising from variation in barometric pressure. It is usually expressed in seconds per day per millimetre of mercury change of pressure.

CLOCK OR WATCH, ESCAPEMENT OF. The mechanism which transmits the power from the weight or spring to maintain the vibration of the pendulum or balance wheel.

CLOCK OR WATCH, MOVEMENT OF. The means by which the timing mechanism is kept in motion. The term is sometimes applied to all moving parts.

CLOCK OR WATCH, QUARTZ. A clock or watch in which the regulating device is an oscillator which is controlled by a quartz crystal kept in resonant vibration by piezo-electric means.

CLOCK OR WATCH, RATE OF. The difference between a time interval measured by a clock or watch and the true time interval as given by a standard clock. It is usually expressed in seconds per day.

CLOCK OR WATCH, TEMPERATURE CO-EFFICIENT OF. The variation in the rate arising from variation in temperature. It is usually expressed in seconds per day per degree.

CLOCK PARADOX. Concerns the difference in time between two identical clocks which are initially synchronized when at rest together, one being accelerated with respect to the other and eventually returning to its initial resting position. Such a difference is a consequence of the special theory of relativity and the paradox involves the question of distinguishing between the two clocks, since all motion is relative within that theory.

CLOCK, PENDULUM. A clock in which the regulating device is a swinging pendulum.

CLOCK, RADIUM. A proposed clock in which the regulating device would be the periodic charging and discharging of an electroscope whose charge is derived from β-rays from a radium source.

CLOCK, RIEFFLER. A pendulum clock in which errors due to interaction between the pendulum and the escapement are minimized by suspending the pendulum in a particular manner.

CLOCK, SCHULER. A clock in which the regulating device is a compound pendulum driven electromagnetically by impulses given to the upper weight. The impulses are controlled photoelectrically by the passage of the lower weights through the central position of swing.

CLOCK, SHORTT. A clock employing two pendulums one of which, the driven pendulum (driven by a falling weight), synchronizes the other pendulum, that of the working clock, electrically.

CLOSED CYCLE. Of a heat engine: a cycle of operation in which the same heat-transfer fluid is used repeatedly.

CLOSED-LOOP SYSTEM. See: *Servomechanism*.

CLOSED SHELL. Of electrons in an atom or molecule: characterizes a system in which all the quantum states of a particular shell are filled. See also: *Electron shell*.

CLOSE-PACKED STRUCTURE. The structure obtained when spheres of equal radii are packed together so as to occupy a minimum volume. Two such structures are possible: face-centred cubic and close-packed hexagonal.

CLOSURE DOMAINS: FLUX-CLOSURE DOMAINS. Small domains so situated and so magnetized as to close the flux path between adjacent (larger) domains without the formation of magnetic poles.

CLOUD. A visible aggregate of small particles of ice or water suspended in free air, formed by condensation when warm moist air rises into cooler regions. Clouds may be classified according to form and height. By form there are two principal classes: the *cumuliform* (heap) clouds and the *stratiform* (layer) clouds. By height there are three groups: the high clouds (usually 5–13 km above sea level)—*cirrus* (*cirrostratus* and *cirrocumulus*), the medium clouds (2–5 km)—*altostratus* and *altocumulus*, and the low clouds (below 2 km)—*stratus* and *stratocumulus*. See other entries under *Cloud*.

CLOUD CHAMBER. An instrument for making the tracks of ionizing particles visible as rows of little droplets, formed by condensation from a supersaturated vapour. In the original form (the *expansion chamber* or *Wilson chamber*) supersaturation is produced, for a short time, by rapid expansion which causes adiabatic cooling. In a later instrument (the *diffusion chamber*) a supersaturated layer exists continuously as vapour diffuses through a temperature gradient from a heat source to a region of lower temperature.

CLOUD, CUMULONIMBUS. A mass of cloud rising to the base of the stratosphere, at which it spreads out horizontally to form an *anvil*. It is the characteristic thunder cloud and hail-producing cloud.

CLOUD, CUMULUS. A dense high detached cloud with sharp outlines and flat bases, the tops resembling domes or cauliflowers.

CLOUDINESS. The fraction, in eighths, of the whole sky covered by cloud, as seen by an observer at the ground. The estimate may be made for low, medium level and high cloud as well as for the total amount.

CLOUD, MAMMATUS. Cloud having sack-like protuberances hanging from its underside.

CLOUD, NACREOUS. Mother of pearl clouds (seen mostly in Scotland and Scandinavia). They are very high (20–30 km) and exhibit striking colour changes before and after sunset.

CLOUD, NIMBUS. A dense layer of dark shapeless cloud with ragged edges. The characteristic rain or snow cloud.

CLOUD, PILEUS. The transient top of detached cloud above cumulus.

CLOUD SEARCHLIGHT. A searchlight apparatus used to measure the height of the cloud base by observation of the patch of light at the cloud base of interest. Radar may be used in the same way.

CLOUDY CRYSTAL-BALL MODEL. See: *Optical model of particle scattering*.

CLUSEC. A unit of inleakage into a gas system. It is 10 cm^3 of gas per second at 1 μm of mercury pressure, i.e. $\frac{1}{100}$ of a *lusec*. See also: *Lusec*.

CLUSTER INTEGRALS. A system of integral–differential equations used by Born in his statistical theory of liquids.

CLUSTER VARIABLES. See: *Cepheid variables*.

CLUTTER. Spurious responses on a radar display caused by reflections from large objects on the ground, rain, storm clouds, etc.

COACERVATION. The production of colloid-rich droplets (coacervates) by the coagulation of a hydrophilic sol.

COAGULA. A precipitate which is formed by coagulation or flocculation of a colloidal solution.

COALSACK NEBULA. A dark nebula lying in the Southern Milky Way, not very far from the star cluster Omega Centauri. See also: *Nebula, dark*.

COANDA EFFECT. The effect whereby a two-dimensional jet of fluid, discharged tangentially along a convex solid surface, may remain attached to the surface for a considerable distance, and may thus be deflected through a large angle. Deflections of up to about 180° have been observed.

COAXIAL LINE. A radio-frequency transmission line consisting of a central copper conductor, embedded in insulating material or supported by insulating beads, surrounded by a cylindrical sheath which may be either a solid metal tube or a woven braid of thin copper wires. Such a line has no external field.

COCKCROFT–WALTON APPARATUS. A device for producing large d.c. voltages of up to about 2 million with output currents of several milliamperes and with good voltage stability. It is used for accelerating particles for research in nuclear physics, for X-ray tubes, high-voltage test equipment, etc. The apparatus is based on a development of the Greinacher voltage doubling circuit.

COEFFICIENT OF PERFORMANCE (C.O.P.). Of a refrigerator or heat pump: the ratio of the quantity of heat extracted to the amount of work expended.

COEFFICIENT OF VARIATION. The ratio of the standard deviation to the arithmetic mean.

COELOSTAT. An instrument for continuously reflecting the same region of the sky into the field of view of a fixed telescope, in spite of the rotation of the Earth. It involves a plane mirror which rotates about an axis parallel to that of the Earth.

COERCIVE FORCE. The magnetic field necessary to reduce to zero the residual intensity of magnetization arising from hysteresis. The value of the coercive force corresponding to magnetic saturation is the *coercivity*.

COGEOID. At a point on the Earth's surface: a surface lying below (or above) the geoid at a distance $\delta V/g$, where δV is the gravitational potential at the point concerned and g is the gravitational acceleration. The potential, δV, refers to the matter (all over the Earth) lying above sea level, plus the defect of mass represented by the oceans, plus variations of rock density (in accordance with any prescribed system of isostasy or other theory). Formerly known as the *compensated geoid*.

COHERENCE. The quality of being coherent. Sometimes defined as the quality that expresses the correlation between two stochastic processes.

COHERENCE TIME. As applied to a typical stochastic process such as a wave travelling through space: the time interval over which appreciable phase correlation exists. The concept can also be applied to particle beams, in terms of the associated de Broglie waves.

COHERENT. As applied to emitted or scattered waves: implies the existence of a definite phase relationship between two or more waves. Where no such relationship exists the waves are said to be *incoherent*.

COHERENT UNITS. Units based on a set of basic units from which all derived units may be obtained by multiplication or division without the introduction of numerical factors. Examples of coherent units are those of the CGS system and the International System (SI).

COHESION. The force between the atoms or molecules of a given material by virtue of which the material resists physical separation.

COHESION PRESSURE. Of a liquid: $\left(\dfrac{\partial E}{\partial V}\right)_T$ where E is the internal energy and V the volume of the liquid. Also known as the *internal pressure*.

COHESIVE ENERGY. Of a solid: the work required to dissociate one mole of the substance into its free constituents.

COINCIDENCE COUNTER. A system of counters which produces an output pulse only when a coincidence occurs, i.e. when two or more particles or photons arrive within the resolving time of the system.

COINCIDENCE COUNTING. The use of a coincidence counter to distinguish particular types of effect from unrelated effects, such as background effects.

CO-LATITUDE. The complement of the latitude.

COLBURN HEAT-TRANSFER FACTOR. A factor introduced into the expression for the heat transfer between two fluids to take account of the fact that the heat-transfer coefficient may vary with temperature.

COLD CATHODE. A cathode of low work function, which emits electrons at room temperature in, for example, an electron tube, a gas discharge tube, or an X-ray tube.

COLD EMISSION. (1) Of electrons: the emission of electrons from a cathode by the application of a sufficiently high-voltage gradient. Sometimes called *auto-emission* or *auto-electronic emission*. (2) Of light: the emission of visible light at ordinary temperatures, as in chemiluminescence, fluorescence, and phosphorescence. See also: *Field emission*.

COLD FRONT. See: *Front (meteorology)*.

COLD OCCLUSION. See: *Front (meteorology)*.

COLD STORAGE. The application of refrigeration to preserve perishable goods for prolonged periods by temperature and humidity control in a closed space. The "closed space" may be a cold-storage warehouse, refrigerator van, railway truck, refrigerated cargo ship, or domestic refrigerator.

COLD WORK. The production of strains and defects in the internal structure of a metal or alloy by mechanical treatment such as bending, rolling, hammering, drawing or filing. The term is also applied to the strains and defects themselves. The original perfection of the structure can generally be restored by annealing at a suitable temperature. See also: *Work hardening*.

COLD WORK, INTERNAL ENERGY OF. The latent energy associated with the defects (e.g. dislocations) produced by cold work. This represents only 10–15% of the total energy associated with cold work, the remainder appearing as heat.

COLLECTING POWER. Of a lens: the property of rendering parallel rays convergent, or reducing the divergence of divergent rays.

COLLECTIVE ELECTRON THEORY. The theory which views the electrons in a solid as being shared by the whole crystal, and is therefore concerned with the problem of finding the appropriate wave functions of individual electrons moving in a periodic potential. It may be contrasted with the *Heisenberg* or *localized electron* theory. See also: *Band theory of solids. Ferromagnetism, Heisenberg theory of.*

COLLECTIVE PARAMAGNETISM: SUPER-PARAMAGNETISM. The phenomenon according to which the magnetization of assemblies of ferromagnetic particles which are so small that each particle consists of only one magnetic domain, in its relation to field and temperature, is comparable with that of paramagnetic atoms or molecules, except that the magnetic moment per particle may be several orders of magnitude larger. See also: *Superantiferromagnetism.*

COLLIGATION. See: *Chemical reaction, homolytic.*

COLLIMATION. The limiting of a beam of radiation to the required dimensions.

COLLIMATION ERROR. (1) In surveying: failure of the line of collimation to be perpendicular to the vertical axis, or to be horizontal when the levelling bubble is central, as the case may be. (2) Of an astronomical telescope: the deviation from 90° of the angle between the optical and mechanical axes.

COLLIMATOR. (1) For light: an assembly involving slits, apertures, lenses, etc., for the production of a parallel beam of light. (2) For X-rays, neutrons, etc.: a contrivance, usually in the form of a tube with terminating slots, which restricts the angular divergence of a beam of radiation, thus producing an approximately unidirectional beam for the study of such processes as scattering and diffraction.

COLLINS MACHINE. A compact unit for the liquefaction of helium and hydrogen. Refrigeration is by isentropic expansion of helium, additional cooling by liquid air or nitrogen being optional.

COLLISION. An interaction between free particles (including photons, atoms and nuclei), aggregates of particles, or rigid bodies, in which they come near enough to exert a mutual influence, with exchange of energy, momentum or charge. Actual contact is not necessarily implied.

COLLISION BROADENING. Of a spectral line or resonance energy level: the broadening of the natural width of a spectral line when the emitting atoms are reduced from an excited state to a lower state by collisions.

COLLISION DENSITY (NUCLEAR TECHNOLOGY). The number of neutron collisions made by a neutron of given energy per unit volume and per unit time. *Partial collision densities* may be defined, characterized by such parameters as energy and direction. The collision density is equal to the neutron flux density divided by the scattering mean free path.

COLLISION DIAMETER. Of molecules in a collision: the distance of closest approach between the centres of two molecules taking part in the collision. The concept is used in the collision theory of chemical reactions.

COLLISION, ELASTIC. A collision in which the total kinetic energy is unchanged. See also: *Restitution, coefficient of. Scattering, elastic.*

COLLISION FREQUENCY. For a given particle in a specified environment at a given temperature and pressure: the number of collisions undergone per unit time.

COLLISION, HEAD-ON. A collision in which the relative motion of the colliding particles is along the same line between centres both before and after "impact".

COLLISION, INELASTIC. A collision in which the total kinetic energy changes. In a *collision of the first kind* one particle loses kinetic energy only, while the second gains not only kinetic energy but ionization or excitation energy. In a *collision of the second kind* one particle loses ionization or excitation energy only, while the second gains in kinetic, ionization or excitation energy. See also: *Restitution, coefficient of. Scattering, inelastic.*

COLLISION MEAN FREE PATH. The average distance traversed by a particle between successive collisions.

COLLISION, PLASTIC. See: *Restitution, coefficient of.*

COLLISION PROBABILITY. For a collision of a particular type between two particles: the ratio of the cross-section for the type of collision in question to the cross-section for all types of collision between the two particles.

COLLISION, REARRANGEMENT. A collision giving rise to the formation of a compound nucleus, usually in a very excited state.

COLLISION, SUPERELASTIC. The collision of an excited atom with a free electron, whereby the atom returns to the ground state and transfers nearly the whole of its excitation energy to the electron.

COLLOID. A substance consisting of very small particles dispersed in a continuous medium. A *colloidal particle* is usually considered to be one whose size lies between 5 Å and 5000 Å. The solution laws do not hold for colloidal "solutions", which are to be regarded as disperse systems. (*Note:* 1 Å = 10^{-10} m.)

COLLOIDAL ELECTROLYTE. An electrolyte in which the ions are of colloidal size. They may be macromolecules or micelles and the term is sometimes restricted to the latter.

COLLOIDAL EMULSION. A colloidal system in which both the dispersing medium and the disperse phase are liquids. The disperse phase is in the form of small droplets and the emulsion is usually stabilized by a third substance: the *emulsifying agent*.

COLLOID, HETEROPOLAR. A colloidal system in which the dispersed particles are polar compounds.

COLLOID, HOMOPOLAR. A colloidal system in which the dispersed particles are non-polar compounds.

COLLOID, HYDROPHILIC. A colloid which readily forms a "solution" in water.

COLLOID, HYDROPHOBIC. A colloid which forms a "solution" in water only with difficulty.

COLLOID, IRREVERSIBLE. One which cannot readily be "dissolved" after separation from the dispersing medium.

COLLOID, LYOPHILIC. A colloid which is readily dispersed in a suitable medium, and may be redispersed after coagulation.

COLLOID, LYOPHOBIC. A colloid which is dispersed only with difficulty, yielding an unstable "solution" which cannot be re-formed after coagulation.

COLLOID MILL. A machine which disintegrates solids into extremely fine colloidal particles.

COLLOID, REVERSIBLE. One which, after separation from the dispersing medium, can be "redissolved" with ease.

COLLOID, SOLID. A system in which a gas, liquid or solid, of colloidal dimensions, is dispersed in a transparent solid, and is thus susceptible to exploration by the methods of colloid chemistry.

COLLOID, STABLE. A colloidal system in which the particles remain constant in size and do not settle.

COLORIMETER. An instrument designed for the measurement of colour specification. See also: *Tintometer*.

COLORIMETRIC PURITY. Of a light stimulus: the ratio of the luminance of a dominant wavelength to the total luminance of the stimulus. If, instead of the luminance, use is made of trichromatic quantities the measure of purity is known as the *excitation purity*. See also: *Dominant wavelength. Light stimulus.*

COLORIMETRY. The measurement of colour specification.

COLORIMETRY, ADDITIVE. The provision of a colour specification in terms of the amounts of any three suitable radiations which, when additively mixed, match a colour under test for the standard observer. Such a specification is known as a *tristimulus specification*, and the individual amounts are termed *tristimulus values*.

COLORIMETRY, SUBTRACTIVE. The provision of a colour specification in terms of the amount of light selectively absorbed ("subtracted") from an initially white beam. Three different spectrally selective filters are needed on account of the trivariance of human colour vision.

COLOSSUS. The first electronic digital computer, proposed secretly in 1943. See: *Turing machine*.

COLOUR. A sensation normally produced in the eye by light, but which may be evoked by pressure, electric or magnetic fields, drugs or pathological conditions. Colour cannot be measured: what a colorimeter measures is colour specification.

COLOUR ANALYSER. A device used to provide information about, or to control, processes that involve colour changes. Whereas a colorimeter furnishes a colour specification on some recognized colorimetric scale, a colour analyser furnishes information that is directly relevant to a process and is in terms of that process. Any of the methods of colorimetry can be used in a colour analyser.

COLOUR ATLAS. A series of material colour standards arranged in a systematic way so as to cover as much of colour space as is practicable. It consists of a collection of charts each of which represents a section through a specified colour space. A common method is to use some form of polar coordinate system, with the black–white axis vertical and a separate chart for each hue or dominant wavelength.

COLOUR CENTRES. Crystal defects in an optically transparent crystal which introduce absorption bands and therefore affect the colour of the crystal. The centres arise from various treatments and at various temperatures, the most common treatments involving exposure to ionizing radiation or heating in a vapour of alkali metal or halogen. Colour centres may be classified as follows: (1) *F-centre:* the best understood colour centre. It consists of an electron trapped at a negative ion vacancy. (2) *F'-centre:* gives rise to a broad band on the long wavelength side of the F-band, arising from low-temperature irradiation in the F-band. It is thought to be an F-centre which has trapped an additional electron. (3) *M- (or C-) centre:* an electron trapped at an L-shaped agglomerate of two negative ion vacancies and a positive ion vacancy. The M-band is on the long wavelength side of the F-band. (4) *R- (or E- or D-) centre:* gives rise to two bands between the F- and M-bands, which are produced by continued irradiation with light in the F-band or continued X-ray exposure at room temperature. (5) *N- (or G-) centre:* gives rise to a faint band on the long wavelength side of the M-band, arising from prolonged exposure to light in the F-band or to X-rays. (6) V_1-*centre:* gives rise to the most prominent band after the F-band, on the short wavelength side of that band. It is thought to consist of a hole trapped at a positive ion vacancy. (7) *Other centres:* U-, V_2-, V_3- and V_4-centres, C_2-centres, H-centres, α- and β-bands, Z-centres, colloid centres and metal impurity centres, all occur under special conditions.

COLOUR COMPARATOR. An instrument to facilitate comparison of coloured samples.

COLOUR CONTENT. Degree of saturation of a colour. See also: *Colour, saturation of.*

COLOUR CONTRAST. The subjective impression of contrast arising from the simultaneous stimulation of different parts of the retina by different colours.

COLOUR COUPLER. A substance which, when added to certain types of photographic developer, will form a colour or dye when "coupling" with the oxidized developer. The use of couplers in combination with one or more developers forms the basis for the production of colour prints and transparencies.

COLOUR, DESATURATED. One which is not a pure spectral colour or a purple formed by the mixture of violet and deep red. It can be considered as formed by a mixture of a spectral colour and white. See also: *Colour, saturation of.*

COLOUR EMISSIVITY. The ratio of the energy radiated in a given (narrow) wavelength band to that radiated by a black body in the same band at the same temperature.

COLOUR FILTER. A layer, film or plate that alters the relative intensities of the component wavelengths of light passing through it. It may operate by simple absorption or by an interference mechanism.

COLOUR, HUE OF. The property by which a colour is identified as tending to resemble the appearance of some particular wavelength of light in the visible spectrum.

COLOUR INDEX. Of a star: formerly the difference between the photographic and visual magnitudes, but now defined in terms of photometric magnitudes. See also: *Stellar magnitude, apparent. UBV photometry.*

COLOUR PHOTOGRAPHY. Photographic reproduction in colour, based upon the Helmholtz theory of colour vision. All processes involve the taking of photographs through red, blue and green colour filters. In *additive processes* a mosaic screen (a minute pattern of red, green and blue filter elements) is used and the primary colours are printed; and in *subtractive processes* (now more usual) an *integral tripack* (three layers of emulsion sensitive to blue, green and red light respectively, coated one above the other) is used, and the complementary colours are printed.

COLOUR RADIOGRAPHY. The display of changes in radiographic intensity as changes in colour, so as to render radiographic contrast more easily detectable, thus adding the qualities of hue and saturation (e.g. degree of "redness") to that of brightness, the only quality displayed by conventional black and white radiographs. A colour radiograph has, of course, no connection with an object's natural colour.

COLOUR RENDERING. Of a light source: a measure of the extent to which the perceived colours of an object or surface illuminated by the source conform to those of the same object or surface when illuminated by a reference source under specified conditions of viewing.

COLOUR, SATURATION OF. The degree of difference of a colour from a neutral (white or gray) sensation, i.e. the vividness.

COLOUR SPACE. A three-dimensional conceptual space used in studying colour phenomena. Any desired attributes of colour may be employed to describe and specify a colour space (e.g. brightness, hue and saturation) and these may be arranged in appropriate coordinate systems. See also: *Colour atlas.*

COLOUR SPECIFICATION. A set of parameters which serve to define uniquely a colour stimulus in terms of some colour system. Only three such parameters are needed for any colour system, on account of the trivariance of human colour vision. A colour specification may be measured directly on a tristimulus colorimeter or may be calculated if the spectral energy distribution of a light stimulus is known. See also: *Colorimetry, additive*.

COLOUR SYSTEM. A system providing a set of data for defining the colour matching characteristics of a "standard" observer. The best known is that of the C.I.E. (Commission Internationale de l'Eclairage).

COLOUR TEMPERATURE. Of a given source of radiation: the temperature of a black body having the same distribution of relative intensities ("colour") as the source. It is assumed that the spectral energy distribution of the source approximates to that of a black body within the visible range of wavelengths. See also: *Astrophysical temperatures*.

COLOUR TRIANGLE. See: *Maxwell colour triangle*.

COLOUR VISION, THEORIES OF. Theories which attempt to explain colour vision in terms of the functions of various receptors in the eye. None are completely satisfactory but most are based on the *Helmholtz* (or *Young–Helmholtz*) theory of colour vision which invokes the existence of three types of receptor in the eye, each of which is, to a greater or lesser extent, stimulated mainly by red, blue, or green light (the three *primary colours*) respectively.

COLPITTS OSCILLATOR. One of the first vacuum tube (triode) oscillators to be devised. It consists essentially of negative reactance (capacitative) paths between the grid and cathode, and between the anode and cathode, and a positive reactance (inductive) path between the grid and anode. It may be considered as the inverse of the Hartley circuit. See also: *Hartley oscillator*.

COMA (OF A COMET). See: *Comet*.

COMA (OPTICS). An aberration in an optical system leading to the formation of a pear-shaped (comet-like) image from an object point.

COMBINATION (MATHEMATICS). The assignment of a group of objects into two or more mutually exclusive sets. The number of ways (or combinations) of selecting r objects from a set of n objects is given by the binomial coefficient $\left(\dfrac{n}{r} \right)$ i.e. $\dfrac{n!}{r!(n-r)!}$. If the r objects are permuted

among themselves, no new combinations are formed but there are $r!$ new arrangements of each combination, i.e. the number of permutations is $n!/(n-r)!$ See also: *Permutation (mathematics)*.

COMBINATION PRINCIPLE OF RITZ. States that the multitude of frequencies shown by the line spectra of atoms or molecules may be expressed in terms of their wave numbers as the differences between relatively few terms taken two at a time in various combinations. Thus, for hydrogen, the wave numbers are given by $\dfrac{R}{x^2} - \dfrac{R}{y^2}$ where R is Rydberg's constant and x and y are integers. Each term corresponds to a state of definite energy. See also: *Hydrogen atom, line spectrum of*.

COMBINATION TONES. Additional tones formed subjectively by the ear when notes of two or more different frequencies are sounded simultaneously. They have frequencies which are the sums (*summation tones*) and differences (*difference tones*) of these notes and are not to be confused with beats. See also: *Beats*.

COMBINATORIAL ANALYSIS. The study of problems concerned with the number of ways in which an operation can be performed under given conditions. The best known examples are the various problems in permutations and combinations.

COMBUSTION, HEAT OF. Another term for gross calorific value. The *net heat of combustion* signifies the net calorific value. See also: *Calorific value, gross. Calorific value, net*.

COMET. A diffuse agglomeration of matter moving under the attraction of the Sun. Cometary orbits exhibit a wide range of eccentricities and inclinations to the ecliptic and the semi-major axes may be very large (up to thousands of astronomical units). The principle visible part of almost all comets is the *coma*. There is a small brighter region within the coma, known as the *nucleus*, and most bright comets are also accompanied by a *tail*, which may extend for more than 10^8 km from the comet itself. Owing, at least in part, to radiation pressure the tail normally extends outwards on that side of the comet remote from the Sun, whatever the direction of motion, but some comets have shown more than one tail, in different directions. (*Note:* an astronomical unit is about $1 \cdot 5 \times 10^8$ km.)

COMET SPECTRA. Cometary spectra consist of a large number of emission bands due to resonance fluorescence excited by sunlight and a weak continuum due to sunlight reflected from the nucleus. There is also a system of bands in the CO^+ spectrum which is exhibited by the tail, and another system in the head, the *Swan bands*, which are defined separately.

COMFORT CONDITIONS. These are largely subjective, but scales of warmth and comfort have been devised based on acceptable values for air temperature, relative humidity, air movement and loss of heat by radiation. These are, of course, influenced by personal factors such as bodily activity, type of clothing worn, state of health etc.

COMMINUTION. The process of breaking up agglomerates of material by machines designed to reduce the lumps to powder form. It is used, for example, in the treatment of ores to increase the surface area so as to improve the efficiency of the extraction treatment.

COMMUNICATION SATELLITE. An Earth satellite forming part of a communication system. It may be *active* (i.e. carry means for transmitting or repeating radiocommunication signals) or *passive* (i.e. designed to transmit signals by reflection). See also: *Satellite*.

COMMUNICATION THEORY. A mathematical theory in which entropy is interpreted as "missing information". The logarithm of the number of possible "states" of a communication channel provides a measure of its handling capacity.

COMMUTATION, ANGLE OF. The difference between the ecliptic longitude of the true Sun and the ecliptic longitude of a planet, as seen from the Earth. For a planet moving in the plane of the ecliptic the angle of commutation is the same as the elongation.

COMMUTATION (MATHEMATICS). See: *Commutative process*.

COMMUTATION RELATIONS. The relations between quantities, typically operators, which do not commute, involving for example the expression of the differences between AB and BA, where A and B are the operators in question. See also: *Commutative process. Commutator*.

COMMUTATIVE PROCESS. Refers to the combination of two or more quantities where the result is independent of the order in which the steps leading to that combination are carried out. Thus if $(a + b) = (b + a)$, or $ab = ba$, the quantities are then said to *commute*. See also: *Commutation relations*.

COMMUTATOR. (1) A mechanical device for periodically changing the connections to a rotating member, or for interchanging the connections of the

leads to an electrical circuit. (2) For operations where commutation does not take place, e.g. for two non-commutative operators for which AB is not equal to BA: the expression $AB - BA$.

COMPACT (POWDER METALLURGY). The compressed mixture produced by hot or cold pressing and shaping in the initial stages of certain techniques. The compact may have a binder incorporated which has a special function in the later fabrication of the product.

COMPARATIVE LIFETIME. In β-decay theory: the product of the half-life and a function (usually represented as f) which expresses the probability per unit time that a β-transition will occur in a given nucleus. For purposes of comparison it is usual to take $\log ft$, where t is the half-life. The comparative lifetime may be taken as characterizing the degree of forbiddenness of a β-process.

COMPARATOR. An instrument for the accurate comparison of lengths. It may be optical (including photoelectric), mechanical, pneumatic, or electrical (including electronic) in principle and may be used, for example, for the comparison of line standards, for the measurement of variations in length, as a sensing and measuring device in more complex instruments, for the control of machine tools, and for the standardization of surveying tapes.

COMPARATOR, VÄISÄLA. An interferometer for measuring large distances (100 m or so) with extreme accuracy (about 1 part in 10^7).

COMPASS, MAGNETIC. An instrument which is employed to establish directions in the horizontal plane, by use of the geomagnetic field, commonly with respect to the magnetic north. It may range from a simple magnetized needle supported on a simple bearing to a system of parallel needles mounted on a card which floats on a suitable liquid. See also: *Gyrocompass. Sky compass*.

COMPENSATED GEOID. See: *Cogeoid*.

COMPENSATOR, OPTICAL. An instrument for measuring the phase difference between the two components of elliptically polarized light. It operates by introducing a known, opposite, phase difference whose magnitude can be varied so as to cancel out (compensate) the phase difference to be measured. The most familiar is the Babinet compensator. See also: *Babinet compensator*.

COMPLEMENTARITY PRINCIPLE. A principle enunciated by Bohr which states that it is impossible to describe quantum-mechanical phenomena in as complete a manner as classical ones, since the pairs of conjugate variables that must be known for an exact description in the latter case are mutually exclusive in the former. Experimentally this signifies that there exist mutually exclusive descriptions of a quantum-mechanical system which are complementary to each other. The best known example concerns the wave corpuscle duality of matter and light. See also: *Conjugate variables. Uncertainty principle: Indeterminacy principle.*

COMPLEMENTARY COLOURS. Pairs of coloured light stimuli which, when combined by additive colour mixture in suitable proportions, can be made to match a specified white. For a given stimulus and specified white there is a range of complementary colours, of approximately constant hue and ranging in saturation from the white to the most pure complementary colour. This last is the monochromatic light known as the *complementary wavelength*. It is the *dominant wavelength* of the pure complementary colour.

COMPLEX BODY. A solid body which, above a critical stress, flows at a rate proportional to the excess of stress over the yield stress. See also: *Fluid, Newtonian. Solid, Hookean.*

COMPLEX (CHEMISTRY). A combination of two or more chemical groups to form a larger stable group.

COMPLEX COMPOUND. A combination of stable groups to form a group in which the valencies are satisfied and which is capable of free existence. Also known as *coordination compound.*

COMPLEX ION. An electrically charged radical or chemical group.

COMPLEX, IONIC. An ionic compound in which the ions are complex.

COMPLEX NUMBER. A number consisting of real and imaginary parts. It is commonly represented in the form $(a + ib)$ where a and b are real numbers and $i = \sqrt{-1}$. A complex number may be represented graphically on an Argand diagram. See also: *Argand diagram. Modulus.*

COMPLEXOMETRIC ANALYSIS: COMPLEXOMETRY. A technique for the quantitative analysis of metals in solution by titration with a complexing agent.

COMPLEX POTENTIAL MODEL OF PARTICLE SCATTERING. See: *Optical model of particle scattering.*

COMPLEX VARIABLE. A complex quantity $(x + iy)$ in which x and y are real variables.

COMPLIANCE. The reciprocal of stiffness. It is the elastic extension or displacement produced by unit load.

COMPLIANCE CONSTANTS. Of a crystal: See: *Elastic constants.*

COMPOUND, COORDINATION. Another name for complex compound.

COMPOUND, COVALENT. A compound in which the bonds are essentially covalent in character. See also: *Bond, covalent.*

COMPOUND, HETEROPOLAR. A covalent compound in which the bonds are essentially heteropolar in character. See also: *Bond, heteropolar.*

COMPOUND, HOMOPOLAR. A covalent compound in which the bonds are essentially homopolar in character. Nowadays, however, the term is often applied to any covalent compound. See also: *Bond, homopolar.*

COMPOUND, IONIC. A compound in which the bonds are essentially ionic in character. See also: *Bond, ionic.*

COMPOUND, MOLECULAR. A compound consisting of two or more molecules held together by weak forces.

COMPOUND NUCLEUS. In Bohr's theory of nuclear reactions: a highly excited atomic nucleus, of short lifetime, formed as an intermediate stage in an induced nuclear reaction.

COMPRESSIBILITY. Of an isotropic material: the reciprocal of the bulk modulus. It is the fractional change in volume produced by unit change of pressure.

COMPRESSIBILITY, LINEAR. For a direction in a crystal: the fractional change in length produced in a given direction by unit change of pressure.

COMPRESSIBLE FLOW. Fluid flow in which changes in fluid velocity are accompanied by changes in fluid pressure and density. When the fluid velocity corresponds to a very low Mach number no appreciable changes in fluid pressure and density occur and the flow is said to be *incompressible.* Four main regions of compressible flow are distinguished, according to the Mach number, M. These are: *Subsonic* $(M < 1)$, *Transonic* $(M \approx 1)$, *Supersonic* $(M > 1)$ and *Hypersonic* $(M \gg 1)$.

COMPRESSIONAL WAVE. In an elastic medium: a form of elastic wave motion in which the particles of the medium are displaced in a direction parallel to the direction of propagation. Also known as *dilatational wave, irrotational wave, longitudinal wave.*

COMPRESSION EFFICIENCY. Of a compressor: the ratio of the work required to compress all the vapour delivered by the compressor, adiabatically and reversibly, to the actual work supplied to the vapour by the piston or blades of the compressor.

COMPRESSION IGNITION. Ignition, as in an internal combustion engine, in which the charge in the cylinder is ignited by the heat of compression alone.

COMPRESSION MODULUS. See: *Bulk modulus: Compression modulus.*

COMPRESSION TEST. A test for the ductility and malleability of alloy or pure metal bars in which a test piece of given dimensions is compressed by a specified amount.

COMPRESSOR. A device for raising the pressure of a gas. It may be of the *positive displacement type*, in which the operation is intermittent (e.g. via a piston) where there is virtually no possibility of flow reversal, or *steady flow (non-positive displacement)* in which the work is imparted to the gas by change of angular momentum (e.g. by moving blades), and in which, away from normal operating conditions, flow reversal or surging can occur.

COMPTON EFFECT. The elastic scattering of a photon by an electron when the electron can be considered to be free and stationary. It is an important mode of scattering for X- and γ-rays. Part of the energy and momentum of the incident photon is transferred to the electron (the *Compton electron* or *recoil electron*) and the remaining part is carried away by the scattered photon (the *modified scatter* or *Compton scatter*). See also: *Modified X- or γ-ray scatter. Unmodified X- or γ-ray scatter.*

COMPTON EFFECT, DOUBLE. A higher order process occurring in the scattering of γ-rays by free electrons in which two scattered quanta appear instead of one, with a relative probability of about $1/137$. The spectrum of the second quantum has a bremsstrahlung-like distribution rising rapidly at low energies.

COMPTON–GETTING EFFECT. The sidereal diurnal variation of cosmic ray intensity which would be expected to occur on account of the rotation of the galaxy (a) if cosmic radiation were of extra-galactic origin and isotropic in inter-galactic space, and (b) if the radiation were unaffected by entry to and passage through the galaxy.

COMPTON WAVELENGTH. The De Broglie wavelength, given by h/m_0c, where h is Planck's constant, m_0 the rest energy of the electron and c the speed of light, which appears in the expression for the increase in wavelength arising from the Compton effect. This increase is $(h/m_0c)(1 - \cos\phi)$, where ϕ is the angle of deflection of the scattered photon.

COMPUTER. A device which, if supplied with numerical information (the "input") in a suitable form, is capable of obeying a series of instructions and thereby yielding numerical information (the "output") derived from the input by a process of logic. The input is usually coded on paper, magnetic tape or punched cards and the output may also be coded, or may be printed out "in clear". See also the following entries.

COMPUTER, ANALOGUE. A computer in which the variables are represented by physical quantities which are made to obey the desired mathematical relations by suitable design and interconnection of the computer elements. See also: *Differential analyser.*

COMPUTER, DIGITAL. A computer, usually electronic, which operates with numbers in their digital form. In its simple form it performs simple arithmetical operations together with certain *logical operations* such as sorting, discrimination, etc. In its more complex form, sometimes described as an *automatic digital computer*, it may operate for long periods of time and do involved calculations without human intervention. A set of instructions and the necessary input data are presented to the machine, which proceeds to work out the results without any further action by the operator. The main requirements of such a computer are: a unit capable of adding, subtracting, multiplying and possibly dividing, and of carrying out certain logical operations; a store to hold instructions; a store to hold numbers; an input device for taking in both instructions and numbers prepared by the user; an output device for recording the results of the calculation; and a control unit to supervise the order of operations.

COMPUTER, DIGITAL LEARNING. A device that processes information without having been programmed to do so, the system having been previously "trained" by being given examples of the processing task to be performed.

COMPUTER HARDWARE. That part of an electronic digital computer that has a physical existence.

COMPUTER LANGUAGE. See: *Computer programme.*

COMPUTER MEMORY. A store in a computer which is capable of holding instructions or numerical data until required.

COMPUTER, ON-LINE. A computer used in such a way that the user communicates directly with it, no intermediate storage media such as magnetic tape or punched cards being used. Information generated by the "user" (which may be a human operator or some piece of equipment connected directly to the computer) is immediately transferred to the memory of the computer. Such a computer may be used for data collection and evaluation or may, in addition, direct the behaviour of the equipment concerned.

COMPUTER PROGRAMME. A sequence of instructions given to a computer for the performance of a particular task or tasks. These instructions are presented to the computer in the form of a code in the *computer programming language* appropriate to the computer concerned.

COMPUTER SIMULATOR. Equipment for checking the performance of computer hardware. It provides and responds to the same signals as the computer and eliminates the need to use the computer itself as a diagnostic tool.

COMPUTER SOFTWARE. The programmes used in an electronic digital computer to organize a user's programme and enable it to run.

CONCENTRATION. Of a solute in a solvent: the amount of solute in a given quantity of solvent commonly expressed as moles per litre.

CONCENTRATION CELL. An electrochemical cell containing electrodes of the same metal immersed in solutions of the same salt but of different degrees of concentration.

CONDENSATION. (1) The process of forming a liquid (the *condensate*) from its vapour. (2) The formation of new carbon–carbon linkages between organic molecules, with the elimination of simple groups such as water and ammonia. See also: *Dropwise condensation. Filmwise condensation.*

CONDENSATION LEVEL. The height to which a volume of surface air would need to be lifted adiabatically in order that it should become saturated.

CONDENSATION NUCLEI. Atmospheric particles in which condensation can occur more or less readily. The majority consist of droplets containing salts or acids in solution, of radius between 0·5 and 1 μm, but some solid insoluble nuclei do occur.

CONDENSATION NUMBER. The ratio of the number of molecules condensing on a solid surface to the total number falling upon it. It is about ten times larger for dropwise than for filmwise condensation. See also: *Dropwise condensation. Filmwise condensation.*

CONDENSATION TRAIL. A visible track in the wake of an aircraft in flight, usually a high-speed aeroplane at high altitude. It is most often due to condensation of water vapour from the engine exhaust but may also arise from condensation from the atmosphere in regions where the temperature has been sufficiently reduced by a local reduction of pressure in the field of flow, e.g. at wing tips or propeller tips.

CONDENSED PHASE. The solid or liquid state.

CONDENSER. (1) A former term for a capacitor. (2) An apparatus used for condensing vapours obtained during distillation. (3) A chamber into which the exhaust steam from a steam engine or turbine is delivered, to be condensed by the circulation or introduction of cooling water. It may be a *surface condenser*, in which the coolant and vapour are separated by a metal or glass surface, at which the necessary heat transfer occurs, or a *jet condenser*, in which the coolant mixes intimately with the vapour as a fine spray or film, both coolant and condensate being discharged together. (4) A lens designed to control the intensity and angular aperture of a beam of light in an optical or electron-optical system; or a mirror designed to perform the same function in an optical system.

CONDUCTANCE. (1) Of a conductor or an electrolyte: the real part of the admittance, i.e. the reciprocal of the resistance. (2) Of a dielectric: the quotient of the conduction current by the applied voltage. (3) Of a component of a vacuum system for a given gas: the throughput of the gas divided by its partial pressure difference across the component, under steady conditions.

CONDUCTANCE COEFFICIENT. The ratio of the equivalent conductance of an electrolyte, at the concentration considered, to its equivalent conductance extrapolated to infinite dilution.

CONDUCTANCE, EQUIVALENT. The electrical conductance of a solution which contains 1 g-equiv weight of solute at a specified concentration, measured when placed between two electrodes separated by 1 cm.

CONDUCTANCE, MUTUAL. A vacuum tube coefficient concerned with the effect of grid voltage on plate current. It is given by $\left(\dfrac{di_p}{de_g}\right)_{E_p \text{ constant}}$, where i_p is the plate current, e_g the grid voltage, and E_p the plate voltage. The mutual conductance is also known as the *transconductance*.

CONDUCTANCE, SPECIFIC. An obsolescent term for electrical conductivity. See also: *Conductivity, electrical.*

CONDUCTION. The transmission of electrical, thermal or acoustic energy via a medium, without movement of the medium itself as such. Thermal conduction is to be distinguished from convection, and electrical conduction from the transmission of electromagnetic radiation.

CONDUCTION BAND. A partially-filled energy band in a solid in which the electrons can move freely, thereby allowing the material to carry an electric current. Although the term is often applied to metals, the distinction between the conduction and other bands is of greatest importance in connection with semiconductors and insulators. See also: *Band theory of solids.*

CONDUCTION ELECTRON. An electron which is free to move under the influence of an external field, e.g. an electron in the conduction band of a solid. Sometimes referred to as a free electron. See also: *Bound electron.*

CONDUCTIVITY, ELECTRICAL. Of a material: the conductance, at a specified temperature, between the opposite faces of a cube of the material having sides of unit length. It is the reciprocal of volume resistivity. Formerly known as *specific conductance.*

CONDUCTIVITY ELLIPSOID. An ellipsoid whose dimensions are such that the magnitude and direction of an electric current may be obtained from it in relation to those of an applied electric field.

CONDUCTIVITY, THERMAL. Of a material: the quantity of heat flowing per second per degree across unit area of the material (the plane of the area being perpendicular to the temperature gradient), for unit temperature gradient. See also: *Wiedemann–Franz law.*

CONDUCTOR. A substance through which electrical or thermal energy flows easily.

CONE. The surface generated by drawing straight lines from a fixed point to every point of a closed curve. Where the closed curve is a circle, and the fixed point is on a line through the centre of the circle and perpendicular to its plane, we have a *right circular cone.* See also: *Quadric.*

CONE FLOW: CONICAL FLOW. Of the steady supersonic flow of a perfect inviscid gas past a conical solid body: refers to the property whereby, in certain regions of the flow field, the principal physical quantities (velocity, pressure, density, and so on) are constant in directions passing through the apex of the cone. For a given body there is a minimum Mach number below which such flow cannot occur.

CONE OF SILENCE. A cone in which the radiation

intensity of an omnidirectional aerial vanishes. It arises from destructive interference between the direct radiation and that reflected from the ground.

CONFIGURATION. (1) Atomic: the spatial arrangement of atoms. (2) Electronic: the distribution of electrons among various states in an atom or molecule.

CONFIGURATION INTERACTION. Interaction between different possible electronic configurations of an atom or molecule so that, for example, the wave function of a spectral term must be taken to include components of both configurations.

CONFOCAL CONICS. A system of conics having the same foci.

CONFORMAL PROJECTION: ORTHOMORPHIC PROJECTION. Of the Earth's surface: one which gives no local distortion, the scale being the same in all directions at a given point, but varying from point to point. One of the best known is the *Mercator projection* for a wide extent of latitude, in which the spheroid is projected on to a cylinder whose axis usually coincides with that of the Earth, although it can be oblique. Another is the *Lambert projection*, which may be visualized as projection on to a cone whose axis also coincides with that of the Earth.

CONFORMAL TRANSFORMATION. A mathematical transformation having the property that an angle is transformed into an equal angle. The customary formulation of the relevant theory is in terms of analytic functions of a complex variable. Typical applications are in aeronautics, for example in considering the flow round an aerofoil, and in map-making, for example in Mercator's projection.

CONGESTION. Of a physical system having a random element: the temporary behaviour of the system as if it were overloaded, even though it is working within its capacity, because of the operation of chance, e.g. in a queuing system.

CONICAL REFRACTION. An effect observed in biaxial crystals whereby a ray incident in a particular direction is spread, on refraction, into a hollow cone of rays.

CONIC: CONIC SECTION. A curve obtained by the intersection of a right circular cone by a plane. It may also be defined as the locus of a point which moves in a plane so that its distance from a fixed point (the *focus*) bears a constant ratio (the *eccentricity*) to its distance from a fixed straight line (the *directrix*). If the eccentricity is unity the curve is a *parabola*; if less than unity it is an *ellipse*; and if greater than unity it is a *hyperbola*.

CONJUGATE: CONJUGATE COMPLEX. Of a complex number: the number obtained by changing the sign of the imaginary part. Thus $(a - ib)$ is the conjugate of $(a + ib)$. The conjugate of a complex number, A, is commonly denoted by A^*, and the product AA^* is always real. It is equal, in the present example, to $a^2 + b^2$.

CONJUGATE DOUBLE BONDS. Double bonds which, in an organic molecule, alternate with single bonds.

CONJUGATE FOCI. Of an optical system: two points such that light rays emanating from one will be brought to a focus at the other.

CONJUGATE IMPEDANCES. Impedances having equal resistance components, and equal and opposite reactance components.

CONJUGATE MOMENTA. See: *Coordinates and momenta, generalized.*

CONJUGATE PLANES. Of an optical system: planes perpendicular to the axis of the system such that any object point near the axis in one plane will be imaged in the other.

CONJUGATE POINTS. Of an optical system: the points of intersection of the principal planes of the system with the optic axis. Any external ray directed towards the first conjugate point will appear to emerge from the second, and will be in a parallel direction if the media on the two sides of the system are optically equivalent. See also: *Lens, cardinal points of.*

CONJUGATE VARIABLES. Pairs of physical variables describing the behaviour of a quantum-mechanical system, either of which, but not both, may be specified precisely at the same time. The most important of these comprise the position and momentum of a particle. See also: *Complementarity principle. Uncertainty principle: Indeterminacy principle.*

CONJUGATION. The delocalization of chemical bonds in some types of molecule, e.g. in aromatic molecules.

CONJUNCTION. (1) Of two celestial bodies: the condition that they have the same celestial (ecliptic) longitude on a celestial sphere with a defined centre, e.g. the centre of the Earth for conjunction between a planet and the Sun. (2) Of two planets: the condition that they have the same longitude on a celestial sphere centred on the Sun.

CONJUNCTION, INFERIOR. For an inferior planet: conjunction with the planet between the Earth and the Sun.

CONJUNCTION, SUPERIOR. For an inferior planet: conjunction with the planet on the side of the Sun remote from the Earth.

CONSERVATION LAWS. For nuclear processes: "laws" expressing the constancy of certain quantitative features of a system of nuclear particles (including photons) when the system undergoes a change. They may be *exact*, i.e. believed to be exact and true for interactions in general, e.g. the total energy of a system (including the energy equivalent of the rest mass), or *approximate*, i.e. based on certain approximate features of the forces responsible for the interaction, e.g. based on the assumption that all the relevant forces are central.

CONSERVATION OF ANGULAR MOMENTUM. A principle stating that, in the absence of external forces, the angular momentum of a system is conserved if both orbital and spin angular momentum are taken into account.

CONSERVATION OF CHARGE. A principle stating that electric charge can neither be created nor destroyed. Its validity is essential for the whole of Maxwell's field theory, and in the context of relativistic electromagnetic theory. See also: *Creation of charge.*

CONSERVATION OF ENERGY. A principle stating that the total quantity of energy in a closed system is constant. Where the conversion of mass to energy, or vice versa, takes place, the energy of the system is understood to include the energy equivalent of all the mass in the system, according to the equation $E = mc^2$ where E is the energy, m the mass and c the speed of light. Thus the principles of the conservation of energy and the conservation of mass are no longer to be considered as distinct. See also: *Mass–energy equivalence.*

CONSERVATION OF LINEAR MOMENTUM. A principle stating that, in the absence of external forces, or if such forces cancel out, the vector of the linear momentum of a system remains constant in magnitude and direction and is not affected by processes occurring within the system.

CONSERVATION OF MASS. A principle stating that the total mass in a closed system is constant. The principle is valid in all conditions only when energy is also taken into account. See also: *Conservation of energy.*

CONSERVATIVE FIELD OF FORCE. A field of force in which the total energy of a moving body is constant. In such a field the work done on the body in moving between any two points is independent of the path taken.

CONSERVATIVE SYSTEM. A system in which only conservative forces are operative.

CONSONANCE. Of musical notes: the production of a sound that is pleasant to the ear when two or more notes are played together. Their frequencies generally bear a simple numerical relation to one another. See also: *Dissonance*.

CONSTANT-BOILING MIXTURE. See: *Azeotropes*.

CONSTANT-DEVIATION PRISM. A prism which, while acting as a normal prism with regard to dispersion and resolving power, has no minimum deviation, the deviation being constant and independent of refractive index and wavelength.

CONSTANT-DEVIATION SPECTROMETER. A spectrometer in which the collimator and telescope axes are fixed, and the wavelength scanning is achieved solely by rotation of the dispersing element.

CONSTANT-POTENTIAL CIRCUIT. A high-voltage generating circuit in which the rectified output voltage is maintained constant, apart from a small ripple.

CONSTITUTION DIAGRAM: EQUILIBRIUM DIAGRAM: PHASE DIAGRAM. A temperature-composition map of the solid and liquid phases of an alloy system. It shows the ranges of composition and temperature over which the various phases exist in equilibrium with each other, and the phase transformations which occur on heating and cooling. The construction of such diagrams is vital to the metallographic study of structures in an alloy system. Binary systems are represented by two-dimensional diagrams, and ternary usually by sections of a three-dimensional figure.

CONSTRAINT. Of a mechanical system of masses: the condition in which the number of degrees of freedom is less than three times the number of masses owing to mechanical or other limitations which prevent the free and independent motion of individual masses. See also: *Least constraint*: *Least curvature*.

CONSTRINGENCE. For an optical glass: the ratio $(n - 1)/\delta n$, where n is the refractive index for a given wavelength (usually that of the helium d-line, 5875·6 Å), and δn is the change of refractive index corresponding to two standard wavelengths (usually those of the hydrogen C- and F-lines, 6562·8 Å and 4861·3 Å). It is a measure of the dispersive power of the glass. Also known as the *reciprocal dispersive power* or *Abbé number*. (*Note:* 1 Å = 10^{-10} m.) See also: *Dispersive power (optical)*.

CONTACT ANGLE. For a liquid drop resting on a solid surface: the angle (measured in the liquid) between the tangent to the surface of the drop and the plane of the solid surface, at the point where the edge of the drop touches the solid, the tangent lying in a plane at right angles to the solid surface. Also called the *wetting angle*. When the angle is zero, complete wetting occurs. If it were 180° (a very rare occurrence) complete non-wetting would occur.

CONTACT, OPTICAL. Contact between two surfaces when their separation is less than about one-quarter of the wavelength of light.

CONTACTOR. A power-operated switch designed for repeatedly making and breaking electric power circuits. It is usually operated by an electromagnet, though pneumatic or mechanical operation is also used.

CONTACT POTENTIAL. More accurately contact potential difference: the potential difference, due to the difference in work function, arising when two solid surfaces are placed in contact.

CONTACT RESISTANCE. The electrical resistance at the point of contact between two conductors.

CONTINENTAL DRIFT. The breaking up and drifting apart of one or more land masses on the Earth to form the present continents. The theory of continental drift has been strongly supported by palaeomagnetic studies. See: *Plate tectonics*.

CONTINENTALITY. The extent to which the climate of a region is influenced by its distance from the sea, a *continental climate* being one with large diurnal and annual ranges of temperature, which arise from the small effective reserve of heat or cold in a land surface.

CONTINUED FRACTION. An expression of the form

$$b_0 + \cfrac{a_1}{b_1 + \cfrac{a_2}{b_2 + \cfrac{a_3}{b_3 + \cdots + \cfrac{a_n}{b_n}}}}.$$

If the fraction terminates with the term $\frac{a_n}{b_n}$ it is *finite*. If it does not terminate it is *infinite*.

CONTINUITY EQUATION. An equation in kinetic theory which relates the local density of a particular kind of particle (n) to the mean current density (j), on the assumption that no such particles are created or destroyed, by the expression

$\frac{\partial n}{\partial t} + \text{div } \mathbf{j} = 0$ where $\frac{\partial n}{\partial t}$ is the increase of n with time. It may be applied, for example, to neutron balance in nuclear reactors, to aerodynamic fluid flow, or to meteorological problems concerned with atmospheric air.

CONTINUITY OF STATE. The transition between two states, as between the gaseous and liquid states (in either direction), without any discontinuity in physical properties.

CONTINUOUS RADIATION. Radiation comprising a continuous range of wavelengths. It is also known as *white radiation* or *heterogeneous radiation*. See also: *Bremsstrahlung. X-rays.*

CONTINUOUS SPECTRUM. A spectrum in which radiation is distributed over an uninterrupted range of wavelengths in contrast to a line or band spectrum. It arises either from radiating matter in a condensed state (which may approximate to a black body) or from the ionization or dissociation of free atoms and molecules or their subsequent recombination.

CONTOUR ACUITY: VERNIER ACUITY. The power of the eye to distinguish a displacement between two parts of a line, as in reading a vernier. It is expressed in terms of the least detectable angular separation between the two parts and varies, for unaided vision, from 3 to 12 s of arc.

CONTOUR FRINGES. Interference fringes at constant optical thickness. Also known as Fizeau fringes.

CONTRAST. The relative brightness of two adjacent areas in a radiograph, photographic reproduction, fluorescent screen image, etc. It depends on many factors, including the granular nature of the film or screen and the nature of the object(s) and processes giving rise to the image. See also: *Colour contrast.*

CONTRAST SENSITIVITY. See: *Fechner law: Weber law.*

CONTROL MECHANISM. A device which causes another mechanism automatically to follow a predetermined behaviour pattern. It may incorporate a servomechanism, but the two terms are not synonymous.

CONVECTION. The transport of heat by a moving fluid in contact with a heated body.

CONVECTION, CELLULAR. A form of convection that may occur when a shallow layer of fluid is heated from below. It is marked by the appearance of a regular array of hexagonal cells, known as *Bénard cells*, the fluid rising in the centres of the cells and falling along the boundaries.

CONVECTION CURRENT. See: *Current.*

CONVECTION, FREE. See: *Convection, natural.*

CONVECTION, FORCED. Convection in which the motion is due primarily to a superimposed flow.

CONVECTION MODULUS: GRASHOF NUMBER. A dimensionless number appearing in the expression for the heat transfer coefficient for natural convection due to the presence of a hot body. It is given by $\frac{l^3 g \alpha \varrho^2 \theta}{\eta^2}$ where l is a typical dimension of the hot body, g is the acceleration due to gravity, α is the temperature coefficient of the fluid density, ϱ is the density and η the viscosity of the fluid, and θ is the temperature difference between the hot body and the fluid. See also: *Dynamical similarity principle.*

CONVECTION, NATURAL. Convection in which the motion is caused by gravitational forces and which arises from density differences due to temperature gradients within a fluid. Natural convection which is not subjected to constraining boundaries is termed *free convection*.

CONVERGENCE LIMIT (SPECTROSCOPY). (1) The long wavelength limit of an ionization continuum, e.g. the limit of a Rydberg series. (2) The limit corresponding to the point at which the separation between successive vibrational bands in an electronic band system decreases to zero.

CONVERGENCE OF MERIDIANS. At a point on a plane representation of the Earth's surface: the angle between the tangent to the line representing the meridian at that point and the tangent to the meridian at the origin of the projection.

CONVERGENT. In nuclear technology: refers to the condition of a neutron-chain-reacting medium for which the effective multiplication factor is less than unity, i.e. to the sub-critical condition. See also: *Chain reaction, neutron. Multiplication factor: Multiplication constant.*

CONVERGENT POINT. Of a moving star cluster: that point on the celestial sphere towards which all the stars appear to be moving.

CONVERSION COEFFICIENT. See: *Internal conversion coefficient.*

CONVERSION ELECTRON. See: *Internal conversion.*

CONVERSION FACTOR: CONVERSION RATIO. Of a nuclear reactor: See: *Nuclear reactor, conversion in.*

CONVERSION LENGTH. The means distance travelled by an energetic photon before its absorption by pair production.

CONVEX POLYHEDRON. A polyhedron of which any two points can be joined by a straight line entirely contained within the polyhedron.

COOLING, NEWTON LAW OF. States that the rate at which a body loses heat is proportional to the temperature difference between the body and the surrounding air. The law is true for quite large temperature differences, if the air is flowing past the body in a forced draught.

CO-OPERATIVE PHENOMENON. A process for which simultaneous interaction between the constituent systems of an assembly (a *co-operative assembly*) is required. Typical examples of co-operative phenomena are the condensation of gases, the melting of solids, order–disorder transformations, and ferromagnetism.

COORDINATE GEOMETRY. Geometry in which the positions of points are specified by means of suitable sets of numbers (*coordinates*) in such a way that, corresponding to geometrical relationships between the points, there are equivalent algebraic relationships between their coordinates, so that theorems concerning geometrical relationships can be proved by a process of algebraic calculation.

COORDINATES AND MOMENTA, GENERALIZED. Coordinates which describe the motion of a mechanical system without their exact nature being specified, and momenta related to these coordinates by the expression $p_i = L/\partial \dot{q}_i$, where $p_1, p_2, ..., p_n$ are the generalized coordinates, $q_1, q_2, ..., q_n$ are the associated (or *conjugate*) momenta, and L is the Lagrangian function for the system.

COORDINATES, CARTESIAN. Coordinates referred to three mutually perpendicular straight lines.

COORDINATES, CENTRE-OF-MASS SYSTEM OF. See: *Coordinates, laboratory system of.*

COORDINATES, CURVILINEAR. Coordinates which define the position of a point as the intersection of curves or of curved surfaces. Latitude and longitude are curvilinear coordinates for the surface of the Earth.

COORDINATES, CYLINDRICAL. Coordinates referred to a cylinder, and specified as the radial distance from the origin, the distance along the cylindrical axis, and the azimuthal angle, respectively.

COORDINATES, ECLIPTIC (ASTRONOMY). Coordinates on the celestial sphere, referred to the ecliptic as equator and to the corresponding poles.

COORDINATES, EQUATORIAL (ASTRONOMY). Coordinates on the celestial sphere, referred to the celestial poles and the celestial equator.

COORDINATES, GALACTIC (ASTRONOMY). Coordinates on the celestial sphere, referred to the galactic poles and the galactic equator.

COORDINATES, GENERALIZED. The minimum number of coordinates necessary to specify the state of a mechanical system subject to constraint (e.g. a diatomic gas with rigid molecules).

COORDINATES, GEOGRAPHICAL. Coordinates of points on the surface of the Earth, referred to the geographical equator and the geographical poles.

COORDINATES GEOMAGNETIC. Coordinates such as latitude and longitude defined with respect to the geomagnetic axis in the same way as *geographical coordinates* are defined with respect to the geographical axis.

COORDINATES, HELIOGRAPHIC. A latitude–longitude system of coordinates, related to the axis of solar rotation, used to define the position on the solar surface of a transient solar phenomenon such as a sunspot.

COORDINATES, HORIZONTAL SYSTEM OF (ASTRONOMY). Coordinates on the celestial sphere, referred to the zenith as pole and the horizon as equator.

COORDINATES, LABORATORY SYSTEM OF. Refers to the coordinate system, with the observer at rest, to which the experimental observation of such phenomena as collision processes is referred. For the theoretical study of these phenomena it is often convenient to use a coordinate system (the *centre-of-mass system*) having as its origin the centre of mass of the system of particles concerned.

COORDINATES, POLAR. Two-dimensional coordinates, comprising the radial distance from the origin and the azimuthal angle.

COORDINATES, SPHERICAL. Coordinates referred to a sphere, and specified as the radial distance from the centre, the angle of longitude, and the angle of latitude.

COORDINATION CHEMISTRY. The chemistry of coordination or complex compounds. See also: *Complex compound.*

COORDINATION NUMBER. (1) In a complex compound: the number of groups associated with

75

the central metal atom. (2) In a crystal: the number of nearest neighbours of a given atom.

C.O.P. See: *Coefficient of performance*.

COPERNICAN SYSTEM. The heliocentric theory of planetary motion, as opposed to the Ptolemaic, or geocentric system.

CORBINO EFFECT. The production of a circumferential electric current in a circular disk, by the action of an axial magnetic field upon a radial current in the disk.

CORIOLIS EFFECT. An effect typified by the deflection, relative to the surface of the rotating Earth, of any object moving with a constant velocity in space. It is usually described in terms of the *Coriolis acceleration* and *Coriolis force*. The deflection is found to be to the right in the northern hemisphere and to the left in the southern. Examples include the scouring of river banks, the motions of air over the Earth and the motions of artificial satellites.

CORNEA. The outermost part of the eye, just in front of the aqueous humour.

CORNER CUBE. A device, made with mirrors or reflecting wires, or cut from the solid, which has the property of reflecting light or radio waves back along the direction of incidence.

CORNER FRACTURE. A term used in the field of explosive metal working to define a fracture surface produced by the interaction of two or more waves of tensile stress.

CORNU PRISM. A prism of quartz, or similarly optically active material, which, used at minimum deviation, overcomes the doubling of an image occasioned by double refraction.

CORNU SPIRAL. A double spiral curve used in computing intensities in the diffraction pattern resulting from Fresnel diffraction.

CORONAE (METEOROLOGY). Coloured diffraction rings round the Moon (or the Sun) when it is seen through light clouds or mists of water droplets. The innermost ring, which appears as a white disk of radius $1-5°$, having an orange or yellow rim, is known as the *aureole*. Outside this are seen two or more rings (of radius up to $10°$ and with the red outside) which, as the Moon (or Sun) is not a point, are not those of a pure spectrum. The radii of the aureole and the various rings vary inversely with the size of the droplets.

CORONAGRAPH. A device which permits the observation of the Sun's chromosphere, prominences and especially its corona, by the production of an artificial eclipse. It incorporates means for removing scattered and reflected light.

CORONA, SOLAR. A bright white halo surrounding the Sun, seen during a total eclipse and also observable by a coronagraph. It emits a characteristic line spectrum. See also: *Solar atmosphere*.

CORPOSANT. Another name for *St. Elmo's fire*.

CORPUSCULAR THEORY OF LIGHT. The view, formerly held, that a beam of light consists of a stream of fast-moving material particles. The modern counterpart of such a particle is the photon.

CORRELATION. The association between two variables. Its magnitude is expressed by the *correlation coefficient*, which lies between 1 (complete correlation) and -1 (complete inverse correlation), zero corresponding to the absence of any relationship at all.

CORRELATION ARRAY. See: *Acoustic array, multiplicative*.

CORRELATION ENERGY. The energy associated with the Coulomb repulsion between electrons. It is part of the reduction in energy of a system of nuclei and electrons arising from the formation of atoms, i.e. from the correlation of the positions of the electrons and nuclei.

CORRESPONDENCE PRINCIPLE. The principle that, in the limit of high quantum numbers, the predictions of quantum theory agree with those of classical mechanics.

CORRESPONDING STATES, LAW OF. Predicts that all substances obey the same *reduced equation of state* $pV/RT = Z(p_R, T_R)$ where p is the pressure, V the volume, R the gas constant, and T the absolute temperature. $Z(p_R, T_R)$ is a universal function of the *reduced pressure* p_R and the *reduced temperature* T_R, i.e. the pressure and temperature expressed as fractions of their respective values at the critical point. See also: *Thermodynamic similarity, principle of*.

CORROSION. The deterioration of a substance (usually a metal) or of its properties, arising from reaction(s) with its environment.

COSINE LAW OF DIFFUSING SURFACE. See: *Light emission, cosine law of*.

COSMIC-RAY ABSORPTION LENGTH. The thickness of matter (usually in grammes per square centimetre) in which the nucleonic secondary component of the cosmic rays is reduced in intensity by a factor e.

COSMIC-RAY CASCADE. (1) *Electron cascade:* the cumulative production of electron pairs from a single high-energy electron via the process of bremsstrahlung emission followed by pair production. (2) *Nucleon cascade:* the cumulative production of nucleons within a single nucleus or in the passage of a nucleon through matter.

COSMIC-RAY EAST–WEST EFFECT. An effect caused by the Earth's magnetic field whereby more cosmic-ray particles arrive from a westerly direction than from an easterly one.

COSMIC-RAY LATITUDE EFFECT. The variation in cosmic-ray intensity with magnetic latitude.

COSMIC-RAY LONGITUDE EFFECT. The variation in cosmic-ray intensity along the magnetic equator, as a function of magnetic longitude.

COSMIC-RAY PENUMBRA. See: *Cosmic rays, allowed and forbidden regions for.*

COSMIC RAYS. Highly energetic extraterrestrial ionizing radiation. The *primary cosmic rays,* i.e. the cosmic rays incident on the top of the Earth's atmosphere, are predominantly composed of fully ionized (i.e. positively charged) atomic nuclei, having energies ranging from a few times 10^6 eV to 10^{18} or 10^{19} eV. The *secondary cosmic rays,* i.e. those produced in the Earth's atmosphere by the primary cosmic rays and other secondary cosmic rays, are complex in character and include nucleons, mesons, electrons, neutrinos, X-rays, γ-rays, etc. (*Note:* 1 eV is about $1 \cdot 6022 \times 10^{-19}$ J.)

COSMIC RAYS, ALLOWED AND FORBIDDEN REGIONS FOR. At a given place and for a given momentum: regions within which cosmic-ray particles arrive at full intensity (*allowed regions*) and those within which the particles cannot arrive at all (*forbidden regions*). There is a *main cone* of allowed directions and a *Störmer cone* of forbidden directions. An intermediate region between the allowed and forbidden regions is known as the *cosmic-ray penumbra.*

COSMIC-RAY SHOWER. The practically simultaneous appearance of secondary particles derived from the more energetic primaries (of energy greater than 10^{13} eV) in groups which contain so many particles as to be recognizable experimentally as separate from the general cosmic-ray flux. In the atmosphere they extend over considerable areas, and are also referred to as *extensive (air) showers* or *Auger showers.*

COSMIC RAYS, N-COMPONENT OF. Refers to all cosmic-ray particles capable of nuclear interaction, i.e. nucleons, heavy mesons and pions.

COSMIC-RAY SPECTROGRAPH. An elaborated cosmic-ray telescope, in which the magnetic deflection of trajectories is determined.

COSMIC-RAY TELESCOPE. A coincidence counter (two-, three- or multifold) used in the detection of cosmic rays.

COSMOGONY. The study of the origin of the solar system. See also: *Astronomy.*

COSMOLOGY. The study of the structure of the universe as a whole, including cosmogony. See also: *Astronomy.*

COSMOTRON. The name given to the proton synchrotron at the Brookhaven National Laboratory. It delivers protons of 3000 MeV energy.

COSOLVENCY. The property whereby a mixture of two or more solvents is capable of dissolving more of a substance than can either solvent separately.

COTTON BALANCE. A balance for determining the intensity of a magnetic field. The field strength is obtained by measuring the force on a current-carrying conductor of special shape placed in the field.

COTTON–MOUTON EFFECT. The double refraction of light in certain pure liquids (such as nitrobenzene) in a transverse magnetic field. The effect is also shown by some isotropic solids.

COTTRELL ATMOSPHERE. A cluster of impurity atoms which have migrated to a dislocation in a crystal. They tend to lock the dislocation in place and cause hardening. See also: *Dislocation.*

COULOMB. The unit of electric charge in MKSA and SI units. It is the quantity of electricity transported in one second by a current of one ampere. See: *Appendix II.*

COULOMB BARRIER. That part of the region of high potential energy through which a charged particle must pass on leaving or entering an atomic nucleus, which arises from the presence of electrostatic forces.

COULOMB ENERGY (1) In a system of particles (e.g. a solid or an atomic nucleus): that part of the binding energy associated with electrostatic interaction between the constituent particles. (2) Of a covalent bond: the electrostatic interaction energy between two or more electron distributions in terms of which the actual electron distribution is described.

COULOMB EXCITATION. The excitation of a nucleus by the electric field of a passing charged particle whose energy is insufficient to penetrate the coulomb barrier.

COULOMB FORCE: COULOMB INTERACTION. The attractive or repulsive force between two electric point charges. It acts along the line joining the two charges and is given, according to *Coulomb's law of force*, by $Q_1 Q_2 / \varepsilon_0 r^2$ where Q_1 and Q_2 are the charges, r the distance between them, and ε_0 is the permittivity of the medium. A negative value corresponds to an attraction and a positive value to a repulsion.

COULOMB INTEGRAL. See: *Exchange integral*.

COULOMB SCATTERING: RUTHERFORD SCATTERING. The scattering of a charged particle by the Coulomb field (electrostatic field) of a nucleus.

COULOMETER: VOLTAMETER. An electrolytic cell used for measuring the quantity of electricity passing through a circuit in a given time by the determination of the amount of metal deposited, or gas liberated, due to the passage of the current. See also: *Silver voltameter*.

COULOMETRY. An analytical process in which a coulometer is used to determine the amount of a constituent present, by measuring the amount of current necessary to achieve a given reaction involving the constituent.

COUNTER. For ionizing radiation: a device which reacts to individual ionizing events, thus enabling them to be counted. The term is also loosely used to describe a complete counting system.

COUNTER, BORON. A counter filled with a gaseous boron compound (e.g. BF_3), or having electrodes coated with boron or a boron compound. It is operated in the proportional region and is used for the detection of slow neutrons by the (n, α) reaction of ^{10}B.

COUNTER, CRYSTAL. A counter depending for its action on a crystal (e.g. diamond) in which the electrical conductivity is momentarily increased as a result of an ionizing event.

COUNTER, EFFICIENCY OF. The fraction of incident particles or quanta that are recorded by a counter.

COUNTER, FOUR PI (4π). A counter which measures the radiation emitted in all directions by a radioactive material.

COUNTER, GEIGER–MÜLLER. A gas-filled ionization chamber, usually consisting of a hollow cylindrical cathode with a fine wire along its axis, which is operated in the Geiger region, and in which each ionizing event is followed by only one terminated discharge.

COUNTER, GEIGER REGION OF. That range of operating voltage in a Geiger–Müller counter over which the output charge is independent of the number of primary ions produced in the initial ionizing event.

COUNTER, HALOGEN-FILLED. A self-quenched Geiger–Müller counter in which the quenching agent is either bromine or chlorine. Its advantages are low operating voltage (400 V) and long life.

COUNTER, LIMITED PROPORTIONALITY REGION OF. The range of operating voltage for a counter, below the Geiger threshold, in which the gas amplification depends on the number of ions produced in the initial ionizing event as well as on the voltage. The counter is then proportional for small initial events but saturates for large ones.

COUNTER, PLATEAU OF. The range of applied potential difference over which the counting rate varies relatively little under constant conditions of irradiation.

COUNTERPOISE WEIGHING: SUBSTITUTION WEIGHING. A method of weighing used to allow for possible differences in the lengths of balance beams. The body to be weighed is placed in one pan and a counterpoise in the other. The body is then replaced by standard weights (including rider) and the counterpoise is weighed. See also: *Double weighing*.

COUNTER, PROPORTIONAL. A counter operating in the proportional region.

COUNTER, PROPORTIONAL REGION OF. The range of operating voltage for a counter in which the gas amplification is greater than unity and is independent of the primary ionization. It follows that, in this region, the pulse size is proportional to the number of ions produced as a result of the initial ionizing event.

COUNTER, QUENCHING OF. The termination of a pulse of ionization in a Geiger–Müller counter. It may be accomplished by the use of an appropriate gas or vapour filling (*internal quenching* or *self-quenching*) or by momentary reduction of the applied potential difference (*external quenching*) by the use of a *quenching circuit*. The gas or vapour used for internal quenching is known as a *quenching agent*.

COUNTER, RECOVERY TIME OF. The minimum time interval after the initiation of a voltage pulse before the next voltage pulse of normal size.

COUNTER, SCINTILLATION. A counter in which the flashes of light produced in a scintillator by ionizing radiation are converted into electrical pulses by a photomultiplier.

COUNTER, SEMICONDUCTOR. A counter in which the electron hole pairs, produced by ionizing particles which penetrate the sensitive volume of a semiconductor, are collected by an applied field to give electrical pulses.

COUNTER, SPARK. A gas counter, having parallel electrodes, in which the primary ionization initiates a very rapid avalanche which appears as a spark. It provides an accurate means of recording both the position and time of an ionizing event. See also: *Spark chamber*.

COUNTING-RATE METER. An instrument which gives a continuous indication of the average rate of arrival of pulses from a counter.

COUNTING STATISTICS. Statistics which provide estimates of certain parameters governing sequences of discrete random events. For example, in the case of a radioactive nuclide, the probability that the observed number of atoms of the nuclide will decay in a given time is given by Poisson's distribution. From this distribution an estimate may be obtained of the decay constant of the nuclide concerned.

COUNTING, WHOLE-BODY. The measurement of the total radioactive content of the human body, whether it originates from natural or man-made sources.

COUPLED CIRCUITS. Usually describes circuits which are coupled, or interconnected, to each other by a common magnetic flux linking an inductor in each circuit (i.e. via a mutual inductance), or by a common capacitor. In general the term refers to any circuits which are related in such a way that a.c. effects are transferred, but steady-state d.c. effects are not.

COUPLING. (1) In general, an interaction between different properties of a system, or between two or more systems. (2) For particles in an atom or nucleus: the *Russell–Saunders coupling* (*L–S coupling*), in which the resultant, L, of the orbital angular momentum of all particles interacts with the resultant, S, of the spins of all particles; or the *j–j coupling*, in which the total angular momenta (orbital plus spin) of the individual particles interact with one another; or a coupling intermediate between these two. (3) In molecular spectra: one of a number

of five ideal ways (*the Hund coupling cases*) of coupling the electron spin angular momentum, S, the electron orbital angular momentum, L, and the nuclear rotation angular momentum, N, to form a resultant, J.

COUPLING CONSTANT. A physical constant expressing the magnitude of the force exerted on a particle by a field.

COVALENT COMPOUND. A compound in which the bonds are essentially covalent in character.

COVALENT RADIUS. See: *Atomic radius*.

COVARIANT EQUATION. One which retains its form under transformation to quantities measured by another observer; more particularly under a Lorentz transformation. See also: *Invariant expression. Tensor analysis.*

CPS TUBE: EMITRON TUBE. A television camera tube in which the optical image is focused on to a photo-emissive mosaic formed on a transparent dielectric and capacitatively associated with a conducting signal plate.

CRAB NEBULA. See: *Supernova*.

CRACKING. The breaking down of hydrocarbons into smaller molecules by heating to about 500–600°C under pressure.

CREATION OF CHARGE. A theory introduced to account, for example, for the Earth's electromagnetic and gravitational fields. Various suggestions have been made for the modification of Maxwell's equations to allow for the creation of charge. See also: *Conservation of charge*.

CREATION OPERATOR. In the quantum theory of fields: an operator which, when applied to a state vector ϕ_n, that describes a system containing n particles, will yield a state vector ϕ_{n+1}, which describes a system having $(n + 1)$ particles. See also: *Destruction operator*.

CREEP. Of a solid: the change of strain with time under the influence of a constant stress. Three types have been distinguished: *primary*, in which the creep rate decreases with time; *secondary*, in which the creep rate is constant with time; and *tertiary*, in which the creep increases with time until fracture occurs.

CREEP, THERMAL. The accelerated development of plastic creep with increase of temperature.

CREP SCALE. See: *Twilight*.

CREPUSCULAR RAYS. (1) Rays which are emitted by the Sun when below the horizon and rendered visible when scattered by clouds or mountains.

(2) Rays from the Sun which pass through openings in the clouds and, on account of scattering, appear to be divergent. See also: *Twilight phenomena.*

CRITICAL. In nuclear technology: refers to the condition of a neutron-chain-reacting medium for which the effective multiplication factor is unity. If such a medium would be critical using prompt neutrons only it is said to be *prompt critical.* If delayed neutrons are necessary to achieve the critical state it is said to be *delayed critical.* See also: *Chain reaction, neutron.*

CRITICAL ANGLE (OPTICS). That value of the angle of incidence above which a pencil of light in a more dense medium, meeting the boundary of a less dense medium, will be reflected back into the more dense medium. At the critical angle itself the light is refracted along the boundary surface. The critical angle is given by $\sin^{-1} \frac{n_2}{n_1}$, where n_1 is the refractive index of the more dense medium and n_2 that of the less dense medium. See also: *Total reflection.*

CRITICAL DENSITY. Of a fluid: the density at the critical point.

CRITICAL FIELD. For a superconductor: See: *Superconductivity, critical field in.*

CRITICAL FLICKER FREQUENCY. See: *Flicker.*

CRITICAL FREQUENCY OF THE IONOS-PHERE. That frequency below which radiation emitted at any angle from a radio aerial on the Earth is reflected back again by the ionosphere.

CRITICAL MAGNETIC SCATTERING. See: *Scattering, critical.*

CRITICAL OPALESCENCE. See: *Scattering, critical.*

CRITICAL POINT. Of a fluid: the point, characterized by pressure and temperature, at which the gaseous and liquid phases become identical and form only one phase.

CRITICAL POTENTIAL. Of an atom in an inelastic collision with an electron: the minimum potential through which the electron must fall to acquire just sufficient energy to raise the atom into an excited state. An atom may have many different critical potentials up to a maximum value equal to the ionization potential.

CRITICAL PRESSURE. Of a fluid: the pressure at the critical point.

CRITICAL SHEAR STRESS. The resolved shear stress required to initiate slip in a given direction of a single crystal along a given crystallographic plane.

CRITICAL TEMPERATURE. (1) Of a gas: the temperature at the critical point. Above this temperature the gas cannot be liquified. (2) Of a solid: the temperature at which a specified transition takes place.

CRITICAL VOLUME. Of a fluid: the volume occupied by unit quantity (usually one mole) at the critical temperature and pressure.

CROOKES DARK SPACE. See: *Cathode fall. Cathode dark space. Discharge, glow.*

CROSS-FIRE TECHNIQUE. See: *Radiotherapy, cross-fire technique in.*

CROSS-LINKING. In a polymerized material: the formation of additional links between the molecular chains.

CROSS OVER. The plane of minimum cross-section through which the beam of electrons from the cathode of an electron gun passes. It may be compared with the exit pupil of an optical system.

CROSS-SECTION. (1) A measure of the probability of the occurrence of a specified interaction between a particular incident radiation and a specified target particle or system of particles. Unless otherwise stated the cross-section (sometimes known as the *microscopic cross-section*) is given by the reaction rate per target particle, for a specified process (e.g. capture), divided by the flux density of the incident radiation. In reactor physics use is also made of the *macroscopic cross-section*, which is the sum of the cross-sections of all the atoms in unit volume and may be considered as a linear absorption coefficient. Cross-sections may also be summed over unit mass or over a specified body. (2) See: *Radar-echo cross-section: Radar echoing area.*

CROSS-SECTION, ABSORPTION. The cross-section for the absorption process. It is the difference between the total cross-section and the scattering cross-section.

CROSS-SECTION, ACTIVATION. The cross-section for the formation of a radionuclide by a specified interaction.

CROSS-SECTION, CAPTURE. The total cross-section for all possible capture processes.

CROSS-SECTION, COLLISION. The cross-section for the collision process.

CROSS-SECTION, DIFFERENTIAL. For an interaction process involving one or more outgoing particles, and for a specified direction or energy: the cross-section per unit interval of solid angle or energy. See also: *Scattering, differential.*

CROSS-SECTION, DÖPPLER-AVERAGED. A cross-section averaged over energy, weighted to take into account the effect of the thermal motion of the target particles.

CROSS-SECTION, FISSION. The cross-section for the fission process.

CROSS-SECTION, GEOMETRICAL. The cross-section corresponding to the geometrical area presented by a target nucleus to the incident beam. It is given by $4\pi a^2$, where a is the radius of the target nucleus.

CROSS-SECTION, MACROSCOPIC. The cross-section per unit volume of a given material for a specified process. It has the dimensions of reciprocal length. See also: *Cross-section.*

CROSS-SECTION, MICROSCOPIC. The cross-section per target nucleus, atom or molecule. It has the dimensions of area and may be visualized as the area normal to the direction of an incident particle which has to be attributed to the target particle to account geometrically for the interaction between the two particles. See also: *Cross-section. Cross-section, geometrical.*

CROSS-SECTION, NON-ELASTIC. The cross-section for all processes other than elastic scattering. It includes inelastic scattering and capture (including fission) as well as charged particle reactions and $(n, 2n)$ reactions.

CROSS-SECTION, SCATTERING. The cross-section for the scattering process. It includes the cross-sections for coherent scattering, incoherent scattering, elastic scattering and inelastic scattering.

CROSS-SECTION, TOTAL. The sum of the cross-sections for all the separate interactions between the incident radiation and a specified target.

CROSS-SECTION, TRANSPORT. The total scattering cross-section multiplied by the average value of $(1 - \cos \theta)$, where θ is the laboratory angle of scattering.

CROSS-TALK. Interference arising between adjacent channels of communication.

CROVA WAVELENGTH. That wavelength, in the spectrum of a radiation at a given temperature, whose intensity varies with temperature at the same relative rate as does the intensity of the total radiation. It is that wavelength for which $\frac{di/dT}{i} = \frac{dI/dT}{I}$, where i is the intensity at the wavelength in question, I the total intensity, and T the temperature.

CRYOGENICS. The study and use of low temperature phenomena. See also: Entries under *Refrigeration. Refrigerator.*

CRYOHYDRATES. Mixtures of ice and a chemical salt to give *freezing mixtures.*

CRYOPHORUS. A laboratory apparatus for demonstrating the principle of a refrigerator.

CRYOPUMPING. The pumping effect associated with the reduction in vapour pressure when the temperature is slightly reduced below that of the triple point of the gas present in a system. Its simplest form is the cold trap used to remove unwanted vapours in conventional vacuum systems but large-scale applications are now possible with the availability of very low-temperature refrigerants on a commercial scale.

CRYOSCOPIC CONSTANT. A constant which, when multiplied by the molality of a solution, gives the freezing point depression of the solution resulting from the addition of one mole of solute per kilogramme. The constant is given by $RT_0{}^2/1000L$, where R is the gas constant, T_0 (absolute) is the freezing point of the pure solvent, and L is the latent heat of fusion per gramme. The expression is valid only for dilute solutions.

CRYOSCOPY. The measurement of the depression of the freezing point of a solvent caused by the addition of a solute. Since, for dilute solutions, this depression is proportional to the molar concentration of the solute, cryoscopy can be used for the determination of molecular weights.

CRYOSTAT. An apparatus in which a low temperature may be maintained.

CRYOTRON. A superconducting switching element, operating in liquid helium, consisting of a short superconducting wire (e.g. Ta), known as the *gate*, in a solenoid of another superconductor (e.g. Nb), known as the *control.* A magnetic field, introduced via the control, sends the gate into the resistive state, although the control itself remains superconducting. The cryotron is of considerable interest to computer designers since it can steer current from one superconductive circuit to another superconductive circuit in parallel, and can therefore be used to perform binary logic. It can also be used to make a computer storage cell.

CRYSTAL. That form of a substance which is characterized by an orderly arrangement of atoms repeated more or less perfectly within a region which is large compared with atomic dimensions. Crystalline material may or may not show macroscopic *crystal faces*, i.e. plane surfaces which are inclined to each other at angles characteristic of the particular substance concerned.

CRYSTAL, BUBBLE MODEL OF. An assemblage of bubbles, about a millimetre in diameter, floating on the surface of, for example, a soap solution and held together by surface tension, whose behaviour is in many ways analogous to that of a two-dimensional metallic structure. Three-dimensional assemblages have also been constructed.

CRYSTAL CHEMISTRY. The study of chemical bonding effects between atoms in solids. It is an essential adjunct to both classical and structural crystallography.

CRYSTAL CLASS. A subdivision of a crystal system in which a crystal may be placed according to its macroscopic symmetry elements (i.e. mirror planes, rotation axes and inversion axes). Each such characteristic group of symmetry elements is known as a *point group*. They are thirty-two in number. See also: *Crystal symmetry, point group in.*

CRYSTAL-CONTROLLED OSCILLATOR. An oscillator with a very high-frequency stability controlled by a quartz crystal which is in mechanical vibration due to the piezoelectric effect. The quartz is cut in the form of a flat plate and the angle at which the plate is cut in relation to the optic axis determines the frequency-temperature coefficient. See also: *Cut.*

CRYSTAL DEFECT. Any departure from the ideal structure in a crystal. See also: *Dislocation. Mosaic structure. Point defect. Surface defect.*

CRYSTAL DIODE. A diode consisting of a semiconducting material with either a point contact or *p–n* junction.

CRYSTAL DIRECTION, INDICES OF. The lattice coordinates of that point on a line in the given direction through the origin whose coordinates may be expressed as the smallest integral multiples of the unit cell sides. In the cubic system these indices are the same as those of the plane to which the direction is perpendicular, but this is not true for other systems. The indices are expressed in square brackets [ ]. See also: *Miller indices.*

CRYSTAL FORM. A group of crystal faces produced by operating on a single face according to one or more symmetry elements of the crystal class. The form may consist of a closed figure, an open group of faces parallel to an axis, a pair of parallel faces or even (in the limit) a single face.

CRYSTAL INDICES. See: *Miller indices.*

CRYSTAL LATTICE. An alternative term for space lattice. See also: *Space lattice: Bravais lattice.*

CRYSTALLINE ELECTRIC FIELD: CRYSTAL FIELD. Refers to the field, experienced by an ion in a crystal, that arises from the influence of neighbouring ions and electric dipoles. The term describes the electric field which, if it acted on an isolated ion, would result in the ion having properties similar to those observed for the same ion in a crystal. Ions of transition group elements are particularly sensitive to crystalline fields.

CRYSTALLITE. A coherent region of a crystalline material, e.g. a mosaic block, which may be much smaller than the particle or grain of which it forms a part.

CRYSTALLOGRAPHIC AXIS. One of a set of three coordinate axes according to which the internal atomic structure or the external form of a crystal can be described. Depending on the symmetry of the crystal, they may be of equal or unequal lengths and may or may not be at right angles to each other.

CRYSTALLOGRAPHY. The study of the form, symmetry, internal structure and properties of crystals.

CRYSTALLOID. (1) A configuration of atoms (or other identical components), finite in one or more dimensions, in a true free-energy minimum, where the units are not related to each other by three lattice operations. It is, in the conditions prevailing, more stable than a crystallite with the same components. (2) (Obsolete) a substance which is soluble in water and is not a colloid.

CRYSTAL MOMENTUM. The product of Planck's constant and a wave vector in a crystal.

CRYSTAL OPTICS. The study of the interaction of crystalline matter with electromagnetic radiation. The term is usually restricted to wavelengths within the visible spectrum.

CRYSTAL PHYSICS. The study of the physical properties of single crystals.

CRYSTAL PLANE. One of a set of parallel equally spaced planes in a crystal structure, characterized by a plane of given Miller indices and a parallel plane through the origin of coordinates.

CRYSTAL PROJECTION. A plane projection of crystal faces or planes in which each is represented by a dot. The projection is obtained in various ways from the points (known as the *poles*) at which normals to the crystal faces or planes cut a sphere (the *reference sphere*) at the centre of which the crystal is assumed to be located. The relative positions of the poles display the fundamental angular relationships between the faces or planes, and afford useful representations of directions in the crystal such as optical directions or structural bonds. See also: *Gnomonic projection. Orthographic projection. Stereographic projection.*

CRYSTAL SPACING. See: *Interplanar spacing.*

CRYSTAL STRUCTURE. The three-dimensional arrangement of atoms in a crystal. It is specified by giving the positions of the individual atoms in the unit cell, by the repetition of which the structure arises. The determination of the structure depends on the application of one or more methods of diffraction.

CRYSTAL SYMMETRY. The totality of the symmetry elements (e.g. plane of mirror reflection, three-fold axis of rotation, etc.) which describe the relationships between a set of macroscopic crystal faces or between the atomic positions in a crystal structure. See also: *Crystal class. Crystal system.*

CRYSTAL SYMMETRY, CENTRE OF SYMMETRY IN. The point about which like faces or edges are arranged in pairs in corresponding positions on opposite sides. Similarly, the point within the unit cell about which the atomic positions are related in the same fashion.

CRYSTAL SYMMETRY, GLIDE PLANE IN. A plane which can be drawn in the crystal structure such that reflection of any atomic position at the plane, followed by a movement of translation, gives an atomic position with identical environment. The translation may be parallel to a crystallographic axis lying in the plane (an *axial glide plane*) or there may be two successive translations parallel to two crystallographic axes (a *diagonal glide plane*). In the former case the translation is half the unit cell edge, and in the latter it is a half or one-quarter.

CRYSTAL SYMMETRY, INVERSION AXIS IN. An axis characterized by rotation through a fraction of a revolution ($^1/_2$, $^1/_3$, $^1/_4$ or $^1/_6$) followed by reflection through a centre of symmetry.

CRYSTAL SYMMETRY, POINT GROUP IN. One of the thirty-two groups of macroscopic symmetry elements which correspond to the thirty-two crystal classes. See also: *Crystal class.*

CRYSTAL SYMMETRY, ROTATION AXIS IN. An axis, characterized by rotation through a fraction of a revolution, which produces equivalent and identical positions in crystal form or structure. The fraction may be a half (*diad axis*), one-third (*triad axis*), one-quarter (*tetrad axis*), or one-sixth (*hexad axis*).

CRYSTAL SYMMETRY, SCREW AXIS IN. An axis of crystal symmetry characterized by rotation through a fraction of a revolution ($^1/_2$, $^1/_3$, $^1/_4$ or $^1/_6$) followed by translation along the axis by a fraction of a unit cell edge.

CRYSTAL SYMMETRY, SPACE-GROUP IN. One of a set of mutually consistent groups of symmetry elements which determine the relative positions in which atoms or groupings of atoms can occur in a unit cell. They are developed from the fourteen Bravais lattices by the application of mirror planes, rotation axes and inversion axes (the *macroscopic symmetry elements*) to give thirty-two crystal classes, and by the application to these classes of glide planes and screw axes (*microscopic symmetry elements*) to give 230 space groups.

CRYSTAL SYSTEM. One of the seven main categories to which a crystal may be assigned according to the symmetry of its external form or internal structure. The systems are: cubic, tetragonal, hexagonal, trigonal, orthorhombic, monoclinic and triclinic (or anorthic).

CRYSTAL WHISKERS. Single crystal filaments of metals and compounds, which possess abnormal properties (e.g. extremely great mechanical strength), often approaching those of a perfect solid.

CRYSTAL ZONE. A set of crystal faces whose intersecting edges are all parallel. Each zone is specified by the direction of the zone axis.

CRYSTAL ZONE AXIS. The direction of the line to which the faces constituting a crystallographic zone are parallel. See also: *Crystal direction, indices of.*

C-STAR. See: *Stars, spectral classes of.*

CUBIC EQUATION. An algebraic equation of the third degree in one or more variables. In its most general form it may be written as $ay^3 + 3by^2 + 3cy + d = 0$, or, in the reduced form, as $x^3 + 3px + q = 0$, where $x = ay + b$, $p = ac - b^2$, and $q = a^2d - 3abc + 2b^3$.

CUBIC SYSTEM. The crystal system with the highest degree of symmetry. The external form and internal structure can be referred to three equal and mutually perpendicular axes.

CULMINATION, LOWER. Of a circumpolar star: the instant when the star's hour angle is 12h.

CULMINATION, UPPER. Of a circumpolar star: the instant when the star's hour angle is zero.

CURIE. A unit of radioactivity, defined as $3 \cdot 7 \times 10^{10}$ disintegrations per second exactly, but originally defined as the amount of radon in equilibrium with 1 g of radium. It is now to be replaced by the SI unit, the becquerel. See: *Becquerel. Appendix II.*

CURIE–CHÉVENEAU BALANCE. A balance in which the specimen is suspended from the arm of a torsion balance between the pole pieces of a magnet. It is designed for comparative dia- and paramagnetic susceptibility measurements at room temperature.

CURIE LAW. States the relationship between the magnetic susceptibility and temperature of a paramagnetic material, where internal interactions can be neglected. It may be expressed as $X_m = C/T$, where X_m is the molar susceptibility, T the absolute temperature, and C is termed the *Curie constant* per mole. See also: *Curie–Weiss law.*

CURIE POINT: CURIE TEMPERATURE. One of four temperatures according to context. (a) *The ferromagnetic Curie point:* the temperature above which permanent or spontaneous magnetization disappears, and ferromagnetic materials become paramagnetic. (b) *The paramagnetic Curie point:* the temperature appearing in the Curie–Weiss law for the susceptibility of a paramagnetic material, in which internal interactions play a part. (c) *The antiferromagnetic Curie point or Néel point:* the temperature above which the antiferromagnetism is destroyed. (d) *The ferroelectric and antiferroelectric Curie points:* points which are analogous to the corresponding ferromagnetic Curie points. See also: *Curie–Weiss law. Néel temperature: Néel point.*

CURIE SCALE OF TEMPERATURE. A scale of temperature used in the range 0 to about 1 K where gas thermometry is impracticable. The temperature on this scale, T^*, is defined by the equation of state of an ideal paramagnetic, i.e. $T^* = CH/I$, where C is the Curie constant, and I the intensity of magnetization in a magnetic field H.

CURIE–WEISS LAW. A modification of the Curie law, applicable where internal interactions play a part. It may be expressed as $X_m = C/(T - \theta)$, where X_m is the molar susceptibility, C the Curie constant, T the absolute temperature and θ the paramagnetic Curie point or temperature.

CURL. The vector product of the differential vector operator, ∇, and a vector. For a vector $\mathbf{V}$, it may be written

$$\nabla \times \mathbf{V} = \mathbf{i} \left\{ \frac{\partial V_z}{\partial y} - \frac{\partial V_y}{\partial z} \right\} + \mathbf{j} \left\{ \frac{\partial V_x}{\partial z} - \frac{\partial V_z}{\partial x} \right\}$$
$$+ \mathbf{k} \left\{ \frac{\partial V_y}{\partial x} - \frac{\partial V_x}{\partial y} \right\},$$

where $\mathbf{i}, \mathbf{j}, \mathbf{k}$ are unit vectors. The curl is also known as the *rotation* (abbreviation *rot*). See also: *Differential vector operator.*

CURRENT. Any ordered motion of electric charge. A *conduction current* is one for which the motion is produced by the application of an electric field: it consists of the flow of free charges. A *displacement current* is that associated with the changing electric field in the dielectric of a capacitor, for example, where it has the same instantaneous value as the charging conduction current in the external circuit. The *polarization current* is that part of the displacement current associated with the movement of bound charges within the dielectric, i.e. with the creation or rotation of dipoles. A *convection current* is one for which no applied field is necessary, e.g. the drift of electrons in a cathode-ray tube, or the drift of charged particles in space under the action of gravitational forces. See also: *Ampère law. Current density.*

CURRENT ALGEBRA. A branch of the theory of elementary particles dealing with the currents measured in electromagnetic and weak interactions.

CURRENT DENSITY. The current flowing in a body per unit area of cross-section taken at right angles to the direction of flow. The term may refer to electric charge, sub-atomic or nuclear particles, or to the flow of a fluid. In the case of nuclear particles the current density is commonly called the *current*.

CURRENT EFFICIENCY. (1) Of an electron gun: that fraction of the total beam current which passes through the beam-limiting aperture. (2) Of an electrochemical process: the ratio of the mass of substance liberated by a given current to that which should be liberated according to Faraday's law.

CURRENT TRANSFORMER. An a.c. transformer for use with a.c. instruments to extend their range, to isolate them for high-voltage circuits, or to operate safety devices in a.c. power installations. The primary is in series with the mains and the instrument is connected to the secondary. The transformer is designed so that the ratio of the primary and secondary currents as well as their phase difference (normally 180°) is accurately maintained.

CURTATE DISTANCE. The distance of a planet

or a comet from the Earth or the Sun, projected upon the plane of the ecliptic.

CURVATURE. A measure of the rate of change of direction of a line either lying in a plane or on a surface.

CURVATURE OF THE FIELD OF A LENS. See: *Lens, curvature of field of.*

CUSEC. A unit of fluid flux, equal to 1 ft³/s (0·0283 m³/s). An abbreviation for *cu*bic foot per *sec*ond.

CUSPS, LUNAR OR PLANETARY. The extremities of the crescent of the Moon, or an inferior planet, when in a crescent phase.

CUT. Of a quartz oscillator crystal: denotes the way in which the crystal plate is cut to achieve various properties. The *AT-cut* refers to a plate including the x-axis and making an angle of about $35·5°$ with the z-axis. The plate has a substantially zero temperature coefficient. The *BT-cut* refers to a plate including the x-axis and making an angle of about $-49°$ with the z-axis. It, too, is characterized by an essentially zero temperature coefficient. The *X-cut* refers to a slab which is normal to the x-axis, with major surfaces parallel to the y- and z-axes. The *Y-cut* refers to a slab which is normal to the y-axis, with major surfaces parallel to the x- and z-axes. The *Z-cut* refers to a slab which is normal to the z-axis with major surfaces parallel to the x- and y-axes.

CUT-OFF FREQUENCY. Of a transmission system: the point at which the system passes from a frequency range of low attenuation to one of high. For a wave guide there is a cut-off frequency for each mode, the lowest of which is known as the *absolute cut-off frequency.*

CUT-OFF VOLTAGE. Of a thermionic valve: the value of that negative voltage which, when applied to a grid in the valve, will just stop the flow of current to an anode. It is approximately equal to the anode voltage divided by the amplification factor.

CUT-OUT. See: *Electrical protective gear or cut-out.*

CYBERNETICS. The study of control and communication in the animal and the machine. It is concerned with the flow of information and involves details of information input and output, feedback, etc.

CYBOTAXIS. The arrangement of the molecules of a liquid into localized crystalline regions, i.e. regions of short-range order. The existence of such regions (*cybotactic groups*) may possibly be a statistical effect in ordinary liquids but probably not in liquids consisting of long-chain molecules.

CYCLE. A series of events in which conditions are the same at the end as they were at the beginning. Usually, but not invariably, a cycle of events is recurrent.

CYCLIC COMPOUND. A compound containing a closed ring of atoms.

CYCLIC STATE. Of a magnetic material: the state in which it responds equally to a magnetic field applied in either of two opposite directions.

CYCLOGRAPH. An instrument producing a closed figure (*cyclogram*) on a screen when a beam of light or cathode rays is subjected to the action of controlling forces acting at right angles to each other.

CYCLOID. The path described by a fixed point on a circle as the circle rolls, without slipping, on a straight line. It is a special case of a trochoid. See also: *Trochoid.*

CYCLONE. A circulatory wind system in the atmosphere, centred on a region of low pressure. In the northern hemisphere the rotation is anticlockwise, and in the southern it is clockwise. A cyclone may cover a horizontal range of a few thousand kilometres, as in the *meteorological depression* or *extra-tropical cyclone*, or it may be confined to a few hundred kilometres as in the *tropical cyclone*. The latter, which is known in the Atlantic Ocean and Caribbean Sea as a *hurricane*, and in the Indian Ocean as a *typhoon*, is associated with very strong winds (100 knots or so) and very low pressures (below 950 mbar). (*Note:* 100 knots is about 180 km/h.) See also: *Whirlwind.*

CYCLONE SEPARATOR. A device for the separation of solid, liquid and gaseous phases or of different materials in the same phase. It is based upon the use of centrifugal acceleration, which is imparted by the translation of fluid pressure energy to rotational kinetic energy. See also: *Centrifugal separator.*

CYCLOTRON. A particle accelerator in which the particles move in a constant magnetic field in a spiral orbit, the energy of the particles being increased by the application of an alternating electric field at constant frequency. See also: *Dee. Synchrocyclotron.*

CYCLOTRON FREQUENCY. The orbital frequency of a charged particle in a magnetic field.

CYCLOTRON RESONANCE. The resonance absorption of energy, by a charged particle moving in a magnetic field, from an impressed alternating electric field with appropriate frequency.

CYLINDER. The surface generated by a straight line which intersects a given curve and moves so as to remain parallel to its original direction. The most common curve is a circle, to the plane of which the straight line is normal, but other common curves are the ellipse, hyperbola and parabola.

CYLINDER GAUGE. A device for checking the diameter and roundness of a cylindrical bore.

CYLINDRICAL WAVE. A wave whose equiphase surfaces form a family of coaxial or confocal cylinders.

CYSTOSCOPE. A particular form of endoscope designed for the interior examination of the bladder.

d

DAGUERROTYPE. The first practical method of photography, involving the development of silver iodide and bromide in mercury vapour and fixing with cyanide or hypo (sodium thiosulphate).

D'ALEMBERT EQUATION. A linear differential equation with constant coefficients, of the form $\frac{dy_i}{dx} + \sum a_{ik}y_k = 0$, where $i = 1, 2, 3, ..., n$, $y_i(x)$ are the n functions to be determined and a_{ik} are the constants.

D'ALEMBERTIAN OPERATOR. A differential operator, represented by $\square$, frequently used in electromagnetic wave theory. Referred to rectangular cartesian coordinates it may be expressed as

$$\square^2 \equiv \frac{\partial^2}{\partial x^2} + \frac{\partial^2}{\partial y^2} + \frac{\partial^2}{\partial z^2} - \frac{1}{c^2}\frac{\partial^2}{\partial t^2},$$

where c is the wave velocity in the medium.

D'ALEMBERT PRINCIPLE. For a moving system of particles connected together by various forms of constraint: states that the external forces acting on the particles, less the effective forces on the particles set up by the constraints, form a system which is in equilibrium.

DALTON LAW OF PARTIAL PRESSURES. States that the total pressure of a mixture of gases under given conditions is the sum of the partial pressures of the individual gases under the same conditions, the partial pressure of a component gas being defined as the pressure that would be exerted by that gas if it were present alone.

DAMPING. Of a mechanical or electrical system: denotes the reduction of motion (usually oscillatory) by dissipation of the energy of motion.

DAMPING CAPACITY: INTERNAL FRICTION. The ability of a material to dissipate mechanical energy internally (i.e. excluding the overcoming of external forces such as friction). A common measure is the fractional energy lost per cycle of oscillation. Others are the logarithmic decrement and the reciprocal of the Q-value. See also: *Q: Q-value.*

DANGER COEFFICIENT. An obsolete term for the mass coefficient of reactivity, i.e. the reactivity change per unit mass of a substance for a standard position in a nuclear reactor.

DANIELL CELL. An electrochemical cell based on the reaction $Zn + Cu^{2+} \rightarrow Zn^{2+} + Cu$. See also: *Electrochemical cell.*

DARK ADAPTATION. See: *Adaptation (vision).*

DARK BURN. Denotes the dark spot which develops in the centre of a television picture caused by the destruction of the luminescent coating.

DARK CURRENT. The small current that flows in photo-emissive and photoconductive detectors in the absence of any incident light.

DARK-FIELD ILLUMINATION. Illumination of an object in a microscope in such a way that only light diffracted by the object reaches the eye. It is applicable only to the examination of small particles or fine lines, which appear as bright images against a dark background.

DARWIN CURVE. In X-ray diffraction: the curve of the intensity of the X-ray reflection from a perfect crystal plotted against angle, as originally calculated by C. G. Darwin.

DATA. Any information obtained or presented in numerical form.

DATA BUOY. A buoy located in a fixed position at sea, which is capable of transmitting to the shore meteorological and oceanographic data, pertaining to that position, for synoptic, operational and research purposes.

DATA PROCESSING. The routine treatment of data, commonly by a digital computer or fixed programme machine, to yield values on which operational decisions can be made, e.g. for the control of industrial processes. See also: *Optical data processing. Optical information processing.*

DATING. See: *Age determination by radioactivity. Archaeomagnetism. Thermoluminescent dating.*

DATING, RADIOACTIVE OR RADIOMETRIC. See: *Age determination by radioactivity.*

DAUGHTER PRODUCT. Of a given radionuclide: that nuclide which is the immediate product of the decay of the radionuclide in question. See also: *Decay product.*

DAUPHINÉ TWIN. A characteristic form of twinning in crystals of quartz, the two components having a common (512) plane.

DAVIS–GIBSON COLOUR FILTER. A two-component filter used, with the appropriate proportions of each component, to convert the spectral energy distribution of an incandescent source of light to that of white light.

DAVISSON–CALBICK FORMULA. An expression for the focal length, f, of a simple electrostatic lens consisting of a single circular aperture in a conducting plate which separates two regions of different potential gradient, V_1' and V_2' respectively. It may be stated as $f = \dfrac{4Vg}{V_2' - V_1'}$, where Vg is the potential of the plate.

DAVY SAFETY LAMP. An oil-burning lamp in which any hot, inflammable gases present are cooled by their passage through concentric layers of copper gauze, on their way to the flame, so that their temperature is below the ignition temperature.

DAY. (1) *Apparent solar day:* the interval between two successive transits of the observed Sun (or *true Sun*) across the meridian. It is not constant. See also: *Time, equation of.* (2) *Mean solar day:* the interval between two successive transits of a fictitious *mean Sun* (which moves at a constant rate equal to the average rate of the true Sun) across the meridian. One mean solar day is equal to 24 *mean solar hours*, each of which equals 60 *mean solar minutes*, each of which equals 60 *mean solar seconds.* See also: *Time, equation of.* (3) *Sidereal day:* the interval between two consecutive transits of the first point of Aries across any selected meridian. Neglecting the small effects of precession, etc., it is the interval between two consecutive transits of the same fixed star. It may be defined for most purposes as the period of rotation of the Earth on its axis, and expressed in mean solar time it is 23 h, 56 min, and 4·0906 s. Like the solar day, the sidereal day is divided into 24 h (*sidereal hours*) each of which is in turn divided into 60 min, each minute being subdivided into 60 s.

DAYLIGHT FACTOR. The ratio of the daylight illumination at any point in a building to the illumination under the open sky outside at the same instant.

D-CENTRE. See: *Colour centre.*

DEAD TIME. (1) *Of a Geiger–Müller counter:* the interval after the initiation of a voltage pulse during which a subsequent ionizing event does not produce a discharge, in the absence of interference by an external circuit. (2) Of a counting system: another term for resolution time.

DEAD-TIME CORRECTION. A correction to the observed counting rate to allow for the probability of the occurrence of events within the dead time.

DE BROGLIE WAVES. The waves associated with a moving particle, as first propounded by De Broglie. The *De Broglie wavelength* is equal to h/p, where h is Planck's constant and p the momentum of the particle.

DEBYE–HÜCKEL THEORY. States that, for dilute solutions, the logarithm of the mean activity is proportional to the product of the cation valence, anion valence and the square root of the ionic strength. To extend the theory to higher concentrations account must be taken of the finite size of the ions and of the attraction between the ions and the dipolar molecules of the solvent.

DEBYE–JAUNCEY SCATTERING. The incoherent background X-ray scattering from a crystal in regions between the Bragg reflections.

DEBYE LENGTH. In a plasma: a characteristic length corresponding to the distance over which an electron will be influenced by the electric field of a given positive ion. It is a measure of the distance over which the electron charge density can differ significantly from the ion charge density.

DEBYE TEMPERATURE: DEBYE CHARACTERISTIC TEMPERATURE. A temperature, θ_D, characteristic of each substance, appearing in Debye's theory of specific heats and given by $\theta_D = h\nu_{max}/k$, where h is Planck's constant, ν_{max} is the maximum frequency of the thermal vibrations of the atoms of the substance, and k is Boltzmann's constant.

DEBYE THEORY OF SPECIFIC HEATS. A theory in which a solid is considered as an elastic continuum whose maximum vibrational frequency is such that the corresponding number of vibrational modes is equal to the total number of degrees of freedom, $3N$, in the solid, where N is Avogadro's number. The specific heat at constant volume, C_v, is given by $C_v = 3R[D(x) - xD'(x)]$, where R is the gas constant and $D(x)$ the Debye function, defined as $D(x) = \dfrac{3}{x^3} \displaystyle\int_0^x \dfrac{t^3\,dt}{e^{t-1}}$, x being equal to θ_D/T and t to $h\nu/kT$, where θ_D is the Debye characteristic temperature, T is the absolute temperature,

h is Planck's constant, v is the frequency, and k is Boltzmann's constant. At low temperatures the theory correctly predicts that C_v is proportional to T^3. See also: *Einstein theory of specific heats. Nernst–Lindemann theory of specific heats. Specific heat theories.*

DEBYE UNIT. A unit of dipole moment equal to 10^{-18} e.s.u. cm. (*Note:* 1 e.s.u. of charge is equal to $10/c$ coulomb, where c is the speed of light in cm/s.)

DEBYE–WALLER TEMPERATURE FACTOR. The factor by which the intensity of a Bragg reflection from a crystal plane is reduced by the thermal motions of the atoms in the crystal. It is e^{-2M}, where $M = 8\pi^2 \overline{u_s^2} \sin^2 \theta / \lambda^2$, $\overline{u_s^2}$ being the mean value of the square of the displacement of an atom along a direction perpendicular to the reflecting plane, θ the Bragg angle and λ the wavelength of the diffracted radiation. The term is sometimes used to denote e^{-M} instead of e^{-2M}. Occasionally it refers to M itself and, in crystallography, it is often used to refer to that part of M which is independent of angle, i.e. $8\pi^2 \overline{u_s^2}$.

DECAY. See: *Radioactive decay: Radioactive disintegration*

DECAY CHAIN. See: *Radioactive series.*

DECAY CONSTANT: DISINTEGRATION CONSTANT. See: *Radioactive decay constant: Radioactive disintegration constant.*

DECAY PRODUCT. Of a given radionuclide: any nuclide that arises from that radionuclide by radioactive decay. See also: *Daughter product.*

DECIBEL. A logarithmic unit used for expressing power ratios. It is one-tenth of a *bel*. The number of decibels corresponding to a power ratio W is $10 \log_{10} W$, the number of bels being $\log_{10} W$. The use of decibels and bels has been extended to cover such quantities as voltage and current ratios.

DECILE. In statistics: one of the nine points (known as the first, second, ..., ninth deciles) which divide a given population into ten equal parts. One-tenth of the population is below the first decile, another tenth between the first and second decile, and so on.

DECLINATION. (1) Of a celestial body: the latitude of the body on the celestial sphere, measured from the celestial equator. See also: *Latitude, celestial: Latitude, ecliptic.* (2) Magnetic: See *Magnetic declination: Magnetic variation.*

DECOMPOSITION VOLTAGE. The minimum voltage which will cause steady electrolysis in a given electrolytic cell.

DEE. Of a cyclotron: one of two electrodes in the form of a hollow semicircular box, which are put together like the two halves of a shallow pill box, and between which a r.f. voltage is maintained in such a way that the circulating particles encounter an accelerating field at every passage.

DEFECT SCATTERING. The scattering of radiation by defects in a crystal. It is of particular importance in scattering by cold neutrons, where Bragg scattering and, to a large extent, absorption may be eliminated, thus permitting the study of the much less intense diffuse scattering arising from the defects, from which something of the nature of the defects themselves may be learned. See also: *Bragg cut-off wavelength.*

DEFECT STRUCTURE. A crystalline structure in which some of the atomic positions are occupied by the "wrong" atoms, or even left unoccupied. In such a structure the compound concerned is usually non-stoichiometric.

DEFICIENT NUMBER. See: *Perfect number.*

DEFLECTION COILS: SCANNING COILS. An assembly of one or more coils around a cathode-ray tube, used for deflecting the electron beam by means of the magnetic field set up when current is passing through the coils. Also known as *Deflection yoke: Scanning yoke.*

DEFLECTION MODULATION. See: *Modulation.*

DEFORMATION. The change in shape or size of a body which accompanies stress.

DEFORMATION BAND. In a metal: a region in a plastically deformed metal crystal which differs in orientation from the rest of the crystal. The band may be a *kink band*, in which the slip direction goes through an S-shaped bend, or a *secondary slip band*, in which the slip system differs from that active in the remainder of the crystal.

DEFORMATION, ELASTIC. Deformation which disappears when the stress is removed.

DEFORMATION, PLASTIC. Deformation which occurs at a level of stress above the elastic limit, and which is therefore permanent.

DEFORMATION TEXTURE. The preferred orientation arising in a polycrystalline metal as a result of a deformation process such as rolling or wire drawing. It is best described in terms of pole figures, which give a projection of the distribution of the various crystallographic planes with respect to some convenient axis system, related to the direction of the deformation process (e.g. rolling direction).

DEGENERACY. (1) Of atomic, electronic, or nuclear energy levels: the existence of an atom, electron or nucleus, as the case may be, in more than one possible state corresponding to a given energy level. Such states are known as *degenerate states*. The term is also used to denote the number of such states corresponding to the given energy level. Degeneracy may be removed by applying a perturbation which results in a separation of the levels for the various states. (2) In classical dynamics: the existence of two or more independent modes of vibration having the same frequency. (2) Of a gas: the condition, occurring at low temperatures (below 20 K), when the molecular heat falls below the value $3/2R$, where R is the gas constant.

DEGREE. (1) A unit of angular measure. (2) A unit of temperature or temperature difference.

DEGREES OF FREEDOM. (1) Of a dynamical system: the minimum number of coordinates required to specify the motion of every particle in the system. Thus a single particle, moving in three dimensions, has three degrees of freedom; a diatomic molecule has six (three translational, two rotational, and one vibrational). (2) In the phase rule: the number of independent variables (pressure, temperature, composition, etc.) required to define a system having a given number of phases and components.

DE HAAS–VAN ALPHEN EFFECT. A periodic variation of magnetic susceptibility with magnetic field strength, observed in a number of metals usually at temperatures below 20 K.

DEIONIZATION TIME. The time taken for the ions formed (for example, in a Geiger–Müller tube) as a result of an ionizing event to disappear by diffusion or recombination.

DEKATRON. A gas-filled multi-electrode tube in which incoming pulses are counted by causing a glow discharge to be transferred sequentially to cathodes arranged in a circle around a central anode.

DEL. See: *Differential vector operator.*

DELAUNAY VARIABLES. A set of canonical variables first used by Delaunay in his theory of lunar motion.

DELAYED COINCIDENCE. The occurrence of a count in a radiation detector at a short, measurable, time after a count in another detector, the two counts being related to the same phenomenon (e.g. two disintegrations of the same nucleus or two pulses due to the same particle going through both counters).

DELAYED NEUTRONS. Neutrons emitted by nuclei in excited states which have been formed in the process of beta decay. The observed half-life is that of the β-decay of the parent nuclide or *delayed neutron precursor*. Most of the delayed neutron precursors originate in nuclear fission.

DELAY LINE. A line or network specifically designed to introduce a desired delay in the transmission of a signal.

DELBRÜCK SCATTERING. The scattering of light by the Coulomb field (electrostatic field) of a nucleus.

DELIQUESCENCE. The absorption of water from the atmosphere (sometimes up to the point of liquefaction) by a chemical salt.

DELTA CONNECTION. One of the two most frequently used ways of connecting a three-phase a.c. circuit, the other being the Y connection. The Δ connection has three coils, arranged in a triangle, each angle of which is connected to one of the three phases. The Y connection has three coils arranged in the form of a Y, each point of which is connected to one of the three phases.

DELTA E (δE) EFFECT. See: *Magnetostriction.*

DELTA FUNCTION. A type of function introduced by Dirac to facilitate the quantum mechanical description of scattering processes.

DELTA RAYS. Electrons ejected from atoms by ionizing radiation, and having sufficient energy to cause further ionization.

DEMAGNETIZATION COEFFICIENT: DEMAGNETIZING FACTOR. See: *Demagnetizing field.*

DEMAGNETIZATION ENERGY. The work done in assembling magnetic poles, in a given configuration, against the field of the poles.

DEMAGNETIZING FIELD. The field acting on a magnetized body in opposition to the magnetizing force and interpreted as being due to free magnetic poles induced on the surface of the body. It is proportional to the magnetic intensity in the body, the proportionality factor being known as the *demagnetizing factor*, or *demagnetization coefficient*.

DEMOIVRE THEOREM. States that, if x is real and n is an integer, positive or negative, $(\cos x + i \sin x)^n = \cos nx + i \sin nx$, where $i = \sqrt{-1}$. If n is a rational number, $\cos nx + i \sin nx$ is only one of the values of $(\cos x + i \sin x)^n$.

DENDRITES. (1) Tree-like crystals which are formed during the solidification of most metals and alloys, and of liquids in general. (2) Tree-like conducting channels in dielectrics which are formed when non-uniform electric fields of sufficient intensity occur.

DENSITOMETER. An instrument used for measuring values of photographic density directly, by photometric comparison. It may measure either the transmission density or the reflection density, and the measurement may be made either visually or by photo-electric means. See also: *Optical density. Photographic density.*

DENSITY. (1) The mass per unit volume of a substance under standard conditions. (2) Optical density (defined separately). (3) Photographic density (defined separately).

DENSITY, APPARENT. The measured density of a substance which contains voids (e.g. powdered materials, which occlude air between or inside individual particles).

DENSITY BOTTLE: SPECIFIC GRAVITY BOTTLE. An accurately calibrated glass bottle used, by comparing the weights of the same volumes of a liquid and a reference liquid, to measure the specific gravity of the first liquid. See also: *Pyknometer.*

DENSITY, DIFFUSE. Of a photographic transparency: the value of the density obtained when (a) the light is incident normally and the whole of the transmitted light is collected and measured, or (b) the incident light is diffused (i.e. incident in all directions) and only the transmitted light emerging normally is measured. See also: *Density, specular. Photographic density.*

DENSITY EFFECT. For stopping power: See: *Stopping power.*

DENSITY MODULATION. The creation of a time-varying space charge density in an electron beam at some surface which is fixed in space, e.g. at the grid of a triode or tetrode.

DENSITY, NUMBER. The number of particles per unit volume or per unit area.

DENSITY, RELATIVE: SPECIFIC GRAVITY. Of a given substance: the ratio of the density of the substance to that of a standard substance, usually water, at a specified temperature. See also: *Atmosphere, relative density of.*

DENSITY, SPECULAR. Of a photographic transparency: the value of the density obtained when the light is incident normally, and only the transmitted light emerging normally is measured. See also: *Density, diffuse. Photographic density.*

DENSITY, TAP. The density of a powdered material after compacting at moderate pressure.

DEOXYRIBONUCLEIC ACID (DNA). The chemical substance occurring in chromosomes which is the carrier of genetic information. See also: *Genetic code.*

DEPLETED URANIUM. Uranium containing a smaller proportion of ^{235}U than is found in nature. The depletion may arise from the irradiation of natural uranium in a reactor (usually the fuel in a thermal reactor) or from the separation of ^{235}U in a diffusion plant. Depleted uranium is used for radiation shielding, as a dense metal, in alloy manufacture, in the ceramics industry, etc.

DEPLETION, ISOTOPIC. The reduction of the concentration of one or more specified isotopes in a material or in one of its constituents.

DEPOLARIZATION. In an electrolyte: the control or limitation of the potential of an electrode, during electrolysis, by the addition of an appropriate substance.

DEPRESSION, METEOROLOGICAL. A region in which the distribution of atmospheric pressure is low and decreases to a minimum at the centre. In the northern hemisphere the winds blow anticlockwise round such a system, and in the southern hemisphere they blow clockwise. See also: *Cyclone.*

DERIVATIVE. Of a function of an independent variable: the instantaneous rate of change of the function with respect to the variable. It is denoted by $\frac{dy}{dx}$ for example.

DERIVATIVE, HIGHER ORDER. The derivative of a derivative. The number of differential operations involved is known as the *order* of the derivative. Thus a derivative of the nth order is the derivative of the $(n-1)$th derivative. It may be designated as $\frac{d^n y}{dx^n}$, $D_x^n y$, $f^{(n)}(x)$, or $y^{(n)}$.

DERIVATIVE, PARTIAL. Of a function of two or more variables: the instantaneous rate of change of the function with respect to a specified variable, the other variables being kept constant. It is denoted by $\frac{\partial u}{\partial x}$ for example.

DERIVATIVE, TOTAL. Of a function of several variables, each of which itself is a function of the same independent variable: the derivative of the function in question with respect to the independent variable, given by

$$\frac{du}{dt} = \frac{\partial u}{\partial x}\frac{dx}{dt} + \frac{\partial u}{\partial y}\frac{dy}{dt} + \frac{\partial u}{\partial z}\frac{dz}{dt} + \cdots$$

where u is a function of $x, y, z, ...,$ and t is the independent variable.

DESALINATION. The purification of sea water or brackish inland water, usually for drinking and commonly in regions where there is a shortage of fresh water, by the removal of salt. It is carried out by distillation, reverse osmosis, electrodialysis, or ion exchange, to name some of the well-known processes, and the use of nuclear energy in connection with one or other of these is increasing.

DE SENARMONT PRISM. See: *Polarizing prism.*

DESENSITIZATION, PHOTOGRAPHIC. The action of rendering a photographic emulsion less sensitive to reactions brought about by light. It should strictly be applied only to the use of reagents which reduce sensitivity without significantly attacking latent images produced by a previous exposure.

DESORPTION. The removal of a substance from a surface on which it is adsorbed.

DESTRUCTION OPERATOR. In the quantum theory of fields: an operator which, when applied to a state vector Φ_n, which describes a system containing n particles, will yield a state vector Φ_{n-1}, which describes a system having $(n-1)$ particles. See also: *Creation operator.*

DETAILED BALANCING PRINCIPLE. States that: in an equilibrium system, any molecular process and its reverse occur with equal frequency. It may be applied to classical or quantum mechanical systems, to chemical reactions, and to irreversible thermodynamics, but is to be distinguished from the principle of microscopic reversibility, which refers to transitions between two states and not to total rates of transition. See also: *Reversibility, microscopic.*

DETERMINANT. A scalar number associated with a square matrix (a_{ik}), which may be written

$$\begin{vmatrix} a_{11} & a_{12} & \cdots & a_{1n} \\ a_{21} & a_{22} & \cdots & a_{2n} \\ \cdots & \cdots & \cdots & \cdots \\ a_{n1} & a_{n2} & \cdots & a_{nn} \end{vmatrix}$$

It is the sum $\Sigma \pm a_{1i_1} a_{2i_2} \cdots a_{ni_n}$, where $i_1, ..., i_n$ is one of the $n!$ permutations of the numbers $1, ..., n$ and the sign $\pm$ is chosen according as the permutation is even or odd. A determinant with n^2 elements is said to be of *order n*. Determinants first arose in connection with the problem of solving n linear equations in n unknowns, and have since found numerous applications in algebra, geometry and analysis.

DETERMINANTAL WAVE FUNCTION. A wave function for an atom or molecule, in the form of a determinant, which takes into account the identity of the electrons and is antisymmetric in all electron pairs.

DETERMINISTIC DATA. Data which vary with time in a way that can be expressed by an explicit mathematical relationship. See also: *Random data.*

DEUTERIDE. A binary compound containing deuterium, e.g. NaD. The deuterides are analogues of the hydrides and have the usual chemical reactions of the hydrides.

DEUTERIUM. The hydrogen isotope ^2_1H, of mass number 2, sometimes represented by the symbol D.

DEUTERON. The nucleus of deuterium. It has a spin of 1, a positive magnetic moment, and a positive electric quadrupole moment.

DEUTERON–DEUTERON REACTION. The reaction between two deuterons. It may proceed in one or two ways, each being equally probable. In the first a helium-3 nucleus is produced, with the emission of a neutron. In the second a tritium nucleus is produced, with the emission of a proton.

DEUTERON STRIPPING. See: *Stripping reaction.*

DEVIATION (OPTICAL). The angle through which a ray of light is bent by refraction or diffraction.

DEW. The deposit of moisture from the atmosphere on a surface whose temperature is below that of the dew point. It may be in the form of tiny drops of water or crystals of ice (*hoar frost*) depending on the temperature.

DEWAR VESSEL: VACUUM FLASK. A double-walled vessel used for the storage of hot or cold liquids or solids, in which the space between the walls is evacuated and the walls themselves may be silvered, to reduce heat flow to (and from) the inside.

DEW POINT. The temperature at which the water vapour present in the atmosphere is just sufficient for the atmosphere to be saturated. Lowering the temperature results in the deposition of drops of water or crystals of ice depending on whether the saturation temperature is above or below the freezing point.

DEXTROGYRIC: DEXTROROTATORY. Refers to an optically active substance that rotates the plane of polarization of a transmitted light beam in a clockwise direction, the observation being made looking through the substance towards the light source. See also: *Laevogyric: Laevorotatory*.

DIAD AXIS. See: *Crystal symmetry, rotation axis in*.

DIAL GAUGE. An instrument for making small linear measurements, in which linear motion imparted to a plunger is converted into rotational motion and displayed by a pointer on the face of a dial.

DIALYSIS. The separation of small molecules and ions from colloids by means of a semipermeable membrane.

DIAMAGNETISM. The property shown by many substances, when they are subjected to a magnetizing force, of becoming magnetized in such a direction as to oppose that force, i.e. exhibiting a negative susceptibility (*diamagnetic susceptibility*). Such substances are said to be *diamagnetic*. See also: *Ferromagnetism. Paramagnetism*.

DIAMAGNETISM, LANGEVIN THEORY OF. A theory that relates the occurrence of diamagnetism to the Larmor precession of the atomic electron orbits. It was originally derived classically, and later corrected quantum-mechanically by Pauli, leading to the following expression for the diamagnetic susceptibility: $\chi = -\dfrac{Ze^2N}{6mc^2}\,\overline{r^2}$, where χ is the susceptibility for an assembly of N atoms per unit volume, Z is the atomic number, e, m and c have their usual significance and $\overline{r^2}$ is the mean square radial distance of the electrons from their nuclei.

DIAPHORIMETER. A device for measuring sweating. There is no standard instrument, but use may be made of the abrupt change in the electrical resistance of the skin which occurs at the onset of sweating, of the change in the colour of an appropriate dye (e.g. iodine dusted with starch), or changes in body weight.

DIAPOSITIVE. A transparency (e.g. a lantern slide) for projection by transmitted light.

DIASTROPHISM. Any process by which the crust of the Earth is relatively rapidly deformed, with the production of continents, ocean basins, plateaux, mountains etc. Examples are: *Epeirogenesis*—the rising or sinking of large areas of the Earth's surface without marked folding, and *Orogenesis*—the process of mountain building, particularly by strong folding of the crust accompanied or followed by uplift and often by metamorphism.

DIATHERMANCY. The property of transmitting radiant heat. *Diathermanous* substances differ as to the energy band which they transmit best.

DIATHERMY. The process by which local heat is produced in body tissues by the application of an oscillating electric current of high frequency (500–100000 kHz). Typical uses of diathermy are in the treatment of arthritis and circulatory diseases, and in fever therapy. Diathermy is also used surgically either for cutting or cautery.

DICHOTIC PRESENTATION: DICHOTIC LISTENING. The simultaneous reception of one message or signal via one ear and another message via the other ear, e.g. by the use of headphones. It is not the same as *binaural presentation* which refers to reception of the same message or signal.

DICHOTOMY (ASTRONOMY). The moment when the Moon, or an inferior planet, appears exactly half-illuminated, i.e. when its phase is 0·5.

DICHROIC MIRROR. A mirror formed by the deposition on a transparent substrate of alternating layers of high- and low-index dielectric material of suitable thickness. In such a mirror the wavelength for maximum reflection varies with the angle of incidence; and, for incident white light, the colours of the transmitted and reflected light are complementary. Dichroic mirrors are used as beam-splitters in colour cameras and colour television cameras.

DICHROISM. The property of exhibiting two colours, especially that of exhibiting one colour in transmitted light and another in reflected light. The name is sometimes restricted to the dichroism of those birefringent crystals which absorb the ordinary and extraordinary rays to different extents. This selective absorption of polarized light is also known as *pleochroism*. The ratio of the two absorption indices is called the *dichroic ratio* and materials which exhibit dichroism are said to be *dichroic*.

DICHROMATISM. (1) An effect shown by some materials in which the colour of transmitted light depends on the thickness. (2) A type of colour blindness in which the eye can distinguish only two colours.

DIELECTRIC. Ideally, a medium in which an electric field, once established, may be maintained without loss of energy. It is an insulator containing no free charges. In practice, however, dielectrics may be slightly conducting and the line of demarcation between an insulator or dielectric and a conductor is not always sharp. Dielectrics may be solid, liquid or gaseous.

DIELECTRIC ABSORPTION. The persistence of electric polarization after removal of excitation by a polarizing field. Substances in which dielectric absorption lasts over long periods are known as *electrets*. See also: *Electret.*

DIELECTRIC AGEING. The change, usually a deterioration, in the properties of dielectrics with time. It is caused by one or more processes of physical, chemical or electrochemical change and is frequently associated with heating.

DIELECTRIC AMPLIFIER. A device in which the non-linear properties of a ferroelectric material are used to amplify a small signal into one of greater power. The source of energy is an a.c. supply working at a frequency that is much higher than the highest frequency at which amplification is required. A dielectric amplifier is analogous to a *magnetic amplifier* in which the non-linear properties of a ferromagnetic material are used in the same way.

DIELECTRIC BREAKDOWN. The failure of a dielectric to act as an insulator. Such a breakdown may arise thermally, from the heat generated by the finite conduction current passing through a dielectric in service; from electric discharges over the surface of a dielectric or in gas-filled cavities; by *tracking*, i.e. the passage of a leakage current in contaminating layers on the surface; or by electrochemical processes which may, for example, produce erosion of the electrodes or cause progressive deterioration of the dielectric itself. When all these factors are absent the occurrence of breakdown will depend on the intrinsic electric strength (usually between 1 and 15 MV/cm), and will take place when this is exceeded.

DIELECTRIC BREAKDOWN, SELF-HEALING. A breakdown in which the breakdown itself produces an insulator, as in electrolytic capacitors, selenium rectifiers, and in capacitors using evaporated metal electrodes.

DIELECTRIC, COMPOSITE. A dielectric consisting of more than one insulating material. Examples are: laminated dielectrics, varnished fabrics, bonded mica, oil impregnated paper.

DIELECTRIC CONSTANT: SPECIFIC INDUCTIVE CAPACITY: RELATIVE PERMITTIVITY. Of a medium: the ratio of the capacitance of a capacitor with the medium between the electrodes to that of a capacitor with a vacuum between the electrodes. Alternatively: the ratio of the electric flux density produced in the medium to that which would be produced in a vacuum by the same electric force. See also: *Permittivity: Absolute permittivity.*

DIELECTRIC DETERIORATION. See: *Dielectric ageing. Dielectric breakdown.*

DIELECTRIC ELLIPSOID. For an anisotropic medium: an ellipsoid whose dimensions are such that the magnitude of the dielectric constant in any given direction may be obtained from it. This mode of representation was first given in terms of the elastic–solid theory of light and the ellipsoid was then called the *ellipsoid of elasticity.*

DIELECTRIC FIELD: DIELECTRIC FIELD, INTERNAL. The average total field which acts on a molecule or group of molecules within a dielectric. It is in general not equal to the external applied field since an additional field is produced by the polarization of the dielectric. It is often defined as the field inside a needle-shaped cavity whose direction is parallel to the polarization.

DIELECTRIC HEATER. An electrical heater in which dielectric material is heated by dielectric loss in a comparatively strong a.c. field of high frequency. Such heaters are widely used in the wood-working and plastic industries.

DIELECTRIC HYSTERESIS. The lag in time between the application of an electric field to a dielectric and the establishment of polarization: in particular the corresponding phase lag when an alternating field is applied.

DIELECTRIC IMPERFECTION LEVELS. Energy levels introduced into the forbidden zone between the conduction and valence bands as a result of imperfections (e.g. impurity atoms, non-stoichiometry, dislocations, mosaic structure) in a dielectric. See also: *Conduction band. Valency band.*

DIELECTRIC LEAKAGE. The very small steady current which flows in a dielectric subjected to a steady electric field.

DIELECTRIC LOSS. The energy lost by absorption when dielectric hysteresis occurs.

DIELECTRIC POLARIZATION. See: *Polarization.*

DIELECTRIC, REFRACTIVE INDEX OF. The refractive index associated with the polarization of a dielectric at optical frequencies. According to the Maxwell relationship the square of the refractive index for a non-polar substance which has no permanent molecular dipole moment is equal to the dielectric constant. The refractive index increases with frequency except in the neighbourhood of a resonance, where anomalous dispersion occurs.

DIELECTRIC STRENGTH: ELECTRIC STRENGTH. The maximum potential gradient which a dielectric or insulator can stand without breaking down, under specified conditions. See also: *Electric strength, intrinsic.*

DIELECTRIC TESTING. The testing of dielectrics to ascertain the following quantities: electric strength, volume and surface resistivity, permittivity and loss angle, resistance to arcing and tracking.

DIESEL CYCLE. A cycle, now modified from Diesel's original conception, involving the compression of air to the ignition temperature, followed by the timed introduction of fuel, expansion, and exhaustion. The ideal cycle is represented thermodynamically on a pressure–volume plot by two parallel adiabatic lines (corresponding respectively to the initial temperature and to the temperature achieved by compression), which are joined by two lines at constant pressure and constant volume (corresponding respectively to expansion during fuel injection and to exhaustion).

DIFFERENCE TONES. See: *Combination tones.*

DIFFERENTIAL. A synonym for *derivative.*

DIFFERENTIAL ANALYSER. A variety of analogue computer possessing a device for performing the mathematical operation of integration. Its main use is in the solution of ordinary differential and integral equations, but partial differential equations may also be treated. See also: *Computer, analogue.*

DIFFERENTIAL EQUATION. An equation involving derivatives of an unknown function.

DIFFERENTIAL EQUATION, DEGREE OF. The highest power of the highest derivative occurring in the differential equation concerned.

DIFFERENTIAL EQUATION, EXACT. An equation of the form $\frac{\partial u}{\partial x} dx + \frac{\partial u}{\partial y} dy = 0$, for a function $u(x, y)$, of which the solution is $u(x, y) = $ constant.

DIFFERENTIAL EQUATION, GENERAL SOLUTION OF. The solution of an ordinary differential equation of the nth order, involving n arbitrary constants.

DIFFERENTIAL EQUATION, LINEAR. A differential equation of the first degree.

DIFFERENTIAL EQUATION, NON-LINEAR. One of a higher degree than the first.

DIFFERENTIAL EQUATION, ORDER OF. The order of the highest derivative occurring in the differential equation concerned.

DIFFERENTIAL EQUATION, ORDINARY. One in which no partial derivatives occur.

DIFFERENTIAL EQUATION, PARTIAL. One in which partial derivatives occur.

DIFFERENTIAL EQUATION, PARTICULAR SOLUTION OF. That solution of an ordinary differential equation which is obtained from the general solution by assigning numerical values to the constants following the imposition of boundary conditions.

DIFFERENTIAL EQUATION, SINGULAR SOLUTION OF. A solution of an ordinary differential equation that cannot be obtained by assigning particular values to the constants in the general solution.

DIFFERENTIAL EQUATION, TOTAL. One in which the derivatives are total.

DIFFERENTIAL GEOMETRY. That branch of the geometry of curves and surfaces embedded in three-dimensional Euclidean space which is studied by means of the differential calculus.

DIFFERENTIAL VECTOR OPERATOR. An operator, written as ∇, sometimes known as *del* or *nabla*, which in Cartesian coordinates is $\left(\mathbf{i} \frac{\partial}{\partial x} + \mathbf{j} \frac{\partial}{\partial y} + \mathbf{k} \frac{\partial}{\partial z} \right)$, where $\mathbf{i}$, $\mathbf{j}$, $\mathbf{k}$ are unit vectors. When applied to a scalar function $\phi(x, y, z)$, $\nabla \phi$ $\left(\text{i.e. } \mathbf{i} \frac{\partial \phi}{\partial x} + \mathbf{j} \frac{\partial \phi}{\partial y} + \mathbf{k} \frac{\partial \phi}{\partial z} \right)$ is known as the *gradient* of the scalar and is also written *grad* ϕ. The vector product of ∇ and a vector $\mathbf{V}$, $\nabla \times \mathbf{V}$, is known as *curl* $\mathbf{V}$ and may be written as

$$\mathbf{i} \left\{ \frac{\partial V_z}{\partial y} - \frac{\partial V_y}{\partial z} \right\} + \mathbf{j} \left\{ \frac{\partial V_x}{\partial z} - \frac{\partial V_z}{\partial y} \right\}$$
$$+ \mathbf{k} \left\{ \frac{\partial V_y}{\partial x} - \frac{\partial V_x}{\partial y} \right\}$$

It is also called the *rotation* (abbreviation *rot*). The scalar product of ∇ and a vector $\mathbf{V}$, $\nabla \cdot \mathbf{V}$, is known as the *divergence* of $\mathbf{V}$ or *div* $\mathbf{V}$. It is $\frac{\partial V_x}{\partial x} + \frac{\partial V_y}{\partial y} + \frac{\partial V_z}{\partial z}$. The operator *div grad*, denoted by ∇^2, is also known as the *Laplace operator*. See also: *Laplace equation.*

DIFFERENTIATION. The process of finding the derivative of a function. See also: *Derivative. Derivative, partial. Derivative, total.*

DIFFERENTIATING CIRCUIT. A circuit which produces an output voltage proportional to the rate of change of the input signal.

DIFFRACTION. The interference of radiation scattered by different elements of a medium. The intensity is increased or reduced in any given direction, for a particular wavelength, according as the scattered waves are in or out of phase with each other. The radiation may be acoustic or electromagnetic or, in general, may refer to any regular wave motion, including that associated with atomic particles. The diffracting elements may, for example, be optical slits, individual atoms in a crystal, or different elements in a radio aerial.

DIFFRACTION ANALYSIS. The study of the atomic arrangement in solids, liquids or gases by the diffraction of X-rays or atomic particles, such as electrons or neutrons.

DIFFRACTION GRATING. A series of parallel lines or reflecting surfaces whose separation is of the same order as the wavelength of light. Most modern line gratings consist of a series of equidistant diamond rulings on a plane or concave metal or glass surface, or a plastic replica of such a set of rulings. A diffraction grating, which may operate with either reflected or transmitted light, gives, for monochromatic light, a series of maxima and minima at different angles of scattering. For polychromatic light a series of spectra are obtained. Diffraction gratings are also available for use with X-rays.

DIFFRACTION MOTTLE. A mottled appearance on a radiograph, which arises from the superposition of diffraction effects on the radiographic image. The presence of large metallic grains in the object being radiographed is a typical reason for such effects.

DIFFRACTION SPLITTING. See: *Interferometer, classification of.*

DIFFRACTOMETER. An instrument used in the study of the atomic arrangement in matter by the diffraction of X-rays, electrons or neutrons. It is usually used with a monochromatic beam of radiation, and permits quantitative measurements to be made of the intensities of the diffracted beams at different angles. An ionization chamber or a particle counter is used for such measurements. See also: *Optical diffractometer.*

DIFFUSER. (1) A duct designed to slow down gas or fluid flow so as to reduce the velocity head and increase the pressure head. Diffusers are used in jet engines, compressors, wind tunnels, etc. (2) A medium which scatters light more or less uniformly in all directions, e.g. glass having an appropriate surface configuration or possessing suitable internal inclusions.

DIFFUSER, IDEAL. A diffuser having a reflection factor of unity, and being, therefore, white and opaque.

DIFFUSER, UNIFORM. A diffuser of light for which the luminance in any direction of observation does not vary with the direction of incidence of the light.

DIFFUSING POWER. Of a medium, for light: is defined by the C.I.E. (Commission Internationale de l'Éclairage) as $\dfrac{L_{20} + L_{70}}{2L_5}$ where L_5, L_{20} and L_{70} are the luminances of the medium when illuminated normally by a narrow beam of light and observed from angles of 5°, 20° and 70° respectively from the direction of incidence.

DIFFUSION. (1) Of light: the random redirection of light by a medium, by numerous small elements which scatter, reflect or refract the light falling on them. A perfect diffuser would be one which redistributed the whole of the incident light so that the luminance was the same in all directions, but no such diffuser has been found. (2) The migration of atoms or molecules in a solid, liquid or gas, normally under the influence of thermal motion and usually in the presence of a gradient of temperature, pressure, electrical or chemical potential, etc. (3) The passage of nuclear particles through matter in such circumstances that the probability of scattering is large compared with that of capture, e.g. neutron diffusion in a nuclear reactor. See also: *Dushman–Langmuir equation. Einstein diffusion equation. Fick diffusion laws.*

DIFFUSION AREA. Of a sub-atomic particle: one-sixth of the mean square displacement (i.e. direct distance travelled irrespective of route) of particles of a given type, between their appearance and disappearance. For thermal neutrons, for example, the distance involved is that between the point at which a neutron becomes thermal and the point at which it is captured.

DIFFUSION CHAMBER. See: *Cloud chamber.*

DIFFUSION COEFFICIENT: DIFFUSIVITY. See: *Fick diffusion laws.*

DIFFUSION DIAMETER. For a gas: the diameter of identical hard spheres which would exhibit the same diffusion as that observed for an actual gas if their motion were treated classically.

DIFFUSION INDICATRIX. The constant in Fick's second law of diffusion. See also: *Fick diffusion laws.*

DIFFUSION LENGTH. Of a sub-atomic particle: the square root of the diffusion area.

DIFFUSION OF NUCLEAR PARTICLES. See: *Neutron diffusion. Transport theory.*

DIFFUSION POTENTIAL: LIQUID JUNCTION POTENTIAL. The potential difference set up across the boundary between electrolytes of different composition. It arises from differences in the rates of diffusion of oppositely charged ions across the boundary.

DILATANCY. The property of certain assemblies, which may be regarded as colloidal solutions, of becoming solid, or setting, under pressure. An example of this is the apparent drying out of wet sand under the pressure of the foot. It arises from dilatation, resulting from pressure, which produces an increase in the volume of voids, into which the water is drawn. Dilatancy is to be distinguished from rheopexy.

DILATATION. The fractional increase in volume of a homogeneous medium.

DILATATIONAL WAVE. See: *Compressional wave.*

DILATOMETER. An apparatus for measuring the (usually thermal) expansion of solids, liquids or gases. It may measure linear or volume expansion and may be based on indirect measurements (as in the Chévenard type), differential measurements, interferometric measurements, the use of line standards and comparators, or X-ray diffraction techniques.

DILUTION EFFECTS IN STELLAR SPECTRA. The occurrence of abnormally strong or weak spectral lines such as those of He, Mg^+ and Si^+. The effects arise in extended stellar atmospheres, or in shell stars or gaseous nebulae, and are a consequence of the "dilution" (a form of self absorption) of the internal radiation by the outer envelope. Dilution effects offer a means of detecting shell stars.

DIMENSION. Of a physical quantity: one of a number of fundamental quantities, such as length, mass and time, in terms of which the physical quantity in question may be expressed. Thus: if L denotes length, M denotes mass and T denotes time, the dimensions of velocity are LT^{-1}, of acceleration LT^{-2}, of force MLT^{-2} and so on. Electrical, magnetic or thermal dimensions may also be used where appropriate.

DIMENSIONAL ANALYSIS. A method based on the principle of the dimensional homogeneity of physical equations—e.g. the two sides of an equation must have the same dimensions—by the use of which it is often possible by a consideration of dimensions only to arrive at the form of an equation connecting physical quantities, or to check the validity of the solution of a physical problem. Dimensional analysis has been particularly useful in the development of aerodynamics and hydrodynamics. See also: *Pi theorem.*

DIMER. A double molecule formed from two single molecules of the same kind. See also: *Monomer. Polymer.*

DIMORPHIC. Refers to a substance which can exist in two different crystalline forms. The phenomenon involved is known as *dimorphism*. See also: *Polymorphism.*

DI-NEUTRON. An unstable system composed of two neutrons, considered to have a transitory existence when a triton collides with a target nucleus so as to produce a proton, with an increase of two in the mass number of the target nucleus.

DIODE. A two-electrode device (thermionic tube or crystal diode) which has markedly unidirectional characteristics.

DIOPHANTICS. See: *Numbers, theory of.*

DIOPTRE. See: *Lens, power of.*

DIOPTRIC. Involving the diffraction of light. See also: *Catadioptric. Catoptric.*

DIORAMA. Stage scenery produced by the projection of enlarged photographs on the back cloth.

DIP CIRCLE, MAGNETIC. An instrument using a magnetic needle (*dip needle*) mounted on a horizontal axis to measure the magnetic dip against a vertical circle. See also: *Magnetic dip: Magnetic inclination.*

DIPOLAR GAS. One whose molecules have a permanent electric moment. The molar electric moment increases with falling temperature in contrast to that of a non-polar gas, which is independent of temperature.

DIPOLE ABSORPTION. The loss of energy experienced by a dipole placed in an electric field which varies with time. It arises from the force which the dipole has to overcome as it strives to take up an orientation in the direction of the applied field.

DIPOLE, CENTRED. A magnetic dipole supposed to be situated at the centre of the Earth by which the main geomagnetic field may be represented. It has a moment of $8 \cdot 06 \times 10^{25}$ gauss cm^3 and has as its axis the line joining 79°N, 70°W to 79°S, 110°E (1954). (*Note:* 1 gauss $= 10^{-4}$ tesla.) See also: *Dipole, eccentric.*

DIPOLE, ECCENTRIC. An imaginary dipole which gives a better representation of the main geomagnetic field of the Earth than the centred dipole. Its position is obtained by moving the latter by a distance of 470 km parallel to itself in the direction 29°N, 151°E. See also: *Dipole, centred.*

DIPOLE, ELECTRIC. A pair of electric charges, equal in magnitude but opposite in sign, and separated by a very small distance. If, in a molecule, the centroids of positive and negative charges present are not coincident, the molecule constitutes a *permanent dipole*. If they are coincident the application of an electric field will cause displacement of the positive and negative charges, giving rise to an *induced dipole*. See also: *Octupole. Quadrupole.*

DIPOLE, MAGNETIC. A magnet with finite magnetic moment, in which the separation of the poles may be considered vanishingly small.

DIPOLE MOMENT. Of an electric dipole: the product of the charge on one element of the dipole, and the distance between the charges. It is a vector quantity. The moment of a magnetic dipole may be defined in an analogous fashion. See also: *Multipole moments.*

DIPOLE RADIATION. Radiation emitted as a consequence of a change in electric or magnetic dipole moment.

DI-PROTON. An unstable system composed of two protons, which occurs in proton–proton scattering experiments.

DIRAC EQUATION. A quantum-mechanical equation introduced to describe the behaviour of the electron and now believed to be capable of describing all elementary particles with spin $\frac{1}{2}$. The equation invokes the concept of states of negative energy from which follows the possibility of the existence of the positron. See also: *Hole theory.*

DIRADICAL: BIRADICAL. A molecule with an even number of electrons of which two are unpaired.

DIRECT CONVERSION OF HEAT TO ELECTRICITY. The generation of electricity by electrogasdynamic, magnetohydrodynamic, thermoelectric or thermionic methods or by the use of solar energy. See also: *Electrogasdynamic generation of electricity. Magnetohydrodynamic generation of electricity. SNAP (Systems for Nuclear Auxiliary Power). Solar energy. Thermionic generation of electricity. Thermoelectric generation of electricity.*

DIRECT CURRENT. An electric current which is continuous, unidirectional and steady.

DIRECTION COSINES. Of a line passing through the origin of a set of rectangular coordinate axes: the cosines of the angles between the line and the positive directions of the three axes. The direction cosines serve to specify the direction of the line.

DIRECTION RATIOS. Of a line passing through the origin of a set of general coordinates axes: the coordinates of the point along the line whose distance from the origin is unity. The direction ratios serve to specify the direction of the line and, for rectangular coordinates, are the same as the direction cosines.

DIRECT MOTION. Of a celestial body: motion such that the longitude (real or apparent) of the body increases with time. See also: *Retrograde motion.*

DIRECT PULSE. In a pulsed radar system: the pulse of energy radiated direct from transmitter to receiver, as opposed to a *returned pulse*, which arrives at the receiver only after reflection from some object illuminated by the transmitter.

DIRECTRIX. Of a conic: See: *Conic. Conic section.*

DIRICHLET CONDITIONS. In the theory of Fourier series: a precise set of conditions formulated to ensure that the Fourier function $f(x)$ shall everywhere converge. The integrals which determine the convergence properties of the Fourier series are known as *Dirichlet integrals.*

DISCHARGE. (1) The removal of charge from a capacitor. (2) The passage of electric current through a gas or "vacuum". (3) The rate of flow of a fluid at a particular point. (4) The process by which a storage battery delivers electrical energy.

DISCHARGE, ARC. See: *Arc.*

DISCHARGE, BRANCHING OF. The branching of a streamer discharge following the simultaneous arrival of two avalanches of electrons at the streamer tip.

DISCHARGE, BRUSH. A form of corona discharge taking place from pointed conductors or fine wires in air.

DISCHARGE COEFFICIENT. For fluid flow through an orifice, etc.: the ratio of the actual rate of flow (expressed as volume per unit time) to the rate of flow of an ideal fluid.

DISCHARGE, CORONA. An electrical discharge between two electrodes, one of which is of such a shape as to cause the electric field at its surface to be much higher than that between the electrodes. It takes place at a potential below that necessary for a spark or arc discharge, and is characterized by a glow on or near the area with the high field, which may be accompanied by faint streamers extending in the direction of the other electrode. Typical sources of a corona discharge are a point and a plane, and a wire on the axis of a cylinder.

DISCHARGE DETECTOR. A test instrument incorporating a sensitive electronic amplifier, used to detect minute electrical transients in voids in solid or liquid insulators.

DISCHARGE, FIELD-INTENSIFIED. See: *Discharge, Townsend.*

DISCHARGE, GLOW. An electrical discharge between cold electrodes in low-pressure gas, characterized by its low current-density (about 10^{-6} to 10^{-1} A cm^{-2}) and occurring at a potential above the ionization potential and below the sparking potential. A series of characteristic regions appears in the discharge between the anode and the cathode and may be displayed in a specially constructed *glow discharge tube.* See also: *Anode dark space. Anode fall. Anode glow. Aston dark space. Cathode dark space. Cathode glow. Cathode layers. Faraday dark space. Negative glow. Positive column.*

DISCHARGE, HOLLOW CATHODE. An electrical discharge in which, by virtue of the hollow shape of the cathode, much higher currents can be sustained by the same voltage than for plane cathodes. The effect of a hollow cathode also occurs when two plane cathodes are brought sufficiently close together for the negative glows to interact. See also: *Hollow cathode.*

DISCHARGE, RING. A ring-shaped electrodeless discharge produced by the high-frequency oscillating electromagnetic field of an external coil. Also known as a *toroidal discharge.*

DISCHARGE, SELF-SUSTAINING. An electrical discharge in which the elementary particles carrying the current are created by the discharge itself; and which does not, therefore, require such adjuncts as irradiation or thermal emission. See also: *Discharge, Townsend.*

DISCHARGE, SPARK. An electrical discharge in which the current is carried by a spark, which constitutes a transient conducting ionized path between the electrodes. After a millisecond or more the characteristics may change to those of an arc.

DISCHARGE, STREAMER. An electrical discharge in which luminous lines ("streamers") occur. The streamers arise from avalanches produced along their length by ionization due to photons.

DISCHARGE, STRIATIONS IN. Streaks or lines occurring in an electrical discharge. Some are periodic in space and time, others in space only. They are produced when an alternating potential difference is applied but may also occur with a fixed potential.

DISCHARGE, SUBNORMAL. See: *Discharge, Townsend.*

DISCHARGE, TIME LAG OF. The time which elapses between the instant at which the voltage applied to a gap exceeds the minimum breakdown voltage of the gap and that at which spark breakdown occurs. It consists of two parts, the *statistical time lag*, namely the time required for a suitably placed electron to appear in the gap to initiate the spark once the breakdown voltage has been exceeded, and the *formative time lag*, namely the time required for the spark, once initiated, to develop across the gap.

DISCHARGE, TOROIDAL. See: *Discharge, ring.*

DISCHARGE, TOWNSEND. An electrical discharge with current density so small that the space charge is negligible, which is not self-sustaining, but requires the presence of an external source to maintain the necessary ionization. The current is amplified owing to ionization by collision. Also known as *field-intensified discharge* or *subnormal discharge.*

DISCHARGE, TRACKING OF. (1) An electrical discharge over the surface of a solid dielectric, caused by the passage of a leakage current in a contaminating layer. (2) The damage produced by such a discharge.

DISCHARGE TUBE. A tube containing gas at low pressure through which an electric current may be passed.

DISINTEGRATION ENERGY: Q-VALUE. For a given nuclear disintegration: the amount of energy released in that disintegration. A negative value signifies that energy is absorbed. See also: *Q: Q-value.*

DISINTEGRATION, NUCLEAR. The transformation of an atomic nucleus, possibly a compound one. It is characterized by the emission of one or more particles or photons, or the splitting of the nucleus into more nuclei.

DISK AREA. Of a system of rotating aerofoils: the area swept out by the tips of the blades.

DISK OF LEAST CONFUSION. Another term for circle of least confusion. See also: *Astigmatism.*

DISLOCATION. (1) A concept in the continuum theory of elasticity involving six types of relative displacement, known as dislocations. These displacements include rotations and pure translations. (2) A type of imperfection in a crystalline solid affecting the regular arrangement of the atoms. It involves displacements of a more restrictive nature than in (1) above in that translations are restricted to complete lattice vectors and rotations to full symmetry operations. An *edge dislocation* refers to a set of translations resulting in a line of lattice misfits (a *dislocation line*) which runs through the crystal in a direction at right angles to the direction of translation. Such a dislocation may be envisaged as lying at the edge of an extra layer of one or more lattice planes inserted into the crystal. In a *screw dislocation* the direction of translation is parallel to the dislocation line. The lattice planes in such a dislocation may be envisaged as constituting a continuous helicoidal surface which screws around the dislocation line. All other dislocations may be resolved into these two types.

DISLOCATION, STRENGTH OF. The magnitude of the Burgers vector associated with a dislocation.

DISORDER. See: *Order–disorder transformation.*

DISPENSER CATHODE. See: *Thermionic cathode.*

DISPERSE SYSTEM. A colloidal system consisting of two phases: the *disperse phase* (consisting of colloidal particles) and the *dispersion medium* (the medium in which the particles are dispersed).

DISPERSION. (1) A suspension of a solid in a liquid. It may range from a colloidal solution to a concentrated paste. (2) The variation of the refractive index of a medium (or, by extension, deviation in general) with wavelength. See also: *Anomalous dispersion. Cauchy dispersion formula.*

DISPERSION FORCE. The force of attraction between molecules possessing no permanent dipole. It arises essentially from the perturbation of the electronic orbits of neighbouring molecules. See also: *Van der Waals forces.*

DISPERSION FORMULAE OF NUCLEAR PROCESSES. Formulae so named because of the similarity of their mathematical structure to the dispersion formulae of optics, energy and cross-section replacing optical frequency and refractive index or atomic polarizability. The simplest and best known of such formulae is the Breit–Wigner formula.

DISPERSION, ROTATORY. The variation with wavelength of the amount by which the plane of vibration of linearly polarized light is rotated by optically active materials.

DISPERSION STRENGTHENING. Of a substance: the reduction of plastic deformation of the substance by the incorporation of a uniform dispersion of particles of another material which limit the movement of dislocations. See also: *Fibre strengthening.*

DISPERSIVE POWER, OPTICAL. (1) Of a particular medium, for light: the ratio of the difference in the deviation of light of two different wavelengths to the deviation of light whose wavelength is an average of the two. (2) For optical glass: the dispersive power is given by $\dfrac{n_F - n_C}{n_D - 1}$ where n_D, n_F and n_C are the refractive indices of the glass for the D, F and C Fraunhofer lines. The *reciprocal* of this dispersive power is known as the *Abbé number* or the *constringence*.

DISPERSIVE POWER, RECIPROCAL. See: *Constringence. Dispersive power, optical.*

DISPERSIVITY: DISPERSIVITY QUOTIENT. Of a medium: the rate of change of refractive index with wavelength.

DISPLACEMENT. (1) In a chemical reaction: the replacement of one kind of atom, molecule or radical by another. (2) Of an atomic or sub-atomic particle: the direct distance travelled by the particle, irrespective of route, between two specified events or types of event. (3) In an electrostatic field (*electric displacement* or *electric induction*): a vector at a point in that field, whose direction is that of the field and whose magnitude is the product of the field strength and the absolute permittivity of the medium in which the field is produced. It may be regarded as the quantity of electricity displaced across unit area and is equal to the electric flux density.

DISPLACEMENT COLLISION. The permanent displacement of an atom from its site in a solid as the result of direct action with a nuclear projectile.

DISPLACEMENT CURRENT. See: *Current.*

DISPLACEMENT LAW. (1) For complex spectra: states that the arc spectrum of an element is similar to the first spark spectrum (singly ionized atom spectrum) of the element one place higher in the Periodic Table, or to the second spark spectrum (doubly ionized atom spectrum) of the element two places higher. The analogy also holds for higher orders. (2) For radioactive decay: one of a set of rules connecting the type of decay with the displacement in the Periodic Table (caused by the change in nuclear charge) of a daughter element relative to the parent.

DISPLACEMENT SPIKE. The zone of displaced atoms produced in a solid by bombardment with high-energy particles. See also: *Fission-spike. Thermal spike.*

DISPROPORTIONATION. A chemical reaction consisting of the conversion of two or more like molecules or ions into two or more unlike molecules or ions. The reverse reaction, the conversion of two or more unlike molecules or ions into two or more like molecules or ions, is referred to as *reproportionation*. Thus: the reaction $2CO \rightarrow C + CO_2$ is an example of disproportionation, while $C + CO_2 \rightarrow 2CO$ is an example of reproportionation.

DISSOCIATION, CHEMICAL. A reversible chemical process which, under suitable conditions (e.g. an increase of temperature or the presence of a given solvent), causes a compound to split up into simpler atomic groups or single atoms or ions. It may be treated by the law of mass action. See also: *Association, chemical.*

DISSOCIATION CONSTANT. An equilibrium constant for a chemical dissociation reaction. It is the ratio of the product of the active masses of (e.g.) the molecules resulting from dissociation to the active mass of the (e.g.) undissociated molecules, when equilibrium is reached.

DISSOCIATION, ELECTROLYTIC. The spontaneous dissociation into ions of opposite sign undergone by acids, bases and salts in solvents of sufficient dielectric constant, e.g. water, the lower alcohols, etc.

DISSOCIATION ENERGY. Of an electronic state in a diatomic molecule: the energy difference between the isolated atoms of the molecule at infinite separation and the lowest rotational vibration level of the electronic state in question. An accurate measurement of the dissociation energy may be obtained from the convergence limit.

DISSOCIATION, HEAT OF. The amount of heat required, per mole, to break a specified interatomic or intermolecular bond in a given chemical compound.

DISSOCIATION, IONIC. Dissociation involving the production of ions.

DISSOCIATION LIMIT. For a molecule: the point at which, as the vibrational quantum number varies, the molecule dissociates into its constituent atoms. It corresponds to the convergence limit.

DISSONANCE. Of musical notes: the production of a sound that is unpleasant to the ear when two or more notes are played together. It is a subjective effect and arises from the existence of beats. See also: *Consonance.*

DISTILLATION. A process for the separation of homogeneous liquid or liquid–solid mixtures in which the liquid is partially vaporized and the resulting vapour subsequently condensed. The condensate is enriched in the more volatile component of the mixture and the residue in the less volatile.

DISTILLATION COLUMN. A series of condenser units arranged in cascade so as to increase the degree of separation of a distillation process, or (more usually) an arrangement of plates, downpipes, etc., in a cylindrical column, designed to achieve the same object.

DISTILLATION, FRACTIONAL. A process for the separation of the components of a liquid mixture into fractions differing from one another on the basis of boiling point or volatility.

DISTILLATION, MOLECULAR. Distillation at a low pressure (e.g. 10^{-3} mm mercury), at which the mean free path of the molecules is of the same order of magnitude as the distance between the heated and cooled surfaces. It permits the distillation of substances which, in a normal distillation process, would be decomposed.

DISTORTED WAVES, METHOD OF. A method used for the solution of problems involving inelastic scattering such as collisions between electrons and atoms. The colliding particle is considered as an incident plane wave which is diffracted by the central field of the scattering centre.

DISTORTION. Of an optical system: the variation of the magnification with the angular distance from the axis. Thus the image of a straight line may appear curved.

DISTORTIONAL WAVE. See: *Shear wave.*

DISTRIBUTION COEFFICIENT: PARTITION COEFFICIENT. The ratio of the equilibrium concentrations of a given substance dissolved in two specified immiscible solvents.

DISTRIBUTION FUNCTION: GENERALIZED FUNCTION. A mathematical function used in bringing symbolic functions such as the Dirac delta function within the scope of a rigorous mathematical treatment.

DISTRICT HEATING. The provision of heat and hot water to a number of buildings in a particular locality by means of a piped service from a central station, the medium of heat transmission being either steam or hot water.

DISTURBED DAYS: ACTIVE DAYS. Five days selected each month as international magnetic "disturbed" or "active" days, on the basis of many magnetic observations. See also: *Quiet days: Calm days.*

DITTUS–BOELTER EQUATION. An equation for the heat transfer from a tube to a viscous fluid flowing through it.

DIURNAL ABERRATION. The aberration of light

101

associated with the rotation of the Earth about its axis.

DIURNAL VARIATION (METEOROLOGY). The diurnal variation of meteorological quantities such as temperature, pressure, atmospheric pollution, wind speed, etc., which arises from the diurnal variation of the solar radiation received at the Earth's surface.

DIVARIANT SYSTEM. A system consisting of one phase only, i.e. solid, liquid or gas, to define the state of which requires a knowledge of both temperature and pressure. The name "divariant" is used instead of the expression "having two degrees of freedom" to avoid confusion with the mechanical degrees of freedom used, for example, in molecular theory. Similarly, instead of using the word "freedom" in the present context, the word "*variance*" or "*variancy*" is commonly used.

DIVERGENCE. Of a vector V: the flux of some physical quantity per unit volume across the surface of a small element at the point x, y, z. It is the scalar product of the differential vector operator, ∇, and the vector V, and may be written

$$\nabla \cdot \mathbf{V} = div \ \mathbf{V} = \frac{\partial V_x}{\partial x} + \frac{\partial V_y}{\partial y} + \frac{\partial V_z}{\partial z}.$$

See also: *Differential vector operator.*

DIVERGENT. In nuclear technology: refers to the condition of a neutron chain reacting medium for which the effective multiplication factor is greater than unity, i.e. to the supercritical condition. See also: *Chain reaction, neutron. Multiplication factor: Multiplication constant.*

DIVERGENT BEAM TECHNIQUE. The use of a divergent beam of X-rays to produce Kossel lines. The technique can be used for the determination of accurate unit cell dimensions (to $\pm 0.001\%$ for cubic crystals), the location of sub-structure boundaries and other variations in crystal texture, and deformation mechanisms in general. See also: *Kossel effect.*

DIVERTER. A resistor shunted across one part of a circuit. It is used, for example, across the field coils of d.c. series motors to control the speed.

DIVIDING ENGINE. An instrument for marking or engraving accurate subdivisions on scales. It consists essentially of a carriage which is adjusted by a micrometer screw controlled by a template, and carring a marking tool. It may produce linear or circular scales and may be automatic.

DIVIDING HEAD. A machine tool attachment for rotating the work through any required angle so as to permit various operations (e.g. machining of faces, drilling of holes) to be performed with an accurate angular relationship.

DNA. See: *Deoxyribonucleic acid.*

DOLDRUMS. A region of calm in equatorial oceans, extending on the average to about 10° north and south of the equator. The region undergoes a displacement with the seasons, moving northwards in June and southwards in December. See also: *Trade winds.*

DOLLAR (NUCLEAR REACTOR TECHNOLOGY). A unit of reactivity equal to the amount of reactivity required to make a reactor critical on prompt neutrons only. One-hundredth of this amount is called a *cent*. See also: *Nuclear reactor, reactivity of.*

DOMAIN, ANTI-FERROMAGNETIC. See: *Domain, ferromagnetic.*

DOMAIN BOUNDARY. The transition layer, which may be as great as $1.2 \ \mu$m, which separates two domains magnetized in different directions and in which the change in spin direction between the two domains is assumed to take place gradually. Also known as a *Bloch wall*.

DOMAIN, FERROELECTRIC. See: *Domain, ferromagnetic.*

DOMAIN, FERROMAGNETIC. A region of spontaneous magnetization in a single direction. Neighbouring domains have different directions of magnetization. Analogous domains of spontaneous polarization also occur in *ferroelectric* materials. In an *anti-ferromagnetic* material some of the magnetic moments are arranged anti-parallel to the direction of magnetization. See also: *Magnetic bubbles.*

DOMAINS, FLUX-CLOSURE. Small domains so situated and so magnetized as to close the flux path between adjacent (larger) domains without the formation of magnetic poles. Also called *closure domains*.

DOMINANT WAVELENGTH. For a given colour: the wavelength of monochromatic light which matches that colour when combined in suitable proportions with a reference light, often taken as the prevailing illumination. See also: *Complementary colours.*

DONNAN DISTRIBUTION COEFFICIENT. A coefficient expressing the distribution of diffusible electrolytic ions on two sides of a membrane in the presence of a non-diffusible ion present on one side only.

DONNAN POTENTIAL. The difference in potential set up across a semipermeable membrane which separates a solution containing a non-diffusible ion from one containing diffusible ions. At equilibrium

(*Donnan equilibrium*) there is also a difference in osmotic pressure.

DONOR. An impurity or imperfection in a semi-conductor, which donates electrons to the conduction band, leading to electron conduction. The energy required to raise a bound electron to the conduction band is the ionization energy of the donor atom. See also: *Acceptor*.

DONOR LEVEL. An electron energy level associated with a donor atom, commonly measured from the bottom of the conduction band.

DOPPLER EFFECT. The phenomenon giving rise to a change in the observed wavelength of a radiation which results from the movement of its source relative to the observer. An approaching source appears to decrease the wavelength and a receding source to increase it. This change in wavelength is known as the *Doppler shift*. The effect is used in astrophysics to study the approach or recession of celestial bodies. It gives rise to the broadening of spectral lines, owing to the thermal motion of the molecules, atoms or nuclei involved; and to the broadening of the observed widths of nuclear energy levels owing to the thermal motion of the target particles. In each of these two instances the effect is known as *Doppler broadening*.

DORN EFFECT. The occurrence of an electrical potential across the ends of a column of liquid when colloidal particles are driven through the column by mechanical forces, as by the action of gravity. It is the converse of electrophoresis, and arises from displacement of the electrical double layers of the particles. See also: *Electrical double layer*.

DOSE, ABSORBED. Of ionizing radiation: the energy imparted to matter per unit mass at a point of interest. The unit is the *rad,* i.e. 10^{-2} J/kg (100 erg/g), which is to be replaced by the SI unit, the *gray.* See: *Gray. Appendix II.*

DOSE EQUIVALENT. Of ionizing radiation: an estimate of the biological effect of an absorbed dose. It is obtained by multiplying that dose by various factors, necessary, for example, to allow for any non-uniform distribution of the radiation and for the difference between the biological effects of the radiation and those of medium-energy X-rays. See: *Rem.*

DOSE EQUIVALENT, MAXIMUM PERMISSIBLE. Of ionizing radiation: the largest dose equivalent received within a specified period which is permitted by a regulatory committee on the assumption that there is no appreciable probability that somatic or genetic injury will result. Formerly known as *permissible dose*, *maximum permissible dose*, or *tolerance dose*.

DOSE, INTEGRAL ABSORBED. Of ionizing radiation: the integral of the absorbed dose over the mass of irradiated matter in a volume of interest.

DOSE, MEDIAN LETHAL: LD50. Of ionizing radiation: the absorbed dose which will kill, within a specified time, 50% of the individuals of a large population of organisms of a given species. Formerly known as *mean lethal dose.*

DOSEMETER: DOSIMETER. An instrument used for measuring or evaluating the absorbed dose, exposure, or similar radiation quantity. It may depend on ionization for its operation or may simply involve the darkening of a photographic emulsion.

DOSE OF IONIZING RADIATION. A general term denoting the quantity of radiation or energy absorbed by a body, and sometimes the quantity to which the body is exposed. For special purposes the word "dose" needs appropriate qualification. The term "dose" has been used with a variety of specific meanings such as "absorbed dose" or "exposure", but such uses are now discouraged.

DOSE, PERCENTAGE DEPTH. Of ionizing radiation: the ratio, expressed as a percentage, of the absorbed dose at any given depth within a body to the absorbed dose at some reference point of the body along the central ray. The location of the reference point depends on the energy of the radiation. For low energy X- or γ-rays it is at the surface, and for high energy at the position of the peak absorbed dose.

DOSE RATE. Of ionizing radiation: the dose per unit time. The nature of the dose should be specified e.g. "absorbed dose", "dose equivalent".

DOSE, VOLUME. Of ionizing radiation: the product of absorbed dose and the volume of the absorbing mass. It is to be distinguished from integral absorbed dose.

DOUBLE DECOMPOSITION. A type of chemical reaction involving an exchange of constituents, as in $AX + BY \rightarrow AY + BX$, where A, B, X and Y may represent atoms, molecules or radicals.

DOUBLE ELECTRON EXCITATION. In spectra: the excitation of two electrons rather than one, giving rise to prominent spectral lines, particularly in the alkaline earths.

DOUBLE QUANTUM TRANSITIONS. Radiative transitions between atomic or molecular states, which involve the simultaneous emission or absorption of two or more photons of low frequency. Such transitions are rare in optical spectra but many have been observed in radiofrequency spectroscopy.

DOUBLE REFRACTION, ALLOGYRIC. In optically active solids: the representation of optical rotation as arising from the difference in velocity (and hence refractive index) between left- and right-hand circularly polarized light.

DOUBLE REFRACTION: BIREFRINGENCE. The property, exhibited by anisotropic crystals, of possessing two refractive indices, i.e. of allowing light to pass through them with two different velocities. A ray of light passing into such a crystal will, unless it is parallel to certain directions (*optic axes*), be split into two rays which pursue different paths and are polarized at right angles. Only one of these, the *ordinary ray*, obeys Snell's law. The other is known as the *extraordinary ray*. The corresponding *refractive indices* are also designated as *ordinary* and *extraordinary* respectively. See also: *Form birefringence.*

DOUBLE SALT. A solid compound formed by mixing solutions of two salts.

DOUBLET, ELECTRIC OR MAGNETIC. An element, often fictitious, comprising two equal and opposite electric or magnetic charges separated by a small distance which may be infinitesimal. The product of the charge and the separation is finite and defines the strength of the doublet. See also: *Dipole, electric. Dipole, magnetic.*

DOUBLET, HYDRODYNAMIC. A source-sink combination analogous to the electric or magnetic doublet, used in the study of flow. The source corresponds to one charge and the sink to the other.

DOUBLET, SPECTRAL. A spectral line with a multiplicity of two. The frequency separation of the two levels, expressed as a wave number, is called the *doublet interval.* See also: *Multiplicity.*

DOUBLE-WEDGE AEROFOIL. See: *Aerofoil, wedge.*

DOUBLE WEIGHING. A method of weighing used to allow for possible inequalities in balance arms. The weight of a given body is measured twice, once with the normal positions of body and weights and once with these positions reversed. The true weight is taken as the average of the two. Also known as the *Gauss method.* See also: *Counterpoise weighing.*

DOUBLING IN MOLECULAR SPECTRA. The splitting of the rotational levels in the spectrum of a diatomic molecule which results from the coupling of the electron orbital and spin angular momenta with the nuclear angular momentum. See also: *Coupling.*

DOUGHNUT: DONUT. The toroidal vacuum chamber of a betatron or synchrotron in which electrons or protons are accelerated.

DOWNWASH (AERODYNAMICS). A vertical component of velocity induced by a lifting wing surface in a direction normal to the direction of flight. An important example is the downwash over the tail surfaces of an aircraft, which reduces the lifting force.

DRACONITIC MONTH. Another name for *Nodical month.* See also: *Month.*

DRAG. Of a body in a flowing fluid: the component of the force exerted by the fluid on the body in the direction of flow. It is thus the resistance to the motion of the body through the fluid. See also: *Lift.*

DRAG, BASE. See: *Base pressure.*

DRAG COEFFICIENT. For a fluid flowing past a submerged rigid body: the ratio $2F/A\varrho v^2$, where F is the longitudinal force on the body, A is the projected area of the body on a plane normal to the direction of flow, ϱ is the density of the fluid and v its velocity.

DRAG, FORM. The drag induced by the lift.

DRAG OF A SPHERE. A quantity much studied theoretically to determine the influence of Reynolds and Mach numbers on drag, and the critical values of these numbers.

DRAG, PROFILE. The sum of the surface drag and form drag.

DRAG, SURFACE. That portion of the drag which can be attributed to surface or skin friction. See also: *Viscous drag.*

DRAPER EFFECT. In a gaseous system exposed to light: the occurrence of a slight expansion followed by a steady state and then by contraction. See also: *Budde effect.*

DRAWING OF METALS. The production of a desired shape (e.g. a wire or tube) by the application of forces which directly produce tensile stresses. Compressive stresses are then developed, however, which effect the actual deformation of the metal.

D-REGION. Of the ionosphere: See: *Ionosphere.*

DRIFT MOBILITY. Of charge carriers diffusing through a solid under the influence of an electric field: the average drift velocity per unit electric field.

DRIFT SPACE. See: *Drift tube.*

DRIFT TUBE. Part of a klystron or a particle accelerator in which charged particles move without being acted upon by magnetic or electric fields. In the drift tube the faster electrons catch up with the slower and a succession of bunches is formed. Where the region of bunching is not tubular the expression *drift space* is used.

DRIFT VELOCITY. Of charge carriers moving in a solid, liquid or gas under the influence of an electric field: the average distance travelled per second in the direction of the field. The drift velocity per unit electric field is known as the *mobility*.

DRIZZLE. See: *Rain.*

DROPPING-MERCURY ELECTRODE. An electrode used in polarography which supplies a regular series of mercury drops within a solution, each with a new clean surface available for polarization. The other electrode, which may consist of a large pool of mercury into which the drops from the dropping electrode fall directly, is non-polarizable. See also: *Polarography.*

DROPWISE CONDENSATION. Condensation in which the condensate forms into drops. It is obtained only when the condensing surface is "contaminated" by an anti-wetting agent, and, other things being equal, it leads to high rates of heat transfer and therefore of condensation. See also: *Condensation. Filmwise condensation.*

DRUDE EQUATION: DRUDE LAW. A relationship between the specific rotation of an optically active substance and the wavelength of the light. It is given by $\alpha = k/(\lambda^2 - \lambda_0^2)$ where α is the specific rotation, λ the wavelength of the light, λ_0 the wavelength of the head of the nearest absorption band, and k is known as the *rotation constant*.

DRY BATTERY. See: *Primary cell.*

DUALITY, WAVE-CORPUSCLE. Refers to the statement that the wave and corpuscular aspects of any physical phenomenon are not contradictory but complementary. See also: *Complementarity principle. Uncertainty principle: Indeterminacy principle.*

DUCTILITY. Of a metal: the property of being readily drawn out into a wire. It is associated with a fairly low yield stress and a high rate of work-hardening which is maintained for a large plastic deformation.

DUDDELL OSCILLOGRAPH. Essentially, a high-frequency moving coil galvanometer, used for investigating the wave form of alternating current or e.m.f.

DULONG AND PETIT LAW. See: *Atomic heat.*

DUPLEXER. A transmission system designed to permit a common aerial to be used for transmission and reception of radio and radar signals. Also known as a *T-R box* or *T-R switch.*

DUPLEX LIQUID FILM. A film of liquid so thick that its two boundary surfaces show no energetic interaction.

DUSHMAN–LANGMUIR EQUATION. An empirical attempt to give a theoretical expression for the constant D_0 in the equation connecting the diffusion coefficient, D, of a solid with temperature, T, in the manner $D = D_0 \exp(-E/RT)$, where E is the activation energy and R the gas constant. The expression given for D_0 is $D_0 = Ed^2/Nh$, where d is the spacing between the crystal planes normal to the diffusion direction, N is Avogadro's constant, and h is Planck's constant.

DUST CORE. A magnetic core composed of pulverized particles held together by a binder. Because of low eddy current losses the cores are suitable for use at high frequencies.

DUST DEVIL. A whirlwind over a sandy area, which picks up dust and sand and may lift them by several hundred feet.

DYE BLEACH PROCESS. A method of making colour photographs by subtractive colour synthesis.

DYNAMICAL SIMILARITY PRINCIPLE. For fluid flow: states that two geometrically similar fluid flows are dynamically similar if the flow field of one may be transformed into that of the other by the same changes in the scale of length and velocity that are necessary to make the boundary conditions identical. If the equations of motion are made non-dimensional by expressing velocities and lengths as fractions of these scales, the equations contain a number of non-dimensional coefficients that determine the character of the flow. The general condition for dynamical similarity is that all these coefficients should be the same for the two flows. The coefficients commonly used are the Reynolds number, Prandtl number, Grashof number, Mach number, and Froude number.

DYNAMICAL THEORY OF ELECTRON AND X-RAY DIFFRACTION. A theoretical treatment of the diffraction process which takes account of the dynamical equilibrium that must be set up between the incident and diffracted beams in a crystal. For example, the amplitude of the incident beam will fall off as it passes through the crystal; the velocity will not be the same in the crystal as in free space; and, more important, multiple reflections will occur and the multiply reflected beams will interfere with the incident beam and with each other. The simple theory, neglecting the above complications, which will still apply in certain cases, e.g. for very small or very thin crystals, is sometimes known as the *kinematic theory*.

DYNAMIC HEAD: DYNAMIC PRESSURE. The impact pressure associated with a moving fluid. The sum of this and the static pressure, i.e. the pressure which would exist in the absence of motion, is known as the *total head*.

DYNAMICS. The study of the behaviour of objects in motion, in particular objects acted on by forces and having variable velocity. Its foundations are Newton's laws of motion. See also: *Kinematics. Mechanics. Newton laws of motion. Statics.*

DYNAMIC TEMPERATURE DIFFERENCE. In a flowing medium: the difference between the static temperature of the medium and the surface temperature at the stagnation point of a heat-insulated body which is immersed in the stream. For other positions than the stagnation point the surface temperature of the body will be below the stagnation value but above the static value.

DYNAMO. Usually denotes an electromagnetic generator which produces current by the rotation of an electrical conductor in a magnetic field. Ideally the machine is reversible and may also be used as a motor. In general, any mechanism which can transform mechanical into electrical energy or vice versa may be described as a dynamo.

DYNAMO EFFECT IN MOVING WATER. The production of an e.m.f. in water which moves in the Earth's magnetic field, by virtue of the fact that water is a conductor. The effect has been used to measure ocean drifts and wind drifts on the surface of deep water.

DYNAMOMETER. (1) A device for measuring force, mainly the brake horse power of a motor. (2) An electrical measuring instrument depending for its action on the e.m.f. between two or more current-carrying coils. Also known as an *electrodynamometer*. It may be used, with appropriate connections, as a voltmeter, ammeter, or wattmeter, and for a.c. or d.c.

DYNATRON. A vacuum tube operated so that its plate characteristic has a negative resistance portion, where the plate current increases as the plate voltage decreases. Its most common application is to the *dynatron oscillator*.

DYNE. The CGS unit of force. It is that force which gives to 1 g an acceleration of 1 cm/sec^2, and is equal to 10^{-5} N.

DYNODE. An electrode in a photomultiplier tube designed to emit secondary electrons when bombarded by a beam of primary electrons.

e

EAGLE MOUNTING. A compact mounting of a concave diffraction grating based upon the principle of the Rowland circle.

EAGRE. See: *Tidal wave.*

EARING. The symmetrical waviness occurring at the brims of cylindrical cups which have been pressed from metal strip. It arises from the presence of plastic anisotropy in the material.

EARTH. The third planet in the solar system in order of distance from the Sun. Its diameter is about 12760 km, its mass about 6×10^{24} kg, its density about 5·52 g cm^{-3}, and its period of rotation about 24 hours. It has one satellite, the Moon. See also: Following entries.

EARTH-AIR CURRENT. The downward electric current observed over most of the Earth, of about 10^{-6} A/km^2.

EARTH, CORE OF. That region of the Earth's interior extending from the mantle to the centre of the Earth. There is an outer core (believed to be liquid) of about 2000 km in thickness and an inner core (believed to be solid) of radius about 1200 km, separated by a transition region of about 140 km in thickness. See also: *Earth, structure of.*

EARTH, CRUST OF. The outer shell of the Earth extending down to a depth of about 35 km under the continents and about 10 km under the ocean floors. It lower boundary is the *Mohorovičić discontinuity.* See also: *Basaltic layer. Earth, structure of.Granitic layer.*

EARTH CURRENTS. Natural electric currents circulating in the ground. They are due to the existence of a natural electric field (the *Earth current field*), which, at a given point, varies continually in magnitude and direction.

EARTH, EXTERNAL POTENTIAL OF. The combined potential of the Earth's attraction per unit mass and the centrifugal force of the Earth's rotation.

EARTH, FIGURE OF. The size and shape of the equipotential surface at mean sea level and its continuation under the continents. This equipotential level is known as the *geoid.* Ideally it would be an ellipsoid of revolution but, owing to mass irregularities of the Earth (both visible and invisible), it departs considerably from this figure.

EARTH INDUCTOR. An instrument for measuring the magnetic dip by means of dynamo-electric action. It depends on the interaction between a hand-operated rotating coil and the ambient magnetic field of the Earth. See also: *Dip-needle.*

EARTH, MANTLE OF. That region of the Earth's interior between the crust and the core. Its upper boundary is the *Mohorovičić discontinuity.* There is an outer mantle of about 375 km in thickness and a lower mantle of about 1900 km in thickness, separated by a transition region of about 590 km in thickness. See also: *Earth, structure of.*

EARTH, MOTIONS OF. (1) *Translation:* the movement about the Sun. The centre of mass of the Earth and Moon, which is about 4000 km from the centre of the Earth, moves about the Sun in approximately an ellipse, the motion being somewhat disturbed by the attractions of other bodies. (2) *Rotation:* the movement of the Earth about its own axis. The time taken per rotation shows slight variations, as does the inclination of the axis. The oscillation of the Earth's pole about the mean position, arising from periodic variations in the plane of the Moon's orbit, is known as *nutation*, and has a period of about 18·6 years. Its effect is to produce small changes in the *precession*, which is the slow revolution of the Earth's axis about the pole of the ecliptic in a period of about 26,000 years.

EARTH, MOVEMENTS IN. Vertical movements of 0·1–10·0 mm per year, the latter figure being rarely achieved except in earthquakes and horizontal movements of uncertain magnitude. Continental drift would represent an extreme case of horizontal movement.

EARTH POTENTIAL. The electric potential of the Earth, usually regarded as zero.

EARTHQUAKE. A violent movement of the Earth's surface which occurs when a large amount of energy (up to as much as 10^{25} erg, i.e. 10^{18} J) is released within a few seconds from a *focal region*. The focal region may have linear dimensions of several kilometres. Its centre, somewhat indefinite, is known as the *focus*. The point on the Earth's surface, vertically above the focus, is the *epicentre*.

EARTHQUAKE INTENSITY SCALE. A scale used to express the observed effects of an earthquake. The most common is the modified *Mercalli scale*, which ranges from I (recorded but not felt) through V (many awakened, some dishes broken) to XII (total damage).

EARTHQUAKE, MAGNITUDE OF. A value, characteristic of a given earthquake, which is a measure of the energy released in the earthquake, and which is based on instrumental records only. The *Richter scale* is concerned with the maximum ground amplitude, and the *Gutenberg–Richter scale* with the relation of this to the period of certain types of earthquake wave, but discrepancies between these (and other) scales led to the *unified amplitude*, given by $\log{(a/T)} + S + Q$, where a is the ground amplitude, T is the period of the waves through the Earth, S is a small constant for each station, and Q is a quantity which depends on the type of wave, the epicentral distance, and the focal depth, which has been tabulated for various conditions.

EARTHQUAKE, PHASES OF. Successive events recorded at a single observatory from a single earthquake. They arise from the arrival of direct P and S waves, from the arrival of reflected and polarized P and S waves, and from the arrival of surface waves, all of which travel at different speeds. See also: *Seismic waves. Seismic waves, primary* (P) *and secondary* (S).

EARTH, ROTATION OF. The rotation of the Earth about its axis. Its period varies slightly by up to 0·002 s/day and the inclination of the axis is also subject to variations, mainly precession and nutation. See also: *Earth, motions of.*

EARTH, STANDARD. A set of geodetic parameters which specify the observed figure of Earth. The data are obtained not only from terrestrial sources but from satellite observations.

EARTH, STRUCTURE OF. The description of the Earth's interior in terms of layers or regions, based upon observations made at the surface by the methods of geology, geophysics and geochemistry. These include seismological observations and measurements of the gravitational field, gravitational constant and mean moment of inertia. The regions into which the Earth is divided consist of the *crust*, extending from the surface to about 35 km (10 km under the ocean floors); the *upper mantle*, extending from about 35 km to 410 km; a transition region from about 410 km to 1000 km; followed by the *lower mantle*, from about 1000 km to 2900 km; the *outer core*, from about 2900 km to 4980 km; a transition region from about 4980 km to 5120 km; followed by the *inner core*, below about 5120 km, and with a radius of about 1200 km. According to these figures the *mean radius* of the Earth is, therefore, about 6320 km. See also: *Earth, core of. Earth, crust of. Earth, mantle of. Mohorovičić discontinuity.*

EARTH, THERMAL CONTRACTION OF. The contraction of the Earth arising from its cooling due to the diffusion of its original heat, modified by the heating up due to the presence of radioactive constituents. The resultant stresses and strains, according to the thermal contraction theory, gave rise to the process of *mountain building* or *orogenesis*.

EARTH TIDES. Periodic elastic deformations in the body of the Earth which are directly analogous to the ocean tides, and, like them, are produced by the gravitational attraction of the Sun and Moon. They cause small but measurable fluctuations in the gravitational acceleration.

EASY GLIDE. In the plastic deformation of a single crystal: the occurrence of considerable deformation with a small increase in stress, in a region following the elastic region (with increasing stress) and preceding the onset of work hardening. It arises from the passage through the crystal, on a single glide system, of many thousands of dislocations.

EBERHARD EFFECT. An effect on the photographic image due to local variations of bromide concentration, during development, over the light and dark parts of the image.

EBULLIOMETER. A device for obtaining the true boiling point of a solution in which the fact that a boiling liquid is slightly superheated is overcome by placing a thermometer in the vapour and spraying it with boiling liquid which cools as it runs down, and reaches the true boiling point.

EBULLIOSCOPIC CONSTANT. Of a solvent: that elevation of the boiling point which would be observed if 1 mole of a solute were dissolved in 1000 g, and if the solution behaved ideally. It is independent of the nature of the solute. Also known as the *molal elevation of the boiling point.*

EBULLIOSCOPY. (1) The measurement of the elevation of the boiling point of a solvent by the addition of a non-volatile solute. (2) The determination of molecular weight by such a measurement. See also: *Ebullioscopic constant.*

ECCENTRIC ANOMALY. Of a planetary orbit the quantity $\cos^{-1}(\alpha - T)/\alpha\varepsilon$, where α is the semimajor axis of the orbit, T is the radius vector from focus to path, and ε is the eccentricity of the ellipse defining the shape of the orbit.

ECCENTRICITY. See: *Conic: Conic section.*

ECCENTRICITY, ENFORCED. Refers to an orbit of a body about its primary, the elliptical shape of which can be attributed to the perturbing force of another body.

E-CENTRE. See: *Colour centres.*

ÉCHELON, ÉCHELLE AND ÉCHELETTE GRATINGS. Dispersive diffraction gratings designed to bridge the gap between conventional gratings and dispersive interferometers as regards resolution, dispersion, and free spectral range. They consist of grooves, rather than lines, formed, for example, by assembling flat plates of glass to form a staircase pattern. The échelle is intermediate in coarseness between the échelette (usually 100–80,000 grooves/cm) and the échelon (10 steps/cm or less). These gratings are to be distinguished from the Fabry–Perot étalon, which operates without angular dispersion.

ECHO BOX. A high-Q resonator used mainly for checking the performance of radar receivers. The cavity is set ringing by a pulse of energy and the receiver performance is checked during the decay of the oscillations.

ECHO PROSPECTING. The location of oil or other mineral deposit by the observation of sound echoes from the discontinuities formed by the deposits.

ECHO SOUNDING. The determination of the range and direction of underwater objects by the observation of pulses of ultrasonic energy reflected from them. The technique may be used for measuring the depth of water below a ship's hull, for locating shoals of fish, or, in the form of *Sonar* (or *Asdic*) for the detection of submarines. In the latter case, provision is also made for the determination of the angle and direction of movement of the target relative to the receiver.

ECLIPSE, LUNAR. The interposition of the Earth between the Sun and the Moon so as to cut off the Sun's light from the Moon's surface. It may be *total*, when all the Sun's light is cut off; or *partial*, when only some of the light is cut off. An analogous state of affairs may also occur for a planet, leading to a *planetary eclipse*.

ECLIPSE, SOLAR. The interposition of the Moon between the Sun and the Earth so as to block out the view of the Sun as seen from the Earth. It may be *total*, when the Moon completely covers the Sun; *partial*, when the Moon partially covers the Sun; or *annular*, when the lunar disk is completely within the solar disk.

ECLIPTIC. The apparent path of the Sun among the stars, i.e. the intersection of the plane of the Earth's orbit about the Sun and the celestial sphere.

ECLIPTIC, OBLIQUITY OF. The angle between the plane of the ecliptic and that of the celestial equator.

ECONOMIZER. In a steam-engine: a heat exchanger in which the waste heat of the flue gases is used to heat the feed water before it enters the boiler.

EDDINGTON THEORY. A theory based on the idea that, in describing a physical system, it is essential to take account of the observer, the nature of the observations, and the processes of measurement, as in relativity and quantum theory. Eddington claimed, with considerable success, that the values of many dimensionless constants of importance in physics (e.g. the proton–electron mass ratio and the fine structure constant) could be worked out from this theory.

EDDINGTON TRANSFER EQUATION. A differential equation describing the transfer of radiative energy in the atmosphere of a star, including effects due to scattering and pure absorption.

EDDY. A macroscopic element of fluid possessing rotational motion.

EDDY CONDUCTIVITY. The apparent thermal conductivity of a fluid in the presence of eddies.

EDDY CURRENT. A local current induced in a conducting medium subjected to a varying magnetic field. Such a current results in energy dissipation known as *eddy current loss*, and a reduction in apparent magnetic permeability. Also known as *Foucault current*.

EDDY CURRENT LAG. The lag in the achievement of the new equilibrium value of the induction when the magnetic flux is changed. With a.c. the induction (and hence the eddy current) decreases from the surface to the centre of the material.

EDDY DIFFUSIVITY. The increase in thermal diffusivity arising from eddy formation in a medium.

109

EDDY VISCOSITY. The apparent viscosity of an incompressible fluid when it is undergoing turbulent motion.

EDGE DISLOCATION. See: *Dislocation.*

EDGE TONE. The tone produced when a sharp edge is placed so as to split an air jet of sufficient speed. It arises from the system of vortices which surround the jet, and which follow each other at regular intervals when the edge is introduced, thereby giving a steady note. The intensity may be increased by resonance effects, as in an organ pipe.

EDGE WAVE. (1) A form of wave generated when an atmospheric disturbance passes over a continental shelf or sloping sea bottom or when an outside wave disturbance is in resonance with a bay or inlet. (2) Another name for *boundary wave.* See also: *Boundary wave.*

EDISON EQUATION. See: *Richardson equation: Richardson–Dushman equation.*

EDSER–BUTLER BANDS. See: *Spectrum, channelled.*

EFFLORESCENCE. The loss of water to the atmosphere by a chemical salt. The opposite of deliquescence.

EFFUSER. A device to convert pressure into velocity, e.g. that part of a wind tunnel, upstream of the working section, in which the air or gas is accelerated.

EFFUSION. The molecular flow of gases through small apertures.

EHRENFEST THEOREM. States that the motion of a quantum-mechanical wave packet will be identical to that of the classical particle it represents if any potentials acting upon it do not change appreciably over the dimensions of the packet.

EHRENHAFT EFFECT: MAGNETOPHOTO-PHORESIS. The (usually) helical movement of fine dust particles in a gas along the lines of force of a magnetic field during their irradiation by light. It is one type of *photophoresis.* See also: *Photophoresis.*

EIGENFUNCTION. Of a differential equation: a solution satisfying specified boundary conditions and corresponding to a particular eigenvalue. Also known as an *eigensolution.* See also: *Schrödinger equation.*

EIGENVALUE. Of a differential equation: one of the special values of a given parameter for which a non-trivial solution (an eigenfunction) exists. See also: *Schrödinger equation.*

EINSTEIN CHARACTERISTIC TEMPERA-TURE. A temperature θ_E, characteristic of each substance, appearing in Einstein's theory of specific heats and given by $\theta_E = h\nu/k$, where h is Planck's constant, ν is the frequency of thermal vibration of the atoms of the substance, and k is Boltzmann's constant.

EINSTEIN–DE HAAS EFFECT: ROTATION BY MAGNETIZATION. The axial rotation of a rod of magnetic material experienced when a magnetic field is applied parallel to the axis. It is of importance in the determination of the gyromagnetic ratio. Sometimes known as the *Richardson effect.* See also: *Barnett effect: Magnetization by rotation.*

EINSTEIN DIFFUSION EQUATION. An equation originally worked out to express the diffusion of atoms in a gas or liquid as a direct consequence of Brownian movement. It is now an important starting point in most theoretical discussions on diffusion in solids, where the Brownian movement of an atom or ion in a crystalline solid may be considered as a succession of jumps from one well-defined lattice site to another. The equation may be written $D = \overline{X^2}/2t$, where D is the diffusion coefficient and X is the net displacement of an atom in time t. See also: *Fick diffusion laws.*

EINSTEIN DISPLACEMENT. The displacement of a ray of light from a star when passing near to the Sun's limb as seen by an observer on the Earth. It is predicted by the general theory of relativity, but there is still some discussion regarding the explanation of such displacements as have been actually observed.

EINSTEIN EQUATION FOR PHOTOELECTRIC EMISSION. An equation giving the kinetic energy of a photoelectron in terms of the quantum energy of the incident photons. It represents one of the earliest successes of the quantum theory, and may be written as $E = h\nu - W$, where E is the kinetic energy in question, h is Planck's constant, ν is the frequency of the incident light, and W is the excess of energy needed by the electron to escape.

EINSTEIN FORMULA IN RADIATION THE-ORY. States that the transition of a system from a higher to a lower quantum state is accompanied by the emission of radiation of frequency $(E_m - E_n)/h$, where E_m and E_n are the two energy levels involved and h is Planck's constant.

EINSTEIN LAW OF PHOTOCHEMICAL EQUIVALENCE. States that there is a simple integral relationship between the primary photochemical yield and the number of light quanta absorbed. The yield may be greater than unity if secondary processes arise, or less when the recombination of products occurs.

EINSTEIN PROBABILITY COEFFICIENTS. Three coefficients introduced by Einstein in his derivation of Planck's radiation formula to take account respectively of spontaneous transitions of an atom from a higher to a lower energy level, induced transitions of an atom in either direction, and the acquisition of energy by absorption.

EINSTEIN THEORY OF SPECIFIC HEATS. A theory in which a solid is regarded as an assembly of bound atoms each of which vibrates independently as a three-dimensional harmonic oscillator, all the atoms vibrating with the same frequency. The specific heat at constant volume C_v is given by

$$C_v = 3Ry^2e^y(e^y - 1)^{-2},$$

where R is the gas constant and $y = \theta_E/T$, θ_E being the Einstein characteristic temperature, and T the absolute temperature. See also: *Debye theory of specific heats. Nernst–Lindemann theory of Specific heats. Specific heat theories.*

ELASTIC AFTER-EFFECT. The time lag in the disappearance of strain following the removal of a previously sustained stress within the elastic limit. It is one of several interrelated phenomena which arise from the operation of a relaxation process.

ELASTIC ANISOTROPY. The variation of the elastic properties of a substance with direction.

ELASTIC CONSTANTS. Of a crystal: the coefficients of the relations by which the components of the elastic strain are expressed as linear functions of the stress components. In general there are twenty-one such coefficients, but the number may be reduced on account of crystal symmetry. When the stresses are written in terms of strains the constants are known as *elastic stiffness constants*. When the strains are written in terms of the stresses another set is obtained, known as the *compliance constants*.

ELASTICITY. The property whereby a body, when deformed by an applied load, recovers its previous configuration when the load is removed. According to *Hooke's law* the strain is proportional to the stress within the elastic limit.

ELASTICITY, ELLIPSOID OF. See: *Dielectric ellipsoid.*

ELASTIC LIMIT. The stress above which a body ceases to deform elastically and at which plastic deformation sets in.

ELASTIC LIQUID. A viscoelastic (or elastico-viscous) liquid, i.e. one which has a "memory" of past deformation and has the ability to store energy when sheared. See also: *Viscoelastic substance.*

ELASTIC MODULUS. For an elastic material: the ratio of the stress to the resulting strain. According as the stress produces elongation, shear, or compression, we speak of the *Young's modulus, shear modulus* (or *modulus of rigidity*), or *bulk modulus* (or *compression modulus*) respectively. Similarly, the *torsional modulus* refers to twisting and the *flexural rigidity* to bending. See also: *Bulk modulus: Compression modulus. Flexural rigidity. Poisson ratio. Stiffness. Torsional modulus.*

ELASTIC WAVES. Any form of wave motion by which energy is transmitted through a body by virtue of the elastic properties of that body. In general there are two types of elastic wave: compressional (or longitudinal) and shear (or transverse). See also: *Compressional wave. Love wave. Rayleigh wave. Shear wave. Surface wave.*

ELASTODYNAMICS. The study of the mechanical properties of elastic waves.

ELASTOMER. A material, usually synthetic, having elastic properties resembling those of rubber.

E-LAYER: HEAVISIDE LAYER: KENNELLY–HEAVISIDE LAYER. The lower of the two main strata or "layers" in the ionized atmosphere, the other being the Appleton or *F*-layer. It occurs at a height of about 100 km. See also: *Ionosphere.*

ELECTRET. A dielectric body that retains an electric moment after the externally applied field has been reduced to zero. It is the electrical analogy of a permanent magnet.

ELECTRET TRANSDUCER. An electrostatic transducer in which the bias is supplied by a semi-permanent electrostatic charge embedded in a dielectric, i.e. an electret. Some applications are to microphones, telephones, pickups etc.

ELECTRICAL ATTENUATION. Of an electrical network: the reduction in output power relative to input. It is usually given in terms of nepers or decibels. Note: 1 neper = 8·686 decibels.

ELECTRICAL ATTENUATOR. An arrangement for introducing a known attenuation into a given circuit. It may take the form of a series of resistors in an artificial line or of coils (or capacitor plates) of which the mutual inductance (or capacitance) can be adjusted. See also: *Piston attenuator.*

ELECTRICAL BATTERY. Two or more electrical cells connected and used as a single unit, as in an accumulator. When it is connected to a load of substantially constant voltage which is just sufficient to keep the battery in a fully charged condition, it is known as a *floating battery*.

ELECTRICAL CABLE. An electrical conductor surrounded by a dielectric and a sheath, or an assembly of conductors so surrounded, used for the transmission of electric power or for communication.

ELECTRICAL CABLE, ARMOURED. A cable which is protected from mechanical damage by one or more spirally wound layers of metallic (usually steel) strip or wire.

ELECTRICAL CABLE, COAXIAL. A cable in which a central cylindrical conducting core is surrounded by a hollow conducting sheath and separated from it by a solid dielectric or inert gas.

ELECTRICAL CABLE, GAS-FILLED. A cable which is filled, inside the sheath, with an inert gas under pressure so as to reduce the effects of voids or spaces of low dielectric strength which may appear in service and limit the working voltage.

ELECTRICAL CABLE, GRADING OF. The use of layers of insulation of different permittivities so as to make the dielectric strength more nearly uniform than that obtained with a homogeneous dielectric.

ELECTRICAL CIRCUIT ANALYSIS: NETWORK ANALYSIS. The use of mathematical techniques to predict the response of an electrical system to a given stimulus. By analogy the behaviour of non-electrical systems (e.g. acoustic, hydraulic, magnetic, mechanical) can also be treated by electrical circuit analysis. See also: *Electrical network*.

ELECTRICAL CIRCUIT, EQUIVALENT. A representation of an actual electrical network, obtained by replacing the network components by fictitious equivalents which, for the particular problem under investigation, express the relevant properties of the network to the desired degree of accuracy.

ELECTRICAL CIRCUIT, THREE-PHASE. A combination of three single-phase circuits in which the three applied voltages successively differ in phase from each other by one-third of a cycle. See also: *Delta connection*.

ELECTRICAL CUT-OUT. See: *Electrical protective gear or cut-out*.

ELECTRICAL CYCLE. One complete set of instantaneous values of an electrical quantity in an indefinite repetition of events. The time of such a cycle is the time between corresponding points in, for example, a voltage or current waveform.

ELECTRICAL DIVERTER. A resistor shunted across one part of a circuit. It is used, for example, across the field coils of d.c. series motors to control the speed.

ELECTRICAL DOUBLE LAYER. At a solid–liquid interface in a colloid system: a composite double layer in the liquid in contact with the solid surface. The inner layer is a bound layer of ions firmly fixed to the surface. The outer layer is diffuse and consists of ions of the opposite sign (*gegenions*) which are subject to thermal agitation. There is a potential difference across the diffuse layer between the bulk of the solution and the bound layer, known as the *electrokinetic potential* or *zeta potential*.

ELECTRICAL ENGINEERING. The practical application of the principles of electricity and magnetism. It is concerned mainly with power or communication.

ELECTRICAL IMAGE. An imaginary charge or dipole introduced to facilitate the solution of problems concerning the evaluation of potential distribution. See also: *Electrostatic image. Fluid flow, image method for*.

ELECTRICAL INVERTER. Any device for converting direct current to alternating current. It may be a rotary machine, a grid-controlled mercury-arc rectifier, or a circuit employing thermionic tubes.

ELECTRICAL LOAD. The component to which the output of an electrical system or network is delivered. In the generalized representation of a system as source–network–sink, the load is the sink. See also: *Balanced load*.

ELECTRICAL LOAD, SYMMETRICAL. A *balanced load*.

ELECTRICAL NETWORK. A combination of electrical elements connected in any way—inductively, conductively, or capacitatively.

ELECTRICAL NETWORK, ACTIVE. A network which contains one or more sources of electromotive force. These sources may arise from a generator or be reactive e.m.f.s due to the components.

ELECTRICAL NOISE. Unwanted electrical effects present in a transmission system or measuring device. See also: *Johnson noise. Shot noise: Shot effect. Temperature fluctuation noise*.

ELECTRICAL POTENTIAL. At a point in an electrostatic field: the work done in moving unit positive charge from infinity (potential zero) to that point. The potential is independent of the path chosen and is a scalar quantity. See also: *Electromotive force*.

ELECTRICAL POWER. The rate at which electrical energy is transformed or delivered. The instantaneous power in an electrical circuit is the product of the instantaneous current through the circuit and the instantaneous voltage across its terminals. See also: *Power factor.*

ELECTRICAL PROTECTIVE GEAR OR CUT-OUT. A relay or group of relays and accessories intended, in case of fault or abnormal conditions, either to isolate a zone of an electrical installation or to actuate a warning signal. The simplest form is the ordinary domestic fuse.

ELECTRICAL SCREENING. The reduction in the interaction between two electrical circuits or circuit elements, achieved, for example, by enclosing circuit elements within closed, earthed, conducting surfaces.

ELECTRIC CURRENT. See: *Current.*

ELECTRIC FIELD STRENGTH. The force experienced by a stationary unit positive charge placed in an electromagnetic field. In MKSA and SI units it is expressed in newtons per coulomb. The electric field strength is sometimes called *electric intensity.*

ELECTRIC FLUX. At a surface in an electrostatic field: the quantity of electricity displaced across the surface and normal to it. The unit is the Coulomb. Also known as the *flux of displacement.*

ELECTRIC FLUX DENSITY. The electric flux per unit area. It is the same as the *displacement*. It is expressed in Coulombs per unit area (cm^2 or m^2).

ELECTRIC INDUCTION. See: *Displacement.*

ELECTRIC INTENSITY. See: *Electric field strength.*

ELECTRICITY. The manifestation of the existence of electric charge, ultimately derived from the ionization of atoms. According as it is concerned with stationary charges or moving charges it is known as *static* or *current* electricity.

ELECTRIC MOMENT. The electric dipole moment. See also: *Dipole moment.*

ELECTRIC MOTOR. A machine which converts electrical energy into mechanical energy.

ELECTRIC STRENGTH, DISRUPTIVE. Another term for the *dielectric strength*, i.e. the maximum potential gradient which a dielectric or insulator can stand without breaking down, under specified conditions.

ELECTRIC STRENGTH, INTRINSIC. The (very high) electric strength attained in a medium at low temperatures where breakdown is believed to arise from cumulative ionization caused by the passage of electrons through the medium. The electrons in the conduction band must gain energy from the field if such a breakdown is to occur.

ELECTRINO. A hypothetical particle, similar to the neutrino, with small or zero rest mass, spin $\pm\frac{1}{2}$, but having a negative charge. It has never been observed.

ELECTROANALYSIS. Analysis by any technique which involves the measurement of an electrical property (e.g. current or potential) of an electrolytic cell.

ELECTROCAPILLARITY. The change in the surface tension at an interface between two phases when an electrical field is imposed. The term refers especially to the change in the position of the meniscus dividing two phases in a capillary tube when a voltage is applied across them.

ELECTROCAPILLARITY CURVE. The curve obtained by plotting surface tension, using a dropping mercury electrode, against the applied potential on the electrode.

ELECTROCARDIOGRAPHY (ECG). The graphical recording of the electrical activity of the heart muscle by means of electrodes attached to the wrists and one ankle. See also: *Bioelectric potentials.*

ELECTROCHEMICAL CELL. A device for the direct conversion of the chemical free energy of a reaction involving electron transfer into electrical energy. Sometimes known as *galvanic cell*. See also: *Accumulator. Concentration cell. Daniell cell. Fuel cell. Primary cell. Secondary cell. Standard cell.*

ELECTROCHEMICAL EQUIVALENT. Of a substance: the mass of that substance liberated by electrolysis by one coulomb of electricity.

ELECTROCHEMICAL MACHINING. The removal of a selected portion of the metal of an electrically conductive work-piece to a desired tolerance by electrochemical action in a suitable electrically conductive solution. It may be thought of as the opposite of electrodeposition.

ELECTROCHEMICAL SERIES. A list of elements arranged in such an order that a given element is negative to any element preceding it in the list, and positive to any element succeeding it, when both are immersed in a suitable solute. An element is said to be *electropositive* or *electronegative* according to its position in the series.

113

ELECTROCHEMISTRY. The study of the behaviour and properties of electrolytes and ions in solution or in the fused state. It includes the effects of the passage of current and the generation of electricity by chemical means.

ELECTRODE. A conductor (commonly solid) by means of which current is passed into or taken out of an electrical system.

ELECTRODE, ANTIMONY. An electrode of antimony, coated with antimonous oxide, used in industry in the measurement of pH for continuous process control.

ELECTRODE, CALOMEL. A reference electrode for the comparison of potentials. It consists of a mercury electrode in a chloride solution of specified concentration which is itself saturated with calomel (mercurous chloride). See also: *Standard electrode.*

ELECTRODELESS-DISCHARGE. A discharge produced by placing a discharge tube in an intense high-frequency electromagnetic field. This device is used for exciting the *spectra* of gases and vapours; and as an *ion source.* In the latter case the discharge is excited by two radio frequency external ring electrodes or by a radio frequency coil surrounding the discharge tube.

ELECTRODEPOSITION. The deposition of a layer of one metal upon another by making one the anode and the other the cathode in an electrolytic cell containing a solution of a salt of the metal to be deposited. See also: *Faraday laws of electrolysis.*

ELECTRODIALYSIS. The removal of electrolytes from a colloidal solution by the application of an e.m.f. across a dialysing membrane. The solution concerned is contained between two such membranes with pure water on either side, and the e.m.f. is applied to electrodes inserted in the water compartments.

ELECTRODYNAMIC INSTRUMENT. An *electrodynamometer,* i.e. an electrical measuring instrument depending for its action on the e.m.f. between two or more current-carrying coils. See also: *Dynamometer.*

ELECTRODYNAMICS. The study of the forces exerted on a conductor carrying a current by a current flowing in another conductor, i.e. of the magnetic effect of one current on another.

ELECTROENCEPHALOGRAPHY (EEG). The graphical recording of the electrical activity of the brain by means of electrodes attached to the scalp. See also: *Bioelectric potentials. Encephalography.*

ELECTROFORMING. The electrolytic deposition of metal on a conducting mould (usually graphite-coated wax) which can subsequently be removed by melting. The process is used, for example, for accurate reproduction or the manufacture of precision tubing.

ELECTROGASDYNAMIC GENERATION OF ELECTRICITY. The generation of electricity by a high-speed hot gas which removes charge (produced, for example, by a corona discharge) from one electrode and deposits it at another. These electrodes take the form, typically, of circular grids. See also: *Direct conversion of heat to electricity.*

ELECTROGRAVIMETRIC ANALYSIS. The quantitative analysis of elements from aqueous solution by electrodeposition, using appropriate solutions and controlled cathode potentials, on the basis of Faraday's laws.

ELECTROHYDRAULIC CRUSHING. The use of repeated underwater electric sparks to cause the breakage of adjacent brittle materials.

ELECTROHYDRAULIC FORMING. The explosive working of metals by the use of underwater electric sparks to generate the explosions.

ELECTROKINETIC PHENOMENA. Denotes the interrelated series of effects having to do with the electrical behaviour of solid–liquid interfaces. The term embraces colloid phenomena (solid phase mobile), on the one hand, and capillary phenomena (liquid phase mobile), on the other.

ELECTROKINETIC POTENTIAL: ZETA POTENTIAL. At a solid–liquid interface in a colloid system: the potential existing across the diffuse portion of the electrical double layer. See also: *Electrical double layer.*

ELECTROLESS DEPOSITION. Electro-deposition which is achieved in the absence of an electric current, the necessary ions for the formation of a metallic film being supplied, for an appropriate metal, by a suitable choice of electrolyte.

ELECTROLUMINESCENT PANEL. A panel, incorporating a phosphor, which provides a uniform illumination all over its surface. By a suitable choice of phosphor the panel may be made to retain its glow for some tens of minutes after the stimulus (e.g. light or X-rays) has been removed. Such an *image-retaining panel* may, therefore, be used with X-rays to examine a radiographic image after the X-rays have been switched off.

ELECTROLYSIS. In general, signifies the passage of an electric current through a liquid containing charged particles or ions. The current flows in both directions and is thus fundamentally different from that in a metallic conductor. The term may, however, also cover phenomena accompanying the passage of current, such as ionic migration, ionic diffusion at an electrode surface, and various electrokinetic phenomena. The most important aspect of electrolysis is that concerning chemical changes at the electrodes, which form the basis of electrochemistry.

ELECTROLYTE. (1) A substance whose solution conducts electricity by the movement of positive and negative ions. (2) The solution itself.

ELECTROLYTE, POLARIZATION OF. See: *Polarization.*

ELECTROLYTIC CONDUCTION. See: *Ionic conduction.*

ELECTROLYTIC DEPOSITION. See: *Electrodeposition.*

ELECTROLYTIC POLISHING. The process of producing a smooth lustrous surface on a metal by making the metal the anode in an electrolytic solution. The protuberances are dissolved away preferentially.

ELECTROLYTIC REDUCTION. The extraction of metals from solution or from fused salts by electrolysis, usually for industrial purposes.

ELECTROLYTIC REFINING. The removal of impurities from metallic material by electrolysis.

ELECTROLYTIC TANK. A shallow tank of electrolyte used for solving various problems for which the tank may be made to give an electrical analogy. It is of constant depth and of geometrical shape appropriate to the problem being investigated. The problems concerned are those which obey the Laplace equation $\nabla^2 \Phi = 0$, where Φ is a potential function; and the physical quantities that may be treated in this way include electric potential in electrostatics, magnetic potential in magnetism, temperature in heat flow, concentration in diffusion, and velocity potential in the irrotational flow of incompressible fluids, among others. The electrolytic tank can also be used for the solution of mathematical problems involving functions of a complex variable. The tank is then a direct simulation of the complex plane.

ELECTROMAGNET. A magnet whose field is produced by an electric current, and which is, to a large extent, demagnetized when the current is cut off. The current flows through a helical coil which often has a core of soft iron or steel to reduce the reluctance and secure a more intense magnetic field for a given current.

ELECTROMAGNETIC FIELD. A field representing the joint interplay of electric and magnetic forces. In free space the relationships in such a field, between the magnetic field strength, magnetic induction, electric field strength, electric induction, electric charge density, and electric current density, are expressed by Maxwell's equations. The direction and magnitude of the energy flow are given by Poynting's vector. According to electromagnetic field theory an electric field is set up only by electric charges and a magnetic field arises only from electric charges in motion. The nature of the field observed, however, depends on the relative motion of the observer, and what appears as a magnetic field to one observer may appear as an electric field to another, or vice versa. See also: *Maxwell equations. Polarization vectors. Poynting theorem.*

ELECTROMAGNETIC GUN. An alternative name for the tubular type of linear motor. See: *Linear motor.*

ELECTROMAGNETIC INDUCTION. The setting up of an electric field by reason of the variation in magnetic flux density with time. Any current so induced is in such a direction as to oppose the change in magnetic flux. See also: *Faraday law of induction.*

ELECTROMAGNETIC MASS. Of a charged particle: the mass arising from the motion of the particle. It corresponds to the difference between the total energy of the electromagnetic field associated with the moving particle and that of the purely electrostatic field of a stationary charge. The latter corresponds to the *rest mass* of the particle.

ELECTROMAGNETIC RADIATION: ELECTROMAGNETIC WAVES. Waves characterized by variations of electric and magnetic fields and propagated through free space with the speed of light. It was shown by Maxwell that the electromagnetic field vectors can propagate as waves of velocity $1/\sqrt{\mu\varepsilon}$, where μ is the permeability and ε the permittivity, and his predictions have been observed at wavelengths ranging from a few kilometres (radio waves) to 10^{-9} cm or less (X- or γ-rays). See also: *Maxwell equations. Slow waves.*

ELECTROMAGNETIC RADIATION, POLARIZATION OF. See: *Polarization.*

ELECTROMAGNETIC SPECTRUM. The complete range of electromagnetic waves from the longest radio waves, through microwaves, infrared, visible, and ultraviolet light, down to the shortest X- and γ-rays, i.e. from several kilometres down to 10^{-12} m or less.

ELECTROMAGNETISM. Denotes all those phenomena concerned with the interaction of electric and magnetic fields.

ELECTROMERISM: ELECTROMERIC EFFECT. The change in the electronic configuration of a molecule as a result of conjugation.

ELECTROMETER. An instrument for detecting or measuring potential difference or electric charge. The earlier types measured the mechanical forces existing between electrically charged bodies, but electronic instruments are now also in service. See also: *Electroscope. Voltmeter, attracted disk.*

ELECTROMETER, ATTRACTED DISK. See: *Voltmeter, attracted disk.*

ELECTROMETER, CAPILLARY: ELECTRO-METER, LIPPMAN. An electrometer in which small changes in potential difference are measured from the changes in level, in a capillary tube, of the surface of separation between a mercury surface and an electrolyte. It is based on the fact that the surface tension at such a surface depends on the potential difference acting across that surface.

ELECTROMETER, COMPTON. A type of quadrant electrometer, having a controlled variable sensitivity.

ELECTROMETER, DOLEZALEK. A very sensitive form of quadrant electrometer.

ELECTROMETER, HOFFMANN. A development of the quadrant electrometer, having two sections instead of four.

ELECTROMETER, LINDEMANN. A form of quadrant electrometer which is insensitive to changes in level and which employs a needle as vane, whose movement is observed through a microscope for measurement.

ELECTROMETER, QUADRANT. The most commonly used type of electrometer. It consists of a shallow cylindrical metal box cut into four quadrants, inside which a metal vane is suspended by a vertical wire or conducting fibre. In the *idiostatic method* of operation one pair of (opposite) quadrants, and the suspended vane, are connected to the potential to be measured, the other pair of quadrants being earthed. The vane turns towards this earthed pair against the torsion of the wire, the deflection of which is observed by the reflection of a beam of light from a mirror mounted on the wire. This deflection is proportional to the square of the potential to be measured. In the *heterostatic method* the vane is given a constant potential (e.g. 100 V), which is large compared with that to be measured. The latter is applied between the two pairs of quadrants. The deflection is now directly proportional to the potential being measured.

ELECTROMETER, STRING. An electrometer in which a very fine conducting fibre is held under tension between two oppositely charged plates. When the fibre itself is charged it shifts towards the oppositely charged plate and its motion is viewed through a microscope with calibrated eyepiece.

ELECTROMETER, VALVE (OR TUBE). An electrometer, employing a triode valve, which operates by the d.c. amplification and measurement of an ionization current.

ELECTROMETER, VIBRATING REED. An electrometer in which a small charge is measured by the use of a variable capacitor, one plate of which takes the form of a vibrating reed which produces an a.c. voltage capable of amplification by a system similar to that used in a chopper amplifier. Such an electrometer may also be used to measure a small ionization current.

ELECTROMETER, WULF. A form of string electrometer in which the oppositely charged electrodes take the form of knife edges with adjustable separation. In the *unifilar* version a single metallized quartz fibre or Wollaston wire is used and its deflection observed. In the *bifilar* version two such fibres or wires are used and their separation measured, the knife edges being uncharged.

ELECTROMOTIVE FORCE (E.M.F.). The electrical force produced by the conversion of some form of energy into electrical energy. The conversion processes which can generate an e.m.f. include the conversion of mechanical energy (as in a dynamo), of heat energy (as in a thermocouple), of magnetic field energy (as in an inductance), and of chemical energy (as in a battery). Electromotive force is to be distinguished from *potential difference*, which is a general term signifying a difference in potential energy, or of voltage, which exists between two points in a circuit or between two electrically charged bodies. See also: *Electrical potential.*

ELECTROMYOGRAPHY (EMG). The graphical recording of the electric currents and action potentials generated in a muscle during its contraction, by means of concentric needle electrodes inserted into the muscle.

ELECTRON. A stable elementary particle having an electric charge of $\pm 1 \cdot 6022 \times 10^{-19}$ C and a rest mass of $9 \cdot 1095 \times 10^{-31}$ kg. When used without specification the term means the negatively charged electron (also called a *negatron* or *negaton*). Its antiparticle is the positively charged electron (*positron* or *positon*). See: *Appendix III.*

ELECTRON ABSORPTION. See: *Absorption of radiation.*

ELECTRON ABSORPTION COEFFICIENT. See: *Absorption coefficient for radiation.*

ELECTRON AFFINITY. (1) The degree of electronegativity of an atom, i.e. the extent to which the atom tends to capture electrons. It is indicated by the position in the electrochemical series. (2) The work required to remove an electron to infinity from a negative ion.

ELECTRON ATTACHMENT. The capture of an electron by an atom or molecule to form a negative ion. The probability of the radiative attachment of electrons is very low, but if the excess energy is removed by a third body or by vibrational excitation or dissociation in the case of a molecule, the *attachment probability* is appreciably increased.

ELECTRON BEAM, GLOW-DISCHARGE: GLOW-DISCHARGE ELECTRON GUN. An electron beam produced by the bombardment of energetic ions in an atmosphere of a suitable gas at a reduced pressure, usually between 20 and 200 millitorr (Note: 1 torr = 133·3 Pa). The ions are produced by a glow discharge between two flat or nearly flat electrodes in the gas, the anode having a hole in the centre and the cathode (which may rotate) serving as a target for the ions. The electrons are accelerated away from the cathode to form the electron beam. The applications of glow-discharge electron guns include single-shot and general purpose welding, the processing of thin coatings or films, vapour deposition and various types of localized heat treatment.

ELECTRON-BEAM MACHINING. The use of electron beams in machining operations. The equipment is similar to that used in electron-beam melting, and is used to carry out such operations as the drilling of holes, the cutting of slots and profiles, and milling in general. The technique is applicable to ceramics as well as to metals.

ELECTRON-BEAM MELTING. The application of electron beams on the industrial scale to such operations as the refining, casting and welding of refractory metals, e.g. tungsten, molybdenum, and zirconium.

ELECTRON-BEAM THERAPY. The treatment of deep-seated malignant tumours by electron radiation. The usual source is radium B or mesothorium.

ELECTRON-BEAM TUBE. An electron tube, the performance of which depends on the formation and control of one or more electron beams.

ELECTRON CAPTURE. (1) Electron attachment. (2) The capture of an orbital electron (usually a *K*-electron) by an atomic nucleus. The vacancy is filled by an electron from an outer shell with the emission of fluorescence X-rays or Auger electrons. See also: *Orbital electron capture.*

ELECTRON CLOUD. Refers to an interpretation of the Schrödinger wave function, ψ, for an atomic electron whereby it is supposed that the electron is spread out in the form of a cloud, the density of which at a given point at a given time is proportional to $|\psi|^2$ at that point. The function $|\psi|^2$ is often called the *probability density function* of the electron cloud. See also: *Schrödinger equation.*

ELECTRON COMPOUNDS. See: *Alloy.*

ELECTRON CONDUCTION. Electrical conduction associated with the movement of electrons as opposed to that of holes or ions. See also: *Acceptor. Donor. Electron hole: Positive hole.*

ELECTRON CONDUCTION OF HEAT. One of the two modes of heat conduction in a metal. The total heat conduction may be considered to be produced by two carriers acting in parallel: electrons and phonons, the latter giving rise to lattice conduction. See also: *Lattice conduction of heat.*

ELECTRON CONFIGURATION. Of an atom or molecule: the distribution of electrons within the atom or molecule. The electron configuration is described in terms of a series of atomic or molecular orbitals arranged in order of increasing energy and classified in terms of atomic or molecular symmetry.

ELECTRON CURRENT. In a given direction: the number of electrons per second crossing unit area at right angles to the direction.

ELECTRON-DEFICIENT MOLECULE. A covalent molecule having fewer valency electrons than are necessary to provide a two-electron covalent bond between each pair of atoms.

ELECTRON DENSITY. (1) In general: the number of electrons per unit mass. (2) In astrophysics: the number of electrons per unit volume.

ELECTRON DETACHMENT. The detachment of an electron from a negative ion to form a neutral atom or molecule. See also: *Electron attachment.*

ELECTRON DIFFRACTION. Interference which occurs when electrons are scattered by different elements (including magnetic elements) of a solid, liquid or gaseous medium. It arises from the wave nature of electrons. See also: *De Broglie waves. Diffraction. Dynamical theory of electron and X-ray diffraction. Neutron diffraction. X-ray diffraction.*

ELECTRON DIFFRACTION, LOW ENERGY (LEED). Electron diffraction at energies of 0·3–1 keV, used for the study of surface structure, e.g. the effects of gas adsorption in catalysis.

ELECTRONEGATIVE. See: *Electrochemical series.*

ELECTRON ENERGY BAND. One of the bands or zones of electron energy levels in a crystal by which the total range of energy states is broken up into alternate allowed and forbidden ranges. The extent of these *allowed* and *forbidden bands* is determined by the nature and structure of the crystal. A *filled band* is an allowed band in which all the energy levels are occupied. An *empty band* is one in which none are occupied. See also: *Band theory of solids. Bloch theory. Brillouin zones. Collective-electron theory. Conduction band. Ferromagnetism, band theory of.*

ELECTRON ENERGY LEVEL. (1) One of the stationary electron energy states in an isolated atom. (2) One of the energy levels in an electron energy band, which band may sometimes be conveniently pictured as arising from the broadening of the energy levels referred to in (1) when the individual atoms are brought into close proximity. See also: *Atomic energy levels. Electron shell.*

ELECTRON GAS. A concentration of electrons whose behaviour is primarily or significantly determined by the interactions between the electrons themselves. The conduction electrons in a metal are themselves sometimes considered to constitute an electron gas.

ELECTRON GUN. An assembly of electrodes which produces an electron beam for use in such equipment as a cathode-ray tube or electron microscope. See also: *Electron beam, glow-discharge. Field emission gun.*

ELECTRON HOLE: POSITIVE HOLE. A vacancy in a previously filled energy band in a solid, caused by the excitation of an electron from that band to one of higher energy. Under an applied field such vacancies are free to move and thus to contribute to the conduction. This type of conduction is known as *hole conduction.* See also: *Acceptor. Semiconductor.*

ELECTRON LENS. A device for focusing an electron beam. It may be *magnetic* (having coils or permanent magnets) or *electrostatic*, or take the form of an aperture in a *diaphragm*. The aberrations of an electron lens are, in general, similar to those of an optical lens—astigmatism, coma, chromatic aberration, spherical aberration, etc. The *aperture* and *cardinal points* of an electron lens are also defined in the same way as for an optical system. See also: *Electrostatic lens. Magnetic lens.*

ELECTRON LENS, IMMERSION. An electron lens in which the object is deeply immersed in the field so that the refractive index varies rapidly in its neighbourhood. The object is usually the cathode, and the purpose of the lens is usually to project an electron optical image of the cathode surface, as in an emission electron microscope or image converter. The significance of the term is thus different from that employed in the context of optical lenses. See also: *Immersion lens, optical.*

ELECTRON MASS. The rest mass of an electron, equal to $9\cdot1095 \times 10^{-31}$ kg. See also: *Electron. Appendix III.*

ELECTRON MICRO-PROBE ANALYSIS. See: *Electron-probe microanalysis.*

ELECTRON MICROSCOPE. An instrument in which a beam of electrons is used to form a magnified image in a manner somewhat analogous to that in a light microscope. The short wavelength associated with electron waves gives, however, a much greater resolution than is attainable with the latter. All electron microscopes have in common a vacuum chamber, some electron-optical system to form the image, and an image detecting or recording system. See also: Following entries.

ELECTRON MICROSCOPE, EMISSION. An electron microscope in which the source of the electron beam is the specimen itself. The first electron-optical element is an electrostatic immersion lens. See also: *Electron lens, immersion.*

ELECTRON MICROSCOPE, HIGH RESOLUTION. An electron microscope in which special attention is paid to the elimination of lens aberrations and to the maintenance of a high degree of stability in the lens currents and the accelerating voltage. The resolution attainable is about 2 Å ($1 \text{ Å} = 10^{-10}$ m) and useful magnifications of 500 000 are possible.

ELECTRON MICROSCOPE, MIRROR. An electron microscope in which a beam of electrons is almost entirely slowed down as it reaches the surface to be examined. The manner in which the electrons are reflected depends on the contact potentials on the surface; and the reflected electrons, after suitable re-acceleration, produce an image of the surface related to the work functions of its constituent parts.

ELECTRON MICROSCOPE, POINT PROJECTION. An electron microscope in which a highly magnified shadow is produced from a real or virtual point source of electrons. See also: *Field emission microscope. Field ion microscope.*

ELECTRON MICROSCOPE, REFLECTION. An electron microscope in which a beam of electrons is reflected from the specimen and observed at an angle between 0 and 45°. The magnification achieved is different in two directions at right angles to each other. The electron-optical system is similar to that of the transmission type instrument. See also: *Electron microscope, transmission.*

ELECTRON MICROSCOPE, SCANNING. An electron microscope in which a fine electron probe scans the surface of the specimen. The secondary electrons are collected, and amplified signals transmitted via a synchronously operated recording system, e.g. a television-type screen.

ELECTRON MICROSCOPE, SCANNING TRANSMISSION (STEM). A transmission electron microscope in which the transmitted electrons are brought back on to the axis of the lens where they can be submitted to energy analysis. All the electrons may be used, or only electrons that have been elastically scattered, or only electrons that have lost a specific amount of energy. See also: *Field emission gun.*

ELECTRON MICROSCOPE, TRANSMISSION. An electron microscope in which a well-collimated electron beam is passed through the specimen, and a highly magnified image is formed on a fluorescent screen or photographic plate. The beam is generated by an electron gun and condenser lens and there are, typically, an objective lens, an intermediate lens and a projector lens in the electron-optical image-producing system. These lenses may be *electromagnetic, electrostatic,* or *magnetostatic* (i.e. having permanent magnets). Appropriate apertures are placed in these lenses to improve the "optical" properties.

ELECTRON MICROSCOPY. The process of making and interpreting highly magnified images produced by electrons. See also: *Replica. Shadow casting.*

ELECTRON MIRROR. A device which, under certain conditions, can reflect back an electron beam and permit, for example, the back-projection of an image. Under other conditions convergent or divergent beams may be produced. A simple type of electron mirror is produced by the electrostatic field in a small gap between two co-linear and co-axial tubes, one of which is at a positive potential, and the other at a negative, with respect to the cathode.

ELECTRON MULTIPLIER: PHOTOMULTIPLIER. A sensitive detector of light in which the initial electron current, derived from photoelectric emission, is amplified by a series of stages of multiplicative secondary electron emission.

ELECTRON NUCLEAR DOUBLE RESONANCE (ENDOR). A combination of electron paramagnetic resonance and nuclear magnetic resonance in which nuclear resonance is detected by the resulting electron paramagnetic resonance. In particular the method is used to observe nuclear resonances which are strongly coupled to the unpaired electrons of paramagnetic centres, in contrast to conventional nuclear magnetic resonance which is usually performed on diamagnetic materials.

ELECTRONOGRAPHY. The acceleration and focusing of photoelectrons to form a high quality electron image with no electron multiplication. The image is directly recorded on a charge-sensitive nuclear emulsion. The image obtained after development is known as an electronograph.

ELECTRON OPTICS. The study of the controlled motion of electrons in free space, and, to some extent, in solids.

ELECTRON-PAIR BOND. See: *Heitler–London theory.*

ELECTRON PARAMAGNETIC RESONANCE (EPR): ELECTRON SPIN RESONANCE (ESR). The resonant absorption of microwaves by a specimen possessing unpaired electrons (and hence a resultant magnetic moment) when the electrons are separated into two or more energy levels by the application of a strong unidirectional magnetic field of appropriate value. It is a branch of microwave spectroscopy. The unpaired electrons may be associated with atomic orbits or with the interatomic binding, and the resultant magnetic moment may be orbital, due to spin only, or intermediate in character. The exact field required for resonance is given by $h\nu/g\beta$, where h is Planck's constant, ν is the microwave frequency, g is the Landé splitting factor (1 for a pure orbital and 2 for a pure spin moment), and β is the Bohr magneton. The electron resonance spectra obtained can be used to give information on, for example, the type of chemical binding, the behaviour of conduction electrons, the values of nuclear spins and moments (when suitable interaction occurs), the symmetry of the crystalline field, the detection (and sometimes identification) of impurities, and the nature of radiation damage (especially biological). See also: *Spectra, radiofrequency.*

ELECTRON PHASE FOCUSING. The process by which a continuous electron stream, moving with a uniform velocity, is passed through a high-frequency field which modulates this velocity so that a variation in electron concentration occurs along the length of the beam, with the formation of maxima and minima. The technique of phase focusing has been applied in high-frequency amplifiers, and ultra-high-frequency oscillators. A well-known example is the klystron.

ELECTRON POLARIZATION. See: *Polarization.*

ELECTRON–POSITRON PAIR. An electron and a positron created simultaneously in an elementary process involving electromagnetic interaction when the energy available exceeds the combined rest energies of the two particles (1·022 MeV). Such a pair may arise, for example, from the interaction with matter of X- or γ-rays, of high energy electrons or of other charged particles; or from electromagnetic transitions in nuclei. (Note: 1 MeV = $1\cdot602 \times 10^{-23}$ J approximately.)

ELECTRON, PRIMARY. In an absorption or scattering process: an incident electron or one which passes through the absorbing or scattering medium unchanged in energy or direction. See also: *Electron, secondary.*

ELECTRON PROBE. A very narrow electron beam, used, for example, for scanning or in electron diffraction.

ELECTRON-PROBE MICROANALYSIS. A technique for the identification and estimation of the component elements in a selected micro-volume at the surface of a solid specimen, by the analysis of the characteristic X-rays emitted from that volume when bombarded by an electron micro-beam. The beam is produced by an electron gun and lens system, and has a diameter of about 1 μm. The X-rays are analysed by crystal spectrometry, pulse-height analysis, or by the use of diffraction gratings. Also known as *electron micro-probe analysis.* See also: *Field ion microscope, atom probe. Nuclear micro-probe analysis. X-ray spectroscopic analysis.*

ELECTRON RADIOGRAPHY. (1) True radiography by photoelectrons. The photoelectrons are generated by a beam of hard X-rays which passes first through a thin sheet (usually of lead) which emits photoelectrons, then through the specimen (which must be very thin) and finally (without appreciable effect) through a fine-grained film, close contact being essential between sheet and specimen, and specimen and film. The resulting radiograph reveals differences in electron absorption. See also: *Radiography.* (2) Autoradiography by electrons emitted from the specimen itself. The electrons are generated by a beam of hard X-rays which passes (without appreciable effect) through a fine-grained photographic film in close contact with the specimen surface. The resulting radiograph reveals differences in electron emitting power.

ELECTRON RADIUS. The value of the radius obtained by equating the electrostatic self-energy of the electron (e^2/r) to its rest mass energy (m_0c^2), on the assumption that the electron is spherical and has the charge distributed uniformly over the surface. Here e is the electronic charge, r the radius of the electron, m_0 its rest mass, and c the velocity of light. The value of the radius so obtained (e^2/m_0c^2) is about $2\cdot82 \times 10^{-15}$ m.

ELECTRON RECOMBINATION. The capture of an electron by a positive ion, with the release of the surplus energy of recombination.

ELECTRON REFRACTION. The refraction suffered by an electron beam in crossing the boundary of two regions which differ in potential. In general the potential changes continuously rather than discontinuously, but it may be simulated by a finite number of discontinuous stages.

ELECTRON REST MASS. The mass of an electron at rest, or that part of the mass of an electron in motion which is associated purely with the electrostatic field of its stationary charge, i.e. that part which is independent of the relativistic mass acquired in motion. It is about $9\cdot11 \times 10^{-31}$ kg. See also: *Electromagnetic mass. Appendix III.*

ELECTRONS, BONDING AND ANTI-BONDING. See: *Orbital, anti-bonding. Orbital, bonding.*

ELECTRON, SECONDARY. In an absorption or scattering process: an electron, other than a primary electron, emerging from the absorbing or scattering medium. See also: *Electron, primary.*

ELECTRONS, EQUIVALENT. Two electrons with the same value of the principal quantum number and the total orbital angular momentum quantum number.

ELECTRON SHELL. A term derived from Bohr's theory of the atom, according to which "planetary" electrons were grouped into a series of shells centred on the nucleus, each shell having a characteristic radius and energy. An electron shell is now taken to signify all those electrons in an atom which have the same principal quantum number. This has the value 1, 2, 3, ..., for K, L, M, ... electrons, respectively, starting at the innermost shell. Each shell may be divided into sub-shells, denoted by s, p, d, f, ..., for which the appropriate orbital angular momentum (or azimuthal) quantum number, l, is 0, 1, 2, 3, ..., respectively. See also: *Atom. Bohr theory. Electron cloud.*

ELECTRONS, K, L, M, ... See: *Electron shell.*

ELECTRON SPIN RESONANCE. See: *Electron paramagnetic resonance (EPR): Electron spin resonance (ESR).*

ELECTRONS, σ, π, δ, ... See: *Quantum number, molecular.*

ELECTRONS, S, P, D, F, ... See: *Electron shell.*

ELECTRON TEMPERATURE. A "statistical" temperature obtained from the Boltzmann equation, $E = \frac{1}{2}kT$ ($E =$ average kinetic energy per degree of freedom, $k =$ Boltzmann's constant, $T =$ absolute temperature), where E refers to the electron energy. The electrons may be in a solid, gas, or plasma, but, since they are not in equilibrium with their surroundings but only (approximately) with themselves, they may have a temperature much higher than, for example, the gas in which they exist. Thus the electron temperature in a fluorescent lamp may be as high as 15000°C. See also: *Energy, kinetic.*

ELECTRON THERAPY. See: *Radiotherapy.*

ELECTRON TRANSFER. The passage of an electron between two constituents of a system. Such a transfer may, for example, be associated with the formation of an ionic bond, or with collision phenomena as in charge exchange.

ELECTRON TRAP. An electron energy level which becomes available by reason of an imperfection in a crystal in which electrons may be trapped. Such traps are of particular importance in luminescence and in semiconductor action.

ELECTRON TUBE: ELECTRON VALVE. A device in which conduction takes place by the passage of electrons or ions between electrodes in a vacuum or a gas.

ELECTRON, UNPAIRED. An electron in a particular shell of an atom to which there is no corresponding electron having the same energy but opposite spin. Such unpaired electrons take part in the formation of covalent bonds, and play a big part in ferromagnetism. See also: *Heitler–London theory.*

ELECTRON, VALENCY. Those electrons in the outer shell of an atom which are involved in chemical combination or certain solid-state processes. See also: *Valency band. Valency: Valence.* Various entries under *Bond.*

ELECTRON VOLT. A unit of energy equal to the change of kinetic energy suffered by an electron when it is accelerated or decelerated through a potential difference of one volt. (Note: 1 eV $= 1\cdot6022 \times 10^{-19}$ J or $1\cdot6022 \times 10^{-12}$ erg.) See: *Appendix II.*

ELECTRON WAVELENGTH. The De Broglie wavelength associated with an electron. For electron energies of 50 or 100 kV the wavelength has about the same value as atomic radii, and beams of such electrons may therefore be diffracted by crystals.

ELECTRON WAVE TUBE. An electron tube in which mutually interacting beams of electrons, travelling with different velocities, cause a signal modulation to change progressively along their length.

ELECTRO-OSMOSIS. The movement of liquid through a porous diaphragm or other permeable solid as a result of an applied electric field. Also known as *electro-endosmosis.* It is the converse of streaming potential.

ELECTROPHIL. A reagent which, in a reaction involving an aromatic compound, preferentially approaches positions at which there is the greatest concentration of π-electrons. See also: *Nucleophil.*

ELECTROPHORESIS: CATAPHORESIS. The migration of charged colloid particles in an electric field. It is the converse of the Dorn effect. One of its principal uses is as an aid to the purification of proteins of various types (enzymes, hormones, viruses, etc.).

ELECTROPHORETIC MOBILITY. In a colloid system: the velocity per unit field acquired by colloidal particles in electrophoresis.

ELECTROPHOTOPHORESIS. The (usually) helical movement of small particles in a gas along the direction of an electric field during their irradiation by light. It is one type of photophoresis. See also: *Photophoresis.*

ELECTROPOSITIVE. See: *Electrochemical series.*

ELECTROSCOPE. An instrument by which small electrostatic charges and hence small electric currents are measured by the observation of the movement of a light mobile member relative to a fixed electrode. The movement is usually observed with the aid of a microscope. See also: *Electrometer.*

ELECTROSTATIC GENERATOR. An instrument or machine for producing a large quantity of electric charge by electrostatic means, e.g. by the use of friction or the discharge of electricity from points. See also: *van de Graaff accelerator.*

ELECTROSTATIC IMAGE. An imaginary point charge or set of point charges on one side of a conducting surface which would produce on the other side of the surface the same field as is produced by the actual electric charge on the surface. Images and image forces are classical concepts which break down at atomic distances. See also: *Electrical image.*

ELECTROSTATIC INDUCTION. The production of electric charge on the surface of a conductor under the influence of an electrostatic field.

ELECTROSTATIC INSTRUMENTS. Instruments for the measurement of potential difference or electric charge which rely on the forces between electric charges to produce some mechanical deflection in the measuring system.

ELECTROSTATIC LENS. A lens for focusing an electron beam, which depends typically on three co-linear apertures in diaphragms at different potentials. See also: *Electron lens. Electron microscope, transmission.*

ELECTROSTATIC PRECIPITATION. The process by which solid particles or liquid droplets are separated from a suspension in a gas by the electrostatic attraction of their surface charges in an electric field. The process is in large-scale industrial use for recovery and purification.

ELECTROSTATICS. The study of the properties of electric charges at rest.

ELECTROSTENOLYSIS. The precipitation of metals in the pores of a membrane by electrolysis.

ELECTROSTRICTION: ELECTROSTRICTIVE EFFECT. A second-order effect shown by all crystals, whereby a strain appears when an electric field is applied. See also: *Piezoelectric effect.*

ELECTROSURGERY. A form of diathermy in which body tissue is cut by a needle electrode carrying high frequency electric current.

ELECTROTONUS. Of a nerve: the state of the nerve when it is being subjected to a steady discharge of electricity. The force experienced by such a nerve is known as the *electronic force.* See also: *Bioelectricity.*

ELECTROVALENT BOND. An alternative name for ionic bond.

ELECTROVISCOUS EFFECT. The decrease in the viscosity of many lyophilic colloids (e.g. protein sols) brought about by small additions of electrolytes.

ELECTROVISCOUS FLUID. A fluid which exhibits an increase in viscosity when subjected to an externally applied electric field. The change in viscosity, which may be so great as to approach the point of solidification, disappears when the field is removed. Electroviscous fluids offer practical means for the transport and control of mechanical power in couplings and hydraulic devices.

ELEMENTARY PARTICLES. Those particles which are (or were) held to be simple, such as neutrons, protons, electrons, mesons, hyperons, and photons. They are characterised by their mass, electric charge, spin, magnetic moment, or interaction properties, as appropriate. Some are stable, e.g. neutrinos, electrons and protons. Others are unstable, e.g. neutrons, mesons, and hyperons. Elementary particles are sometimes known as *fundamental particles.* See: *Hadrons. Interactions between elementary particles. Leptons. Quarks.*

ELEMENTARY PARTICLES, INTERACTIONS BETWEEN. See: *Interactions between elementary particles.*

ELEMENT, CHEMICAL. A species of matter consisting of atoms with the same atomic number.

ELEVATION HEAD. The energy possessed per unit mass of fluid due to its elevation above some datum. Also known as *potential head.* For a perfect fluid in steady flow: Pressure head + velocity head + elevation head = constant, according to the Bernoulli equation where all are expressed as heights. See also: *Kinetic head: Velocity head. Pressure head.*

ELINVAR. An alloy, containing typically 36% nickel and 12% chromium, whose modulus of elasticity is almost invariable with change of temperature.

ELLIPSE. See: *Conic: Conic section.*

ELLIPSOMETER: ELLIPTIC POLARIMETER. An instrument for the measurement of the elliptical polarization of light. It is usually associated with the determination of the thickness and optical constants of surface films on metals and dielectrics.

ELLIPTIC FUNCTION. The inverse of an elliptic integral.

ELLIPTIC INTEGRAL. An integral of the type $\int f(x, \sqrt{R})\,dx$, where f is a rational function of its two arguments and R is a third- or fourth-degree polynomial in x, with no repeated roots.

ELONGATION. Of a celestial object: the angle between the direction of the object and a chosen reference direction, e.g. the angle between the direction of a planet and that of the Sun.

ELUTION. A sedimentation process for the separation of the finer grains of a powder. The heavier grains are allowed to settle in a liquid and the finer grains are poured off with the upper layer of the liquid.

ELUTRIATION. A sedimentation process for

separating and grading finely divided materials, e.g. clays, pigments, soils, which cannot be sieved on account of their size, in an upward jet of fluid (usually water) or of air.

EMANATING POWER. The ratio of the rate of emission of radioactive emanation from the surface of a material to the rate of production within the material. See also: *Hahn emanation technique.*

EMANATION: RADIOACTIVE EMANATION. The inert radioactive gases, consisting of seventeen isotopes of atomic number 86, and having mass numbers ranging from 204 to 223. The best known are *radon* itself ($^{222}_{86}Rn$), *thoron* ($^{220}_{86}Rn$) and *actinon* ($^{219}_{86}Rn$) which are decay products of radium, thorium, and actinium respectively. See also: *Emanating power.*

EMERGENT RAY POINT. See: *Beam direction indicator (radiotherapy).*

EMERSION. In Astronomy: the reappearance of the Moon or other celestial body after its eclipse or occultation.

EMISSION, SPECIFIC. Of a radiating surface: the emissive power for a given wavelength per unit area per unit time.

EMISSIVE POWER. Of a body losing energy to its surroundings by radiation: (1) The total energy emitted per second per unit area of surface, known as the *total emissive power* or *emittance*. (2) The quantity e_λ in the expression $e_\lambda \, d\lambda$ for the rate at which energy is lost over a small range of wavelengths λ to $\lambda + d\lambda$, known as the *emissive power for the wavelength* λ and sometimes the *monochromatic emissive power*. (Note: the total emissive power is $\sum e_\lambda \, d\lambda$.)

EMISSIVITY. Of a body losing energy to its surroundings by radiation: the ratio of its emissive power to that of a black body at the same temperature. It is known as the *total emissivity* when the total heat is taken into account, and the *spectral emissivity* when related to some selected region of the heat spectrum.

EMITRON TUBE: CPS TUBE. A television camera tube in which the optical image is focused on to a photo-emissive mosaic formed on a transparent dielectric and capacitively associated with a conducting signal plate.

EMITTANCE. The total emissive power. See also: *Emissive power.*

EMULSIFICATION. The dispersion of one liquid within another to form an emulsion.

EMULSIFYING AGENT. A substance which ensures the stabilization of an emulsion of one liquid dispersed in another. It acts by forming a protective film around each droplet.

EMULSION. (1) A system containing two immiscible liquids, one of which (the *internal* or *disperse* phase) is subdivided as fine droplets within the other (the *external* or *continuous* phase). (2) A photographic emulsion.

EMULSION GRAIN DENSITY. See: *Nuclear emulsion, grain density of.*

EMULSION STAR. See: *Track of ionizing particle.*

ENANTIOMORPHISM. The occurrence of two crystals or crystal forms which are related to each other as are right and left hands, i.e. one is the mirror-reflection of the other.

ENANTIOTROPY. The existence of two structural or crystalline forms of a substance one of which is stable below a definite transition temperature and the other above. Such forms are known as *enantiotropes* and the change from one to the other is said to be *enantiotropic*. See also: *Allotropy.*

ENCEPHALOGRAPHY. The radiographic examination of the brain, the cavities and spaces of which have previously been filled with air. See also: *Electroencephalography.*

ENDOSCOPE. An optical instrument used for the medical inspection of body cavities and having an optical system similar to that of a periscope but on a much smaller scale. See also: *Cystoscope. Fibreoptics.*

ENDOTHERMIC. (1) Of a process or reaction: one in which the system concerned absorbs heat from its surroundings. (2) Of a chemical compound: one which is formed by an endothermic reaction.

END RADIATION. Another term for *quantum limit*. See also: *X-rays.*

ENERGY. The capacity of a body or a system for doing work. It may, for example, be thermal, electrical, mechanical, chemical or nuclear. See also: *Erg. Joule. Work.*

ENERGY BANDS, ALLOWED OR FORBIDDEN. Zones of allowed or forbidden energy levels in a crystal. See also: *Band theory of solids. Brillouin zones.*

ENERGY, INTERNAL. That part of the energy of a system which is determined only by the state of the system. It does not include, therefore, the kinetic energy of bulk motion or the potential energy in an external field of force. The internal energy, however, does include the kinetic energies of the individual molecules or atoms of the system and their potential energies in each other's fields of force. The change of internal energy in any process is the difference between the energy gained and the external work done.

ENERGY, KINETIC. Energy stored in a system by virtue of the velocities of various moving masses within the system. The kinetic energy of a body of mass m and velocity v is $\frac{1}{2}mv^2$. See also: *Equipartition of energy.*

ENERGY LEVEL DIAGRAM. A graphical representation of the various transitions which an atom, molecule, nucleus, etc., may undergo. Typically the energy levels are shown as a series of horizontal lines plotted on a vertical scale of energy or wave number. Transitions are then shown by lines drawn between the states.

ENERGY LEVELS: ENERGY STATES. (1) The set of discrete energy states in which an individual atom, molecule, nucleus, etc., may exist. (2) The energy values associated with these states.

ENERGY LEVEL, SYMMETRICAL. An energy level for which the sign of the associated wave function remains unchanged under a given operation. If the sign changes it is said to be *unsymmetrical.*

ENERGY LEVEL WIDTH: ENERGY LEVEL BREADTH. The energy spread associated with a given energy level which, according to the uncertainty principle, is inversely proportional to the mean life of the atom, molecule, nucleus, etc., in the corresponding energy state. This breadth is reflected in the breadth of the spectral lines emitted in a transition between two states. If several transitions are possible each may be assigned a *partial width.*

ENERGY LOSS PER ION PAIR. The mean energy lost by all processes per ion pair produced in a given material by a particular type of radiation under specified conditions.

ENERGY MOMENTUM TENSOR. For an electromagnetic or material field: a set of quantities which specifies the energy density, momentum density, and stress density, and which transforms as a tensor.

ENERGY, POTENTIAL. Energy stored in a body or system by virtue of the position of the body or the configuration of the system. It is the work done in bringing the body from some reference position to the position of interest, or in bringing the system from some reference configuration to the configuration of interest; and is expressed or implied with respect to that reference position or configuration.

ENERGY PRODUCT. Of a magnetic material: the maximum attainable value of the product BH,

where **B** is the magnetic induction and **H** the magnetic field. This product (**BH**)$_{max}$ is a criterion (the *Evershed criterion*) of the "quality" of a magnetic material.

ENERGY STATE, NEGATIVE. An unexpected concept originally invoked by Dirac's theoretical treatment of the electron. Negative energy states are normally all occupied, so that no transition to them can occur, a condition which represents a vacuum. But if a particle in a negative energy state absorbs sufficient energy to raise it to a positive energy state it appears as an electron and the hole left behind in an otherwise occupied negative "sea" appears as a positron. See also: *Dirac equation. Hole theory.*

ENGINEERING. The science and technology of the application of physical and chemical principles to the exploitation of natural resources and phenomena, and the design, construction, and maintenance of the equipment associated with this.

ENSKOG THEORY. An assumption as to the form of the distribution function (or probable density distribution) f in the Boltzmann equation. The assumption is that f can be expressed in the form $f = f^{(0)} + \varepsilon f^{(1)} + \varepsilon^2 f^{(2)} + \cdots$, where $f^{(0)}$, etc., are successive approximations.

ENTHALPIMETRY. The use of an analytical technique in which the heat change in the analytical reaction is used to measure the extent or degree of advancement of the reaction.

ENTHALPY. Signifies heat content. It has the dimensions of energy and has a value, for any system in a definite state, that depends only on that state, and not upon the previous history of the system. For a thermodynamical system which may exchange heat and work with its surroundings the enthalpy H is given by $H = E + PV$, where E is the internal energy, P is the pressure and V the volume.

ENTRANCE PUPIL. The image of the aperture stop or diaphragm of an optical system as seen from the object.

ENTROPY. A quantity which is characteristic of the thermodynamic state of a system, whose significance usually relates to changes in its value, rather than to the value itself. Various aspects may be considered: (1) In a reversible process the change in entropy, dS, is given by dQ/T, where Q is the quantity of heat taken in at absolute temperature T. For an irreversible process (e.g. a physical system which is left to itself and allowed to distribute its energy in its own way) the total entropy of the system always increases, and entropy may be regarded as a measure of the unavailability of energy. (2) The entropy of a system in a given state is related to the statistical probability of finding it in that state by the equation $S = k \ln W$, where S is the entropy, k is Boltzmann's constant, and W is the probability in question. Here again the principle of increase of entropy applies. (3) In information theory the concept of entropy has been taken over to signify a measure of the uncertainty of our knowledge. See also: *Nernst heat theorem*.

ENTROPY, PARTIAL. That part of the entropy of a system related to a particular process or configuration. Thus we may have the entropy of activation, of magnetization, of mixing, or of solution. Other examples are lattice entropy, nuclear entropy and stellar entropy.

ENTRY AND EXIT PORTAL. See: *Radiotherapy, entry and exit portal in.*

ENVELOPE. (1) A curve that is tangential to each of a given family of curves. (2) An analogous surface relating to a family of planes or to a family of surfaces.

ENVIRONMENTAL TESTING. The testing of materials, devices or equipment to ensure that they are able to resist the environment in which they are to be operated.

ENZYME. A colloidal protein of high molecular weight produced by the living cell, which acts as a biological catalyst. An enzyme is designated by adding the suffix "-ase" to the substance on which it acts, or to the type of reaction it provokes, e.g. esterase, oxidase.

EÖTVÖS RULE. A relation between the surface tension and temperature of a liquid. It states that the molar surface energy $S(M/\varrho)^{2/3}$, where S is the surface tension, M the molecular weight, and ϱ the density, is proportional to $T_c - T$, where T_c is the critical temperature and T the temperature of the liquid.

EÖTVÖS TORSION BALANCE. A torsion balance, used extensively for differential measurements in geo-gravitational surveying (e.g. for prospecting), in which two masses attached to a beam are spaced apart vertically as well as horizontally, permitting the measurement of the local distortion of the Earth's gravitational field arising from the presence of geological features such as hidden mineral deposits. See also: *Balance, torsion.*

EPEIROGENESIS. See: *Diastrophism.*

EPHEMERIS. A table giving the predicted astronomical coordinates of a body at specific times, based upon the theory of the body's motion.

EPICADMIUM NEUTRONS. Neutrons of kinetic energy greater than the cadmium cut-off energy.

EPICENTRE. See: *Earthquake.*

EPITAXY. The growth of one crystal on the face of another in such a way that the orientations of the two crystals bear a definite relationship to each other. By extension the term may also designate a process of manufacture of transistors and integrated circuits in which a thin layer is formed on a silicon chip.

EPITHERMAL NEUTRONS. Neutrons of kinetic energy greater than that of thermal agitation. The term is often restricted to energies just above thermal, i.e. energies comparable with those of chemical bonds.

EQUALIZER. A reactive or dissipative network inserted in a communication system to overcome distortion.

EQUATION OF STATE. An equation (sometimes known as a *characteristic equation*) relating such variables as pressure, volume and temperature, which describes the state of a solid, liquid, or gaseous system. The best known is perhaps the equation of state for an ideal gas $pV = RT$, where p is the pressure, V the volume per mole, R the gas constant per mole, and T the absolute temperature. Some of the well-known equations of state are given in the following entries. See also: *Corresponding states, law of.*

EQUATION OF STATE, CLAUSIUS. A modification to the van der Waals equation which corrects the pressure–correction term for its variation with temperature. In this term V^2 is replaced by $T(V + c)^2$, where V is the volume per mole, T the absolute temperature, and c a function of the constants relating to intermolecular forces and finite molecular size. See also: *Equation of state, van der Waals.*

EQUATION OF STATE, GRÜNEISEN. Refers to a solid under the action of hydraulic pressure. It is often written in the form $PV + V\phi'(V) = \gamma E$, where P is the pressure, V the volume, $\phi'(V)$ the derivative with respect to volume of that part of the free energy which depends only on V, E the thermal energy, given by $E = \int_0^T C_v \, dT$, where T is the absolute temperature, C_v is the specific heat at constant volume, and γ is the *Grüneisen constant*. This is an important equation in the study of thermal expansion. See also: *Grüneisen law.*

EQUATION OF STATE, VAN DER WAALS.
An equation of state for a real gas, which takes into account finite molecular size and intermolecular attractive forces. It may be written

$$\left(p + \frac{a}{V^2}\right)(V - b) = RT,$$

where p is the pressure, V the volume per mole, R the gas constant per mole, and T the absolute temperature, as in the equation of state for an ideal gas; a and b are constants depending on the gas, a relating to intermolecular attractive forces and b to finite molecular size. See also: *Equation of state, Clausius.*

EQUATOR. (1) *Celestial:* the great circle in which the plane of the Earth's equator cuts the celestial sphere. (2) *Terrestrial:* the great circle on the surface of the Earth which is perpendicular to the axis of rotation of the Earth. It is also known as the *geographical equator.*

EQUATORIAL MOUNTING. Of an astronomical telescope: a mounting in which one axis of rotation is parallel to the axis of rotation of the Earth, enabling the telescope to follow a given star by a single movement of rotation.

EQUATORIAL RING CURRENT. An east–west current of charged particles flowing around the Earth in an approximately equatorial plane at a distance of a few Earth radii. See also: *Zones of radiation, extraterrestrial.*

EQUILIBRIUM. The state in which the observable properties (chemical, mechanical, thermodynamic, etc.) no longer change with time. See also: *Chemical equilibrium. Thermodynamic equilibrium.*

EQUILIBRIUM CONSTANT. For a closed system in a state of chemical equilibrium at a given temperature: a constant expressing the relation between the activities, fugacities, concentrations or mole fractions of the various species of which the system is comprised.

EQUILIBRIUM DIAGRAM: PHASE DIAGRAM: CONSTITUTION DIAGRAM. A temperature–composition map of the solid and liquid phases of an alloy system. It shows the ranges of composition and temperature over which the various phases exist in equilibrium with each other and the phase transformations which occur on heating or cooling. The construction of such diagrams is vital to the metallographic study of structures in an alloy system. Binary systems are represented by two-dimensional diagrams, and ternary usually by sections of a three-dimensional figure.

EQUILIBRIUM OF A RIGID BODY. The state in which the vector sum of all the external forces acting on the body is zero (*translational equilibrium*), or in which the algebraic sums of the torques about each of three mutually perpendicular axes are zero (*rotational equilibrium*).

EQUILIBRIUM, RADIOACTIVE. In a radioactive series: a quasi equilibrium condition in which the products decay approximately with the period of the parent. This occurs if the half-life of the parent is very much longer than that of any of the products and is known as *secular equilibrium.* If, however, the half-life of the parent is not much longer, the condition is known as *transient equilibrium.*

EQUILIBRIUM SHAPE. Of a crystal: See: *Wulff theorem.*

EQUINOCTIAL COLURE. The great circle passing through the First Points of Aries and Libra and through the celestial poles. See also: *Solstitial colure.*

EQUINOX, AUTUMNAL. (1) That point of intersection of the celestial equator and the ecliptic where the declination of the Sun is changing from positive (northerly) to negative (southerly). This point is known as the *First Point of Libra*, $\simeq$. (2) The time at which the Sun is in the direction of the First Point of Libra (on or about 22 September).

EQUINOX, VERNAL. (1) That point of intersection of the celestial equator and the ecliptic where the declination of the Sun is changing from negative (southerly) to positive (northerly). This point is known as the *First Point of Aries*, ϕ. (2) The time at which the Sun is in the direction of the First Point of Aries (on or about 21 March).

EQUIPARTITION OF ENERGY. In a system consisting of a large number of particles, which is in equilibrium and can be treated classically: the principle, enunciated by Boltzmann, that the mean kinetic energy per particle is the same for each degree of freedom and is equal to $\frac{1}{2}kT$, where k is Boltzmann's constant and T the absolute temperature. The mean kinetic energy of each kind of particle is then $\frac{1}{2}skT$, where s is the number of degrees of freedom of one particle.

EQUIPOTENTIAL SURFACE. (1) *Electrical:* a surface on which every point is at the same potential and to which the lines of force are perpendicular. (2) *Geodetic:* a surface on which the Earth's external potential is constant, and to which its resultant direction is perpendicular. The equipotential surface which coincides approximately with mean sea level is known as the *geoid*. See also: *Earth, external potential of. Earth, figure of.*

ERG. The CGS unit of work and energy. It is the work done (i.e. the energy transferred) when the point of application of 1 dyne is displaced by a distance of 1 cm in the direction of the force. It is equal to 10^{-7} J exactly. See: *Appendix II.*

ERGODIC THEORY. A theory which justifies the probability postulates employed in constructing the theory of statistical mechanics by linking them directly to dynamical laws.

ERICHSEN TEST. A cupping test to measure the ductility of metal strip or sheet.

ERIOMETER. A device for measuring the diameter of small particles or thin fibres by observing the diameter of the optical diffraction pattern which they produce.

ERROR FUNCTION (ERF). For measurements described by a Gaussian distribution: a function giving the probability that the error of a single measurement will lie between given limits. See also: *Gaussian distribution: Normal distribution.*

ERROR, PROBABLE. That deviation from the mean value of a series of observations above and below which half of the observed values lie.

ERROR, RANDOM. A small error present in a measurement, which, for a sufficient number of measurements, may be treated statistically by the theory of errors.

ERROR, STANDARD. A measure of the accuracy of the mean of a series of observations. For a normal distribution, for example, the standard error of the mean of N measurements is given by σ/N where σ is the standard deviation. See also: *Significance. Standard deviation.*

ERROR, SYSTEMATIC. An error in a measurement which is not random, being caused, for example, by an error in calibration. An accumulation of measurements involving systematic errors cannot be treated by any general theory.

ERYTHEMA. The reddening of the skin as a result of exposure to ultraviolet radiation or to radiations of higher energy such as X-rays.

ESCAPEMENT. See: *Clock or watch, escapement of.*

ESCAPE PROBABILITY. Of neutrons in a nuclear reactor: the proportion of neutrons produced which escape without being absorbed. Separate probabilities can be defined, for example, for thermal neutrons, or neutrons in the process of being slowed down; but the actual values assumed will depend on the particular approximation to neutron behaviour which is being used. See also: *Neutrons, resonance escape probability for.*

ESCAPE VELOCITY. Of a body from a planet: the minimum initial velocity needed by the body to overcome the gravitational attraction of the planet and escape into space. For Earth it is 6·9 miles (11·1 km)/s, for the Moon 1·5 miles (2·4 km)/s and for the Sun 384 miles (618 km)/s.

ESTER. A neutral organic compound formed when the hydroxylic hydrogen atom in oxygenated organic or inorganic acids is replaced by an alkyl or an acyl group.

ÉTALON, FABRY-PEROT. An instrument in which interference fringes are produced by the passage of approximately monochromatic light through a pair of plane, parallel, half-silvered glass plates separated by a few millimetres of air. It is used, in combination with a spectrograph as a monochromator (to select the spectral region of interest), for high-resolution spectroscopy.

ETCHING. The preferential removal of portions of the surface of a crystal (often metallic) to reveal its macro- or micro-structure. The etching process may be chemical, electrolytic or thermal, or may involve cathodic bombardment or heat tinting.

ETCH PITS. Depressions on the surface of a crystal which has been etched for a sufficiently long time. The shape of the pits is related to the symmetry of the crystal.

ETHER. (1) A neutral oxygenated organic compound. It may be aliphatic or aromatic. (2) A hypothetical non-material fluid (the *luminiferous ether*) formerly supposed to permeate all space, and having the property of propagating electromagnetic waves and permitting action at a distance.

ETTINGSHAUSEN EFFECT. In an electrical conductor in a transverse magnetic field: the appearance of a temperature gradient at right angles to both the electric current and the magnetic field.

EUCLIDEAN GEOMETRY. The familiar two-dimensional geometry of the drawing board, ruler and compasses, and its extension into three dimensions. See also: *Non-Euclidean geometry.*

EUCLIDEAN SPACE. A mathematical formulation of the intuitive geometrical properties of physical space.

EUCOLLOIDS. Natural colloids such as rubber, starch, albumen, collagen, and all other macromolecular substances. See also: *Hemicolloid. Mesocolloid.*

EULER CONSTANT. Is defined to be

$$\lim_{n \to \infty} \left(1 + \frac{1}{2} + \frac{1}{3} + \dots \frac{1}{n} - \log n \right).$$

It is approximately equal to 0·5772 ...

EULER EQUATION. States that the condition that the integral

$$\int_{x_1}^{x_2} I(x, y, y') \, dx$$

shall have a stationary value is

$$\frac{\partial I}{\partial y} - \frac{d}{dx} \frac{\partial I}{\partial y'} = 0.$$

A solution $y = f(x)$ that satisfies this equation is known as an *extremal* and is a maximizing or a minimizing curve. See also: *Calculus of variations.*

EULER EQUATIONS FOR FLUID FLOW. See: *Fluid flow, equations of motion for.*

EULER EQUATIONS OF MOTION: EULER DYNAMICAL EQUATIONS. For a mechanical system which consists of connected particles: a series of equations which may be expressed in a variety of ways (e.g. relative to the centre of mass or to a fixed point), all of which are basically forms of the two vector equations

$$\frac{d\mathbf{p}}{dt} = \mathbf{R} \quad \text{and} \quad \frac{d\mathbf{h}}{dt} = \mathbf{G},$$

where $\mathbf{p}$ and $\mathbf{h}$ are the linear and angular momenta of the system referred to an inertial frame of reference with a stationary origin, $\mathbf{R}$ is the vector sum of the forces applied to the system, and $\mathbf{G}$ is the sum of the vector moments of these applied forces about the origin.

EULER FORMULA. For a *convex polyhedron* (i.e. a polyhedron of which any two points can be joined by a straight line entirely contained within the polyhedron): states that $V - E + F = 2$, where V is the number of vertices, E the number of edges, and F the number of faces.

EULERIAN ANGLES. A set of three angles of rotation used in describing the position of a body which moves about a fixed point (with Cartesian axes fixed in the body), relative to another Cartesian coordinate system fixed in space. Such angles are particularly useful in the analysis of the dynamics of precession and nutation.

EULERIAN INTEGRAL OF THE FIRST KIND. A beta function. See also: *Beta function.*

EUPATHESCOPE. A device for estimating the heat lost by an individual in an enclosure.

EUTECTIC. In a binary phase diagram: (1) the composition and temperature at which two descending portions of the liquidus curve meet; (2) material having the eutectic composition. Ternary eutectics also occur.

EUTECTOID. In a binary phase diagram: is entirely analogous to a eutectic, the change in name signifying only that all the phases involved are solid.

EVAPORATION. The escape from the surface of a liquid (or solid) of those molecules whose energy of thermal agitation is greater than the work function of the surface.

EVAPORIMETER. A device for measuring the rate of evaporation.

EVECTION. An inequality in the mathematical expression of the Moon's orbital motion, associated with the variable eccentricity of the orbit, which arises from solar perturbations.

EVERSHED CRITERION. See: *Energy product.*

EVERSHED EFFECT. Refers to the general motion of the gases in the penumbral regions of sunspots: on the basis of spectrographic evidence this motion is believed to be radially outwards from the sunspot centre.

EXCHANGE. The postulated continuous exchange of charged particles between two similar quantum-mechanical systems. The interaction between two such systems gives rise to *exchange energy* and *exchange forces*. The simple wave functions descriptive of the two systems must be combined with a function which takes account of the fact that the particles have actually changed places, the resulting energy being significantly different from that of the state in which no exchange is supposed to occur. A typical example is that of electron exchange, which may be used to explain covalent bonding, ferromagnetism and antiferromagnetism. A parallel explanation has also been suggested to account for *nuclear forces* by the exchange of π-mesons. The magnitude and sign of an exchange effect depends on the value of the exchange integral.

EXCHANGE DEGENERACY. Exchange degeneracy is said to occur when the total energy of two systems in which exchange takes place is the same as if no exchange had occurred.

EXCHANGE INTEGRAL. For two electrons: one of two types of integral, of which the first (known as the *Coulomb integral*) represents the mutual electrostatic energy of two charge clouds $\phi^*\phi$ and $\psi^*\psi$, where ϕ and ψ are the two independent electronic wave functions. The second (the exchange integral proper) represents the mutual energy of two charge clouds $\phi^*\psi$ and $\phi\psi^*$.

EXCITATION. The addition of energy to a system to bring it into an excited state.

EXCITATION ENERGY. For a given atomic or nuclear state: the difference between the energy of the given state and that of the ground state. Also known as *excitation potential*.

EXCITATION FUNCTION. A function expressing the relationship between the cross-section for an atomic or nuclear excitation reaction and the energy of the incident particle or photon causing the excitation.

EXCITATION POTENTIAL. See: *Excitation energy*.

EXCITATION PURITY. See: *Colorimetric purity*.

EXCITED STATE. Of an atomic or nuclear system: any stationary state with an energy above that of the ground state.

EXCITON. An excited state of a crystal which does not give rise to or increase its electrical conductivity. It may be regarded as a state of atomic excitation which moves with infinite mean path through the crystal, affecting a series of atoms in turn and giving rise to the transport of energy but not of charge.

EXCLUSION PRINCIPLE. A principle enunciated by Pauli which states that no two identical fermions (i.e. particles of spin $\frac{1}{2}$) in any system may occupy states which have the same set of quantum numbers. This principle holds, therefore, for electrons, protons and neutrons, but not for photons or π-mesons.

EXIT POINT (RADIOTHERAPY). See: *Beam direction indicator (radiotherapy)*.

EXIT PUPIL. The image of the aperture stop or diaphragm of an optical system as seen from the image.

EXOTHERMIC. (1) Of a process or reaction: one in which the system concerned evolves heat. (2) Of a chemical compound: one which is formed by an exothermic reaction.

EXOTIC ATOM. See: *Mesonic atom: mesic atom*.

EXPANDING UNIVERSE. A hypothesis, based on spectroscopic observations of the red shift, according to which all extragalactic objects are receding from each other. Those which are farthest away are moving fastest, and it has been estimated that all the distances concerned are doubled in a period of about 10^9 years. See: *Universe*.

EXPANSION. The increase of a given mass of substance in length, area or volume, as a result of a change in physical conditions, commonly a change in temperature. See also: *Thermal expansion coefficient*.

EXPANSION CHAMBER. See: *Cloud chamber*.

EXPANSION COEFFICIENT. See: *Thermal expansion coefficient*.

EXPANSION WAVE. A sound wave through a region of a gas in which the density and velocity are increasing in the direction of propagation. See also: *Shock wave*.

EXPLODING WIRE. A fine wire, fired electrically by the energy stored in a capacitor, which may be substituted for a chemical explosive when an extremely high energy density is needed. It may be used, for example, as a light source in spectroscopy, for plasma formation, for shock-wave generation, for fast switching, for welding or plating, or for the production of high temperature. It is mainly used industrially as a low-voltage fuse for conventional explosives and in electroforming processes.

EXPLOSION. A rapid increase of pressure accompanied by heat, light, sound, and mechanical shock, arising as a result of a sudden release of energy. The source of the energy may be chemical, electrical, mechanical, or nuclear.

EXPLOSION LIMITS. Limits which specify the range of concentration over which a gas mixture can be caused to ignite explosively. They are usually expressed as the volume percentage of the combustible gas. Thus, for hydrogen in air, the lower limit is $4 \cdot 1\%$ and the upper 75%.

EXPLOSIVE WORKING. Of metals: any operation which involves the use of explosive energy to change the shape of a metal part, to effect the displacement, removal or joining of metals, to produce cutting and shearing or to effect changes in the engineering and metallurgical properties.

EXPONENTIAL EXPERIMENT. In nuclear technology: an experiment performed with a subcritical assembly of reactor materials which contains, or is in close apposition to, a neutron source, in order to determine the neutron characteristics of a configuration of these materials. The assembly is termed exponential because there is one direction in which the neutron flux density decreases exponentially with distance.

EXPONENTIAL FUNCTION. A transcendental function of a real variable denoted by e^x or $\exp x$

and defined by the convergent infinite series

$$\exp x = 1 + x + \frac{x^2}{2!} + \frac{x^3}{3!} + \frac{x^4}{4!}$$
$$+ \ldots + \frac{x^n}{n!} + \ldots.$$

For $x = 1$, $\exp x = 2.718218 \ldots$, which is the base of the system of natural logarithms, and is known as e.

EXPONENTIAL INTEGRAL. Denotes the function $ei(x)$ defined by

$$ei(x) = \int_x^\infty \frac{e^{-u}}{u} \, du, \quad (x > 0)$$

This is equal to

$$-\gamma - \log x + x - \frac{x^2}{2 \cdot 2!} + \frac{x^3}{3 \cdot 3!} - \ldots,$$

where γ is Euler's constant.

EXPONENTIAL TRANSMISSION LINE. A two-conductor transmission line whose characteristic impedances vary exponentially with electrical length along the line.

EXPOSURE. (1) Of X-rays or γ-rays at a given place: a measure of the amount of radiation, based on its ability to produce ionization in air. See: *Röntgen*. (2) In general: the incidence of radiation on matter (living or inanimate), either by accident or intent.

EXTENSION. Of a metal: an abbreviation for *"extension to fracture"*. It is usually expressed as the percentage increase in length between gauge marks on a specimen of standard dimensions when extended to fracture.

EXTENSIVE SHOWER. See: *Cosmic-ray shower*.

EXTENSIVE VARIABLE. One which is proportional to the amount of matter being considered. Thus, volume, entropy and total energy are typical extensive variables. An extensive variable is also known as *extensive quantity* or *extensive parameter*. See also: *Intensive variable*.

EXTENSOMETRY. The measurement of the deformation of a specimen under tensile loading, usually in connection with the tensile stress–strain diagram of a material.

EXTERNAL BALLISTICS. See: *Ballistics*.

EXTINCTION. In optics: is synonymous with optical density. The extinction per unit path in a medium is termed the *optical extinction coefficient*. See also: *Bouguer–Lambert law of absorption: Bouguer–Beer law. Light, extinction coefficient for. Optical density*.

EXTINCTION. In diffraction by crystals: (1) the attenuation of a beam of radiation by diffraction, as distinct from absorption. *Primary extinction* occurs with perfect crystals and arises from interference between the direct and multi-reflected beams, so that only a "skin" of crystal is able to reflect. *Secondary extinction* occurs with imperfect crystals and arises from the shielding of underlying mosaic blocks by identically oriented blocks nearer the crystal surface. (2) the total absence of Bragg reflections from certain sets of crystal planes, as a consequence of the symmetry of the space group. Such absences are also known as *space-group extinctions*.

EXTRACORPOREAL CIRCULATION. The circulation of blood in the body by means of an external agent, e.g. a heart–lung machine. Venous blood is removed from the body, oxygenated, and pumped back as arterial blood.

EXTRAORDINARY RAY. See: *Double-refraction: Birefringence*.

EXTRAORDINARY REFRACTIVE INDEX. See: *Double-refraction: Birefringence*.

EXTRAPOLATED BOUNDARY. In reactor physics: a hypothetical surface formed outside a neutron chain-reacting assembly, whose distance from the assembly is at all points equal to the linear extrapolation distance.

EXTRAPOLATION. The estimation of the value of a function of an independent variable for a value of the variable which is outside the range for which the function has been determined. See also: *Interpolation*.

EXTRAPOLATION DISTANCE. In reactor physics, using the one-group theory of neutron transport: the distance beyond the boundary of a neutron chain-reacting medium to a point at which the asymptotic neutron flux density would go to zero if it were represented by the same function as within the boundary. See also: *Neutron flux density, asymptotic*.

EXTRAPOLATION DISTANCE, LINEAR. In reactor physics, using the one-group theory of neutron transport: the distance beyond the boundary of a neutron chain-reacting medium to a point at which the tangent to the curve of asymptotic neutron flux density at the boundary goes to zero. See also: *Neutron flux density, asymptotic*.

EXTREMAL. A value of a function which makes a given definite integral of that function either a maximum or a minimum.

EXTRUSION. A process whereby a metal billet is forced by compression to flow through a suitably shaped aperture in a die to give a product usually of a smaller, uniform cross-sectional area. Pressure is usually applied by a piston, known as a *ram*.

EXTRUSION, HYDROSTATIC. A method of forcing a metal billet through a die in which the conventional ram is replaced by a pressurized fluid. For brittle materials it is arranged that the product of extrusion emerges into a pressurized fluid, for example by immersing the whole extrusion apparatus. Extrusion involving the use of such a back pressure is known as *Differential pressure extrusion* or *Fluid-to-fluid extrusion*.

EYEPIECE. A lens or lens system which produces, for presentation to the eye, a virtual image of an intermediate image in an optical system.

EYEPIECE, GAUSS. An eyepiece suitable for an autocollimating telescope. It has a mirror between the eye lens and the graticule, which directs light coming from a lateral source into the axis of the telescope.

EYEPIECE, HUYGENS. An eyepiece consisting of two thin lenses of the same sort of glass, separated by a distance equal to one half the sum of their focal lengths. It is free from chromatic aberration but is not suitable for use with a graticule.

EYEPIECE, KELLNER. See: *Eyepiece, Ramsden*.

EYEPIECE, ORTHOSCOPIC. A complex eyepiece with a wide field of view. The distinction between field- and eye-lens is here not clear cut.

EYEPIECE, RAMSDEN. An eyepiece suitable for use with a graticule, consisting of a combination of two converging lenses of equal focal length. When an achromatized doublet eye-lens is used, the eyepiece is known as a *Kellner eyepiece*.

f

FABRY LENS. A lens placed between the image of a star and a light receptor (photographic plate and photoelectric cell) so as to provide at the latter a more uniform image, and one more suited to photometric measurement than that offered by the primary image.

FACE-CENTRED CUBIC STRUCTURE. One of the two types of close-packed crystal structure, the other being close-packed hexagonal. Atoms are positioned at the corners and centres of the faces of the unit cube. See also: *Close-packed structure.*

FACE-CENTRED LATTICE. A particular type of space lattice and not a crystal structure. See: *Space lattice: Bravais lattice.*

FACSIMILE TELEGRAPHY. A transmission system by which still pictures or other forms of graphic material may be transmitted by line or radio and received in a form which is essentially an exact copy of the original material.

FACULAE, SOLAR. Bright features observed on the surface of the Sun. They are seen at the solar poles, in the Sun's chromosphere (upper part of the solar atmosphere) or its photosphere (lower part). Chromospheric faculae are sometimes known as *flocculi.*

FADING. Of radio waves: a variation in the received signal strength arising from variation with time in the conditions of propagation. Fading which affects unequally the components of a modulated wave is known as *selective fading.*

FAIRING. A secondary structure added to smooth the flow in an aerodynamic or hydrodynamic system.

FAJANS RULES. A series of rules, now superseded, put forward by Fajans in 1923, which explained the transition from ionic to covalent bonding in a related series of compounds in terms of the mutual deformation or polarization of ions.

FALCATED. Refers to the Moon, or an inferior planet, when its phase is less than 0·5 (i.e. when the illuminated portion is crescent-shaped). See also: *Gibbous.*

FALL-OUT. From a nuclear explosion: the radioactive debris from a nuclear explosion, which, after being forced upwards into the atmosphere by the explosion, condenses and falls to the ground. It consists of fission products, unexpended fissile material, and the products of neutron activation.

FARAD. The unit of capacitance in MKSA and SI units. It is the capacitance of a capacitor between the plates of which there appears a difference of potential of 1 volt when it is charged by a quantity of electricity equal to 1 coulomb. See: *Appendix II.*

FARADAY BALANCE. A torsion balance designed for the measurement of the susceptibilities of para- and diamagnetic materials. It has also been developed for measuring the susceptibilities of gases.

FARADAY CAGE. A room or part of a room which is screened so as to keep out external electromagnetic fields. Metal (often wire) screening is usually used, the screening being taken directly to earth; and often two separately earthed screens are employed, one inside the other.

FARADAY DARK SPACE. In a glow discharge tube: a relatively dark region immediately adjacent to the positive column. See also: *Discharge, glow.*

FARADAY DISK. See: *Homopolar induction.*

FARADAY EFFECT. The rotation of the plane of polarization produced when plane-polarized light is passed through a homogeneous medium in a magnetic field, the light travelling along the direction of the field. For a given substance the rotation is proportional to the path length of the light in the substance and to the magnetic field strength. The constant of proportionality is known as the *Verdet constant.*

FARADAY: FARADAY CONSTANT. The quantity of electricity that will liberate 1 mole from solution. It is the product of Avogadro's number and the electronic charge and is equal to about 96 500 coulomb. See: *Appendix III.*

FARADAY LAW OF INDUCTION. States that the e.m.f. induced in a closed circuit is equal to the rate of decrease of magnetic flux through the circuit, or $E = -d\phi/dt$, where E is the e.m.f. and ϕ is the magnetic flux linking the circuit. See also: *Electromagnetic induction*.

FARADAY LAWS OF ELECTROLYSIS. State that: (a) the amount of chemical action produced by a current is proportional to the quantity of electricity passed; and (b) the masses of different substances deposited or dissolved by the same quantity of electricity are proportional to their chemical equivalents.

FAR POINT. Of the eye: the furthest point of clear vision when the eye is relaxed. It is also known as *punctum remotum*.

FAST FISSION FACTOR. For nuclear fission in an infinite medium: the ratio of the mean number of neutrons produced by fissions due to neutrons of all energies, to the mean number of neutrons produced by thermal fissions only.

FATA MORGANA. See: *Mirage*.

FATIGUE. Of a material subjected to an alternating stress: a phenomenon leading to failure after repeated applications (thousands or millions of cycles) of a stress which would not have caused failure had it been maintained at a steady value. Fatigue failure is accelerated in the presence of *corrosion* or *fretting* (the rubbing together of surfaces), or at increased temperatures.

FATIGUE, THERMAL. (1) Failure occurring on the repeated heating and cooling of anisotropic materials owing to their anisotropic thermal expansion. (2) Failure caused by stresses induced in materials when free thermal expansion is not permitted.

FATTY ACIDS. A term denoting the whole range of monobasic aliphatic carboxylic acids, i.e. acids in which the acid group is —CO(OH).

FAULT, GEOLOGICAL. A fracture of the Earth's crust, with a displacement in the surface of the fracture. This surface, the *fault-plane*, is specified by its *dip* (angle with the horizontal) and *strike* (line of intersection with the horizontal).

F-CENTRE AND F'-CENTRE. See: *Colour centres*.

F-CURVE. See: *Atomic f-curve*.

FEATHER METHOD: FEATHER PLOT. A method for making a rough analysis of the energy spectrum of β-rays from a radioactive source, usually one of low specific activity, from a comparison of the absorption of the β-rays from the source with that of the β-rays from a reference body, using the same material as absorber in each case.

FECHNER LAW: WEBER LAW. A general law of human sensation which states that the increase of stimulus necessary to cause an increase of sensation which is just perceptible, is a constant fraction of the whole stimulus. In the case of vision the *contrast sensitivity* (the smallest detectable difference in brightness divided by the background brightness) is known as the *Fechner fraction*.

FEEDER. (1) A cable or overhead line through which power is supplied. (2) The transmission line from a radio transmitter to the aerial.

FEL(D)SPARS. Silicates containing various cations but having in common a structure which is based on a three-dimensionally continuous framework of composition $(Al, Si)_4O_8$, formed from linked tetrahedral SiO_4 and AlO_4 groups.

FERMAT LAST THEOREM. States, but does not prove, that the equation $a^n + b^n = c^n$ has no solution in integers if n is greater than 2. Attempts to prove this famous theorem have given rise to a vast amount of new mathematics. See also: *Numbers, theory of*.

FERMAT LAW: FERMAT PRINCIPLE. Stated that the path of a ray of light from one point to another, through one or more media, is such that the time taken is a minimum—the *principle of least time*. It is now more usually expressed as the *principle of stationary time*, i.e. the path of the ray is the path of least or greatest time.

FERMI AGE. The value of the *slowing-down area* for a neutron as calculated by Fermi age theory. It has, thus, the dimensions of area, not of time, and is a measure of the distance travelled by neutrons under specified conditions. More precisely, it is one-sixth of the mean square displacement of neutrons, in an infinite homogeneous medium, from their points of origin to the points where they have been slowed down from the initial energy to a specified energy. The square root of the slowing down area is known as the *slowing down length*.

FERMI AGE EQUATION. An equation, due to Fermi, by which the slowing-down density may be calculated. It is usually written as

$$\nabla^2 q = \frac{\partial q}{\partial \tau},$$

where q is the slowing-down density and τ the Fermi age.

FERMI AGE THEORY. An approximate method, due to Fermi, for calculating the slowing-down density in a medium by means of the Fermi age equation and also for calculating the Fermi age.

It assumes that neutrons, in being slowed down, lose energy continuously and not in finite discrete amounts.

FERMI CONSTANT. The constant in the Fermi theory of beta decay which denotes the strength of the interaction between the proton, neutron, and electron–neutrino field.

FERMI–DIRAC STATISTICS. The form of quantum statistics, having to do with the distribution of particles among various allowed energy values, applicable to an assembly of particles in which only one particle is allowed in each state. Their wave functions have symmetry properties such that the wave function changes sign if any two particles are interchanged. Such wave functions are said to be *antisymmetric*. At sufficiently high temperatures, where a large number of energy levels are excited, Fermi–Dirac statistics, like Bose–Einstein statistics, reduce to the classical Maxwell–Boltzmann statistics. See also: *Bose–Einstein statistics. Fermion.*

FERMI ENERGY: FERMI LEVEL. In the electron theory of metals: the energy of the highest energy level occupied by electrons in a metal. See also: *Metals, electron theory of.*

FERMI GAS. (1) The "free" electrons contributed by the atoms of a solid metal to provide electrical conductivity. They are considered to form a dense electron gas obeying the Pauli exclusion principle and Fermi–Dirac statistics. (2) The aggregate of nucleons in a nucleus of sufficiently large mass number, which may be treated as a statistical "gas".

FERMI INTERACTION. A direct action between four Dirac fields.

FERMION. A particle to which Fermi–Dirac statistics apply. The Pauli exclusion principle holds for such particles. Electrons, protons, neutrons, and all particles with spin $\frac{1}{2}$ are fermions. See also: *Boson.*

FERMI PLOT: KURIE PLOT. Of a β-particle spectrum: a graph in which a suitable function of the observed intensity is plotted against the particle energy, the function being chosen so that the graph is a straight line for allowed β-transitions. It is used in determining the character of the β-transition and the maximum energy.

FERMI SURFACE. In the electron theory of metals: the surface, in momentum space, formed by electrons having the Fermi energy. See also: *Metals, electron theory of.*

FERMI THEORY OF BETA DECAY. A theory giving the probability of the ejection, in unit time, of a β-particle whose energy lies between E and $E + dE$. It involves the assumption of the creation of leptons in the decay process, the specification of appropriate selection rules, and the conservation of leptons.

FERRIC INDUCTION. Defined as $B - H$, where B is the magnetic induction in a magnetic field H. It is also called *intrinsic induction.*

FERRIMAGNETIC RESONANCE. See: *Ferromagnetic resonance.*

FERRIMAGNETISM. The existence of a state analogous to the ferromagnetic state but with neighbouring spins anti-parallel instead of parallel. It differs from anti-ferromagnetism in that the individual magnetic moments are not all equal and the vector sum of all spins is not zero, i.e. spontaneous magnetization occurs. The saturation magnetization is, however, much lower than the sum of the magnetic moments present.

FERRITES. Non-metallic compounds with the formula MFe_2O_4, where M is a bivalent metal. Most of them show ferromagnetic properties, and, because of their high resistivity, have widespread technological applications, e.g. in microwave circuits.

FERROELECTRICITY. The phenomenon whereby certain dielectric crystals exhibit a spontaneous dipole moment. Ferroelectric materials exhibit hysteresis, possess one or more electric Curie temperatures, and are analogous in behaviour to ferromagnetic materials. See also: *Antiferroelectric*

FERROHYDRODYNAMICS. The study of the mechanics of strongly magnetizable fluid media in the presence of applied magnetic fields.

FERROMAGNETIC RESONANCE. The resonant absorption of energy from an external radiofrequency field by ferromagnetic materials when a strong polarizing field is applied. This resonance occurs at frequencies in the microwave region. *Ferrimagnetic resonance* is a similar phenomenon, exhibited by ferrimagnetic materials. See also: *Spectra, radiofrequency.*

FERROMAGNETISM. The phenomenon whereby certain materials exhibit a high degree of magnetism in weak fields and possess a very high permeability. Ferromagnetic materials exhibit spontaneous magnetization, residual magnetism, and are subject to hysteresis. Their behaviour is markedly dependent on whether the temperature is above or below that of the Curie point. See also: *Antiferromagnetism. Diamagnetism. Ferrimagnetism. Paramagnetism.*

FERROMAGNETISM, BAND THEORY OF. A theory according to which ferromagnetism may be attributed to the electrons in the unfilled bands.

The equilibrium magnetization depends on the number of electrons, the form of the band, the magnitude of the exchange interaction, and the temperature, and must be calculated on the basis of Fermi–Dirac statistics. See also: *Collective electron theory. Ferromagnetism, Heisenberg theory of.*

FERROMAGNETISM, EWING THEORY OF. An elaboration of the "molecular" theory of magnetism put forward by *Weber* according to which each molecule of a magnetic substance contained one or more small permanent magnets which were free to turn.

FERROMAGNETISM, HEISENBERG THEORY OF. A theory according to which the explanation of ferromagnetism is given in terms of the exchange forces between electrons, and the electrons themselves are localized on individual atoms rather than shared by the whole crystal as in the collective electron theory. See also: *Collective electron theory. Ferromagnetism, band theory of.*

FERROMAGNETISM, WEBER THEORY OF. See: *Ferromagnetism, Ewing theory of.*

FERROMAGNETISM, WEISS THEORY OF. A theory according to which ferromagnetism is attributed to an ensemble of independent molecular magnets which are subject, on the one hand, to the aligning effect of an applied magnetic field and, on the other, to the disorientation effect of thermal agitation.

FERTILE. (1) Of a nuclide: capable of being transformed, directly or indirectly, into a fissile nuclide by neutron capture. (2) Of a material: containing one or more fertile nuclides.

FÉRY SPECTROGRAPH. A spectrograph in which the only optical element is a back-reflecting prism with cylindrically curved faces. The curvatures of these faces are such that rays diverging from a point on a circle are focused on that circle.

FEYNMAN DIAGRAM. A graphical scheme for representing the interactions between elementary particles and fields, including a pictorial representation of intermediate states. The elements in a Feynman diagram give a graphical picture of matrix elements. For example, an outgoing electron is represented by a line along the time axis, a positron by a line in the opposite sense, and a photon by a dotted (or sometimes wavy) line.

FIBONACCI SEQUENCES. See: *Numbers, theory of.*

FIBRE DIAGRAM. The characteristic X-ray diffraction pattern given by a group of crystallites (such as those in natural fibres) which tend to have one crystallographic direction parallel to a common direction but are otherwise randomly arranged. The diagram is intermediate between a powder diagram and that of a rotating crystal photograph. The arrangement of crystallites concerned is known as a *fibre structure* or *fibre texture.*

FIBRE OPTICS. (1) The use of glass fibres for the transmission of optical images as, for example, in endoscopes for the examination of the human body. (2) The use of glass fibres for the transmission of information by light, sometimes over considerable distances. See also: *Optical communication systems.*

FIBRE STRENGTHENING. Of a substance: the reduction of plastic deformation of the substance by the incorporation of fibres of a strong material which are so distributed as to carry elastically a large part of the applied load. See also: *Dispersion strengthening.*

FICK DIFFUSION LAWS. *First law:* states that the rate of diffusion of particles across a given area, i.e. the flux of particles per unit area per unit time, perpendicular to the gradient of concentration of the particles, is proportional in amount to that gradient and opposite in sign. It may be written as

$$J = -D \frac{\partial x}{\partial c},$$

where J is the flux in question, D the *diffusion coefficient* or *diffusivity*, and $\frac{\partial x}{\partial c}$ the gradient of concentration. This law holds only when this concentration is independent of time. When this is not so the *Second law* must be applied. This may be written

$$\frac{\partial c}{\partial t} = D \frac{\partial^2 c}{\partial x^2},$$

where $\frac{\partial c}{\partial t}$ is the rate of change of concentration with time. For both laws it is assumed that D is independent of concentration. Both laws may be extended to three-dimensional diffusion by replacing $\frac{\partial c}{\partial x}$ by ∇c and $D \frac{\partial^2 c}{\partial x^2}$ by $\nabla \cdot (D \nabla c)$ respectively. See also: *Einstein diffusion equation.*

FIELD. (1) Qualitatively: a region in which a physical force is operative. Quantitatively: the description of the way in which energy, momentum, etc., are carried in such a field. Examples are the electromagnetic field of Maxwell and the gravitational field of Einstein. Fields may be *scalar*, as exemplified by the distribution of temperature or electrical potential, or *vector*, as exemplified by the distribution of velocity or magnetic field strength. (2) An abbreviation for field strength. See also: *Action at a distance. Electromagnetic field. Gravitational field.*

FIELD EMISSION. (1) The emission of *electrons* from the surface of a conductor, which is produced by a strong electric field (i.e. cold emission). (2) The emission of *ions* from a positively charged surface under the influence of an extremely high field.

FIELD EMISSION GUN. An electron gun that produces electrons by field emission from a source of about 5 nm (50 Å) radius. Acceleration and focusing produces a very narrow beam (semi-angle about $8\frac{1}{2}°$) which is one million times brighter than that produced by a conventional electron gun. One of the main applications of the field emission gun is to the scanning transmission electron microscope.

FIELD EMISSION MICROSCOPE. An instrument which produces an image of the surface of a metal tip, or of material adsorbed by the tip, by the projection of field emission electrons on to a fluorescent screen. The resolution is some 20 or 30 Å.

FIELD-ION MICROSCOPE. An instrument similar to the field emission microscope in which the image from the surface of a metal tip is produced by gaseous ions which originate close to that surface. By *field evaporation* lower and lower depths of the original tip may be examined. The resolution is good enough to permit the visualization of individual atoms.

FIELD-ION MICROSCOPE, ATOM-PROBE. A development of the field-ion microscope which permits the analysis of a single particle selected from the atomically resolved image of a specimen. See also: *Electron-probe microanalysis.*

FIELD OF VIEW. The area of an object which can be "seen" by an optical system. It is expressed as an angle or can also be a diameter in suitable circumstances.

FIELD STRENGTH. The magnitude of the force experienced in a field by unit charge, mass, etc., as appropriate.

FIELD THEORY. A theory of the way in which the field potentials, be they electromagnetic, gravitational or nuclear, account for the propagation of a field and the consequent transfer of energy, momentum, etc.

FIELD THEORY, QUANTUM. A field theory in which all the physical observables of a system are represented by appropriate operators which obey certain commutation relations. The quantized field can be considered as an assembly of particles each of which is characterized by its own energy, momentum, charge etc., the total energy, momentum, etc., of the field being built up additively from the individual contributions of the particles present. Any particle may thus be considered as a "quantum" of a corresponding field.

FIELD THEORY, UNIFIED. Formerly any field theory which attempts to unify Maxwell's electromagnetic theory with general relativity theory; but now refers to any field theory which attempts to unite two or more physical theories. See also: *Gauge theories. Neutral currents.*

FIGURING. Of optical components: the grinding or polishing of optical surfaces in order to remove aberrations, for example, in the production of large mirrors for astronomical telescopes. See also: *Lens, aspherical.*

FILAMENT. (1) The resistive element in a thermionic device or electric lamp. (2) A prominence seen on the disk of the Sun.

FILM BALANCE. An instrument, incorporating a torsion balance, for measuring the two-dimensional pressure of surface films spread on water or other liquids.

FILM BOILING. See: *Heat transfer, boiling.*

FILM COOLER. A type of recuperator in which the cold fluid is a film of liquid (usually water) flowing down the outside of the wall which contains the hot fluid.

FILMWISE CONDENSATION. Condensation in which the condensing surface is insulated by a film of condensate. It usually leads to relatively low rates of heat transfer and therefore of condensation. See also: *Condensation. Dropwise condensation.*

FILTER. (1) A device for eliminating or reducing certain frequencies in the transmission of electrical energy, sound, radio, light, X-rays, etc. (2) A device for separating out solid particles from a liquid or gaseous phase.

FILTER, BAND-PASS. An optical or electrical filter which transmits a single band of frequencies. Such a filter may be classified as broad or narrow according to the range of frequencies transmitted. The frequency range within which transmission takes place without appreciable attenuation is termed the *pass-band.* See also: *Stop band.*

FILTER, CHRISTIANSEN. A selective light filter which may operate in the infrared, visible, or ultraviolet region of the spectrum. Such a filter comprises small solid crystals suspended in a liquid having a different dispersion but with a refractive index equal to that of the crystals at the wavelength at which maximum transmission is required.

FILTER, COLOUR. A layer, film, or plate that alters the relative intensities of the component wavelengths of light passing through it. It may operate by simple absorption or by an interference mechanism.

FILTER, INTERFERENCE. An optical filter which depends on interference in a thin film to give selective transmission of a narrow waveband.

FILTER, NARROW-CUT. An optical filter which changes abruptly from complete absorption to high transmission over a short wavelength region.

FILTER NETWORK. A system for filtering a.c. voltages or currents involving either the interconnection of various circuit elements (inductors, capacitors or resistors) or the use of mechanically resonant elements (coaxial lines, waveguides, or piezoelectric crystals).

FILTER, NEUTRAL DENSITY. An optical filter to reduce the illumination irrespective of colour, i.e. one with a non-selective absorption.

FILTER, ROTARY VACUUM. A device for the continuous filtration of slurries and suspensions.

FILTER, THORAEUS. A filter made of tin, copper, and aluminium used in radiotherapy to permit the passage of certain wavelengths only.

FILTER, WAVE. An electrical network whose insertion loss is low (and approximately constant) over a band or bands of frequencies and high over other bands, and can thus act as a filter.

FILTER, X-RAY. See: *K-beta filter: Beta filter.*

FILTRATION. The separation of solids from a fluid phase, either liquid or gaseous. It may be achieved by gravitational action (as in the clarification of water by filtration through beds of sand), by the use of a vacuum or pressure, or by centrifugation.

FINENESS RATIO. Of an axisymmetric body: the ratio of the length of the body to its maximum thickness.

FINE-STRUCTURE CONSTANT. A non-dimensional constant appearing in the theory of the fine structure of the atomic states of a single electron. It is equal to $2\pi e^2/hc$, where e is the electronic charge, h is Planck's constant, and c is the speed of light. Its numerical value is about $\frac{1}{137}$, or 7.3×10^{-3}.

FINITE DIFFERENCES, METHOD OF. A method of evaluating, for a function $f(x)$ whose numerical value is known only at a limited number of points x_i ($i = 0$ to n), (a) the value of the function for some x other than the x_i's, (b) the derivative of $f(x)$ at any point, (c) the definite integral of $f(x)$ between two specified values of x, and (d) the solution of other similar problems.

FINITE STRAIN THEORY. A theory of elasticity which is applicable to high compressions and is therefore appropriate for studying the physics of the Earth's interior.

FINITE TRANSFORM. A transform analogous to a Fourier transform but used when the coordinate to be transformed has a finite range. Instead of using an integral inversion formula as in a Fourier transform it is necessary to use a series inversion in the form of a sum of orthogonal functions. See also: *Fourier transform. Integral transform.*

FIREBALL. A very bright meteor, also known as a *bolide.* It is not to be confused with ball lightning, which is purely electrical in origin.

FIRST SOUND. In liquid helium II: ordinary sound which results when the superfluid and normal components of liquid helium II oscillate in phase with each other to produce periodic density variations. See also: *Helium II, two-fluid model for. Second sound.*

FISCHER–TROPSCH PROCESS. A catalytic process for the synthesis of liquid hydrocarbon fuels from purified water gas. It is also known as the *synthine* process (from the words "*synth*etic" and "gaso*line*").

FISSILE. (1) Of a nuclide: capable of undergoing nuclear fission by interaction with slow neutrons. (2) Of a material: containing one or more fissile nuclides. See also: *Fertile. Fissionable. Nuclear fission.*

FISSIONABLE. (1) Of a nuclide: capable of undergoing nuclear fission by any process. (2) Of a material: containing one or more fissionable nuclides. See also: *Fissile. Nuclear fission.*

FISSION CHAMBER: FISSION COUNTER. An ionization counter containing fissile material which is used to detect and measure neutrons by the ionization produced by fission fragments.

FISSION FRAGMENTS. The nuclei resulting from nuclear fission.

FISSION PRODUCTS. Nuclides produced either directly by nuclear fission or by the radioactive decay of such nuclides.

FISSION SPIKE. A displacement spike in a solid resulting from damage by fission fragments. See also: *Displacement spike. Thermal spike.*

FISSION YIELD. The fraction of nuclear fissions leading to fission products of a given type.

FITZGERALD–LORENTZ CONTRACTION. See: *Lorentz contraction.*

FIZEAU EXPERIMENT. An experiment carried out to measure the ether drag in moving water by measuring the velocity of light up and down stream, using an interferometric method. It was a predecessor of the Michelson–Morley experiment. The results were inconclusive.

FIZEAU FRINGES. Interference fringes at constant optical thickness. Also known as *contour fringes.*

FLAME. A region in which chemical interaction occurs between gases, which is associated with the combustion of a substance and is accompanied by the evolution of light and heat.

FLAME, BURNING VELOCITY OF: FLAME SPEED. The speed at which a plane flame front advances into the unburned gas in a direction normal to its surface.

FLARE SPOT. A diffuse bright patch in an optical image produced, for example, by multiple reflections at lens surfaces.

FLASHING. Of an optical system: the apparent filling of a parabolic mirror, for example, with light when viewed from a distance, as a result of the production of a parallel beam from a small source of light at the focus. The intensity produced is proportional to the product of the area flashed and the source brightness and will clearly be at a maximum when the mirror is fully flashed.

FLASHOVER. A form of surface breakdown between two electrodes on a dielectric surface. See also: *Dielectric breakdown.*

FLASHPOINT. The lowest temperature at which the vapour or decomposition products of an oil will become flammable.

FLASH RADIOGRAPHY. The production of radiographs of moving objects by the use of intense bursts of X-rays lasting for microseconds or less. A succession of such flashes will permit ciné-radiography.

FLASH SPECTRUM. The chromospheric emission spectrum of the Sun, visible for a few seconds just before and just after totality in a solar eclipse. This spectrum furnishes evidence of the great abundance of helium in the Sun.

FLASH TUBE. (1) A glass or quartz tube filled with inert gas which, when discharged across condensers, produces a brief flash of white light which can be used for high-speed photography. (2) A tube, containing neon to which a suitable mixture of gases has been added, in which the passage of a particle is indicated by the production of a flash. It has been used for the detection of cosmic ray particles, particles produced by accelerators, and gamma rays.

FLATNESS. In metrology: one of the two fundamental qualities in precision engineering, the other being straightness. A flat surface is one for which all lines lying in the surface are co-planar. See also: *Optical flat. Straightness.*

FLAW SENSITIVITY. In the industrial radiography of a specified homogeneous material, for a given technique of examination: the smallest fractional change in thickness which is detectable. It is usually expressed as a percentage of the total thickness.

F-LAYER: APPLETON LAYER. The higher of the two main strata or "layers" in the ionized atmosphere, the other being the Kennelly–Heaviside layer or E-layer. It occurs at above about 150 km. See also: *Ionosphere.*

FLETCHER INDICATRIX: OPTICAL INDICATRIX. A geometrical construction for the determination of propagation velocities and vibration directions in a doubly refracting crystal. It takes the form of an ellipsoid, the semi-axes of which are proportional to the principal indices of refraction, and the directions of which are parallel to the corresponding vibration directions. The required velocities and vibration directions can be obtained from the axes of the ellipse passing through the centre of the ellipsoid and at right angles to the direction of the wave normal. See also: *Fresnel ellipsoid: Ray ellipsoid. Normal surface. Ray surface.*

FLETTNER ROTOR. See: *Magnus effect.*

FLEXURAL RIGIDITY. Of a thin elastic bar clamped at one end: the force at the other end required to produce unit displacement of that end.

FLICKER. A flickering sensation experienced when the eye is stimulated by regular pulses of light, under certain conditions and at appropriate frequencies. Above a certain frequency, the *critical flicker frequency* or *flicker-fusion frequency*, the flickering effect ceases and a uniform sensation is experienced. This is known as *persistence of vision.*

FLICKER EFFECT. Of a thermionic valve or tube:

minute variations in the cathode current caused by random changes in cathode activity or positive ion emission.

FLOATING BATTERY. See: *Electrical battery.*

FLOCCULATION The coalescence of a finely divided precipitate into larger particles.

FLOCCULI, SOLAR. See: *Faculae, solar.*

FLOOD WAVE. Another name for a storm surge. See: *Storm surge: storm tide.*

FLOTATION PROCESS. A process for removing particles suspended in water by attaching air bubbles to them to form a *froth* or *foam*. Aggregates of particles and bubbles with a density less than the suspending medium rise to the surface where they may be retained as a stable foam. The process is extensively used in the treatment of mineral ores.

FLOUQUET THEOREM. States that the general solution of Mathieu's equation

$$\frac{d^2y}{dx^2} + (\lambda + \gamma \cos \omega t)\, y = 0,$$

where λ and γ are constants, is of the form

$$y = c_1 \phi(x)\, e^{\mu x} + c_2 \phi(-x)\, e^{-\mu x},$$

where c_1 and c_2 are arbitrary constants, $\phi(x)$ is a periodic function of x with period $2\pi/\omega$, and μ is a constant depending on λ and γ.

FLOW BIREFRINGENCE. The birefringence of a flowing liquid arising from the orientation of anisotropic molecules which occurs as a result of the flow, or of certain particles carried in suspension. See also: *Form birefringence.*

FLOW CURVE. (1) *For a fluid:* a graphical representation of the flow, e.g. a plot of total shear against time. (2) *For a solid:* the plastic section of the stress–strain curve.

FLOW VISUALIZATION. Any means of making visible the flow in a liquid or gas. Such means may involve the use of changes in refractive index, the injection of smoke or dye, the employment of wool tufts to serve as "markers", or the evaporation of suitable surface finishes to reveal turbulent regions. See also: *Schlieren method.*

FLUENCE, PARTICLE. See: *Particle flux density.*

FLUID. Any substance that flows and which offers no permanent resistance to changes of shape induced by pressure. Only a uniform isotropic pressure can be supported without distortion. A fluid may be compressible (a gas) or practically incompressible (a liquid).

FLUID DYNAMICS. The mathematical study of the motion of perfect fluids.

FLUID FLOW, COMPRESSIBLE AND INCOMPRESSIBLE. See: *Compressible flow.*

FLUID FLOW, EQUATIONS OF MOTION FOR. (1) *Euler equations:* describe the motion in terms of the fluid as a whole, i.e. they are essentially statistical. They apply to a homogeneous, incompressible fluid, of constant density. (2) *Lagrange equations:* describe the motion in terms of individual moving particles. The fluid is assumed to be ideally inviscid and incompressible. (3) *Navier–Stokes equations:* describe the motions of an incompressible Newtonian fluid.

FLUID FLOW, FRICTIONLESS. Of a fluid: flow in which shear stresses are entirely absent, i.e. non-turbulent flow of a fluid with zero viscosity. It is also known as *inviscid flow.*

FLUID FLOW, HOMENERGIC. Flow in which the sum of the kinetic energy, potential energy, and enthalpy is the same throughout the fluid and remains constant.

FLUID FLOW, HOMENTROPIC. Flow in which the entropy is the same throughout the fluid and remains constant.

FLUID FLOW, IMAGE METHOD FOR. A device for the study of fluid flow at a solid boundary in which the flow is simulated by extending the region of flow across the boundary and introducing into the extended region an additional body or bodies so arranged that the boundary becomes a stream line. The "bodies" may be represented by sources, doublets, and vortices and the additional "bodies" are images of the original ones, but not necessarily mirror images. See also: *Electrical image. Electrostatic image.*

FLUID FLOW, INVISCID. Another term for *frictionless flow.*

FLUID FLOW, ISENERGIC. Flow in which the sum of the kinetic energy, potential energy, and enthalpy of each element of the fluid remains constant.

FLUID FLOW, ISENTROPIC. Flow in which the entropy of each element of fluid remains constant.

FLUID FLOW, LAMINAR. Flow in which there is no macroscopic mixing between adjacent layers. It is the same as *stream line flow.*

FLUID FLOW, ONE-DIMENSIONAL. Flow whose behaviour can be approximated by assuming that the rate of change of fluid properties across the stream lines is negligible compared with that along the stream lines.

FLUID FLOW, ROTATIONAL. Flow in a region of the fluid in which vorticity exists.

FLUID FLOW, SECONDARY. A flow which is additional to the main flow and arises from local effects within the fluid.

FLUID FLOW, SEPARATION OF. The detachment of the flow of the fluid from a solid surface with which it was previously in contact. See also: *Boundary-layer separation.*

FLUID FLOW, SLENDER BODY THEORY OF. The simplest form of the general theory of compressible inviscid fluid flow past a body, which requires that the body shall have a pointed nose and a pointed base, that the ratio of its maximum thickness to its length shall be much less than one, that the angle between the tangent plane to the body and the direction of motion shall be small, that the rate of change of this angle along the direction of motion shall also be small, and that the body shall be smooth along its length.

FLUID FLOW, STREAM LINE. See: *Fluid flow, laminar.*

FLUID FLOW, TURBULENT. Flow in which small irregular fluctuations (with time) are superposed on a mean flow.

FLUID FLOW, TWO-DIMENSIONAL. Flow whose behaviour can be approximated by considering two-dimensional space only.

FLUID FLUX. The flow of fluid across a surface and normal to it. It is usually expressed as the volume of fluid flowing in unit time.

FLUIDICS. The technology of the use of gaseous or liquid fluids in motion for purposes such as switching, sensing, and amplification. The fluid energy is manipulated in a manner similar to that of electricity in electronic devices, with the advantage that there are no moving parts.

FLUID, IDEAL. Usually refers either to an inviscid incompressible fluid, or to a perfect gas.

FLUID, INVISCID. A fluid which exerts no shearing stress, flows freely past any boundary without energy dissipation, preserves a constant density, and whose pressure at any point is independent of direction.

FLUIDIZED BED. A bed of powdered solids which is maintained in a state of non-circulating suspension by a flow of fluid through it at right angles to the bed. Its technological advantages stem from its fluidity and from the very high area of contact between the solid particles and the liquid or gas in which the suspension is maintained.

FLUID, NEWTONIAN. A fluid in which the shear stress is proportional to the shear rate. See also: *Complex body. Solid, Hookean.*

FLUIDYNE ENGINE. An engine, based on the Stirling cycle, in which the gas is replaced by a liquid (e.g. water). See: *Stirling engine.*

FLUORESCENCE. The excitation of a substance so as to excite radiation of a wavelength, or wavelengths, characteristic of that substance. If the emitted radiation is in the visible or near-visible regions of the spectrum the process is *optical fluorescence*, as distinct from *X-ray fluorescence* which occurs at much shorter wavelengths. When the emitted radiation is of the same wavelength as that of the incident radiation it is termed *resonance radiation*. Substances which continue to emit characteristic radiation after the excitation has ceased are termed *phosphorescent*. See also: *Anti-Stokes emission. Phosphorescence. Stokes law* (2).

FLUORESCENCE MICROSCOPY. The microscopic examination of fluorescent material to establish its position (e.g. in biological tissue) and to reveal fine structure which might not otherwise be visible. Fluorescent dyes may also be used as "stains" in such work.

FLUORESCENCE X-RAYS. Characteristic X-rays arising from the absorption of X- or γ-rays. See also: *Characteristic radiation. Fluorescence. X-rays.*

FLUORESCENCE YIELD. For a given type of characteristic X-rays emitted by a specified type of atom: the fraction of the ionizing events involving the ejection of photoelectrons which also result in the emission of the characteristic X-ray photons, rather than of Auger electrons.

FLUORESCENT SCREEN. A suitably mounted layer of material which fluoresces in the visible region of the spectrum when irradiated by X-rays or other ionizing radiation. See also: *Fluorography. Fluoroscopy.*

FLUORIMETRY. (1) The measurement of the fluorescent properties of materials in general. (2) An analytical technique involving the photometric measurement of fluorescence to detect, identify, and estimate fluorescent material. Also known as *fluorometry* and *fluorophotometry*.

FLUOROGRAPHY. The photography of a fluoroscopic image.

FLUOROSCOPY. The production of a visual image on a fluorescent screen by X-rays or other ionizing radiation and its use for medical diagnosis, the examination of materials, or for other purposes.

FLUTTER. An oscillatory instability that can occur on aircraft when the forward speed is greater than the critical flutter speed. It is usually characterized by a structural vibration of which two or more components of different phase are coupled by a powerful aerodynamic force.

FLUX. Refers to the flow of physical entities such as energy, charge, radiation, or atomic particles across a given area or in a given direction. The precise units of flux employed depend on the types of physical quantity involved. See also: *Electric flux. Fluid flux. Luminous flux. Magnetic flux. Neutron flux. Particle flux density. Radiant flux: Radiance.*

FLUX DENSITY. The flux per unit area. Its precise definition depends on the type of physical quantity involved. See also: *Flux. Particle flux density.*

FLUXGATE MAGNETOMETER. An electronic saturable-core magnetometer in which the degree of saturation by an external magnetic field is used as a measure of the strength of the field. It was developed during World War II and used as an airborne detector of submarines.

FLUX METER. An instrument used for measuring magnetic flux. It is essentially a moving-coil galvanometer with its coil mounted in such a manner that the suspension exerts negligible restoring torque, and in which no damping occurs other than electromagnetic. It is sometimes known as a *gauss meter.*

FLUX OF DISPLACEMENT. Another term for *electric flux.*

FLUX UNIT (ASTRONOMY). See: *Jansky.*

F-NUMBER: STOP NUMBER. Of a lens or lens system: the ratio of the focal length to the diameter of the entrance pupil. Since the amount of light passing through the system decreases as the square of this diameter the conventional *f*-numbers used in photography increase by steps of $\sqrt{2}$ (1, 1·4, 2, 2·8, 4, 5·6, etc.) so that each rise in the series requires a doubling of the exposure time, other things being equal.

FOAM. An agglomeration of gas bubbles separated from each other by thin liquid films. The films may also contain solid particles (as in the flotation of ores) and the foam is then said to be a *three-phase foam.*

FOCH SPACE. An infinite vector space in which the state of a quantum-mechanical system may be represented by an infinite set of wavefunctions each of which refers to a fixed number of particles. Each wavefunction may be regarded as a coordinate in the Foch space.

FOCOMETER. An instrument for measuring the focal length of an optical system.

FOCUS. (1) The position of convergence and divergence of a beam of radiation (particulate, electromagnetic or acoustic). The focus is usually a point or a line, but may take other forms. See also: entries under *Lens.* (2) Of a conic. See: *Conic: Conic section.* (3) Of an earthquake: See: *Earthquake.*

FOCUS, DEPTH OF. Of an optical or electron–optical system: the range, in a direction normal to the focal plane, within which the image of a point object remains in focus. It has been defined as the distance at which the central intensity of the Airy pattern image of a point is reduced to 80% of its value at the exact focus. Other definitions exist but give similar results.

FOCUSED COLLISION SEQUENCE. Refers to the "focusing" of cascades of inter-atomic collisions produced by recoil atoms arising from the bombardment of a crystal with high-energy particles. The cascades are transmitted by collision along closely packed rows of atoms, so that energy is transferred in specific directions, i.e. appears to be focused.

FOG. A term usually taken to mean a cloud of small water drops near ground level, which is sufficiently dense to reduce horizontal visibility to below 1 km (1100 yards). It may, however, refer to clouds of other composition (e.g. smoke fog, ice fog), cloud at other levels (e.g. high fog), or cloud of other densities (e.g. thin fog, dense fog).

FOG BOW. A colourless (or almost so) white "rainbow" associated with fog or mist, its pallid appearance being due to the very small size of the droplets involved. Sometimes the edges of a fog bow are seen as slightly tinted.

FOG, PHOTOGRAPHIC. The darkening of a developed photographic emulsion or a portion of such an emulsion, when the portion affected had not been exposed to light or other radiation.

FÖHN. A warm, dry wind which blows down the leeward slopes of a mountain.

FOOT. One third of an Imperial yard, and therefore equal to 0·3048 m exactly. See also: *Length, units of.*

FOOT-CANDLE. A unit of illumination equal to 1 lumen/ft² or 10·764 lux. See also: *Illuminance: Illumination.*

FOOT-CANDLE, EQUIVALENT. The term formerly used for *foot-lambert*. See also: *Luminance.*

FOOT-LAMBERT. A unit of luminance. It is the luminance of a uniform diffuser emitting 1 lumen/ft² (or 3·426 cd/m²). See also: *Luminance.*

FORBUSH DECREASE. The decrease in cosmic-ray intensity associated with a large magnetic storm.

FORCE. The name given to the physical agent which, acting upon a body, tends to cause a change in its momentum or to produce an internal strain. By extension the term is used to denote such physical agencies as electromotive force, magnetomotive force, coercive force, etc. Force is a vector quantity, unit force being that which produces unit acceleration of unit mass. In CGS units this is the *dyne*. In MKS and SI units it is the *Newton*. In the Imperial (pound–foot–second) system it is the *poundal*.

FORCE, CENTRIFUGAL. An apparent force experienced by a body in a reference frame which is rotating with respect to a stationary frame. For a body of mass m, moving in a circle of radius r with angular velocity ω, this force is equal to $m\omega^2 r$ and its direction is outwards along the radius. It is, mathematically, an imaginary inertial force introduced to make valid the use of Newton's third law of motion in a moving frame of reference, and may be regarded as the reaction to the centripetal force.

FORCE, CENTRIPETAL. The force which acts on a body which is moving about a fixed point so as to deflect that body from a straight path. For a body of mass m, moving in a circle of radius r with angular velocity ω, this force is equal to $m\omega^2 r$ and its direction is inwards along the radius; the corresponding *centripetal acceleration* is $\omega^2 r$.

FORCE, CORIOLIS. See: *Coriolis effect.*

FORCED VIBRATIONS. The vibrations induced in a system (electrical, mechanical, nuclear, etc.) by an external periodic force. The system oscillates at a frequency equal to that of the applied force and the amplitude is at a maximum when this frequency is the same as the natural frequency of the system, when *resonance* is said to occur. See also: *Oscillation, forced.*

FORCE, INERTIAL. An imaginary force introduced to permit Newton's laws of motion to be applied to a moving frame of reference. For example, the equation of motion for a body of mass m, subject to a force P, and acquiring an acceleration f, is $P = mf$. When written as $P - mf = 0$ it appears as the equilibrium condition for two forces P and $-mf$. The latter force, $-mf$, is the inertial force.

FORCES, POLYGON OF. A geometrical diagram by which the resultant of two or more forces acting simultaneously at a point is represented by the successive vectorial addition of the forces involved. The best-known example is the *parallelogram (or triangle) of forces.*

FORGING. The process by which hot metal is altered in shape or form as a result of being struck by a hammer, squeezed by a press, or forced through a die.

FORMAL CHARGE. On the atoms of a molecule: the average charge per atom calculated on the assumption that the shared electrons are divided equally between bonded atoms.

FORMANT. A specific set of resonances by which a vowel sound is characterized. For the production of a given vowel the vocal chords resonate at the formant frequencies in the cavities of the mouth, both vocal chords and mouth cavities being adjusted accordingly. See also: *Voice mechanism.*

FORM BIREFRINGENCE. Birefringence which arises not from the intrinsic anisotropic properties of a substance as in normal birefringence, but from the regular arrangement within the substance of rod-shaped particles of which the thickness and distance apart are small compared with the wavelength of light, and of which the refractive index is different from that of the medium in which they exist. See also: *Flow birefringence.*

FORM FACTOR. (1) *Atomic:* the ratio of the atomic scattering factor to the atomic number. It approaches unity as the Bragg angle approaches zero. (2) *Electrical:* the ratio of the root-mean-square value of a periodically varying wave to the mean value taken over half a period from zero to zero.

FORTRAT DIAGRAM. For the spectral lines of a diatomic molecule: a series of parabolae obtained by plotting wave number against the whole numbers which serve to identify successive lines, from which, by projection on the wave number axis, may be obtained the positions of the spectral lines in each branch.

FORWARD SCATTER. (1) The scattering of a beam of particles or radiation through angles less than 90°. (2) The actual scattered particles or radiation. See also: *Back scatter.*

FOSSIL REACTOR. See: *Oklo phenomenon.*

FOSTER REACTANCE THEOREM. A theorem giving the generalized impedance of a low-loss electrical network.

FOUCAULT CURRENT. Another name for eddy current.

FOUNTAIN EFFECT. For liquid helium II: the spraying of liquid helium II above the top of its container when the container is connected to another (also containing liquid helium II) by a capillary or porous plug, and when slight heat is applied to the first container. It arises from the flow of liquid helium II towards a region of higher temperature.

FOUR-COLOUR PROBLEM. The problem of proving the generally held belief that no more than four colours are essential to colour any map, drawn on a plane surface or the surface of a sphere, so that no two regions with a common border will have the same colour. It has been shown that four colours are sufficient for a map of less than thirty-nine regions. Proofs of the generalized problem have recently been put forward but have not so far been confirmed.

FOUR-FACTOR FORMULA. See: *Nuclear reactor, four-factor formula for.*

FOURIER ANALYSIS. The analysis of a periodic function into its simple harmonic components.

FOURIER EQUATION. A differential equation giving the temperature at any place and time when heat transport is by conduction only. See also: *Fourier law.*

FOURIER INTEGRAL. The limit of a Fourier series when the period tends to infinity.

FOURIER LAW. States that the amount of heat flowing per unit time, $\dfrac{dQ}{dt}$, across an infinitesimally small area, δA, is given by $-K\,\delta A\,\dfrac{d\theta}{ds}$, where K is the thermal conductivity of the material involved

and $\dfrac{d\theta}{ds}$ is the temperature gradient normal to the area.

FOURIER NUMBER. A number appearing in the theory of similarity for heat transport by conduction only. It is defined as xt/d^2, where x is the thermal diffusivity, t is the time, and d is a linear dimension characteristic of the geometrically similar systems under study.

FOURIER SERIES. A series

$$f(x) = \sum_{n=1}^{\infty} a_n \sin nx + \frac{1}{2} b_0 + \sum_{n=1}^{\infty} b_n \cos nx$$

which can be used to represent a single-valued function, continuous except possibly for a finite number of finite discontinuities in the interval $-\pi$ to π, and having a finite number of maxima and minima in that interval. The coefficients a_n and b_n are given by

$$a_n = \frac{1}{\pi} \int_{-\pi}^{\pi} f(x) \sin nx\, dx$$

and

$$b_n = \frac{1}{\pi} \int_{-\pi}^{\pi} f(x) \cos nx\, dx.$$

If the function is periodic, with period 2π, the Fourier series represents it for all values of x.

FOURIER SYNTHESIS. A technique applied in diffraction analysis to form a graphical representation of a crystal structure from the measured diffracted intensities. It is based on the representation of the periodic distributions of electrons, nuclei, or electric potential, as the case may be, by Fourier series, the coefficients of which are measured by the appropriate diffraction method (i.e. X-ray, neutron, or electron diffraction).

FOURIER TRANSFORM. The Fourier transform of a Fourier series,

$$F(x) = \frac{1}{\sqrt{2\pi}} \int_{-\infty}^{\infty} e^{-ixy} f(y)\, dy,$$

is given by

$$f(y) = \frac{1}{\sqrt{2\pi}} \int_{-\infty}^{\infty} e^{-ixy} F(x)\, dx,$$

subject to certain restrictions.

FOURIER TRANSFORM SPECTROSCOPY. Denotes the means of finding a spectrum by recording a two-beam interference pattern as a function of path difference within an interferometer and then performing a numerical Fourier transform on this function, from which the spectrum can be constructed. The technique offers the greatest advantage over conventional methods in the infrared region.

FOVEA. The central portion of the retina of the eye where vision is most distinct. The fovea of a normal eye subtends an angle of about 2°, but under low illumination the most sensitive portion of the retina extends to about 15° around the fovea. This area is called the *parafovea*. See also: *Purkinje effect*.

FRACTURE. The separation of a solid into parts, usually by the growth of a crack. *Brittle fracture* is fracture that occurs without a previous stage of plastic deformation. *Ductile fracture* is fracture that occurs after plastic deformation has taken place.

FRACTURE, CLEAVAGE. The separation of a crystalline material parallel to some particular crystallographic plane. It may or may not be preceded by plastic deformation. Cleavage fracture occurs when the resolved part of the applied stress normal to the fracture attains a critical value. This criterion is known as the *Sohnke law*.

FRANCK–CONDON PRINCIPLE. For electronic transitions in a molecule or crystal: a quantum mechanical criterion for the probability of an electronic transition which, in classical terms, can be understood by supposing that, for a transition to occur with a reasonable probability, it must take so little time that neither the positions nor the velocities of the heavy particles can change appreciably during the transition.

FRANCK–RABINOWITCH HYPOTHESIS. A hypothesis which explains the decreased quantum efficiencies of certain photochemical reactions in the liquid or dissolved state, compared with those of the same reactions in the gas phase, according to which the solvent molecules form a cage round a molecule which has been excited by the absorption of a quantum.

FRANKLIN. That electric charge which exerts on an equal charge, at a distance of 1 cm *in vacuo*, a force of 1 dyne. It is the unit of charge in the CGS electrostatic system of units, and is equal to $10/c$ coulomb, where c is the speed of light *in vacuo*, expressed in centimetres per second.

FRANK–READ SOURCE. A self-regenerating configuration of dislocations from which, under the action of applied stress, successive concentric loops of dislocation can be thrown off in a glide plane, thereby providing a mechanism for continuing slip in a crystal and an increase in its dislocation content, essentially without limit, through plastic deformation.

FRAUNHOFER DIFFRACTION. Diffraction of a parallel beam at an aperture, observed effectively at infinity, i.e. when the wave fronts are plane. See also: *Fresnel diffraction*.

FRAUNHOFER LINES. Dark lines in the solar spectrum, produced mainly by selective absorption in the relatively cool gaseous envelope surrounding the incandescent photosphere but partly by similar absorption in the Earth's atmosphere. There are thousands of such lines, to the most prominent of which Fraunhofer assigned reference letters *A–H*. The elements and wavelengths to which they refer are as follows: *A*, terrestrial oxygen, 7594 Å (extreme red); *B*, terrestrial oxygen, 6867 Å (red); *C*, hydrogen, 6563 Å (red); D_1 and D_2, sodium, 5896 Å and 5890 Å doublet (yellow); *E*, iron, 5270 Å (green); *F*, hydrogen, 4861 Å (blue); *G*, iron and calcium (group), 4308 Å (violet); *H*, calcium, 3968 Å (extreme violet). (Note: $1 Å = 10^{-10}$ m.) See also: *Reversing layer. Telluric lines*.

FRAZIL ICE. Ice formed in rapidly flowing streams. It is in the form of spikes and plates which do not coalesce.

FREAK WAVES. See: *Waves, episodic*.

FREE-AIR DISPLACEMENT. Of a mechanical vacuum pump: the volume of air passed in unit time, at atmospheric pressure.

FREE ELECTRON. An electron which is free to move under the influence of an external field. The term may refer to electrons *in vacuo* or to electrons in the conduction band of a solid. See also: *Bound electron. Conduction electron*.

FREE ELECTRON MODEL. For an electronic system containing more than one electron: a model for the solution of the quantum mechanical equations for the system, according to which the electrons concerned are assumed to have no interactions with each other. The model has been very successful in the study of the electrical conductivity of metals.

FREE ENERGY. That portion of the energy of a system which is available for conversion to work: (1) *Helmholtz free energy:* given by $U - TS$ where U is the internal energy, S the entropy, and T the temperature. It is especially useful in describing systems where the independent variables are temperature, volume, and composition. It is also known as the *thermodynamic work function:* (2) *Gibbs free energy:* given by $U - TS + pV$, where p is the pressure and V the volume. It is especially useful in describing systems where the independent variables are temperature, pressure and composition. See also: *Surface energy*.

FREE MOLECULE FLOW. See: *Rarefied gas laws*.

FREE RADICAL. A molecule which possesses an incomplete shell of valency electrons, does not carry an electrical charge, and is not a free ion. Such a radical usually possesses considerable additive properties and is very reactive. The free radical is one of the most important intermediates in many organic reactions.

FREE-RUNNING GENERATOR. See: *Pulse generator*.

FREEZE COATING. The application of a coating of one material on to another in a continuous process involving the removal of latent heat from the coating material by using the substrate as a heat sink.

FREEZE-DRYING. The pumping away of water vapour sublimed from a frozen specimen (usually biological). The process is used, for example, in drying blood serum, which remains in a sponge-like skeleton, readily stored at normal temperatures, but capable of rapid re-solution when water is added.

FREEZING. The process of solidification from the liquid state. It is believed that the long-range order characteristic of a crystalline solid is preceded in the liquid by some form of short-range order. See also: *Melting*.

FREEZING MIXTURE. See: *Cryohydrates*.

FREEZING POINT. Of a liquid: the temperature at which it solidifies at atmospheric pressure.

FREEZING-POINT DEPRESSION. See: *Cryoscopic constant*.

FREMONT TEST. See: *Impact tests*.

FRENKEL DEFECT. See: *Point defect*.

FREON. One of a series of compounds consisting of ethane or methane in which some or all of the hydrogen has been substituted by fluorine or by fluorine and chlorine. Typical uses are as refrigerants and as dispersal fluids for insecticides.

FREQUENCY. (1) Of a periodic phenomenon: the number of oscillations or cycles occurring in unit time. For a progressive wave motion the frequency is given by the speed divided by the wavelength. (2) Of a class in statistical distribution: the number of members of that class. See also: *Frequency distribution*.

FREQUENCY CONVERTER. A heterodyne device for changing an incoming radio signal from its original carrier frequency to a fixed intermediate carrier frequency.

FREQUENCY DISTRIBUTION. The description, by a simple mathematical expression or smooth curve, of the way in which a particular property is distributed among the members of a group, e.g. the number of men whose heights fall within given intervals of height. The number falling in such an interval is known as the *frequency* appropriate to that interval. See also: *Average: Mean. Gaussian distribution: Normal distribution. Poisson distribution. Standard deviation*.

FREQUENCY, FUNDAMENTAL. Of a periodic quantity (e.g. a musical note or electrical waveform): the frequency of the lowest harmonic component of that quantity.

FREQUENCY MODULATION. The process of imparting information to a carrier wave by causing the frequency of the carrier to vary in accordance with an input signal. See also: *Modulation*.

FREQUENCY MULTIPLIER. An electrical network for generating harmonics from a sinusoidal input and selecting a higher harmonic as its output.

FREQUENCY, NATURAL. The frequency of free oscillation. For a resonant electrical circuit the frequency, known as the *resonant frequency*, is given by $1/(2\pi \sqrt{LC})$, where L is the inductance and C the capacitance of the circuit.

FRESNEL BIPRISM. A single prism of very obtuse angle that can be considered as two prisms of very acute angle placed base to base. It is used for producing two coherent images from the same source.

FRESNEL DIFFRACTION. Diffraction at an aperture, when either the source of radiation or the point of observation, or both, are at a finite distance from the aperture, i.e. when the wave fronts are spherical. See also: *Fraunhofer diffraction*.

FRESNEL ELLIPSOID: RAY ELLIPSOID. A geometrical construction for the determination of propagation velocities and vibration directions in a doubly refracting crystal. It takes the form of an ellipsoid, the semi-axes of which are proportional to the reciprocals of the principal indices of refraction, and the directions of which are parallel to the corresponding vibration directions. See also: *Fletcher indicatrix: Optical indicatrix. Normal surface. Ray surface*.

FRESNEL EQUATIONS. Equations expressing the fraction of incident radiant energy reflected at a surface boundary in terms of the plane of polarization of the radiation, the angles of incidence and refraction, and the refractive indices of the media on each side of the boundary. Suitably modified they are also applicable to metallic reflection.

FRESNEL–HUYGHENS PRINCIPLE. See: *Huyghens principle*.

FRESNEL INTEGRALS. Integrals which occur in problems of wave motion, defined by:

$$C(x) = \int_0^x \cos\left(\tfrac{1}{2}\pi u^2\right) du, \qquad S(x) = \int_0^x \sin\left(\tfrac{1}{2}\pi u^2\right) du.$$

145

FRESNEL MIRROR: BIMIRROR. A pair of plane mirrors slightly inclined to one another, used for producing two coherent images in interference experiments.

FRESNEL RHOMB. A glass rhomb used for obtaining circularly polarized light from plane-polarized light by total internal reflection.

FRESNEL ZONES: HALF-PERIOD ZONES. Annular elements in which it is convenient to divide a wave front when determining the amplitude at a given point resulting from Fresnel diffraction. The zones are such that radiation from one zone reaching the point is one half-period out of phase with that from the adjacent zones. The resultant amplitude is half the amplitude that would result if all but the first (axial) Fresnel zone were blocked out. If a screen (known as a *zone plate*) is interposed so as to obstruct alternate zones, the intensity at the point of interest can be made very large.

FRETTING CORROSION. The wear or surface damage occurring when two surfaces rub over each other cyclically with a "normal" load and very small amplitude. It can cause serious damage to machinery.

FRICTION. The resistance offered to the relative motion of two bodies in contact. It acts in a direction tangential to the surface of contact. *Static friction* refers to bodies which are motionless relative to each other, and is greater than *kinetic* or *sliding friction*, which refers to sliding surfaces. This in turn is greater than *rolling friction*, which refers to one body rolling on the other. See also: *Friction, fluid. Surface friction: Skin friction.*

FRICTION, ANGLE OF. The angle whose tangent is equal to the coefficient of static friction. Also known as the *limiting angle of friction*. See also: *Friction, cone of.*

FRICTION, BOUNDARY. Friction between two surfaces which are neither perfectly dry nor separated completely by a lubricant. They may be covered, for example, by a thin film which is only a few molecules thick.

FRICTION, COEFFICIENT OF. (1) For kinetic or sliding friction between two surfaces: the ratio of the tangential force which is required to sustain motion without acceleration of one surface with respect to the other, to the normal force at the surface of contact. (2) For static friction between two surfaces: the ratio of the tangential force which is required initially to produce motion of one surface with respect to the other, to the normal force at the surface of contact.

FRICTION, CONE OF. A conical surface containing the resultant of the force of friction between two surfaces and the normal force at the surface of contact. Its semi-angle is the angle of friction.

FRICTION, FLUID. Friction appertaining to the relative movement of a solid and a fluid. It is proportional to the speed of motion and the area of the surface of contact, but not to the normal force, and must therefore be distinguished from solid (dry) friction.

FRICTION, STICK-SLIP. A friction process in which static and kinetic friction alternate.

FRIEDEL LAW. A law originally applied to X-ray diffraction by crystals, but also valid for electron or neutron diffraction. It states that the intensities of reflection from opposite sides of the same crystal planes (i.e. the *hkl* and *hkl* planes) are the same, so that certain asymmetries (e.g. polarity) are not revealed and the presence or absence of a centre of symmetry is left open. The law is not valid where anomalous dispersion occurs, e.g. in the neighbourhood of an X-ray absorption edge.

FRIGORIE. A unit of heat equal to 10^{-3} thermie. It is 10^3 cal$_{15°}$, i.e. $4 \cdot 1868 \times 10^3$ J. See also: *Thermie.*

FRINGING FIELD. The field in an electron microscope produced by scattered electrons. Confusion of the final image by this is prevented by a suitable stop.

FRONT (METEOROLOGY). A surface of discontinuity between air masses at different temperatures, especially at the intersection of the surface with the surface of the Earth. A *polar front* separates polar air from tropical air. A *cold front* signifies the advance of cold air to displace warm, the latter being pushed forwards and upwards to produce rain, thunderstorms and squalls, followed by clearer weather. A *warm front* signifies the advance of warm air to replace cold, the latter being pushed forwards and the former sliding up and over it to produce steady rain. An *occluded front* develops when a cold front overtakes a warm front, the advancing cold air overtaking the preceding cold air (albeit usually at a different temperature) and pushing the warm air upwards with the production of rain. When the overtaking cold air is at a lower temperature than the preceding cold air the occlusion is known as a *cold occlusion*. When the opposite is the case the term *warm occlusion* is used.

FRONTOGENESIS. In meteorology: the formation or intensification of a front.

FRONTOLYSIS. In meteorology: the weakening or disintegration of a front.

FROST POINT. The temperature at which the water vapour present in the atmosphere will pass directly into the solid state without passing through an intermediate liquid state. It is always below 0°C.

FROUDE METHOD. A method of extrapolating the resistance of a ship model to that of a full-scale vessel, based on the division of ship resistance into viscous and wave-making resistance. See also: *Ship resistance.*

FROUDE NUMBER. A number of importance in considering the dynamical similarity of fluid flows. It is sometimes defined as V^2/lg and sometimes as $V/\sqrt{lg}$, where V is the velocity, l is a typical length of a body belonging to a family of geometrically similar bodies, and g is the acceleration due to gravity. See also: *Dynamical similarity principle.*

F-STAR. See: *Stars, spectral classes of.*

FUEL CELL. An electrochemical cell in which the energy of reaction between a conventional fuel and oxygen (preferably from air) is converted directly and continuously into low-voltage direct-current electrical energy. It involves an anode (at which fuel is oxidized), a cathode (at which oxygen is reduced), an external circuit and an electrolyte. Charging is not necessary, all that is required being a supply of fuel and oxygen.

FUEL SPIKE: FUEL SEED. Nuclear fuel containing more highly enriched material than the majority of the other fuel in a reactor. See also: *Nuclear reactor fuel, enriched.*

FUGACITY. A fictitious pressure which, when substituted for the partial pressure, enables some of the thermodynamic equations for a real gas to be written in the same simple forms as those for a perfect gas. The fugacity p^* of a gas in a real gas mixture may be defined by

$$\ln p^* = \frac{\mu - AT}{RT} \, ,$$

where μ is the chemical potential of the gas, A is a constant, T is the temperature, and R is the gas constant.

FULCHER BANDS. A system of spectral bands of molecular hydrogen which are preferentially excited by a low-voltage electron discharge. They consist of a number of regularly spaced lines in the red and green and were the first systems of molecular hydrogen to be analysed.

FULL RADIATION. Synonym for black-body radiation. See also: *Black body.*

FULL RADIATOR. A black-body radiator.

FULL-WAVE CIRCUIT. A high-voltage circuit in which alternate half-cycles are reversed.

FUNCTION. A mathematical expression describing the relationship between variables. The function takes on a definite value or set of values when values are assigned to one or more independent variables. Thus, a dependent variable y may be expressed as $y = $ function of x or $f(x)$ in the case of one independent variable, or, if there are several independent variables, as $y = f(x_1, x_2, ..., x_n)$. If, to each value of x, or of the set of values $x_1, ..., x_n$, there is a unique value of y, y is said to be a *single-valued function* of x or of the set in question. If there are several values of y, y is said to be a *many-valued function.*

FUNCTION, HOMOGENEOUS. A function $f(x)$ which, for any parameter t, satisfies the relation $f(x) = f(tx)/t^n$, $f(x)$ signifying $f(x_1, x_2, ..., x_n)$. The exponent n is the *degree* or *order* of the function.

FUNCTION, MONOTONIC. A function which changes continuously in the same sense with a continuous increase of the independent variable (and hence has no maxima or minima), i.e. the derivative always has the same sign.

FUNDAMENTAL FREQUENCY. Of a periodic quantity (e.g. a musical note or electrical wave form): the frequency of the lowest harmonic component of that quantity.

FUNDAMENTAL INTERVAL. Of a thermometer: the difference between the reading at the ice point and the steam point. A platinum-resistance thermometer, for example, may have a fundamental resistance of about 10 ohms.

FUNDAMENTAL PARTICLES. See: *Elementary particles.*

FUNDAMENTAL PHYSICAL CONSTANTS. Constants which occur naturally in the field of physics, such as the electronic charge, the velocity of light, the Avogadro number, Planck's constant, the Rydberg constant, the fine structure constant, the gravitational constant, and so on. With the apparent exception of the gravitational constant, the fundamental constants are interdependent, and a precise knowledge of only a small number is all that is needed to establish the values of the entire group. See: *Appendix III.*

FUNDAMENTAL THEOREM OF ALGEBRA. States that: if $f(z)$ is a polynomial of degree m in z, then the equation $f(z) = 0$ possesses at least one root, which may be complex.

FUNDAMENTAL THEORY OF EDDINGTON. A theory based on the idea that, in describing a physical system, it is essential to take account of the observer, the nature of the observations, and the processes of measurement, as in relativity and quantum theory. Eddington claimed, with considerable success, that the values of many dimensionless constants of importance in physics (e.g. the proton–electron mass ratio and the fine structure constant) could be worked out from this theory.

FURNACE. Any device in which materials may be subjected to continuous intense heat. The source of heat may be the *combustion* of liquid, gaseous, or solid fuel, or be derived from the use of *electrical*, *solar*, or *nuclear* energy, or from bombardment by an *electron beam*.

FURNACE, ARC. An electric furnace in which the heat is produced by means of an electric arc between suitable electrodes or between an electrode and the furnace charge. In the *arc-image furnace* the heat is focused optically at the point of application.

FURNACE, BLAST. Commonly refers to a furnace in which iron-ore is smelted to produce pig-iron. It derives its name from the use of a blast of air in the melting process. Similar furnaces are also used in the melting of copper, lead, antimony, tin, cobalt, and nickel ores.

FURNACE, INDUCTION. An electric furnace used for the melting of metals, in which the heat is produced by currents induced in the charge itself.

FURNACE, MUFFLE. A furnace in which heat is applied to the outside of a refractory chamber within which the charge is contained.

FURNACE, OPEN HEARTH. A gas-heated furnace in which the material being heated rests on a hearth. The flame and heating gases pass over the surface of the hearth.

FUSED-RING COMPOUND. A compound containing two or more closed rings of atoms and having two adjacent atoms in each ring in common with the next ring.

FUSIBLE PLUG. A plug with a core of a low-melting-point alloy, used in the lid of a pressure vessel to release the pressure if overheating occurs. By a suitable choice of metals (usually a mixture of bismuth, lead, tin, cadmium, etc.) melting temperatures ranging from 60° to 180°C can be obtained.

FUSION. See: *Melting*.

g

GAIN. (1) A general term used to denote an increase in signal power in transmission from one point to another. (2) The ratio of output to input, e.g. voltage or current gain. See also: *Aerial gain.*

GALACTIC CLUSTER: OPEN CLUSTER. See: *Star cluster.*

GALACTIC EQUATOR. The great circle on the celestial sphere which is 90° from the galactic poles.

GALACTIC HALO. A diffuse halo of stars, with a radius of some 30 kiloparsecs, which surrounds the Galaxy. It is composed of *RR Lyrae* variables, subdwarf stars, globular clusters, bright red giants, and—in the galactic polar regions—blue stars. The galactic halo exhibits a marked concentration towards its centre. (Note: 1 parsec $\approx 3.0857 \times 10^{16}$ metres.)

GALACTIC PLANE. The plane of the Galaxy defining the galactic equator.

GALACTIC POLES. The poles defined on the celestial sphere by the intersection with the sphere of a line through its centre drawn normal to the plane of circular symmetry of the Galaxy. A less ambiguous definition of this plane is provided by the concentration of hydrogen gas in the Galaxy into a single galactic plane, observed from the emission of 21 cm wavelength radio waves.

GALAXIES, EMISSION OF RADIO AND X-RAYS FROM. See: *Radiogalaxy. X-ray stars and galaxies.*

GALAXIES, RECESSION OF. The recession of the extragalactic nebulae (which themselves constitute galaxies), which is deduced from observed displacements of their spectra towards the red end, these displacements being interpreted as manifestations of the Doppler effect, corresponding to radial motion away from the Earth. See also: *Nebula, extragalactic. Hubble law.*

GALAXY. A flattened system of stars, dust, and gas, which includes the Sun and at least all stars that are visible to the naked eye. It has a diameter of about 25 kiloparsecs (kpc) and a central thickness of about 5 kpc, with a mass at least 10^{11} times that of the Sun. The Galaxy is popularly known as the *Milky Way*. By extension, any similar system of stars, etc. is known as a galaxy. (Note: 1 parsec $\approx 3.0857 \times 10^{16}$ metres) See also: *Nebula, extragalactic.*

GALAXY, FIELD. An isolated galaxy that does not belong to any particular cluster. See also: *Stars, field.*

GALAXY MACHINE. An acronym for *G*eneral *A*utomatic *L*uminosity and *XY*, referring to a machine which can extract the vast amount of stellar information recorded on photographic plates at a rate comparable with the rate at which the information is produced. The machine finds the stellar images and measures their positions, sizes, and densities, all automatically, at a rate of 1000 images an hour, the size and density being a measure of the strength of such an image.

GALAXY, MAGNETIC FIELD OF. A field which has been postulated to account for a number of phenomena, including the polarization of starlight, the origin of cosmic rays, and the emission of synchrotron radiation from the Galaxy and from extragalactic sources.

GALAXY, NUCLEUS OF. The concentration of mass at the centre of the Galaxy.

GALAXY, ROTATION OF. The rotation of the Galaxy about a distant centre. It accounts for the observed motions of the stars and is such that the rotational velocity at the distance of the Sun from the centre is about 200–300 km/s, corresponding to a period of revolution of about 2×10^8 years.

GALAXY, VIOLENT (ALSO EXPLOSIVE OR ACTIVE). A galaxy releasing large quantities of energy, i.e. about 10^{51}J compared with a typical supernova release of 10^{42}J.

GALLON. (1) The fundamental unit of volume in the Imperial system of measurement. By definition, it is the volume of 10 Imperial pounds of distilled water at 62°F at a barometric pressure of 30 in. (2) The United States unit of volume, defined as the volume of 231 in^3. One Imperial gallon is 1·20094 US gallons, and one US gallon is 0·83267 Imperial gallons. (Note: 1 Imperial gallon = 4·5461 dm³.) See also: *Litre. Appendix II.*

GALVANIC CELL. An older name for electrochemical cell.

GALVANIZING. The process of coating iron or steel with a layer of zinc to prevent atmospheric corrosion.

GALVANOMAGNETIC EFFECTS. Phenomena in which a potential difference is developed as a result of the flow of electric current in the presence of a magnetic field, as in the Hall effect, which is defined separately. See also: *Thermoelectric effects. Thermomagnetic effects.*

GALVANOMETER. An instrument for measuring electric current by its magnetic effect, in which observations are made of the deflection produced by the torque exerted between an electric circuit in a coil and a magnet. In an *astatic* galvanometer the coil is stationary and the magnet turns. In a *moving-coil* galvanometer the reverse is the case.

GALVANOMETER, BALLISTIC. A galvanometer designed to measure the total quantity of electricity in a transient current. The period of swing of its moving element is long compared with the time of discharge so that the amplitude of the first oscillation of the galvanometer is proportional to the quantity of electricity to be measured.

GALVANOMETER, D'ARSONVAL. The best-known form of moving-coil galvanometer. It is practically undamped and if fitted with a heavy coil can also be used as a ballistic galvanometer. It is, however, normally fitted with a light coil. The current to be measured passes through the coil via the suspension.

GALVANOMETER, TANGENT. A galvanometer in which a small magnetic needle is mounted at the centre of a circular coil, the current being proportional to the tangent of the angle of deflection.

GALVANOMETER, VIBRATION. A galvanometer for used with a.c. The movement is undamped but tunable to have natural frequencies ranging from 5 to 1000 Hz.

GALVANOPLASTY. The production of exact copies of articles by electrolytic means, a coating of copper, for example, being deposited on a mould which is subsequently removed.

GAMES, THEORY OF. The theoretical study of the outcome of a game of strategy, as opposed to one of pure chance, in which are taken into account, at each stage of the game, the various courses of action open, the amount of information available, and the probabilities of chance events. Criteria for deciding the end of the game and the "amount" received or paid out by each player are also included. The theory of games may be applied to military, economic and political problems.

GAMMA. (1) Of a photographic emulsion: See: *Photographic emulsion, characteristic curve of.* (2) A unit of magnetic flux density equal to 10^{-9} tesla (10^{-5} gauss). See: *Appendix II.*

GAMMA-DISPERSION. In body tissues: See: *Body tissues, dielectric dispersion in.*

GAMMA FACTOR. In the theory of the fine structure of spectral lines: the interaction energy of two quantum vectors. It is of special importance in the Zeeman effect.

GAMMA FUNCTION. This is defined as

$$\int_0^\infty e^{-x} x^{n-1}\, dx,$$

where n is a real number greater than zero. See also: *Beta function.*

GAMMA-RAY ABSORPTION COEFFICIENT. The fraction of a beam of γ-rays which is removed by absorption on passing through unit thickness of material, measured for very small thicknesses. It is less than the attenuation coefficient. See: *Absorption coefficient for radiation. Attenuation coefficient for radiation.*

GAMMA-RAY CONSTANT, SPECIFIC: GAMMA-RAY EMISSION, SPECIFIC. For a radioactive nuclide: the exposure rate at unit distance from a unit point source. It is also known as the *k-factor.*

GAMMA-RAY HEATING. A method of supplying a controlled amount of heat to a thermally isolated sample at a temperature below 1 K, by γ-ray irradiation. It is used, for example, in low-temperature calorimetry. Gamma-rays from cobalt-60 are commonly used.

GAMMA-RAY RADIOGRAPHY. Radiography by means of γ-rays emitted by radioactive sources. Formerly radium or radon was employed but with the advent of nuclear reactors, radioisotopes such as ^{60}Co are in more common use. See also: *Radiograph.*

GAMMA-RAYS. Electromagnetic radiation emitted in the process of nuclear transition or particle annihilation. See also: *Electromagnetic spectrum.*

GAMMA-RAYS, DELAYED. Gamma-rays emitted in the decay of fission products. They represent about 3% of all γ-rays emitted as a result of fission.

GAMMA-RAY SPECTROMETER. An instrument designed for measuring the distribution of energy and intensity in γ-ray spectra. Such an instrument usually involves the use of scintillation techniques, proportional counters, or spectroscopy by crystal diffraction. See also: *Pair spectrometer.*

GAMMA-RAY SPECTRUM. The emission spectrum of sharp lines, corresponding to well-defined quantum energies, resulting from nuclear transitions between two or more energy states.

GAMMA-RAYS, PROMPT. Gamma-rays accompanying fission without measurable delay.

GAS. A state of matter in which the molecules move freely, so that a gas will completely fill the region in which it is contained. A substance which is in the gaseous state but below the critical temperature is generally described as a *vapour.*

GAS AMPLIFICATION: GAS MULTIPLICATION. (1) The process whereby, in a sufficiently strong electric field, the ions produced in a gas by ionizing radiation produce additional ions. (2) The factor by which the initial ionization in (1) is multiplied.

GAS ANALYSIS. See: *Katharometer. Infrared analysis of gases.*

GAS CONSTANT: MOLAR GAS CONSTANT. The constant of proportionality R in the equation of state for a perfect gas, $pv = RT,$ referred to one mole of gas. In this equation p is the pressure, v the volume, and T the absolute temperature. R is equal to $8 \cdot 314 \, \text{J/K}$, or $1 \cdot 985 \, \text{cal/K}$. It may be interpreted as being two-thirds the total kinetic energy due to transitional motion of the molecules of an ideal gas at 1 K. See: *Appendix III.*

GAS-DENSITY BALANCE. A detector designed for use in gas–liquid chromatography systems. It is based on the comparison of the density of the gaseous chromatographic effluent with that of a gaseous reference standard.

GAS DISCHARGE. See: *Discharge, glow.*

GAS DYNAMICS. The study of the dynamics and thermodynamics of flowing gases and vapours.

GASES, ANOMALOUS DISPERSION IN. Exceptional optical dispersive effects shown by some gases (e.g. sodium) in the neighbourhood of certain spectral lines, although gases in general exhibit very small dispersive powers for light. Such anomalous effects have important applications in geophysics.

GAS FOCUSING. The focusing of an electron beam, e.g. in a cathode-ray tube, by the action of ionized gas.

GAS, PERFECT: GAS, IDEAL. A gas which behaves as an assembly of perfectly elastic spherical "point" masses in random motion which exert no forces on each other except during collisions. The behaviour of such a gas is described by seven well-recognized "laws", the *perfect-gas laws.* See also: *Avogadro law. Boyle law. Charles law: Charles–Gay-Lussac law. Dalton law of partial pressures. Gay-Lussac law. Joule law of energy content. Maxwell distribution law. Maxwell–Boltzmann distribution law.*

GAS SCRUBBING. The treatment of a gas with a liquid, usually water, to remove particulate or gaseous components.

GAS TURBINE. An internal combustion engine in which mechanical energy is obtained from the expansion of a high-temperature, high-pressure gas stream provided by the combustion of a suitable fuel with compressed air. The expanded gas drives the compressor via a turbine, and may either drive a second turbine to provide shaft power, or may be expanded to atmospheric pressure in the exhaust system to provide a jet of high-velocity gas for propulsion. In the *turbo-jet* both methods are used, with the emphasis on the first. Gas turbines are used for aircraft propulsion; marine, locomotive, and automobile propulsion; stationary power plants; and as ancillaries in industrial plant. See also: articles beginning *Turbine.*

GAUGE. (1) An instrument or device for measuring dimension, pressure, or other physical property. (2) The condition imposed upon the vector and scalar potentials of an electromagnetic field to achieve symmetry (*gauge invariance*).

GAUGE THEORIES. (1) Modern gauge theory: a development of quantum electrodynamics. (2) Gauge theories with spontaneous symmetry breaking: a development of gauge theory which has been successful in linking the strong and weak interactions and has led to the prediction and discovery of neutral currents and the charmed quark. See also: *Field theory unified. Neutral currents.*

GAUSS. The CGS unit of magnetic induction or flux density. It is equal to 1 line (or Maxwell) per cm^2 and to 10^{-4} tesla. See: *Appendix II.*

GAUSSIAN DISTRIBUTION: NORMAL DISTRIBUTION. The most important theoretical distribution in statistics. It is that which gives, among other things, the distribution of errors of observation. It is described by the equation

$$p(x) = \frac{1}{\sigma \sqrt{(2\pi)}} \exp\left(-\frac{(x-\mu)^2}{2\sigma^2}\right),$$

where $p(x)\,dx$ gives the probability that x lies

151

between x and $x + dx$, μ is the mean of the data, and σ the standard deviation. The probability that a randomly selected value of x shall have a value greater than X is given by

$$\int_x^\infty p(x)\,dx = \frac{1}{\sqrt{(2\pi)}} \int_u^\infty \exp\left(-\frac{1}{2}u^2\right) du,$$

where $u = (X - \mu)/\sigma$. See also: *Error function.*

GAUSSIAN IMAGE POINT. For a system with spherical aberration: the image point to which the paraxial rays converge.

GAUSS METER. See: *Flux meter.*

GAUSS METHOD OF WEIGHING. See: *Double-weighing.*

GAUSS THEOREM. States in general that the total flux of a vector field through a closed surface is equal to the volume integral of the divergence of the vector taken over the enclosed volume. It may be applied to surfaces drawn in electric and magnetic fields, and to gravitational, magnetostatic, and fluid-velocity fields. Thus, for example, the electric flux from an enclosed electric charge q is $4\pi q$ and the magnetic induction is zero.

GAY–LUSSAC LAW. For a perfect gas: states that the volumes in which gases combine chemically bear a simple whole-number relation to each other and to that of the resulting product if that is also gaseous, and if measured at the same temperature and pressure.

G-CENTRE. See. *Colour centres.*

GEGENIONS. See: *Electrical double layer.*

GEGENSCHEIN: COUNTERGLOW. A region associated with the zodiacal light which is brighter than the neighbouring regions. It is opposite to the Sun and is about 10° in diameter.

GEIGER LAW. A relationship between the initial velocity of α-particles and their range, applicable to ranges from 3 to 7 cm. It states that $v^3 = Ar$, where v is the initial velocity, r the range, and A a constant.

GEIGER–NUTTALL RULE. A rule originally relating the lifetime of an α-emitter to the range of the α-particles, but now usually expressed in terms of the decay energy as follows: $\log \lambda = a + b \log E$, where λ is the decay constant, E the kinetic energy of the emitted α-particle, and a and b are constants appropriate to each of the three naturally occurring radioactive families.

GEIGER REGION. In a Geiger–Müller counter: that range of operating voltage over which the output charge is independent of the number of primary ions produced by the initial ionizing event. See also: *Counter, Geiger–Müller.*

GEIGER THRESHOLD. The lowest operating voltage in the Geiger region.

GEISSLER TUBE. An early form of discharge tube, containing air or ether at a pressure of about 1% of an atmosphere, and exhibiting marked fluorescent effects.

GEL. A colloidal system in which the dispersed substance forms a continuous, branching, cohesive network. It may contain a large proportion of liquid and yet still possess solid-like properties.

GENE. A constituent part of a chromosome. The nature and arrangement of genes govern the transmission of hereditary characteristics. See also: *Chromosome. Genetic code.*

GENERALIZED FUNCTION: DISTRIBUTION FUNCTION. A mathematical function used in bringing symbolic functions such as the Dirac delta function within the scope of a rigorous mathematical treatment.

GENERALIZED SECONDARY IONIZATION COEFFICIENT. See: *Ionization coefficient, secondary.*

GENETIC CODE. Refers to the relationship between the genetic information stored in the deoxyribonucleic acid (DNA) of the chromosomes and that manifested in protein structure. It is concerned with the way in which combinations of the four different kinds of nucleotide found in DNA are associated with the subsequent production of one or more of the twenty different kinds of amino acid present in proteins. See also: *Chromosome. Gene.*

GEOCENTRIC. In astronomy: refers to any system or mathematical construction which has the centre of the Earth as its reference point.

GEOCHRONOLOGY. The study of physical methods for the dating of events in the Earth's history, e.g. by radioactive methods. See also: *Age determination by radioactivity.*

GEODESIC. That curve connecting two fixed points which has an extreme length (maximum or minimum). It is a straight line in plane Euclidean geometry and part of a great circle on the surface of a sphere. In non-Euclidean geometry geodesics are of priimportance, e.g. in the general theory of relativity.

GEODESY. The study of the size and shape of the Earth. It includes such things as triangulation and survey, measurement of height above sea-level, observations of the direction and intensity of gravity, determination of the figure of the Earth, magnetic measurement, tidal measurements, and—sometimes—seismic measurements.

GEODESY, FOUR-DIMENSIONAL. The study of changes in the configuration of the Earth with time, e.g. of changes in sea-level, wandering of shorelines, movement of the poles, etc. Four-dimensional geodesy involves latitude, longitude, elevation, and time.

GEODETIC TRIANGLE: SPHEROIDAL TRIANGLE. Three points on a spheroid together with the three geodesic lines which join them.

GEODIMETER. An instrument for measuring a large distance by measuring the time taken for light to travel that distance. Conversely, knowing the distance, it can be used to measure the speed of light. Distances of up to 35 km have been measured with an accuracy of 1 part in 10^6. See also: *Laser applications* (3). *Mekometer. Tellurometer.*

GEOID. That equipotential surface of the Earth which coincides approximately with mean sea-level. Its size and shape and its continuation under the continents is known as the figure of the Earth. See also: *Earth, figure of. Equipotential surface. Spheroid.*

GEOID, ISOSTATIC. A surface lying above or below the spheroid at the same distance as the geoid lies above or below the co-geoid.

GEOMAGNETIC ALBEDO. See: *Albedo* (3).

GEOMAGNETIC AXIS. The axis of the centred dipole of the Earth. See also: *Dipole, centred.*

GEOMAGNETIC FIELD. The magnetic field of the Earth. It may be represented by a magnetic dipole supposed to be situated at the centre of the Earth, but the major part almost certainly arises from electric currents flowing inside the Earth. See also: *Dipole, centred. Magnetopause.*

GEOMAGNETIC FIELD, ANNUAL VARIATION OF. The variation of the monthly mean value of the geomagnetic field with a period of one year. The variation is about 10^{-4} oersted, i.e. about 79.6×10^{-4} A/m.

GEOMAGNETIC FIELD, REVERSALS OF. The reversal of the polarity of the mean dipole field at several periods in the geological past. The evidence for such reversals comes from palaeomagnetic measurements.

GEOMAGNETIC FIELD, VARIATIONS OF. The variations of the mean value of the geomagnetic field. They may be daily, annual, or transient.

GEOMAGNETIC POLES. The intersections of the geomagnetic axis with the Earth's surface. See also: *Magnetic poles.*

GEOMAGNETIC STORMS. Intense world-wide variations in the Earth's magnetic field. They are associated with the occurrence of sun-spots and solar flares.

GEOMAGNETISM. The study of the Earth's magnetic field and its secular and transient variations.

GEOMETRICAL OPTICS. The treatment of light paths as a series of rays. Diffraction, interference, and polarization effects are ignored and there is straight-line propagation in a homogeneous medium. See also: *Physical optics. Rays.*

GEOMETRIC ATTENUATION. See: *Attenuation of ionizing radiation.*

GEOMETRODYNAMICS. The study of the geometry of curved empty space and the evolution of this geometry with time. It is based on the theory of general relativity and is a broad, but often speculative, investigation of the extent to which physical phenomena can be interpreted as aspects of the geometry of space-time.

GEOMETRY. In radiation experiments: a term used colloquially to signify the arrangement in space of the various components, as in, for example, *plane geometry*, 2π *geometry*. *Good geometry* and *bad geometry* denote the suitability or otherwise of this arrangement for some particular purpose.

GEON. A postulated entity consisting of a bundle of electromagnetic energy held together for a finite time by its own gravitational attraction. It has not been observed. A similar bundle of gravitational energy has been termed a *gravitational geon*. It also has not been observed.

GEOPHYSICS. That branch of physics which correlates and explains certain natural phenomena occurring in the Earth and its environs. The term is also used in a restricted sense to mean the physics of the solid Earth.

GEOPOTENTIAL. At a given point in the atmosphere: the gravitational potential energy of unit mass located at that point.

GEOSYNCLINE. An elongated large-scale depression of the Earth's surface, generally wholly or partially filled with sediment and frequently subjected to strong folding.

GEOTHERMAL ENERGY. Energy derived from the utilization of the geothermal gradient, especially in places where this is high. Geothermal wells have been sunk in a number of places in the U.S.A., and geothermal energy has been used in Iceland for a long time.

GEOTHERMAL GRADIENT. The rise in temperature of the Earth's crust with increasing depth.

GETTER. A material introduced into a vacuum enclosure to establish and maintain a very low pressure of residual gas. Materials commonly used are phosphorus, zirconium, titanium, and charcoal.

G-FACTOR. See: *Gyromagnetic ratio. G-value. Landé g-factor: Landé splitting factor.*

GHOSTS. (1) Spectral ghosts. (2) Spurious images seen on a television screen and arising from echoes.

GIANT PLANET. See: *Planet, major.*

GIANT STAR. A star which is brighter than the main sequence stars of the same spectral class. See also: *Hertzsprung–Russell diagram. Stars, spectral classes of. Star, supergiant.*

GIBBOUS. Refers to the appearance of the Moon or a planet as less than a circular disk but greater than a half-disk. The condition for this is that the Moon shall be between quadrature and opposition, an inferior planet shall be between maximum elongation and superior conjunction, and a superior planet shall be at any point of its orbit other than opposition or conjunction. See also: *Falcated.*

GIBBS–DUHEM EQUATION. Relates the chemical potentials in a mixture to the constitution of the mixture at constant temperature and pressure.

GIBBS FREE ENERGY. See: *Free energy.*

GIBBS–HELMHOLTZ RELATION. The reformulation of the expression for the Helmholtz free energy used to facilitate calculations of changes in the energy or enthalpy of a system. See also: *Free energy.*

GIBBS THEOREM. States that when two perfect gases, each occupying the same volume, are mixed isothermally so as to form a perfect gas mixture also occupying the same volume, the change in entropy is zero.

GILBERT. The unit of magnetomotive force in the CGS system. It is equal to $4\pi/10$ ampere-turns.

GIPSY PARTICLE. See: *J/Psi particle.*

GLACIER. A moving ice mass on the surface of the Earth which survives from year to year.

GLADSTONE–DALE LAW. A relationship between the changes in refractive index induced by compression or change of temperature. The form of the law is: $(n + 1)/\varrho =$ constant, where n is the refractive index and ϱ the density. See also: *Lorentz–Lorenz formula.*

GLARE. (1) The visual discomfort experienced by an observer in the presence of a visible source of light. (2) The visual disability arising from the presence of visible sources or areas of luminance in the field of view.

GLARE, VEILING. The reduction in contrast of an optical image (e.g. in a telescope) owing to the superposition of scattered light.

GLASS. A substance, commonly a complex of metallic oxides and silicates, which has been cooled from the liquid state without acquiring any crystalline properties other than a certain amount of short-range order. Glasses are normally transparent and may exhibit flow, although the coefficient of viscosity is extremely high. They may devitrify into the crystalline state after standing for very long periods of time.

GLASS CERAMICS. Polycrystalline solids made by the controlled crystallization of glasses.

GLASSY METALS. Non-crystalline alloys based largely either on iron, nickel and phosphorus, or on iron, phosphorous and carbon, which show remarkable combinations of high strength and corrosion resistance.

GLAUERT NUMBER. The number $(1 - M^2)^{-1/2}$, where M is the Mach number.

GLIDE. See: *Slip.*

GLIDE PLANE. A plane which can be drawn in the crystal structure so that reflection of any atomic position at the plane, followed by a movement of translation, gives an atomic position with identical environment. The translation may be parallel to a crystallographic axis lying in the plane (an *axial glide plane*) or there may be two successive translations parallel to two crystallographic axes (a *diagonal glide plane*). In the former case the translation is half the unit cell edge, and in the latter it is a half or one quarter.

GLOBAR. A rod, consisting mainly of silicon carbide, which at white heat radiates almost like a black body. See also: *Nernst glower.*

GLOBULAR CLUSTER. See: *Star cluster.*

GLORY (METEOROLOGY). See: *Brocken, spectre of.*

GLOSS. Of a surface: the ratio of the amount of light reflected specularly by the surface to the total amount of reflected light. The photometer used for measuring gloss is known as a *glossmeter.*

GLOW-DISCHARGE ELECTRON GUN. See: *Electron beam, glow discharge.*

GLOW DISCHARGE TUBE. See: *Discharge, glow.*

GLUCOSIDE. A compound formed by the union of a molecule of sugar and one of a hydroxylic non-sugar, with the elimination of a molecule of water.

GLYCEROL: GLYCERINE. A trihydric alcohol of formula

$$CH_2OH \cdot CHOH \cdot CH_2OH.$$

The esters of glycerol are known as *glycerides.*

GLYCOL. A dihydric alcohol of formula

$$CH_2OH \cdot CH_2OH.$$

It is the common name for *ethylene glycol.*

GNOMONIC PROJECTION. A plane projection used for displaying the positions of the poles of a crystal. These poles are projected radially from the centre of the reference sphere on to any specified non-diametral plane, usually taken as the plane tangential to the sphere. See also: *Orthographic projection. Stereographic projection.*

GOLAY CELL. A detector of radiation (ultra-violet to micro-wave) which registers the rise in temperature of a gas contained in a small transparent cell by recording it as a change in pressure.

GONIOMETRY. The measurement of angles, usually those between flat reflecting surfaces. The term is commonly applied to measurements on crystal faces, optical prisms, and angle gauges. It has also been used in radio direction finding.

GOUY BALANCE. A balance for determining para- and diamagnetic susceptibility. The specimen is mounted on one arm of a chemical balance and the apparent change in weight is noted when a uniform magnetic field is applied to its lower end.

GRADIENT. Of a scalar function ϕ: the direction of maximum rate of change of the scalar. It may be written

$$\nabla \phi = \text{grad } \phi = \mathbf{i} \frac{\partial \phi}{\partial x} + \mathbf{j} \frac{\partial \phi}{\partial y} + \mathbf{k} \frac{\partial \phi}{\partial z},$$

where ∇ is the differential vector operator, and $\mathbf{i}$, $\mathbf{j}$, $\mathbf{k}$ are unit vectors. See also: *Differential vector operator.*

GRADIENT WIND. Wind whose motion is in the direction of the isobars, with the coriolis, pressure, and centrifugal forces, all normal to the motion, being in balance.

GRADIOMETER. A simplified form of Eötvös torsion balance, which measures only the horizontal changes in the gravitational field.

GRAETZ NUMBER. A non-dimensional coefficient used in problems of heat transfer. It is defined as $\omega \gamma F C_p / \lambda l$, where ω is the velocity of flow, γ the density of the medium, F the surface area, C_p the specific heat at constant pressure, λ the thermal conductivity, and l a characteristic length.

GRAIN BOUNDARY. The boundary between two coherent crystals which differ in orientation, in lattice dimensions, in crystal structure, in composition, or in any combination of these. An ordinary grain boundary is believed to consist of a layer in which the atoms are exposed to forces from both crystals at once and it approximates to a thin amorphous layer. At a boundary between identical crystals, if the difference in orientation is only a few degrees, the boundary consists of dislocations. A *tilt* boundary consists of edge dislocations and a *twist* boundary of screw dislocations. Most dislocation boundaries are a mixture of the two.

GRAIN BOUNDARY, LARGE-ANGLE. A grain boundary between adjacent crystals whose edges make an appreciable angle with each other.

GRAININESS. Of a processed photographic emulsion: the visible lack of homogeneity arising from the particulate nature of the photographic image. See also: *Granularity.*

GRAM-EQUIVALENT WEIGHT. One mole of an element or a radical, divided by its valency. In the case of multi-valent substances there will, of course, be more than one possible value of the gram-equivalent weight.

GRAM-MOLECULE. Formerly the amount of a pure substance whose weight in grams was numerically equal to the molecular weight. It has now been superseded by the mole. See also: *Amount of substance. Mole.*

GRAM-RAD. The former unit of integral absorbed dose. One gram-rad is 100 ergs or 10^{-5} J. See: *Gray.*

GRANITIC LAYER. The upper of the two layers that may occur in the Earth's crust, the lower being the *basaltic layer.* The granitic layer is absent from the floor of the deep ocean basins.

GRANULARITY. Of a processed photographic emulsion: a measure of the graininess of a photographic image, based upon observations of the variation in photographic density. See also: *Graininess.*

GRAPH. (1) A diagrammatic expression of a function of two variables in which one variable is plotted against the other. (2) Any record produced by physical methods.

GRASHOF NUMBER: CONVECTION MODULUS. A dimensionless number appearing in the expression for the heat-transfer coefficient for natural convection due to the presence of a hot body. It is given by

$$\frac{l^3 g \alpha \varrho^2 \theta}{\eta^2}$$

where l is a typical dimension of the hot body, g is the acceleration due to gravity, α is the temperature coefficient of the fluid density, ϱ is the density and η the viscosity of the fluid, and θ is the temperature difference between the hot body and the fluid. See also: *Dynamical similarity principle.*

GRASSMANN LAWS OF COLOUR VISION. State that: (a) the eye can distinguish only three types of variation; luminance, dominant wavelength and purity for the stimulus; and brightness, hue, and saturation for the sensation: (b) if one component in a two-component mixture is steadily changed, the other remaining constant, the colour of the mixture steadily changes: and (c) lights of the same colour produce identical effects in mixtures, regardless of their spectral composition.

GRASSOT FLUX METER. A type of flux meter in which the restoring couple of the suspended coil is negligible in comparison with its electromagnetic damping, the ends of the coil being connected to an outside search coil of known area. See also: *Flux meter.*

GRATICULE. (1) An engraved or ruled scale or a set of cross-wires or similar measuring marks, situated at the common focal plane of an objective and eyepiece. (2) A scale on the light shield of a cathode-ray oscilloscope for the measurement of patterns on the screen. (3) The pattern of lines representing parallels of latitude and meridians of longitude on a map or chart.

GRAVIMETER. An instrument for the measurement of gravity differences between observation points. It may be *dynamic* (involving the measurement of time, generally vibrations of a mechanical system), *stable* (involving the balancing of weight against some other force—springs or gas pressure), or *astatized* (any gravimeter in which the deflection from equilibrium is increased by the application of an impressed force, often in the form of a spring).

GRAVIMÈTRE. An instrument for the determination of the density of a liquid by the use of a total-immersion float which is attached to the bottom of the containing vessel by a spiral spring. The extension of this spring gives a measure of the density of the liquid.

GRAVITATIONAL ACCELERATION. The acceleration towards the centre of mass of the Earth experienced by a body of negligible mass (compared with that of the Earth) moving *in vacuo*. Its value (usually denoted as g) varies from place to place but has a mean value of about $9\cdot81$ m/s^2 or $32\cdot2$ ft/s^2. Various formulae have been proposed for calculating g at any point on the Earth's surface, the results of which all differ to some extent from the actual value observed. This difference at a given point is known as the *gravitational anomaly* at that point. See also: *Units, gravitational.*

GRAVITATIONAL CONSTANT. (1) The constant G appearing in Newton's law of gravitation. (2) The *Gaussian constant of gravitation*, equal to $\sqrt{G}$.

GRAVITATIONAL FIELD. The field in which gravitational forces are operative. According to Einstein's general theory of relativity this field may be described in terms of *gravitational waves* and quanta which are analogous to the waves and quanta of the electromagnetic field.

GRAVITATIONAL POTENTIAL. At a point in the gravitational field of an isolated spherically symmetric mass: the quantity Gm/r, where G is the gravitational constant, m is the mass concerned, and r is the distance of the point from the centre of mass. It has the dimensions of the square of a velocity.

GRAVITATION, NEWTON'S LAW OF. States that every particle of matter in the Universe attracts every other particle with a force acting along the line joining the particles. For two particles of masses m_1 and m_2, separated by a distance d, the force between the two particles, known as the *gravitational force*, is given by Gm_1m_2/d^2, where G is the *gravitational constant*. Its value is about $6\cdot67 \times 10^{-8}$ dyn cm^2/g^2, or $6\cdot67 \times 10^{-11}$ Nm^2/kg^2.

GRAVITON: GRAVITINO. The quantum of the gravitational field as postulated by Einstein. It has a spin of 2.

GRAVITY WAVES. Surface waves whose motion is controlled by gravity and not by surface tension. They may be small wave systems in the atmosphere or surface waves in water. See also: *Waves, surface.*

GRAY. The SI unit of absorbed dose (of ionizing radiation), which is to replace the rad. It is equal to 1 joule per kilogram, hence 1 rad is equal to 10^{-2} gray. See: *Appendix II.*

GREAT CIRCLE. The intersection of the surface of a sphere by a plane passing through the centre of the sphere. See also: *Small circle.*

GREAT RIFT. A conglomeration of superposed dark nebulae in the Milky Way, stretching from Deneb (in Cygnus) past Altair to Scorpius and Sagittarius. See also: *Nebula, dark*.

GREEN FLASH. A green coloration sometimes observed at sunset and sunrise. It arises from the removal of light of shorter wavelength than the green by Rayleigh scattering.

GREEN THEOREM. The vector form of the Gauss theorem.

GREENWICH TIME. See: *Time*.

GREINACHER CIRCUIT: LATOUR CIRCUIT. A constant-potential voltage-doubling circuit in which two capacitors connected in series are each charged in alternate half-cycles.

GRENINGER CHART. A chart for reading directly, from a back-reflection X-ray diffraction photograph, the angular relations between planes and zones.

GRENZ RAYS. Very soft X-rays, of 15 kV or less.

GREY BODY. A body which, for a given temperature, emits radiant energy which, at every wavelength, is a constant fraction of the energy emitted by a black body at the same wavelength.

GRID. (1) In an electron tube: a metallic electrode, located between a source of electrons and their final collecting point, whose function is to exercise control over the electron flow. (2) In geodesy: lines of a Cartesian coordinate system plotted on a map or projection. (3) In an ionization chamber: an electrode, located between the plates of a parallel-plate ionization chamber, which is used to ensure that the charge appearing on the collector is equal to the total ionization produced by the ionizing particles.

GRIFFITH THEORY OF BRITTLE FRACTURE. States that the relatively low strength of brittle materials is due to the presence of small cracks (*Griffith cracks*). According to the theory, fracture occurs at a stress level such that, if a crack grows by an infinitesimal amount, the increase in surface energy associated with the longer crack is supplied by the decrease in elastic energy plus the applied stresses as the specimen elongates. For a crack of length $2c$ in a sheet with Young's modulus E, and surface energy γ, the crack will grow when the externally applied stress f reaches a value given by $f^2 = 2\gamma E/\pi c$.

GROTTHUS–DRAPER LAW. In photochemistry: states that only light which is absorbed is effective in producing chemical change.

GROUND STATE. The state of lowest energy of an atomic, nuclear, or molecular system.

GROUP. A finite or infinite set of algebraical elements satisfying certain conditions, e.g. the product of two elements of a group is a member of the same group. The algebraic theory of groups finds important applications in the analysis of symmetry in atomic, molecular, and nuclear systems.

GROUP VELOCITY. The velocity of propagation of a disturbance or pulse, e.g. the crest of a group of interfering waves where the component wave trains have slightly different wavelengths. It is the velocity of propagation of the associated energy and, in a dispersive medium, differs from the phase velocity. See also: *Phase velocity*.

GRÜNEISEN CONSTANT. See: *Equation of state, Grüneisen*.

GRÜNEISEN FIRST RULE. An equation, derived from the Grüneisen equation of state for a solid, on the assumption that the interatomic potential energy is a function of the volume only and is independent of temperature. It may be written as $\Delta V = \gamma K_0 E$, where V is the volume, K_0 the isothermal compressibility at absolute zero, E the thermal energy, and γ the Grüneisen constant.

GRÜNEISEN LAW: GRÜNEISEN SECOND RULE. An equation derived from the Grüneisen equation of state on the assumption that the Grüneisen constant is independent of temperature. It may be written as $\beta = \gamma K_0 C_{v/V}$, where β is the volume coefficient of thermal expansion, K_0 the isothermal compressibility at absolute zero, C_v the molar specific heat at constant volume, V the volume, and γ the Grüneisen constant. Note: The Grüneisen constant varies from 1·5 to 2·5 at medium temperatures.

G-STAR. See: *Stars, spectral classes of*.

GUARD RING. A ring surrounding an electrically charged or heated body to ensure an even distribution of potential or heat by eliminating edge effects.

GUARD VANE. A body of aerofoil shape inserted in a flowing medium to produce a change in direction of flow with minimum loss.

GUARD WIRE. An earth wire used on an overhead transmission line. It is so arranged that, should a conductor break, it will immediately be earthed by contact with the wire.

GUILLEMIN EFFECT. A magnetostrictive effect whereby a bent ferromagnetic bar is observed to straighten when magnetized.

GUINIER–PRESTON ZONES. See: *Age hardening: Precipitation hardening.*

GUNN EFFECT. The occurrence of coherent microwave current oscillations in certain homogeneous, *n*-type, semiconductors subject to high electric fields. Gunn oscillators provide simple, low-voltage, solid-state sources of microwave power.

GUST. A relatively rapid and random change of air speed or direction. It is a manifestation of atmospheric turbulence. The gust vector is usually resolved into longitudinal, lateral, and vertical components.

GUTENBERG–RICHTER SCALE. A scale of earthquake magnitude. See also: *Earthquake, magnitude of.*

G-VALUE. In a radiochemical process: the number of specified changes in an irradiated substance produced per 100 eV of energy absorbed from ionizing radiation. Examples of such changes are: cross-linking, production of a particular type of molecule, and production of free radicals. (Note: $1 \text{ eV} \approx 1 \cdot 602 \times 10^{-19}$ J.)

GYPSUM PLATE. A thin plate of gypsum used for determining the sign of birefringence of crystals in a polarizing microscope by observing the change in the interference colour of the plate due to the additive or subtractive effects of a superimposed crystal section.

GYROCOMPASS. A compass which relies on the principles of the rigidity of a gyroscope for its direction-finding ability. It embodies a spinning rotor or flywheel driven at high speed, together with some means of controlling its position relative to the meridian plane. Once set, the gyrocompass will continue to point in the same direction irrespective of its attitude. It forms the basis of all types of automatic pilot and of various navigational systems.

GYROMAGNETIC FREQUENCY. See: *Larmor precession frequency.*

GYROMAGNETIC RATIO. In an atomic or nuclear system: the ratio of the magnetic moment to the angular momentum. Thus, for an atomic nucleus the gyromagnetic ratio (usually denoted by g) is equal to μ/I, where μ is the nuclear magnetic moment and I the nuclear angular momentum quantum number.

GYROSCOPE. A rapidly spinning, symmetrical, well-balanced wheel, with its mass distributed as far as possible from the axis of spin, and having bearings possessing the minimum of friction. It possesses the property of maintaining the direction of its spin axis in space (i.e. with reference to the "fixed" stars), except when a force acts on it to cause a torque around an axis other than that axis, in which case it precesses about the spin axis.

GYROSTABILIZER. A device used on ships to prevent excessive rolling. It may employ a large and heavy gyroscope to counteract the roll, or a small one which senses the roll and by a servomechanism, actuates stabilizing fins below water-level.

h

HABIT. Of a crystal: one or other of the various crystal forms shown by the crystal. The habit gives a clue to the underlying atomic arrangement in many instances.

HABIT PLANE. The matrix plane on which a precipitated plate lies when a supersaturated solid solution precipitates a second phase.

HADRONS. Elementary particles which take part predominantly in strong interactions but also, to some extent, in electromagnetic, weak and gravitational interactions. Hadrons include the heavy mesons (pions and heavier) and the baryons: there are many hadrons of many masses. See also: *Baryon. Interactions between elementary particles. Quantum number, internal. Quarks.*

HAEMODYNAMICS. The physical laws describing the circulation of the blood.

HAHN EMANATION TECHNIQUE. A technique for the measurement of emanating power, used as a qualitative measure of the changes occurring in solids under various treatments, a small quantity of radon or thoron being incorporated into the material being examined, and the changes in rate of evolution measured. See also: *Emanating power.*

HAIDINGER BRUSHES. Faint yellow brushes seen when looking at a bright surface through a nicol prism. They are believed to be due to an effect at the fovea of the eye.

HAIDINGER FRINGES. Interference fringes seen with thick, flat plates near normal incidence. The fringes of the Fabry–Perot interferometer are of this type.

HALATION. The formation of haloes on photographic images of bright objects due to light reflected back into the emulsion from the rear of the supporting film.

HALF-LIFE, BIOLOGICAL. Of a particular substance in a biological system: the time required for the amount of that substance to be reduced to one-half of its value by biological processes, when the rate of removal is approximately exponential.

HALF-LIFE, EFFECTIVE. Of a particular radionuclide in a system: the time required for the amount of that radionuclide to be reduced to half its value as a consequence both of radioactive decay and other processes such as biological elimination or burn-up, when the rate of removal is approximately exponential.

HALF-LIFE, RADIOACTIVE. Of a single radioactive decay process: the time required for the activity to decrease to half its value. It is the time taken for half of the nuclei to decay.

HALF-PERIOD ZONES: FRESNEL ZONES. Annular elements in which it is convenient to divide a wave front when determining the amplitude at a given point resulting from Fresnel diffraction. The zones are such that radiation from one zone reaching the point is one half-period out of phase with that from the adjacent zones. The resultant amplitude is half the amplitude that would result if all but the first (axial) zone were blocked out. If a screen (known as a *zone plate*) is interposed so as to obstruct alternate zones, the intensity at the point of interest can be made very large.

HALF-SHADOW DEVICE. A device used to divide a field of view into two parts to facilitate the detection by the eye of small changes in a dark field of view by balancing one half against the other, reference, half. Such devices are used extensively in measurements made with polarized light and also in photometers.

HALF-VALUE LAYER: HALF-VALUE THICKNESS. Of a specified substance for a given beam of ionizing radiation: that thickness which, when introduced into the path of the beam, reduces the value of a specified radiation quantity by one-half. It is sometimes expressed in terms of mass per unit area.

HALF-WAVE CIRCUIT. A circuit in which current flows during alternate half-cycles only.

HALF-WAVE PLATE. A plate of doubly refracting material which, for a given wavelength, resolves a beam of linearly polarized light which traverses the plate, into two components with a path difference of half that wavelength, i.e. a phase-difference of π. See also: *Wave plate: Retardation plate.*

HALF-WIDTH. Of a peak on a distribution curve, e.g. a spectral line: the full width of the peak measured at half the height.

HALIDE. A compound formed as a result of combination with a halogen.

HALL ANGLE. For the Hall effect: the angle between the direction of current flow and that of the total electric field (i.e. that of the vector sum of the applied and Hall electric fields.

HALL COEFFICIENT. The coefficient R in the expression for the Hall e.m.f., $RBJT$, where B is the magnetic flux density, J the current density, and T the thickness of the specimen in the direction of the e.m.f. It is the potential gradient for unit flux density and unit current density.

HALL EFFECT. The development of a transverse electric potential difference (the *Hall e.m.f.*) across a current-carrying conductor or semiconductor when a magnetic field is applied at right angles to the current. The potential gradient, electric current, and magnetic field are mutually perpendicular. See also: *Nernst effect.*

HALL MOBILITY. A measure of carrier mobility afforded by the Hall effect. It is the ratio of the Hall angle to the magnetic flux density.

HALO. (1) In atmospheric optics: a coloured circle of light apparently surrounding a light source. It may arise from the reflection (either external or internal) of light from the Sun or Moon at the surfaces of ice crystals, from the concentration of light in certain directions due to refraction at minimum deviation of light passing through the crystals, or from a similar concentration arising from the preferred orientation of the crystals. (2) In X-ray and electron diffraction: one of the broad rings appearing in the X-ray or electron diffraction pattern of a material which cannot be regarded as truly crystalline. Such haloes occur with liquids and gases as well as with solids.

HALOGEN. One of the elements of Group VIIʙ of the Periodic Table, consisting of fluorine, chlorine, bromine, iodine, and astatine.

HAMILTONIAN FUNCTION. (1) In classical particle mechanics: a function defined by

$$H = \sum_{i=1}^{n} (p_i \dot{q}_i) - L,$$

where H is the Hamiltonian function, which refers to a system of n generalized coordinates q_i, and momenta p_i, L is a Lagrangian function, and $\dot{q}_i$ is the first time derivative of q_i. Where L is time independent, H is equal to the total energy of the system. H then satisfies the canonical equations of

motion. (2) In quantum theory: an operator H which gives the equation of motion for the wave function in the form $i\hbar\,\delta\psi/\delta t = H\psi$ and $H\phi = E\phi$. See also: *Canonical equations of motion. Schrödinger equation.*

HAMILTON PRINCIPLE. Concerns the potential energy V and the kinetic energy T of a mechanical system. For a conservative system the principle postulates that the integral

$$\int_{t_1}^{t_2} (T - V)\,dt,$$

where t is the time, shall have a stationary value. The integrand $(T - V)$ is known as the *Lagrangian function.* The Hamilton principle is a more general form of the principle of least action and Fermat's principle.

HAMMER TRACK. In a nuclear emulsion: a T- or hammer-shaped track formed when a nuclear particle comes to rest and decays into two fragments which travel in opposite directions. Such tracks are most commonly formed by the decay of the excited $^{8}_{4}$Be nucleus into two α-particles, but other causes are also not unusual.

HANKEL FUNCTIONS. See: *Bessel equation.*

HANKEL TRANSFORM. Provided that certain conditions are satisfied, the Hankel transform of $F(x)$ is $f(y)$, where

$$F(x) = \int_0^\infty J_n(xy)\,f(y)\,y\,dy$$

and

$$f(y) = \int_0^\infty J_n(xy)\,F(x)\,x\,dx,$$

where J_n is a Bessel function.

HARD AND SOFT RADIATION. Qualitative terms describing the penetrating power of the radiation, "hard" denoting a relatively high degree of penetration and "soft" a relatively low degree. What is hard in one context may be soft in another.

HARD MAGNETIC MATERIAL. Generally taken to mean a material which has a coercivity of more than about 100 oersteds or 7.96×10^3 A/m. The associated value of the product of field strength and flux density is the quantity of interest for permanent magnets. See also: *Soft magnetic material.*

HARDNESS. Of a solid: its resistance to local deformation by such means as indentation or scratching. It is customarily expressed as: (a) *Scratch hardness*, which is based on the ability of one solid to scratch another. Here the *Mohs scale* (defined separately) may be used in which ten minerals are selected as standards, from talc (hardness 1) to diamond (hardness 10): (b) *Indentation hardness*, to determine which permanent deformation is made by an indenter and the hardness expressed in terms of the load and area of indentation formed. The indenter may be in the shape of a ball (*Brinell* and *Rockwell* hardnesses), cone (*Rockwell* hardness), or pyramid (*Vickers* hardness): or (c) *Dynamic* or *rebound hardness*, in which the hardness is expressed in terms of the indentation made by a bouncing indenter falling under gravity, or of the height to which the indenter rebounds under specified conditions (as in the *Herbert* pendulum and *Scleroscope* tests). See also: *Meyer hardness number. Work hardening.*

HARDNESS, HOT. The hardness of a metal at high temperatures.

HARMONIC. (1) A continuous function which satisfies Laplace's equation, and whose first derivative is also a continuous function. (2) A sinusoidal quantity having a frequency which is an integral multiple of the fundamental frequency of a periodic quantity of which it is a component. (3) In musical terms, a partial whose frequency is an integral multiple of the fundamental frequency. See also: *Subharmonic.*

HARMONIC ANALYSIS. The analysis of a periodic function into its Fourier components. This may be done automatically from a graphical representation of the periodic function by a *harmonic analyser.*

HARMONIC MOTION. A vibratory motion represented by

$$x = A \cos \omega t + B \sin \omega t,$$

where x is the displacement of a given point, A and B are constants, and ωt is the phase at time t. Simple examples are the oscillations of a mass at the end of a coiled spring, the elastic displacement of a beam from its equilibrium position, and a sinusoidal wave motion.

HARMONIC OSCILLATOR. An oscillator in which the restoring force varies linearly with the displacement of the system from its equilibrium position.

HARMONIC SERIES. The series $1 + \frac{1}{2} + \frac{1}{3} + \frac{1}{4} + \dots + 1/n.$

HARTLEY OSCILLATOR. One of the first vacuum-tube (triode) oscillators to be devised. It consists essentially of a parallel resonant circuit connected between the grid and anode, the cathode being connected to a tapping point along the coil. See also: *Colpitts oscillator.*

HARTMANN DISPERSION FORMULA. For a prism: the equation

$$dn/d\lambda = - C/(\lambda - \lambda_0)^2,$$

where n is the refractive index, λ the wavelength, and C and λ_0 are constants.

HARTMANN OSCILLATOR. A device, due to Hartmann and Trolle, for generating high-power sound waves in fluids.

HARTMANN TEST. A test for aberration and coma in an optical system in which the images formed by different parts of the field are compared in position and sharpness.

HARTMANN WAVELENGTH FORMULA. For a prism: the equation

$$\lambda = \lambda_0 + C/(d_0 - d),$$

where λ is the unknown wavelength of a spectral line on a spectroscopic plate, λ_0, C, and d_0 are constants (found by calibration), and d is the distance of the spectral line measured along the plate from a fixed point.

HARTREE METHOD: SELF-CONSISTENT FIELD METHOD. An iteration method used in the determination of solutions of the Schrödinger equation for atomic problems. See also: *Slater method.*

HARVARD–DRAPER SEQUENCE. See: *Stars, spectral classes of.*

HAZE. (1) *Dry haze:* reduced visibility in the atmosphere arising from very small (less than 1 μm) particles of smoke, dust, or salt. It is yellowish in colour. (2) *Wet haze* or *mist:* reduced visibility in the atmosphere arising from small water droplets. It is grey in colour. See also: *Fog.*

HAZE FACTOR. For adjacent black and white objects as seen through a diffusing medium: the quantity $L_B/(L_W - L_B)$, where L_B and L_W are the apparent luminances of the black and white bodies respectively.

HEAD AMPLIFIER: PRE-AMPLIFIER. A unit containing the input stage of an amplifying system and designed to be mounted integrally with, or close to, the detector.

HEAD, TOTAL. The *head* (i.e. the height of a liquid column necessary to develop a given pressure at the base) measured by a pitot tube, i.e. the sum of the hydrostatic pressure and the kinetic head.

HEAD WAVE. The shock wave emanating from the front of a body in supersonic motion.

161

HEALTH PHYSICS. The branch of medical physics that is concerned with radiation protection as applied to man. See also: *Biophysics. Hospital physics. Medical physics.*

HEARING LOSS. For a given individual at a specified frequency, the difference, in decibels, between the threshold of hearing of that individual and the normal (standard) threshold. The hearing loss is also known as the *sensation level* for the relevant frequency.

HEAT. A form of energy associated with the motion of individual atoms or molecules of a body. It is to be distinguished from temperature, which is a measure of the degree of hotness.

HEAT CAPACITY. Of a system, entity, or substance at a given temperature: the amount of heat required to raise the temperature by one degree. It may be defined as the limit, as ΔT approaches zero, of the ratio $\Delta Q/\Delta T$, where ΔQ is the amount of heat that must be added to raise the temperature from T to $T + \Delta T$. The heat capacity per unit mass of a substance is the *specific heat* of that substance.

HEAT CONDUCTION. See: *Thermal conduction.*

HEAT CONTENT. See: *Enthalpy.*

HEAT ENGINE. A device for obtaining work from a cycle of operations involving the expansion and compression of a fluid (the "working" fluid). This fluid may be gas or vapour, which, in expanding, exerts pressure either on a piston moving in a cylinder, or on a curved, rotating turbine blade. In ideal cycles the fluid is permanent, but in real cycles it is constantly being replaced. See also: *Carnot cycle. Diesel cycle. Joule cycle. Otto cycle. Stirling cycle.*

HEAT ENGINE, EFFICIENCY OF. The ratio of the work obtained to the heat absorbed.

HEAT EXCHANGER. Any device in which heat is transferred from one fluid to another. If both fluids flow continuously through separate channels, divided by a wall, it is termed *recuperative*. If the two fluids flow alternately through the same channels it is termed *regenerative*. And if both fluids flow continuously through the same channels the heat exchanger is said to be of the *direct-contact* variety. See also: *Heat-transfer coefficient.*

HEAT FLOW, RATE OF. Across a structure or through a solid: the quantity of heat flowing per unit time. It is also known as the *thermal transmission*.

HEAT FLOW, SPECIFIC RATE OF. The quantity of heat flowing through unit cross-section of a body per unit time.

HEAT FLOW, STEADY STATE. Refers to the situation in which the quantity of heat flowing into a body is equal to that flowing out of the body. Although the temperature may differ at different points within the body, the temperature at a given point will remain constant.

HEAT OF ABSORPTION, ADSORPTION, COMBUSTION, CONDENSATION, CRYSTALLIZATION, DECOMPOSITION, DISSOCIATION, FORMATION, IONIZATION, MELTING, SOLIDIFICATION, SOLUTION, TRANSFORMATION, WETTING, VAPORIZATION, ETC. The amount of heat lost or gained when 1 mole of a substance is subjected to the appropriate process.

HEAT PIPE. A sealed tube containing a small quantity of working fluid (typically water) which provides a very fast and effective way of transferring heat. The working fluid boils off at the hot end, absorbing heat, and condenses at the cold end, giving up the heat again. The cycle is completed by the capillary action of a wick or wire wool or sintered metal which lines the inside of the pipe.

HEAT PUMP. A device which absorbs heat from an external low-temperature source, and rejects the whole as heat at a higher temperature. It is essentially a reversed heat engine, of which a refrigerator is an example. See also: *Coefficient of performance (COP).*

HEAT SWITCH. A device for stopping or starting the flow of heat between two regions.

HEAT TRANSFER. The transfer of heat by conduction, convection, radiation, or a combination of these. See also: *Heat pipe.*

HEAT TRANSFER, BOILING. Refers to the cooling of a hot surface by a boiling liquid. The two types of boiling of most significance are *nucleate boiling*, in which the bubbles formed at the hot surface burst into vapour at that surface, and *film boiling*, in which the vapour forms a continuous film at the hot surface, thus leading to reduced heat transfer.

HEAT TRANSFER COEFFICIENT. The amount of heat emitted per second by unit area of a surface maintained at a temperature of one degree above its surroundings.

HEAT TREATMENT. Any sequence of temperature changes to which a metal or alloy may be subjected in order to modify its properties. Most heat treatments are designed to alter mechanical properties but it is not unusual for other properties (e.g. the magnetic properties) to be the main consideration.

HEAVISIDE LAYER: KENNELLY–HEAVISIDE LAYER: E-LAYER. The lower of the two main strata or "layers" in the ionized atmosphere, the other being the Appleton or *F*-layer. It occurs at a height of about 100 km. See also: *Ionosphere.*

HEAVISIDE MUTUAL-INDUCTANCE BRIDGE. See: *Bridge, electrical.*

HEAVY PARTICLE. An elementary particle with a mass greater than that of the π-meson. The term includes *K*-mesons, nucleons, and hyperons.

HEAVY WATER. Water in which the hydrogen isotope of mass number 1 is replaced by deuterium, of mass number 2. The term is often applied to water containing a high proportion of deuterium.

HEIGHT, BAROMETRIC. The height of a point above sea-level as determined by barometric measurements.

HEIGHT, DYNAMIC. The height of a point above the surface of the Earth in terms of the equipotential surface on which it lies. Thus all points with the same dynamic height lie on the same equipotential surface but may not have the same orthometric height.

HEIGHT, ORTHOMETRIC. The vertical height of a point above mean sea-level.

HEITLER–LONDON THEORY. Of covalent bonding: a theory based on consideration of the exchange forces between two atoms in which the two interacting electrons are assumed to be in atomic orbitals about each of the nuclei, the spins of the electrons being paired, i.e. anti-parallel. The theory explains the origin of a *two-electron* (or *electron-pair*) bond. See also: *United atom method.*

HELIACAL RISING OR SETTING. The simultaneous rising or setting of a celestial body and of the Sun. In ancient times heliacal risings and settings were used as a basis for solar calendars.

HELICITY. Of an elementary particle: the component of its spin along the direction of its motion. It may be positive (right-handed) or negative left-handed).

HELICONS. Low-frequency circularly polarized electromagnetic waves propagated in a metal in the presence of a constant external magnetic field. Such waves are valuable in the investigation of many properties of solids.

HELIOCENTRIC. Describes a quantity such as latitude or longitude which is referred to a hypothetical observer located at the centre of the Sun.

HELIOSTAT. An arrangement of mirrors driven automatically and used to reflect a beam of sunlight in a constant direction. Originally used for signalling,

it is now used in some solar furnaces to direct the Sun's rays continuously on to a fixed parabolic mirror. A similar instrument used as a beacon by surveyors is known as a *surveyor's heliotrope.*

HELIOTROPE, SURVEYOR'S. See: *Heliostat.*

HELIUM DILUTION REFRIGERATOR. See: *Refrigerator, dilution.*

HELIUM FILM. The film which covers any surface which is in contact with helium II. Such a film, formed on the wall of a container of helium II, will creep over the walls of the container to find its own level until the level inside is the same as that outside or until the container is empty, according to circumstances. The *creep rate* of this film rises to a maximum (the *critical creep rate*) which is almost independent of the pressure head or the length of the film. This film is also known as the *Rollin film.*

HELIUM, LIQUID. (1) He-4 in the form of helium I: the form of liquid helium stable above the lambda-point (about 2·19 K). (2) He-4 in the form of helium II: the form of liquid helium stable between the lambda-point and absolute zero. It exhibits superfluidity, and has an extremely high thermal conductivity and extremely low viscosity. (3) He-3, present in natural helium in minute proportions. It has no lambda-point transition but possesses superfluid properties. See also: *Fountain effect. Helium film. Zero sound.*

HELIUM II, MECHANOCALORIC EFFECT IN. The increase in temperature of helium II in a tube when the helium II flows through a porous plug. Conversely, the cooling of helium II when some of it flows into a tube, closed by a porous plug, which is plunged into the liquid. The temperature change in each case is about 0·01 degrees.

HELIUM II, THERMOMECHANICAL EFFECT IN. The flow of helium II towards a region of higher temperature. See also: *Fountain effect.*

HELIUM II, TWO-FLUID MODEL FOR. A model in which it is assumed that helium II consists of two freely interpenetrating fluids, one a normal fluid and the other a superfluid, the latter having zero entropy, zero viscosity, and zero thermal conductivity. See also: *First sound. Second sound.*

HELIX. A curve drawn on a cylindrical or conical surface so that all the generators of the surface are cut at the same angle.

HELLMAN–FEYNMANN THEOREM. A theorem used in calculating the forces on the nuclei in molecules or solids which are described by the Born–Oppenheimer approximation. It states that these forces are those which would arise electrostatically if the electron probability density were treated as a static distribution of negative electricity.

HELMHOLTZ COILS. Two circular parallel coils of the same radius separated by a distance equal to that radius. They produce a region of constant magnetic field on either side of the midpoint between the two coils.

HELMHOLTZ FREE ENERGY. See: *Free energy.*

HELMHOLTZ RESONATOR. An acoustic resonator in which an enclosed cavity is open to the air only via a small opening or neck. It resonates at a single frequency which depends on the shape and size of the resonator. Such a resonator may form part of an acoustic filter (as in a sound-absorbing panel), or may appear as the body of a musical instrument such as a cello.

HELMHOLTZ THEORY OF COLOUR VISION. See: *Colour vision, theories' of.*

HEMICOLLOID. A colloid composed of very small particles, up to about 250 Å (0·025 μm) in length. See also: *Eucolloids. Mesocolloid.*

HENRY. The unit of inductance in MKSA and SI units, equal to 10^9 CGS units. It is the inductance of a passive circuit in which an induced e.m.f. of 1 volt is produced when the inducing current changes at the rate of 1 ampere per second. See: *Appendix II.*

HENRY LAW. States that the mass of gas dissolved by a given volume of liquid at a given temperature is proportional to the pressure of gas with which it is in equilibrium. It is strictly true only for ideal solutions.

HERMITE EQUATION. The second order differential equation $y'' - 2xy' + 2vy = 0$, where v is a constant, but not necessarily an integer. The solutions to this equation are known as *Hermitian polynomials.*

HERSCH CELL. An oxygen cell comprising a metallic silver cathode partially immersed in KOH solution, used for measuring oxygen in solutions or in gases. The current afforded is proportional to oxygen concentration over the range 0·1–50 ppm O_2.

HERTZ. The unit of frequency in the International System of units (SI). It is equal to 1 cycle per second. See: *Appendix II.*

HERTZIAN DIPOLE. A radiating oscillatory electric doublet. Although it is not physically realizable it has formed the starting point of many investigations of radiation and diffraction phenomena. See also: *Dipole, electric.*

HERTZIAN VECTOR. A single vector describing an electromagnetic wave, given by $\mathbf{Z} = \int \mathbf{A}\, dt$,

where $\mathbf{Z}$ is the Hertzian vector and $\mathbf{A}$ is the vector potential. The electric and magnetic intensities, E and H, are given by

$$E = \text{grad (div } \mathbf{Z}) - \frac{1}{c^2}\frac{\partial^2 \mathbf{Z}}{\partial t^2}$$

and

$$H = \frac{1}{c}\,\text{curl}\,\frac{\partial \mathbf{Z}}{\partial t},$$

where c is the velocity of light.

HERTZSPRUNG–RUSSELL (H–R) DIAGRAM. A diagram in which the luminosity of the stars is plotted against their spectral type, and which shows that they fall mostly into two categories: the main sequence stars (for which the luminosity is directly related to the spectral type) and the giant stars. Other categories represented are the super-giants, sub-giants, sub-dwarfs, and white dwarfs. See also: *Stellar luminosity. Stellar magnitude. Stellar populations I and II. Stars, spectral classes of.*

HESS DIAGRAM. A diagram in which the luminosity of the stars is plotted against their absolute magnitude. The stars of each of the stellar classification groups 0, B, A, ..., fall on different curves, there being one curve for each group.

HESS LAW. States that the net change of energy in a chemical reaction or series of reactions depends only on the initial and final states, and is independent of the intermediate steps.

HETERODESMIC STRUCTURE. See: *Homodesmic structure.*

HETERODYNE. Denotes the phenomenon of beats at radio frequency. The heterodyne principle is used in radio reception to bring the frequency of the incoming signals into the audio range. An intermediate stage, often employed, in which the new frequency is supersonic is described as supersonic heterodyne or *superheterodyne.* See also: *Beats.*

HETEROGENOUS RADIATION. See: *Continuous radiation.*

HETEROLYSIS. See: *Chemical reaction, heterolytic.*

HETEROPOLAR COMPOUND. A covalent compound in which the bonds are essentially heteropolar in character. See also: *Bond, heteropolar.*

HEUSLER ALLOYS. A series of alloys of the type Mn–Cu–Al and Mn–Cu–Sn which, although their constituents are all non-magnetic, exhibit well-defined ferromagnetic properties.

HEXAD AXIS. See: *Crystal symmetry, rotation axis in.*

HEXAGONAL CLOSE-PACKED STRUCTURE. The close-packed structure having hexagonal symmetry. See also: *Close-packed structure.*

HEXAGONAL SYSTEM. The crystal system which is characterized by the possession of a single axis of sixfold symmetry, either of simple rotation or inversion. Crystals belonging to this system may be referred to three equal co-planar axes, making angles of 120° with each other, and a fourth axis perpendicular to their plane. The *trigonal system* (characterized by a single threefold axis of rotation or inversion) is sometimes considered as a subdivision of the hexagonal system. See also: *Trigonal system.*

HIGH POLYMER. See: *Polymer.*

HILBERT SPACE. An abstract space which is important in many branches of mathematics and theoretical physics, e.g. in quantum mechanics. It is a special type of Banach space.

HILDEBRAND RULE. States that the entropy of vaporization of a mole of liquid, when the density of the vapour phase above the liquid is constant, is itself constant. For a vapour-phase density of 0·05 mole/l, the change of entropy is about 130 J/mole. The rule is not applicable to quantum liquids, nor to liquids in which molecular association occurs.

HILTNER–HALL EFFECT. The phenomenon of the polarization of the light from stars at large distances. It is believed to arise from the presence of interstellar grains which are preferentially aligned, and manifests itself as a reduced absorption of that light whose polarization vector is approximately parallel to the galactic plane.

HINDERED ROTATION. Rotation of a chemical molecule or part of a molecule, which is opposed by internal forces.

HITTORF DARK SPACE. See: *Cathode dark space. Cathode fall. Discharge, glow.*

HOAR FROST. See: *Dew.*

HODOGRAPH. A curve relating to a particle moving in two dimensions in which the independent variables are the components of the velocity along the x and y axes (rather than the x- and y-coordinates) or, alternatively, the modulus of the velocity and the angle between the velocity and the x-axis. Such curves permit the acceleration of the particle to be determined directly and are extensively applied to problems of compressible and incompressible flow.

HODOSCOPE. An array of radiation counters used in studying the paths of cosmic-ray particles.

HOHLRAUM. A uniform temperature cavity which acts as a black body.

HOLE CONDUCTION. See: *Electron hole: Positive hole.*

HOLE THEORY. A theory leading to the prediction of the existence of the positron and, by extension, to that of other anti-particles. According to this theory the states of negative energy required by the Dirac equation are normally all occupied so that no transition to them can occur. If sufficient energy is applied, however, a particle may be excited from a negative to a positive energy state and will appear as an electron. The hole left behind in an otherwise occupied negative sea, i.e. the absence of an electron with certain values of energy, momentum and spin, will then appear as the presence of a particle with the same mass and spin, but opposite charge—the positron. See also: *Dirac equation.*

HOLLOW CATHODE. (1) The name given to an electron emission system (e.g. an electron gun) which produces, under certain conditions, a beam of electrons with zero current density along the axis and peak density away from it. (2) The name given to the effect in a discharge tube whereby, when two plane cathodes are brought close enough together for their negative glows to interact, a marked increase in the current density is obtained without the need for altering voltage or pressure.

HOLOGRAPHY. The technique of recording an optical image of an object, which preserves the amplitude, wavelength, and phase of the light reflected, diffracted or transmitted by that object; and the use of this record to produce a three-dimensional image of the object itself. The object is illuminated by a laser beam, part of the beam being separated off into a phase-coherent reference beam which is allowed to interfere with the light reflected, diffracted, or transmitted by the object so that the phase information is manifest as an intensity variation. A photograph of the resultant pattern, without focusing, then carries information about both intensity and phase. This photograph (known as a *hologram*), when illuminated with the original coherent reference beam (or one bearing some simple relationship to it), will yield a diffracted wave which is identical in amplitude and phase distribution with the original wave from the object, so that a three-dimensional image of the object can be obtained, which can either be photographed or viewed directly.

HOLOGRAPHY, ACOUSTIC. Holography in which sound waves are used instead of light waves.

HOLOHEDRAL CRYSTAL. One exhibiting the maximum symmetry of the crystal system to which it belongs.

HOLOMORPHIC FUNCTION: ANALYTIC FUNCTION: REGULAR FUNCTION. A function $f(z)$ of the complex variable $z = x + iy$ is holomorphic, analytic or regular at a point on the z-plane if the function and its first derivative are finite and single-valued there. If this property applies to all points within a given region of the complex plane, $f(z)$ is a holomorphic (analytic or regular) function throughout the region. Points at which there is no derivative are known as *singular points* or *singularities*.

HOMODESMIC STRUCTURE. A crystal structure in which only one type of bond (e.g. ionic or covalent) occurs, as opposed to a *heterodesmic* structure, in which two or more different types of bond are in operation.

HOMOLOGOUS SERIES. A set of organic compounds in which successive members differ from each other by a CH_2 group. Individual members are known as *homologues*.

HOMOLYSIS. See: *Chemical reaction, homolytic.*

HOMOPOLAR COMPOUND. A covalent compound in which the bonds are essentially homopolar in character. Nowadays, however, the term is often applied to any covalent compound. See also: *Bond, homopolar.*

HOMOPOLAR INDUCTION. The generation of an induced e.m.f. of constant polarity by the motion of a circuit element through a magnetic field of constant strength. A simple example is the *Faraday disk,* a metal disk rotating between the poles of a magnet, which generates a constant e.m.f. between the rim and axis of the disk and from which a direct current may be obtained between brushes placed in these two areas. In recent commercial machines, using liquid metal brushes, currents of the order of 10^5 amp can be obtained.

HOMOTHETIC CURVES. A set of curves such that any straight line through a given point cuts all members of the set at the same angle. The set is then said to be homothetic to that point.

HOOKE LAW. For an elastic solid: states that the strain is proportional to the stress within the elastic limit.

HOPE APPARATUS. An apparatus for showing the variation of the density of water with temperature in the neighbourhood of the freezing point, and for determining approximately the temperature of maximum density.

HORIZON, ARTIFICIAL. (1) On land: usually a shallow basin of mercury, used in the determination of the altitudes of celestial bodies with a sextant, when no natural true horizon is visible. (2) At sea or in the air: a device presenting a visual display of the horizon, which may involve the use of a gyroscope. See also: *Sextant.*

HORN. (1) Acoustic: a tube intended for the transmission and radiation of sound, of which the cross-sectional area increases progressively from the throat to the mouth. (2) Radio: an elementary aerial consisting of a waveguide in which one or more transverse dimensions increase towards the aperture.

HORSE LATITUDES. The regions of comparative calm in the sub-tropical anticyclone belts between about 30° North and 30° South of the equator.

HORSE POWER. (1) A British unit of power defined as a rate of work of 33 000 ft-lb/min. It is equal to 745·700 W. (2) A metric unit defined as a rate of work of 75 m-kg/s. It is equal to 735·5 W. See: *Appendix II.*

HORSE POWER, BRAKE. The power actually delivered and measurable by a brake or dynamometer, as distinct from the *indicated horse power,* i.e. the theoretical horse power in the absence of friction and other losses in the mechanism.

HOSPITAL PHYSICS. The branch of medical physics that deals mainly with problems arising from the application of ionizing radiations and radioactive tracers in hospitals. Hospital physics is, however, often extended beyond this to overlap with other parts of medical physics. See also: *Biophysics. Health physics. Medical physics.*

HOUR. See: *Day.*

HOUR ANGLE. Of a celestial body for an observer: the angle made at the celestial pole between the meridian of the body and that of the observer. It is measured positively westwards from the observer's meridian, in hours, minutes, and seconds.

HOVERCRAFT. A jet-driven vessel which "flies" over land or water on a cushion of air. The air is drawn in vertically by a fan and discharged circumferentially downwards and inwards below the flat undersurface of the craft. A flexible curtain or "skirt" serves to minimize any damage arising from contact with the sea or the ground.

H-THEOREM OF BOLTZMANN AND LORENTZ. A theorem which shows that a non-equilibrium distribution of a system of weakly interacting particles will eventually approach the Maxwell–Boltzmann distribution. It may be stated in the form: the entropy of a closed system will never decrease; and may therefore be thought of as the kinetic counterpart of the second law of thermodynamics.

HUBBLE LAW. States that distant galaxies recede from the Earth at velocities directly proportional to their distances. The constant of proportionality is known as the *Hubble constant,* and is about 10^{-10} per year, or 3×10^{-17} per second.

HUBBLE RADIUS. The radius of the observable universe, upwards of 10^{24} km.

HUE. Of a colour: The property by which a colour is identified as tending to resemble the appearance of some particular wavelength of light in the visible spectrum.

HÜFNER RHOMB. A single rhomb of glass that can be considered as two prisms of large angle placed base to base. It is used for bringing two separated beams of light into juxtaposition for photometric comparison.

HUGONIOT CURVE. The curve connecting pressure rise and density variation in a compression shock.

HUGONIOT EQUATION. The relation between the cross-section and velocity of a compressible fluid in one-dimensional flow.

HUME-ROTHERY RULE. For alloys of copper, silver and gold with metals of the B subgroups: states that there is a tendency for intermediate phases to be formed at compositions corresponding to valence-electron/atom ratios of $\frac{3}{2}, \frac{21}{13}$ and $\frac{7}{4}$. Each ratio corresponds to a particular crystal structure or group of structures. The rule has now been extended to cover alloys between all the transition metals and metals of the B subgroups.

HUMIDITY. The presence of water vapour in the atmosphere. Various ways of specifying the degree of humidity are given in the following entries.

HUMIDITY ABSOLUTE. The mass of water vapour per unit volume of moist air, usually expressed in grammes per cubic metre.

HUMIDITY, RELATIVE. The ratio of the partial pressure of the water vapour in the atmosphere to the saturation vapour pressure at the same temperature. The ratio of the amount of water vapour present to the amount present in saturated air at the same temperature is known as the *saturation ratio*.

HUMIDITY, SPECIFIC. The mass of water vapour per unit mass of air.

HUMPHREYS SERIES. A series in the line spectrum of the hydrogen atom. See also: *Hydrogen atom, line spectrum of.*

HUM TONE. The lowest note emitted by a bell.

HUND COUPLING CASES. See: *Coupling.*

HUND RULES. Empirical rules for the sequence of the energy states of atomic terms arising from one electronic configuration. Suitably reformulated they have some application to molecules.

HUNTER SHORT BASE. For surveying: a light, portable apparatus for rapidly laying out a baseline over rough or sloping ground. It takes the form of a catenary with intermediate supports.

HUNTING. Of an automatic control system: the swinging of the controlled variable about the desired value without any settling down to that value.

HURRICANE. See: *Cyclone.*

HUYGENS PRINCIPLE. States that every element of an advancing wavefront acts as a source of secondary disturbances, i.e. as a source of a new train of waves. A modification, the *Fresnel–Huygens principle*, states that the secondary disturbances interfere with one another in accordance with the usual principle of superposition.

HYBRIDIZATION. The combination of two or more constituent orbitals of an atom to form equivalent and directed orbitals (*hybrids*). These hybrids are described, according to the directions of their orbitals, as *digonal, trigonal, tetrahedral, octahedral,* or *tetragonal.*

HYDRACID. An acid that does not contain oxygen in its molecule.

HYDRATED COMPOUND. A chemical compound in which a definite proportion of water is taken up to form a stable compound, the water being loosely held as *water of crystallization*. A well-known example is $CuSO_4 \cdot 5 H_2O$.

HYDRATION. The combination of a substance with water, but not necessarily to form a hydrate, e.g. in solvation.

HYDRAULIC COUPLING: HYDRODYNAMIC COUPLING. Coupling between rotating shafts consisting of two rotating units immersed in a viscous fluid. The first of these (the primary) acts as a pump, setting the fluid in motion. The second (the secondary) is then caused to rotate by turbine action.

HYDRAULIC RADIUS. The ratio of the cross-sectional area of flow to the wetted perimeter in a pipe or channel.

HYDRAULICS. The science and engineering of the flow of fluids. It embraces the flow of fluids in closed boundaries, e.g. flow in pipes, pumps, and turbines, and the flow of liquids with a free surface within rigid or loose boundaries, e.g. flow in rivers and estuaries, and wave motion. It may also include the flow of water in saturated and partly saturated soils.

HYDRIDE. A compound of an element with hydrogen.

HYDROCARBON. An organic compound containing carbon and hydrogen only.

HYDRODYNAMICS. The study of the motion of incompressible fluids. See also: *Hydromechanics.*

HYDROFOIL. A lifting wing, analogous to an aerofoil, used in a liquid, usually water.

HYDROGENATION. The chemical reaction of a substance with molecular hydrogen in the presence of a catalyst.

HYDROGEN ATOM, LINE SPECTRUM OF. Consists of several series of spectra, named after their discoverers, whose wave numbers, v, are given by the following expressions:

Lyman series:

$$v = R\left(\frac{1}{1^2} - \frac{1}{n^2}\right), \quad n = 2, 3, 4, \ldots$$

Balmer series:

$$v = R\left(\frac{1}{2^2} - \frac{1}{n^2}\right), \quad n = 3, 4, 5, \ldots$$

Paschen series:

$$v = R\left(\frac{1}{3^2} - \frac{1}{n^2}\right), \quad n = 4, 5, 6, \ldots$$

Brackett series:

$$v = R\left(\frac{1}{4^2} - \frac{1}{n^2}\right), \quad n = 5, 6, 7, \ldots$$

Pfund series:

$$v = R\left(\frac{1}{5^2} - \frac{1}{n^2}\right), \quad n = 6, 7, 8, \ldots$$

Humphreys series:

$$v = R\left(\frac{1}{6^2} - \frac{1}{n^2}\right), \quad n = 7, 8, 9, \ldots$$

In these expressions R is the Rydberg constant. See also: *Rydberg series.*

HYDROGEN BOND. A bond or linkage between a molecule such as OH, NH, or FH, and an electronegative atom such as O, N, or F, which operates via the hydrogen atom. The binding is mainly electrostatic. Hydrogen bonds play a vital role in influencing crystal structure. They are also known as *hydrogen bridges* and *hydroxyl bonds.*

HYDROGEN ELECTRODE. The electrode formed on platinum metal coated with platinum black, immersed in an aqueous electrolyte, when gaseous hydrogen is supplied to it. It is used as a standard zero of potential and for measuring pH values.

HYDROGEN ION CONCENTRATION. See: *pH*

HYDROGEN, METALLIC. Hydrogen which, by subjection to high pressure (a few megabars: upward of 10^{11} Pa) and low temperature (4·2 K), becomes a conducting metal.

HYDROGEN I AND II. The hydrogen of interstellar space. H I is neutral hydrogen and is detected solely by its emission of the 21-cm radio line; and H II is ionized hydrogen, emitting both visible and radio radiation.

HYDROGEN SCALE OF TEMPERATURE. A temperature scale based on the variation in pressure of hydrogen at constant volume. It was accepted internationally in 1887 but has now been superseded. See also entries under: *Temperature scale.*

HYDROLOGIC CYCLE. The circulation of water from the oceans, through the atmosphere, and back to the oceans; or from the oceans, to the land, and back to the oceans by overland or subterranean routes.

HYDROLOGY. The study of the movement and storage of ground water.

HYDROLYSIS. A chemical reaction in which water, as a reactant, produces bond rupture in an inorganic compound, or in which an organic compound takes up an OH group.

HYDROMECHANICS. The science of the mechanics of fluids. It comprises hydrostatics and hydrodynamics.

HYDROMETEOROLOGY. The study of the atmospheric and land phases of the hydrologic cycle with emphasis on the interrelationships involved.

HYDROMETEORS. Bodies of liquid or solid water falling to the ground as a result of condensation, i.e. rain, snow, drizzle, sleet, freezing rain and drizzle, hail.

HYDROMETER. An instrument for measuring the density of a liquid. A common form, the *constant-mass* type, consists of a floating bulb, weighted at the lower end so that it floats in the vertical position, the depth of immersion being inversely proportional to the density. The *constant-displacement* type, e.g. the *Nicholson* hydrometer, which incorporates scale pans above and below the bulb, can also be used to determine the density of a solid.

HYDROPHILIC. Of a colloid or sol: signifies one which readily forms a "solution" in water.

HYDROPHOBIC. Of a colloid or sol: signifies one which forms a "solution" in water only with difficulty.

HYDROPHONE. A sound detector for use under water. It is used for echo-sounding, for sonar, and for communications.

HYDROSTATIC PRESSURE. Describes a system of forces acting on a body to compress it equally in all directions. It was originally applied to the pressure imposed on a body in a liquid due to the weight of the liquid.

HYDROSTATICS. The study of the equilibrium of fluids at rest. See also: *Hydromechanics*.

HYDROXIDE. An inorganic compound containing one or more *hydroxyl groups* (i.e. OH groups).

HYGROGRAPH. An instrument for continuously recording the relative humidity of the air.

HYGROMETER. Any instrument for measuring the humidity of the atmosphere. The commonest is the wet and dry bulb thermometer or psychrometer. See also: *Hygrometer, absorption. Hygrometer, dew point (or frost point). Hygrometer, diffusion. Hygrometer, hair. Psychrometer.*

HYGROMETER, ABSORPTION. A hygrometer in which the moisture in a measured volume of gas is absorbed and weighed.

HYGROMETER, DEW POINT (OR FROST POINT). A hygrometer which involves the measurement of the dew point (or frost point) by the optical observation of a surface condensate.

HYGROMETER, DIFFUSION. A hygrometer which involves the measurement of gas transfer through a porous plate.

HYGROMETER, HAIR. A hygrometer in which is measured the extension of grease-free human hair under tension. The humidity-extension relationship is, however, somewhat variable.

HYPERBOLA. See: *Conic: Conic section.*

HYPERBOLIC FUNCTIONS. Combinations of e^x and e^{-x} with properties analogous to those of the trigonometric functions. They are:

$$\sinh x \left[= \tfrac{1}{2}(e^x - e^{-x}) \right], \quad \cosh x \left[= \tfrac{1}{2}(e^x + e^{-x}) \right],$$

$$\tanh x \left[= \frac{\sinh x}{\cosh x} \right], \quad \coth x \left[= \frac{1}{\tanh x} \right],$$

$$\operatorname{sech} x \left[= \frac{1}{\cosh x} \right], \quad \text{and} \quad \operatorname{cosech} x \left[= \frac{1}{\sinh x} \right].$$

HYPERBOLOID. See: *Quadric.*

HYPERCHARGE. Of a high-energy particle: the sum of the strangeness and the baryon number.

HYPERCONJUGATION. Refers to the delocalization of chemical bonds in a molecule, of which the bonding is described entirely in terms of resonance.

HYPERFINE STRUCTURE. See: *Spectrum, hyperfine structure in.*

HYPERFRAGMENT: HYPERNUCLEUS. A short-lived nucleus, emitted in nuclear disintegration, which contains a bound hyperon, usually the lambda hyperon Λ^0.

HYPERMETROPIA: HYPEROPIA. A condition of the eye in which parallel rays incident upon the lens are focused at a point behind the retina. Commonly known as long sight.

HYPERON. A term formerly denoting one of a group of unstable elementary particles which are heavier than a neutron, but now extended to cover any baryon with non-zero strangeness. Such particles, which (in order of increasing mass) are lambda (Λ^0), sigma (Σ^0, Σ^+, and Σ^-), cascade or xi (Ξ^0 and Ξ^-) and omega (Ω^-) particles, eventually decay to a proton or a neutron. See also: *Particle symmetry.*

HYPERSONIC. Refers to flow fields, phenomena and problems appearing at speeds far greater than that of sound, usually taken as five times that speed or greater. See also: *Compressible flow.*

HYPO –. A prefix indicating that the central atom of an acid or compound is in a lower valency state than normal.

HYPOTHERMIA. The reduction of body temperature, used therapeutically when the oxygen demand of the body is too high to be satisfied normally.

HYPSOCHROMIC SHIFT. The shift of an absorption band to a region of shorter wavelength. The opposite of bathochromic shift. See also: *Bathochromic shift.*

HYPSOMETER. An instrument used for determining the boiling point of water, either to determine altitude or correct the upper fixed point of a thermometer.

HYSTERESIS. The phenomenon exhibited by a system whose properties depend on its previous history. The term is commonly applied to magnetic properties, but hysteresis may, for example, be *mechanical* (*dynamic* or *static* according as the number of cycles is greater than or equal to unity) or *dielectric*. The term is also used to denote the difference between the response of a *physical instrument* or system to an increasing signal and the response to a decreasing signal.

HYSTERESIS HEATING. A method of supplying a controlled amount of heat to a thermally isolated sample at a temperature below 1 K, involving taking the sample through a magnetic hysteresis loop. The sample must be paramagnetic and near or below its Curie point.

HYSTERESIS LOOP. See: *Magnetic hysteresis.*

HYSTERESIS LOSS. See: *Magnetic hysteresis.*

i

ICE. Water in the solid state. It is crystalline and birefringent and exists in seven polymorphic phases.

ICE BLINK. A whitish glare in the sky arising from an extensive snow-field (inland ice or snow-covered sea ice). In comparison open water shows a dark sky.

ICELAND SPAR. A clear transparent form of calcite which cleaves into perfect rhombohedra and is widely used in constructing Nicol prisms for the polarizing microscope.

ICE-POINT. The temperature at which pure ice and water can exist together in equilibrium at a pressure of 760 mm of mercury (133·322 Pa). It was formerly used to define 0°C on the International Temperature Scale, but has now been superseded as a fixed point by the triple point of water (0·10°C above the ice-point). See also: *Steam point. Temperature scale, international practical.*

ICONOSCOPE. An early form of television camera tube in which the optical image of the scene to be transmitted is projected on a photosensitive target mosaic which is scanned by a high-speed electron beam. See also: *Camera tube.*

IDLE COMPONENT. See: *Reactive component.*

IGNITION. Of an electrical discharge: See: *Accommodation time.*

IGNITRON. A single-anode mercury vapour rectifier in which current passes as an arc between the anode and a mercury pool cathode. It is used for the control of currents between about five and many thousands of amperes, for such things as resistance welding, voltage regulation and switching duty.

ILLUMINANCE: ILLUMINATION. At a point of a surface: the luminous flux received per unit area. The SI unit is the *lux* and is equal to an illumination of 1 lumen/m². It was formerly known as a *metre-candle*. The *phot* is an illumination of 1 lumen/cm² and the *foot-candle* 1 lumen/ft² (= 10·764 lux). See also: *Luminance. Appendix II.*

ILLUMINATION, UNIFORMITY RATIO OF. Over a given area: the ratio of the minimum illumination to the maximum illumination over that area.

ILLUMINATION, UTILIZATION FACTOR OF. Over a given surface: the ratio of the total luminous flux reaching the surface to the total luminous flux of the source. It is also known as the *utilization coefficient.*

IMAGE CONVERTER. A device producing a visible display of an image formed by radiation outside the visible spectrum or by ultrasound. In the most usual type the image is formed on a photocathode, photoelectrons being accelerated from this to form a visible image on a fluorescent screen. However, the X-ray fluoroscope is also a type of image converter. See also: *Image intensifier. Thermograph. Vidicon.*

IMAGE CONVERTER, SHUTTERED. An image converter employing a photocathode, in which the photo-electrons may be rapidly switched off to permit a stationary camera to record the image on the fluorescent screen, e.g. by means of a grid. Shuttered image converters are capable of giving exposure times as short as 10^{-9} s. See also: *Image converter.*

IMAGE DIPOLE. See: *Electrical image.*

IMAGE FORCE. The force on a charge due to the charge or polarization that it induces on neighbouring conductors or dielectrics. See also: *Electrostatic image.*

IMAGE IMPEDANCE. The impedance which terminates a network without introducing reflection losses at the junction, so implying image termination at the other end of the network.

IMAGE INTENSIFIER. An image converter which produces an optical image at a brightness greater than that of the initial image. The need for such a device first arose in X-ray fluoroscopy, especially for medical diagnosis, because of the insufficient brightness of fluorescent screens; but similar devices are now available for neutrons.

IMAGE METHODS. See: *Electrical image. Electrostatic image. Fluid flow, image method in. Magnetic image.*

IMAGE, REAL. In an optical system: an image at which rays of light are actually brought to a focus, i.e. one which can be formed on a screen placed in the image plane. See also: *Image, virtual.*

IMAGE-RETAINING PANEL. See: *Electroluminescent panel. Lamp, ceramic.*

IMAGE SPACE. Of an optical system: the space which light rays traverse after passing through the refracting system. See also: *Object space.*

IMAGE STORAGE. The accumulation of sufficient image information to present a useful picture. Examples are provided by the storage tube used in television and radar, and by the electroluminescent panel. However, the simplest example is the use of a photographic film for optical, X-ray, electron or neutron photography. See also: *Electroluminescent panel. Storage tube.*

IMAGE, ULTRASONIC. Usually signifies the image produced when an ultrasonic beam passes through an object, which is analogous to a radiographic image. Various detection methods are possible but the most sensitive is probably the use of piezoelectric or electrostrictive transducers combined with a scanning technique. Presentation may be by direct recording on film or paper, or on an oscilloscope.

IMAGE, VIRTUAL. In an optical system: an image from which rays of light appear to diverge (as with a diverging lens) but which does not actually exist and cannot be formed on a screen placed in the image plane. See also: *Image, real.*

IMAGINARY NUMBER. A real number multiplied or operated on by $\sqrt{-1}$. See also: *Complex number.*

IMBIBITION. The absorption of a liquid by a solid or gel. It is often accompanied by swelling, e.g. in a colloidal system.

IMMERSION. In Astronomy: the disappearance of the Moon or other celestial body at eclipse or occultation.

IMMERSION LENS, OPTICAL. The term usually denotes that the space between the object and objective of a microscope is filled with a medium (usually oil) of higher refractive index than air. This increases the numerical aperture and hence improves the resolution. In an electron lens the term has a different significance. See also: *Electron lens, immersion.*

IMPACT. The collision between two bodies when relatively large contact forces exist during a very short interval of time.

IMPACT PARAMETER. For a collision between two particles: the least distance at which the centres of the two particles would have passed each other if they had both continued with their original direction and speed.

IMPACT STRENGTH: TOUGHNESS. Of a material: the maximum energy of deformation which may be produced by impact loading. The *static toughness* is given by the area under the true stress–strain curve; and an index of the *dynamic toughness* is given by the energy absorbed when a specimen of specified dimensions is subjected to a blow sufficient to produce fracture. See also: *Impact tests.*

IMPACT TESTS. Tests made to determine the dynamic toughness. In nearly all the standard tests some form of *notched bar* is introduced, but notch specifications vary. The *Charpy* test and the *Izod* test each employ a pendulum, and the energy absorbed in fracture is obtained from the height to which the pendulum rises. Similar tests are the *Fremont* and *Mesnager* tests. See also: *Impact strength: Toughness.*

IMPATT DIODE MICROWAVE GENERATOR. Refers to a variety of *p–n* (or more complex) semiconductor diodes which, when reverse biased, generate power at frequencies above 100 MHz. The term IMPATT stands for "*Imp*act" *a*valanche *t*ransit *t*ime.

IMPEDANCE. The (complex) ratio of voltage to current in an alternating current circuit. The real part of the impedance is the *resistance* and the imaginary part is the *reactance*. Analogous quantities occur in acoustics and mechanics. See also: *Acoustic impedance. Mechanical impedance.*

IMPEDANCE, EXTERNAL AND INTERNAL. See: *Impedance, load. Impedance, source.*

IMPEDANCE, INPUT. The impedance presented by a transducer to a source.

IMPEDANCE LOAD. The impedance external to the source in an electric circuit. It is also known as the *external impedance.*

IMPEDANCE, MATCHING. The adjustment of a load to a source so that the maximum power is accepted, i.e. so that there is no reflection loss due to mismatch. To achieve this the impedances on either side of the junction should be equal, or rendered so by the insertion of a matching device.

IMPEDANCE, OUTPUT. The impedance presented by a transducer to a load.

IMPEDANCE, SOURCE. The impedance of the source in an electric. circuit. It is also known as the *internal impedance.*

IMPELLER. The rotating member of a centrifugal pump or blower, which imparts kinetic energy to the fluid which is being pumped.

IMPERFECT CRYSTAL. A crystal in which the regular periodicity of the structure is impaired, on account of strain, distortion, or other defects, except over limited regions; i.e. a crystal which has a mosaic structure. A crystal with a mosaic structure in which the mosaic blocks can be regarded as negligibly small is known as an ideally *imperfect crystal*. See also: *Extinction. Mosaic structure. Perfect crystal.*

IMPERIAL SYSTEM OF UNITS. A system based on the Imperial Standard Yard as a standard of length and the Imperial Pound as a standard of mass. The unit of volume, the gallon, is defined in terms of the volume of 10 lb of distilled water under specified conditions. It was developed in Great Britain and adopted, with some modification, in the U.S.A. See also: *Gallon. Units, Imperial. Appendix II.*

IMPULSE APPROXIMATION. For a collision between a particle and a (bound) target particle: the neglect, for purposes of calculation, of the binding forces on the target particle during the collision.

IMPULSE ELECTRIC STRENGTH. Of an insulator for a voltage applied for a very short time (microseconds): the breakdown voltage divided by the thickness of the insulator.

IMPULSE GENERATOR. An arrangement for producing high voltage pulses, usually by charging capacitors in parallel and discharging them in series. See also: *Marx circuit.*

IMPURITY CENTRE. A foreign atom or group of atoms in a crystal. It often serves to activate a phosphor. See also: *Acceptor. Donor.*

INCH. See: *Length, units of. Appendix II.*

INCOHERENT. As applied to emitted or scattered waves: implies the absence of a definite phase relationship between two or more sets of waves. Where a definite phase relationship does exist the waves are said to be *coherent.*

INCOMPRESSIBLE FLOW. See: *Compressible flow.*

INDEPENDENT DAY NUMBERS. Quantities appearing in the expressions for the changes in the coordinates of a star due to precession and nutation.

INDETERMINACY PRINCIPLE: UNCERTAINTY PRINCIPLE. A principle enunciated by Heisenberg which states that no measurement can determine both the position of a particle and its momentum (or any other pair of conjugate variables, such as time and energy) so accurately that the product of their errors is less than Planck's constant. It is a necessary result of the quantum theory. See also: *Complementarity principle. Conjugate variables.*

INDICATOR. A substance used to indicate the completion of a chemical reaction, often by a change in colour. Such substances are widely used in volumetric analysis.

INDICATOR DIAGRAM. A graphical representation of the pressure existing in the cylinder of a reciprocating engine for any position of the piston.

INDUCTANCE. (1) The property of an electrical circuit which gives rise to the phenomenon of electromagnetic induction. (2) A quantitative measure of this given by the e.m.f. produced per unit rate of change of current. The MKSA and SI unit is the henry (equal to 10^9 cm). (3) An inductor. See also: *Henry.*

INDUCTANCE, DISTRIBUTED. The inductance that exists along the entire length of a conductor as opposed to that which is concentrated in an inductor.

INDUCTANCE, MUTUAL. For electromagnetic induction between two circuits (e.g. two, parallel wires): the e.m.f. induced in one circuit when the current in the other is changing at unit rate.

INDUCTANCE, SELF. For electromagnetic induction within a single circuit (e.g. a coil): the back e.m.f. induced in a circuit when the current in the same circuit is changing at unit rate.

INDUCTION. (1) Electric induction, or displacement. (2) Electromagnetic induction. (3) Electrostatic induction. (4) Magnetic induction. (5) Nuclear induction.

INDUCTION, COEFFICIENTS OF. The coefficients which, together with the coefficients of capacitance, specify the charges on a system of conductors in terms of their potentials.

INDUCTION COIL. A primitive form of transformer for obtaining a relatively high intermittent voltage from a low, steady voltage such as that supplied by a battery. It consists essentially of a primary coil, of few turns, a secondary coil wound on a soft iron core, and a simple interrupter in the primary circuit.

INDUCTION HEATER. An electrical heater in which material or its container is heated by current induced by a primary inductor. Such heaters are widely used for soldering and brazing in mass production, and for case hardening.

INDUCTION, INTRINSIC: MAGNETIC ENERGY. Defined as $B - H$, where B is the magnetic induction in a magnetic field H. It is also called *ferric induction*. The CGS unit is the gauss-oersted and the SI unit the tesla-ampere per metre ($= 40\,\pi$ or about 125·6 gauss-oersted).

INDUCTION, MOTIONAL. Electromagnetic induction in which the variation of magnetic flux density with time arises from the motion of a circuit and not from a change in current.

INDUCTION MOTOR: ASYNCHRONOUS MOTOR. An a.c. motor in which currents in the primary winding induce currents in the secondary winding (usually the rotor) which interact with the flux set up by the primary winding to produce rotation.

INDUCTION, MUTUAL. Electromagnetic induction between two circuits, e.g. two parallel wires.

INDUCTION, SELF. Electromagnetic induction within a single circuit, e.g. a coil.

INDUCTOMETER. A variable inductor, used for varying the inductance in a network, e.g. for balancing a bridge.

INDUCTOR. (1) A device which has an inductance which is large compared with its resistance and self-capacitance. (2) One of the rotating masses of magnetic material used to produce the necessary changes in magnetic flux density in an inductor generator.

INDUCTOR GENERATOR. An electric generator in which the field and armature windings are fixed relative to each other, the necessary changes in magnetic flux density being produced by rotating masses of magnetic material.

INEQUALITY. In astronomy: denotes any departure from uniformity in orbital motion, either periodic or secular.

INERT GAS: NOBLE GAS: RARE GAS. Any element of Group 0 of the periodic system, namely helium, neon, argon, krypton, xenon and radon. The electron shells are completely filled and hence these gases are practically devoid of chemical properties, but a number of crystalline compounds have been prepared with fluorine, oxygen and certain metals.

INERTIA. The resistance offered by a body to acceleration. See also: *Moment of inertia. Momentum. Newton laws of motion.*

INERTIAL FRAME OF REFERENCE. A frame of reference that remains constant, usually a coordinate system based on the "fixed" stars.

INERTIAL GUIDANCE. A navigational technique employing the inertial properties of masses to measure the magnitude and direction of motion. Accelerometers and gyroscopes are used, operating with respect to an inertial frame of reference, usually associated with the "fixed" stars. Inertial guidance systems are used in missiles, ships (including submarines) and aircraft.

INFINITE SERIES. A collection of numbers $a_1 + a_2 + a_3 + \cdots + a_n + \cdots + a_\infty$ with real or complex terms. It is said to be *convergent* if a_n converges to a finite sum as n tends to infinity. Otherwise, i.e. if a_n diverges, the series is *divergent*.

INFLECTION, POINT OF. A point on the curve $y = f(x)$ where the derivative $\dfrac{dy}{dx}$ changes sign but where neither a maximum nor a minimum occurs, the curve crossing the tangent and bending away from it in opposite directions on opposite sides of the point of contact.

INFORMATION THEORY. Concerns the separation and measurement of information content from other aspects of information such as semantic value. In this theory the concept of entropy has been taken over to signify a measure of uncertainty.

INFRARED ANALYSIS OF GASES. The quantitative determination of known elements and simple compounds in a mixture by their differential absorption of particular infrared wavelengths.

INFRARED CATASTROPHE. The theoretical infinity in the emission of low frequency radiation by a charged particle, which appears as a consequence of classical electrodynamics. The difficulty disappears when appropriate quantum corrections are applied. See also: *Bloch–Nordsieck method. Ultraviolet catastrophe.*

INFRARED GALAXIES. Galaxies which are strong emitters of infra-red radiation at wavelengths ranging from 1 to 1000 μm. The energy output of such a galaxy can be as much as 100–1000 times that of the brightest optical galaxy.

INFRARED RADIATION. That part of the electromagnetic spectrum which extends from the long-wave limit of visible radiation (about 7500 Å or 0·75 μm) to a wavelength sometimes taken as 350 μm, since this represents the extreme wavelength studied by conventional heat radiation methods, and sometimes taken to include in addition the microwave range up to 10000 μm or 0·1 cm.

INFRASONIC. See: *Sound.*

INFRASONIC WAVES: INFRASOUND. Sound waves having frequencies which are too low for audibility. Such sound may give rise to unpleasant or even dangerous physiological and psychological effects. See also: *Sound.*

INHOUR. In nuclear reactor technology: a unit of reactivity equal to that increase in reactivity of a critical nuclear reactor which would produce a reactor time constant of one hour. The term is short for inverse hour. See also: *Nuclear reactor, reactivity of.*

INJURY POTENTIAL. In bioelectricity: the electrical potential between an intact part of a nerve or muscle cell and a damaged part of the same cell.

INNER BREMSSTRAHLUNG: INTERNAL BREMSSTRAHLUNG. Bremsstrahlung which may arise from the emission or absorption of a charged particle by an atomic nucleus. See also: *Bremsstrahlung.*

INNER EFFECT. See: *Outer effect.*

INNER POTENTIAL. The mean value, taken over the volume of a crystal, of the electrostatic potential arising from the regular arrangement of atoms within the crystal. The inner potential varies between about 5 and 25 V. It is to be distinguished from t1e work function.

INSERTION GAIN OR LOSS. The gain or loss of an electrical network when it is connected between a generator and a load. It is the increase or decrease in signal strength when the network is inserted, and is usually expressed in decibels.

INSOLATION. (1) Radiation received from the Sun. (2) The energy flux density of this radiation. See also: *Radiant flux density.*

INSTRUMENT DESIGN, PASSIVITY IN. The design of instruments in such a way that the measuring devices incorporated in the instruments do not interfere with the phenomena being measured.

INSULATION. (1) Dielectric material having permanent low electrical conductivity and high dielectric strength, used to separate conducting materials which are at different electrical potentials. (2) Material having low thermal conductivity, used to retain or exclude heat.

INSULATOR. Refers to a material which does not conduct electricity or heat as the case may be.

INSULATOR, ELECTRICAL. Refers, apart from its general meaning, to an appliance designed to support or locate an uninsulated current-carrying conductor maintained at a different potential from that of its surroundings.

INTEGER. A whole number, either positive or negative.

INTEGRAL, ELLIPTIC. An integral of the type $\int f(x, \sqrt{R})\,dx$, where f is a rational function of its two arguments and R is a third or fourth degree polynomial in x, with no repeated roots.

INTEGRAL EQUATION. An equation containing an integral which involves an unknown function. The equation is solved when this function is determined. It is termed *linear* if it is linear in the unknown function, otherwise it is *non-linear.*

INTEGRAL, LEBESGUE. Defines an integral similar to that of Riemann but which is more general in its applications.

INTEGRAL, RIEMANN. Defines an integral which gives an exact analytical form to the idea of the area under a curve.

INTEGRAL TRANSFORM. The integral transform $\mathbf{f}(p)$ of a function $f(x)$ is defined by the relation $\mathbf{f}(p) = \int_{0}^{\infty} f(x)\,K(p, x)\,dx$, where $K(p, x)$ is a given function of p and x, called the *kernel* of the transform. When the range of integration is replaced by a finite range the transform is said to be *finite.* The Fourier transform is a well-known example of an integral transform. See also: *Kernel.*

INTEGRAPH. An instrument used for drawing a graph of the integral of a function from that of the function itself.

INTEGRATED CIRCUIT. An electronic circuit which is so designed and constructed that it may be considered to be an indivisible whole both for use and servicing. The applications are most numerous in computing, and data processing and control, but linear amplifier circuits are also available. Either thin films or silicon "chips" are used for the circuit elements. Where both are used in the same circuit it is known as a *hybrid circuit.* See also: *Microcircuit. Monolithic circuit.*

INTEGRATED REFLECTION. Of X-rays or neutrons diffracted by a crystal: a measure of the intensity of a crystal reflection given by the area under the *reflection curve* (i.e. the curve of reflecting power versus angular displacement of the crystal). For monochromatic radiation, it is given by $E\omega/I$, where E is the total energy reflected by a crystal which is bathed in an incident beam of energy I per second, when the crystal is rotated through the reflecting position with an angular velocity of ω. It is, of course, independent of ω. For non-monochromatic radiation (i.e. a beam of "white" X-rays) a corresponding expression exists.

INTEGRATING CIRCUIT OR NETWORK. A circuit or network whose output waveform is the time integral of the input wave form. Such a circuit or network if it precedes a phase modulator produces a frequency modulator. Following a frequency detector it produces a phase detector.

INTEGRATING FACTOR. A multiplying factor which converts a differential equation into an exact differential.

INTEGRATING PHOTOMETER: PHOTOMETRIC INTEGRATOR. An instrument which enables the luminous flux emitted by a source to be determined by a single measurement. The source is placed inside a large sphere, cube, or globe, the interior surfaces of which are painted matt white. The light is thus diffused throughout the sphere and some is allowed to escape through an aperture on to a photometer, usually photoelectric.

INTEGRATION, NUMERICAL. The process of calculating an integral of a function from a set of its approximate values. A well-known example is Simpson's rule for estimating the area of an irregular figure.

INTENSIFICATION, PHOTOGRAPHIC. The chemical treatment of a processed emulsion when under-exposed or under-developed, to increase the overall contrast or density or both.

INTENSIFYING SCREEN. A layer of material which, when placed in close contact with a photographic emulsion, adds to the photographic effect of incident radiation. Such screens are widely used in radiography. The screen may be a *salt screen*, consisting of a chemical salt which fluoresces when struck by X-rays (for example), or a *metal screen*, which emits secondary radiation under such bombardment.

INTENSITY. A quantitative measure of the "strength" of a physical effect. For a beam of radiation (heat, light, sound, sub-atomic particles) it is usually taken as the amount of radiant energy incident on or crossing unit area per second, but in other contexts different concepts are employed, e.g. for electric and magnetic intensity. See, however, under the various effects involved.

INTENSITY MODULATION. See: *Modulation.*

INTENSIVE VARIABLE. One which is independent of the amount of matter considered. Thus, temperature, density and pressure are typical intensive variables. They are opposed to extensive variables such as total energy, mass and volume. An intensive variable is also known as an *intensive quantity* or *intensive parameter*. See also: *Extensive variable.*

INTERACTION REPRESENTATION. A representation of the equations of motion, especially in quantized field theory, in which time dependence is carried partly by the state vector and partly by the operators.

INTERACTIONS BETWEEN ELEMENTARY PARTICLES. Four types of interaction which occur between elementary particles. Their relative strengths are: *strong* (about 1), *electromagnetic* (about 10^{-2}), *weak* (about 10^{-15}), and *gravitational* (about 10^{-39}). See: *Hadrons. Leptons. Quarks.*

INTERATOMIC DISTANCE. The distance between atoms in a molecule or crystal. It is usually taken to be the same as the internuclear distance. See also: *Atomic radius. Ionic radius.*

INTERCOMBINATION LINES. See: *Spectrum, atomic, intercombination lines in.*

INTERCRYSTALLINE FAILURE. See: *Intergranular fracture.*

INTERFACE. A surface which forms the boundary between two phases or systems.

INTERFACIAL ANGLES. Of crystals: the angles between the normals to pairs of crystal faces.

INTERFACIAL FILM. The non-homogeneous region between two phases, of different composition from either.

INTERFACIAL TENSION. The free energy per unit area of the interface between two phases.

INTERFERENCE. (1) In aerodynamics: the aerodynamic influence of one body upon another. (2) In a communications system: the disturbance caused by unwanted signals. (3) Of a progressive wave motion: the effects resulting from the superposition of coherent wave trains of the same or nearly the same frequency, as in diffraction, for example. The interference may be *constructive* or *destructive* according as the waves are in or out of phase.

INTERFERENCE FILTER. An optical filter which depends on interference in a thin film to give selective transmission of a narrow wave band.

INTERFERENCE FRINGES. A pattern of light and dark bands arising from the constructive and destructive interference of wave trains.

INTERFERENCE, ORDER OF. The integral number of wavelengths by which the paths of two wave trains differ when constructive interference occurs. Thus the first order refers to a path difference of one wavelength, the second order to a path difference of two wavelengths and so on.

INTERFEROMETER. Originally an instrument employing the interference of light to obtain knowledge of the light itself, as in spectroscopy, or to perform physical measurements with known light, such as metrological measurements or measurements of refractive index. It consists essentially of a beam splitter, which divides the light from some suitable source into two or more beams, and a recombination system, which brings the beams together again in a condition suitable for interference to occur. The term is now also applied to instruments involving the interference of other forms of wave motion, e.g. sound, microwaves, and X-rays.

INTERFEROMETER, OPTICAL, CLASSIFICATION OF.
This depends on the type of beam splitting employed. In *aperture splitting,* as in the *Rayleigh interferometer,* light from a narrow slit is split by passing through two other slits, and combined by a lens. In *amplitude splitting,* as in the *Michelson, Twyman and Green, and Mach–Zender interferometers,* light is split by reflection and transmission at a partly silvered surface; in the *Fabry–Perot interferometer* multiple reflections are employed and in the *Dyson interferometer* use is made of polarization. A third form of splitting, namely *diffraction splitting,* is known but rarely used.

INTERFEROMETER, MICROWAVE.
Microwave interferometers work on the same principles as optical ones, although dispersion effects are more troublesome. The best known are microwave analogues of the Michelson and Fabry–Perot types.

INTERFEROMETER, STELLAR.
A device employing interference (of light or other electromagnetic waves) to measure small angles, typically angular diameters of stars.

INTERGRANULAR FRACTURE.
Fracture in which cracks run along grain boundaries but do not cross the grain. It is the usual mode of fracture in high-temperature creep tests of metals and is sometimes found when stress corrosion occurs. It is also known as *intercrystalline failure.*

INTERMEDIATE STATE.
(1) In nuclear theory: a state via which a forbidden transition may occur, the transitions to and from the intermediate state being allowed. (2) Of a superconductor: the state of the superconductor when it is in an external magnetic field which approaches the critical field. It is characterized by the existence of domains of alternately normal and superconducting regions.

INTERMETALLIC COMPOUND.
A compound of two or more metallic elements possessing a characteristic crystal structure and a specific composition or composition range.

INTERMOLECULAR POTENTIAL.
The work done in separating two molecules by an infinite distance.

INTERNAL COMBUSTION ENGINE.
A heat engine in which a fuel (oil, gas or coal) is burnt with air to form the working fluid. By extension the indirectly heated gas turbine is usually classed as an internal combustion engine.

INTERNAL CONVERSION.
The process in which a nuclear transition, which would otherwise result in γ-ray emission, imparts the full available energy to one of the electrons in the atom (termed a *conversion electron*) which is ejected from the atom with the emission of an X-ray photon or Auger electron.

INTERNAL CONVERSION COEFFICIENT.
The ratio of the number of conversion electrons produced to the number of γ-ray photons emitted, i.e. the fraction of nuclear transitions of a given type which are accompanied by internal conversion.

INTERNAL FRICTION: DAMPING CAPACITY.
The ability of a material to dissipate mechanical energy internally (i.e. excluding the overcoming of external forces such as friction). A common measure is the fractional energy lost per cycle of oscillation. Others are the logarithmic decrement and the reciprocal of the Q-value.

INTERNAL PRESSURE.
(1) Of a liquid: $\left(\dfrac{\partial E}{\partial V}\right)_T$ where E is the internal energy and V the volume of the liquid. Also known as the *cohesion pressure.* (2) Of a solid: See: *Simon melting equation.*

INTERNAL ROTATION.
The rotation of one part of a chemical molecule about a line joining it to the rest of the molecule.

INTERNATIONAL SYSTEM OF UNITS.
See: *SI Units.*

INTERPLANAR SPACING.
In a crystal: the perpendicular distance between adjacent members of any family of planes drawn through the three-dimensional structure of a crystal. These distances are of the order of 1 Å and can be accurately determined by X-ray diffraction methods. (*Note:* $1 \text{ Å} = 10^{-10} \text{ } \mu\text{m.}$)

INTERPLANETARY MATTER.
The matter and radiation lying between the planets of the solar system. It consists of solar corpuscular streams, interplanetary plasma, cosmic rays, and interplanetary "dust" (including meteors and comets). Interstellar molecules have now been observed, most of which are organic, but molecules of SiO have been detected. See also: *Interstellar dust. Zodiacal light.*

INTERPOLATION.
The estimation of the value of a function of an independent variable for a value of the variable which is within the range for which the function has been determined. See also: *Extrapolation.*

INTERSTELLAR DUST.
Small grains in the interstellar gas, consisting primarily of silicates, which cause the dimming and reddening of starlight. Numerous organic molecules have also been identified. See also: *Interplanetary matter.*

INTERSTELLAR GAS.
The gas, consisting mostly of hydrogen, which occupies interstellar space, and amounts to about 1% of the total mass of the Galaxy. See also: *Hydrogen I and II.*

INTERSTITIAL ATOM. See: *Point defect*.

INTERSTITIAL COMPOUND. A metallic compound in which atoms of a non-metallic constituent occupy the spaces between the metallic atoms. The metallic atoms are usually those of the transition metals, and the (small) non-metallic atoms are typically those of C, N, B, H, and sometimes O, S, Si or P.

INTERSTRATIFICATION, RANDOM. The succession of crystalline layers of two or more types in a random fashion. It is of importance in clay minerals but also occurs in many substances with a lamellar structure as well as in some metals.

INTERTROPICAL CONVERGENCE ZONE. A belt of cloud and disturbed weather, up to some hundreds of kilometres in width, which occurs in low latitudes between the trade winds of the two hemispheres.

INTRINSIC PROPAGATION CONSTANT. For an electrical transmission line: another term for *propagation coefficient*.

INTRINSIC PROPERTIES. Of a material: those properties, such as specific heat or heat of sublimation, which are not seriously changed by the presence of imperfections or impurities. See also: *Structure-sensitive properties*.

INTRINSIC RESISTANCE (ACOUSTICS). See: *Acoustic impedance, characteristic*.

INVARIANT EXPRESSION. One which remains constant when a transformation, such as a translation or rotation of coordinate axes, is made. See also: *Covariant equation*.

INVARIANT SYSTEM. One with no degrees of freedom.

INVERSE FUNCTION. Of a function $f(x)$, equal to y: a function $g(y)$, equal to x.

INVERSE SQUARE LAW. Describes (a) the law which states that the intensity of radiation (acoustic or electromagnetic) falls off inversely as the square of the distance from a point source of radiation; or (b) the law which states that the force between two points (e.g. the gravitational force, the force between two electrical charges or two magnetic poles) is inversely proportional to the square of the distance between them.

INVERSION. (1) In mathematics: the mathematical transformation of all points outside a circle or sphere into points inside, and vice versa. (2) In meteorology: refers to an increase of temperature with height as opposed to the usual decrease. (3) In stereochemistry: the transformation of an optically active substance into one having the opposite rotatory effect, without essential change of chemical composition. (4) In communications: a form of speech scrambling which essentially inverts the frequency spectrum of the original signal.

INVERSION AXIS. In crystallography: an axis characterized by rotation through a fraction of revolution ($\frac{1}{2}$, $\frac{1}{3}$, $\frac{1}{4}$ or $\frac{1}{6}$) followed by reflection through a centre of symmetry.

INVERTER: INVERTED RECTIFIER. Any device for converting direct current to alternating current. It may be a rotary machine, a grid-controlled mercury arc rectifier, or a circuit employing thermionic tubes.

IODESIN TEST. A test of the ability of optical glass to retain its polish and transparency. It is based on the measurement of the amount of alkali released from the glass by fracture, using iodesin, which is a very sensitive indicator.

IODIDE. A salt of hydroiodic acid HI.

ION. An atom, or aggregate of atoms, which is not electrically neutral. In certain circumstances an electron may also be described as an ion.

ION BURN. The deactivation of a small spot of the screen of a cathode ray tube caused by the bombardment by heavy negative ions in the beam.

ION DENSITY. The number of ion pairs per unit volume.

ION DOSE. For any ionizing radiation: the total amount of electric charge produced per unit mass of air by the ions of one sign.

ION EXCHANGE. A process involving reversible exchange between ions in solution and ions of specific solid electrolytes. Ion exchange is extensively used, for example, in water-softening devices.

ION GUN. An assembly of electrodes which produces, from a suitable ion source, a beam of ions for use in such equipment as mass spectrometers, proton microscopes, or accelerating devices for nuclear investigations.

IONIC COMPOUND. A compound in which the bonds are essentially ionic in character.

IONIC CONDUCTION. Strictly, any electrical conduction which is brought about by the movement of positive or negative ions. The term is, however, usually restricted to such conduction in solids, the equivalent process in molten salts or aqueous solution being more often described as *electrolytic conduction*.

IONIC MOBILITY. (1) For gaseous ions: the drift velocity per unit electric field. (2) For ions in solution: the quotient of the ion conductance and the Faraday constant. It is different for positive and negative ions.

IONIC POLARIZATION. See: *Polarization.*

IONIC RADIUS. The effective radius of an ion, which is an important factor in deciding the type of crystal structure and the interatomic distances in ionic solids.

IONIC RECOMBINATION. In gases: the mechanism by which the charge of an ion is neutralized in a collision with an oppositely charged particle.

IONIC SOLID. A solid in which the preponderant type of bond is ionic. See also: *Born–Haber cycle. Born–Mayer equation. Lattice energy.*

IONIC STRENGTH. Of a solution: is given by $\frac{1}{2}\Sigma c_i z_i^2$ where c_i is the ionic concentration, z_i is the ionic charge and the summation is over all the ionic species present in the solution. For dilute solutions the concentration c_i may be expressed in terms of mol/litre.

ION IMPLANTATION. The controlled injection of energetic ionized atoms (or molecules) into solid targets so as to alter the properties of the material either by introducing a different atomic constituent or displacing atoms of the target material.

IONIZATION. Any process which results in the formation of ions.

IONIZATION BURST. A sudden pulse of ionization recorded in an ionization chamber.

IONIZATION BY COLLISION. The removal of one or more electrons from an atom or particle by collision with another atom or particle.

IONIZATION CHAMBER. A gas-filled enclosure for the detection or measurement of ionizing radiation. It contains in its simplest form two electrodes one of which may be the chamber wall. The electrodes are at different potentials and one (the *collector electrode* or *collector*) is connected to an instrument which detects or measures the ionization current which flows when ionizing radiation enters the chamber. See also: *Counter, Geiger–Müller.*

IONIZATION CHAMBER, AIR-EQUIVALENT OR AIR-WALL. See: *Air-equivalent material.*

IONIZATION CHAMBER, CONDENSER. An ionization chamber which is charged before irradiation, the residual charge (measured by an electrometer) indicating the dose received.

IONIZATION CHAMBER, EXTRAPOLATION. An ionization chamber of adjustable volume which permits the estimation of the limiting value of the ionization current per unit volume as the volume becomes vanishingly small.

IONIZATION CHAMBER, FREE-AIR. The basic standard instrument for absolute measurements. By the establishment of artificial boundaries to the sensitive volume, effected by means of guard rings, the chamber walls are in effect composed of air.

IONIZATION CHAMBER, THIMBLE. An ionization chamber in which the outer electrode has a size and shape comparable with those of a thimble.

IONIZATION COEFFICIENT, PRIMARY: TOWNSEND FIRST IONIZATION COEFFICIENT. For ionization in a gas: the number of ion pairs produced per electron per unit displacement in the direction of the field (i.e. per unit distance of drift). See also: *Ionization, specific.*

IONIZATION COEFFICIENT, SECONDARY. (1) For ionization in a gas: the *Townsend second ionization coefficient*, i.e. the number of electrons liberated from the cathode per single primary ionizing collision in the gas. (2) The *generalized secondary ionization coefficient*, which is a coefficient representing all possible contributions to the electron current. It is the linear sum of the Townsend second ionization coefficient and coefficients representing the number of photoelectrons emitted from the cathode by ionizing particles produced in the gas, the number of electrons liberated from the cathode or produced in the gas by the incidence of excited atoms (including metastable atoms), together with photo-ionization of the gas and ionization arising from ion–atom collisions within the gas or at the cathode. See also: *Ionization, specific.*

IONIZATION CONSTANT. In a solution: describes the degree of ionization of a solute in a given solvent. It is given by $[A^+][B^-]/[AB]$, where the brackets indicate concentration and dissociation occurs according to the equation $AB \rightleftharpoons A^+ + B^-$.

IONIZATION, CUMULATIVE. The multiplication of ionizing events in a gas by successive collisions in cascade, which occurs when the applied potential is sufficiently high. The result is often known as an *avalanche*. According to the magnitude of the applied potential the cumulative ionization may be directly proportional to that produced by the original ionizing event or, as the multiplication reaches saturation, may be independent of the initial ionization. See also: *Counter, Geiger region of. Counter, proportional region of.*

IONIZATION GAUGE. For the measurement of small pressures: (a) A thermionic triode connected to the vacuum to be measured, the pressure being obtained from measurement of the grid current. (b) A wire ring anode between two metal plate cathodes, all placed within the vacuum to be measured. The pressure is obtained from measurement of the ionization current produced when the gas molecules are ionized by electrons ejected from the plates and caused to take a spiral path by the application of an outside magnetic field. This type of gauge is the *Penning gauge*.

IONIZATION POTENTIAL: IONIZATION ENERGY. (1) Of a particle such as an atom, molecule or ion: the energy required completely to remove the most weakly bound electron from the particle in its ground state so that the resulting positive ion is also in its ground state. This is the *first ionization potential*, successive stages of ionization corresponding to the *second, third, ..., ionization potentials*. (2) Of a semiconductor: the energy required to release an electron from a donor level to a conduction band, or from the valence band to the acceptor level.

IONIZATION, SPECIFIC. The number of ion pairs produced by an ionizing particle per unit length of its path. The *primary specific ionization* excludes ionization due to δ-rays, but these are included in the *total specific ionization*. The specific ionization is sometimes known as the *ionization coefficient*. See also: *Ionization coefficient, primary: Townsend first ionization coefficient. Ionization coefficient, secondary*.

IONIZATION, SPORADIC-E. Radio reflections from the E-layer of the ionosphere which are not predictable by the theory of ionization by solar photons.

IONIZATION TEMPERATURE. Of a gas having a Maxwellian distribution: the temperature at which the gas molecules would have an average kinetic energy equal to the ionization energy.

IONIZATION, THERMAL. The ionization in a gas induced by collisions between the atoms or molecules moving with thermal energy.

IONIZING PARTICLE. An elementary particle which produces ion pairs in its passage through a medium. It may be *directly ionizing* (e.g. an electron, proton or photon) or may achieve ionization only by the production of directly ionizing particles, as does a neutron, when it is described as *indirectly ionizing*. Photons have sometimes been classified, erroneously, as indirectly ionizing particles, presumably since they are uncharged.

IONIZING RADIATION. Any radiation, either electromagnetic or particulate, which is capable of producing ions, directly or indirectly, in its passage through a medium. For some purposes, e.g. in connection with nuclear energy legislation, the term is not held to include visible or ultraviolet radiation.

IONOSPHERE. The region of the Earth's upper atmosphere which exhibits appreciable electrical conductivity due to the ionization of the air. This ionization is caused mainly by the effect of ultraviolet radiation from the Sun on the constituent gases of the upper atmosphere. The ionosphere consists of two main strata, the lower being the E-layer (or *Kennelly–Heaviside layer*) at about 100–150 km and the upper the F-layer (or *Appleton layer*) above about 150 km. There is a lower stratum (the D region) at 70–90 km which occurs only during the daytime. See also: *Atmosphere*.

IONOSPHERE, PLANETARY. A region in the atmosphere of a planet other than the Earth, in which ionization occurs. Ionospheres have so far been detected in the atmospheres of Mars and Venus.

IONOSPHERIC WAVE: SKY WAVE. A radio wave that is reflected by the ionosphere before reception.

ION PAIR. A positive and a negative ion or electron formed from the same neutral atom by the process of ionization.

ION PUMP. A pump for the production of a very low vacuum in which the residual gas is ionized (usually by electron bombardment) and the positive ions removed to the cathode.

ION SOURCE. A device in which material is ionized, and from which the ions are extracted, to form the initial stage of the production of an ion beam. See also: *Ion gun*.

ION SPECTRUM. The distribution in energy, momentum, or velocity of a particular species of ion in an ion beam or ion atmosphere.

ION TRAP. A device to prevent the ions present in a cathode-ray tube from producing blemishes on the phosphor coating.

IRIDESCENCE. The exhibition of rainbow-like colours by a surface composed of thin layers, e.g. mother-of-pearl, which arise from the interference of light reflected from the front and back surfaces of the layers.

IRIS. A mechanism in the eye permitting the pupil to alter in aperture.

IRIS DIAPHRAGM. An adjustable diaphragm for optical use which operates in a fashion analogous to that of the iris of the eye.

IRON LOSSES. In a.c. and d.c. machinery: the losses due to hysteresis and to eddy currents.

IRRADIANCE: IRRADIANCY. The radiant flux per unit area, also known as *radiant flux density*.

IRRADIATION. Exposure to radiation, either by accident or intent. The term is often restricted to ionizing radiation, as in the field of radiotherapy, radiation protection and nuclear energy.

IRRATIONAL NUMBER. A number which cannot be expressed as the ratio of two integers, but can be interpolated on a scale of rational numbers, e.g. $\sqrt{2}$.

IRREVERSIBLE PROCESS. A process at the conclusion of which it is not possible to return the system involved to its original thermodynamic state. By the second law of thermodynamics all natural physical processes are irreversible.

IRROTATIONAL MOTION. Of a fluid: flow of a fluid at all points of which the vorticity is zero. See also: *Fluid flow, rotational*.

IRROTATIONAL WAVE. See: *Compressional wave*.

ISALLOBAR. A contour line on a weather chart joining points at which the rate of change of atmospheric pressure is constant.

ISENTHALPIC PROCESS. One carried out at constant enthalpy.

ISENTROPIC CHANGE. A change that is accomplished at constant entropy.

ISENTROPIC CHART. See: *Adiabatic chart*.

ISOBARIC CHANGE OR PROCESS. See: *Polytropic change or process*.

ISOBARIC SPIN. A quantum number which, on the supposition of the charge independence of nuclear forces, serves to distinguish between neutrons and protons.

ISOBARS. (1) In meteorology: contour lines on a weather chart joining points at which the atmospheric pressure is the same. (2) In atomic physics: nuclides having the same mass number but different atomic numbers, i.e. the opposite of isotopes, e.g. $^{40}_{19}K$ and $^{40}_{20}Ca$.

ISOCANDLE DIAGRAM: ISOCANDELA DIAGRAM. A diagram showing the distribution of light from a lighting system (e.g. a row of street lamps). It gives the loci of directions of equal luminous intensity by means of a suitable projection.

ISOCHORE. A graph representing the state of a system as a function of two variables (e.g. temperature and pressure), the volume remaining constant.

ISOCLINIC LINES. Lines joining points of the Earth's surface at which the magnetic dip is the same.

ISOCOLLOID. One in which the dispersed phase and the dispersion medium have the same chemical composition.

ISO-COMPOUNDS. Compounds which have the same total formula, but in which the atomic arrangement is different.

ISODESMIC STRUCTURE. An ionic crystal structure in which all bonds have the same strength, so that no distinct atomic groups occur.

ISODOSE CHART. For ionizing radiation in a body: a chart showing *isodose lines*, i.e. lines along which the absorbed dose is the same.

ISODYNAMIC LINES. Lines joining points on the Earth's surface at which the total magnetic intensity is the same.

ISOELECTRIC POINT. (1) The pH value at which the charge on a colloid is zero. (2) The pH value at which the ionization of an ampholyte is at a minimum.

ISOELECTRONIC PRINCIPLE. States that any two or more molecules which are isoelectronic (i.e. possess the same number of electrons distributed over a similar molecular framework) will have similar molecular orbitals.

ISOELECTRONIC SEQUENCE. In spectroscopy: a sequence of ions having the same number of extranuclear electrons, and hence having spectral terms of the same types. Such a sequence starts with a neutral atom and proceeds through the singly ionized atom which follows it in the Periodic Table, and so on.

ISOGEOTHERM. The locus of points at depths below the surface of the Earth which are at the same temperature.

ISOGONIC LINES. Lines joining points on the Earth's surface at which the magnetic declination is the same.

ISOGYRES. The dark brushes which appear in the interference figure of a biaxial crystal when observed in convergent light between crossed nicols in a polarizing microscope. They correspond to directions in the crystal for which the vibration directions are parallel to those for the analyser and polarizer.

ISOMAGNETIC LINES. Lines joining points on the Earth's surface at which some given property of the Earth's magnetic field is the same.

ISOMERIC STATE. Of an atomic nucleus: an excited nuclear state having a mean life that is long enough to be observable.

ISOMERISM: See: *Isomers.*

ISOMERS. (1) Chemical: compounds having the same percentage composition and molecular weight but having different chemical and physical properties. This may arise from a difference in interatomic linkage (*structural isomerism*), or from differences in the relative positions of the atoms (*stereoisomerism*). In the latter case the isomerism may be *optical* (due to an enantiomorphic arrangement, with a consequent effect on polarized light) or simply *geometrical* (or *cis–trans*) isomerism. See also: *Tautomerism.* (2) Nuclear: nuclides having the same mass number and atomic number, but occupying different nuclear energy states.

ISOMETRIC PROCESS: ISOVOLUMIC PROCESS. The process of heating or cooling a gas, keeping the volume constant and allowing the pressure to change. See also: *Polytropic change or process.*

ISOMORPHIC: ISOMORPHOUS. Refers to crystals having similar chemical and physical properties, and being *isostructural*, i.e. possessing the same or very nearly the same crystal structure. Isostructural compounds, however, are not necessarily isomorphic. The phenomenon involved is known as *isomorphism.*

ISOPERIMETRIC PROBLEM. In the calculus of variations: the problem arising when subsidiary conditions are imposed on the argument function which relate to its entire course. Boundary conditions do not belong to this type of condition.

ISOPIESTIC PROCESS. The process of heating or cooling a gas, keeping the pressure constant and allowing the volume to change.

ISOPORIC LINES. Lines joining points on the Earth's surface at which the annual change in some given property of the Earth's magnetic field is the same.

ISOPYCNIC CHART. A chart showing the different heights in the atmosphere at which, at a given

time, the atmospheric density is the same. Surfaces of equal specific volume are called *isosteres.*

ISOSPIN. See: *Quantum number, internal.*

ISOSTASY. Of the Earth: a principle according to which, at a sufficient depth, the masses of all columns of equal cross-section are the same, whether the column be topped by mountain, lowland or ocean. Thus the mountains are compensated by masses of smaller density beneath them and the oceans by masses of higher density. This phenomenon is known as *isostatic compensation.*

ISOSTERE. See: *Isopycnic chart.*

ISOSTRUCTURAL. See: *Isomorphic: Isomorphous.*

ISOTENISCOPE. An apparatus for measuring vapour pressure. It consists of a U-tube, one end of which is connected to a manometer and the other to a closed vessel, the manometer pressure being adjusted so that the liquid levels in the U-tube are equal.

ISOTHERM. A line joining points on the Earth's surface at which the temperature is the same.

ISOTHERMAL. Qualifies anything (e.g. lines on a graph, processes, layers of the atmosphere) for which the corresponding temperature is constant. See also: *Polytropic change or process.*

ISOTONES. Nuclides having the same neutron number but different atomic numbers, e.g. $^{39}_{19}K$ and $^{40}_{20}Ca$, each of which has twenty neutrons.

ISOTOPE-DILUTION ANALYSIS. A method of analysis by which the amount of a given element in a specimen is found by observing how the isotopic abundance in that element is changed by the addition of a known amount of one of the isotopes of the element. The method can be used with stable isotopes (using the mass spectrometer) or with radioactive isotopes.

ISOTOPE EFFECTS. Differences in chemical properties, optical spectra, lattice dimensions, molecular vibrations and so on, exhibited by different isotopes of the same element. Such effects are mostly very small but for light elements may be appreciable.

ISOTOPES. Nuclides having the same atomic number but different mass numbers, e.g. $^{235}_{92}U$ and $^{238}_{92}U$. See also: *Isobars.*

ISOTOPE SEPARATION. The separation of isotopes of the same element from each other. As they are chemically the same, physical methods have to be employed in almost all instances. The principal methods involve gaseous diffusion, electromagnetic separation, thermal diffusion and centrifugation. The one chemical method of note involves isotopic exchange reactions.

ISOTOPIC ABUNDANCE. Of a particular isotope in a mixture of isotopes of the same element: the fractional amount of that isotope. When expressed as a percentage it is known as the *relative abundance*.

ISOTOPICALLY LABELLED COMPOUNDS: LABELLED COMPOUNDS. Compounds containing isotopic tracer(s).

ISOTOPIC EXCHANGE REACTION. A reaction in which isotopic atoms in different valency states or in different molecules, or in different sites in the same molecule, exchange places. Sometimes known as a *chemical exchange reaction*.

ISOTOPIC INCOHERENCE IN NEUTRON DIF-FRACTION. The background incoherent scattering arising from the difference in scattering length between different isotopes of the same element.

ISOTOPIC TRACER. A radioactive or stable isotope of some component of a system, which is introduced into the system to trace the behaviour of that component. The system may be physical, chemical or biological. Radioactive isotopes are detected and measured by normal radioactivity methods, and stable isotopes by mass spectrometry or, where feasible, activation analysis.

ISOTROPY. The exhibition by a medium of uniformity in physical properties whatever the direction of measurement. A medium may be isotropic in respect of some physical properties but anisotropic in respect of others. See also: *Anisotropy*.

ISOVOLUMIC PROCESS. See: *Isometric process: Isovolumic process*.

ITERATION METHOD. A method of successive approximation used in the numerical solution of algebraic equations, transcendental equations and differential equations, and for interpolation, etc.

ITERATIVE STRUCTURE. A number of identical passive networks connected in series, e.g. a wave filter or artificial transmission line.

IZOD TEST. See: *Impact tests*.

j

JACOBIAN. In the theory of partial differentiation: one of a number of determinants whose elements consist of partial derivatives, which play a part analogous to that of single derivatives in the theory of functions of a single variable.

JACOBI ELLIPSOID. A tri-axial ellipsoid which represents the form taken by a rapidly rotating liquid body, arranged in shells of uniform density, and exposed to no forces but its own gravitation.

JACOBI POLYNOMIAL. A solution of the differential equation $x(1 - x) y'' + [c - (a + 1) x] y'' + n(a + n) y = 0$, where a and c are real, $c > 0$, $a > (c - 1)$, and n is an integer.

JADERIN WIRE. A thin wire some 24 m in length used in surveying for the measurement of distance.

JAHN–TELLER RULE. States that, if a particular symmetric configuration of a non-linear molecule gives rise to an orbitally degenerate ground state, then that configuration is unstable with respect to one of lower symmetry which does not give rise to such a state.

JANSKY. In Astronomy: a unit of flux density (symbol Jy) equal to 10^{-26} W m^{-2} Hz^{-1}. It was adopted by the International Astronomical Union in 1973.

JET. A stream of fluid issuing from a nozzle or orifice, and having free boundaries. An example is the exhaust stream from a jet engine.

JET DISINTEGRATION. The destruction of the jet structure and forward momentum owing to mixing which takes place at the jet boundaries and spreads into the jet core causing an outward spread of the flow.

JET, EMULSION. A jet-like appearance in a nuclear emulsion arising from a nuclear interaction in which the incident particle has a very high energy (>100 GeV). (*Note:* $1 eV \approx 1 \cdot 6021 \times 10^{-19}$ J.)

JET ENGINE. A gas turbine used to provide a propulsive jet for aircraft, consisting of compressor, combustion system, and turbine. See also: *Rocket.*

JET FLAP. A thin sheet of air or gas discharged at high speed close to the trailing edge of a wing so as to induce lift over the whole wing independently of wing incidence. A *shrouded jet flap* is one in which the angle of discharge is controlled by means of a small flap along the trailing edge.

JET (JOINT EUROPEAN TORUS) A development of the tokamak system which has non-circular (D-shaped) plasma cross section, a high magnetic field and a "tight" torus. It is expected to have a nominal plasma current of 3×10^6 A, which is within a factor of 3–6 of that envisaged in a reactor. It is not aimed at achieving reactor conditions but should show how big an experimental reactor will be necessary. JET is a joint venture by the European Community and is located at Culham near Oxford, England. See also: *Thermonuclear reaction, controlled. Tokamak.*

JET, PULSE. A jet propulsion engine using a pulsating flow of atmospheric air into which fuel is injected and burned to produce a pressure build-up. Such an engine was used in the German V-1 bomb.

JET, RAM. The simplest form of jet propulsion engine, using a steady flow of atmospheric air with ram (i.e. piston) compressions. It consists of intake diffuser, combustion chamber, and propelling nozzle, and is of particular value for flight at supersonic speeds but produces no static thrust.

JET STREAM. A variable region of strong westerly wind occurring in the upper troposphere with maximum speeds of 50 or 100 m/s (100 or 200 mph). Two main jet streams occur, one in the northern hemisphere (the *circumpolar jet stream*) at up to about 10 km above sea level, and the other (the *sub-tropical jet stream*) extending round both hemispheres at a height of about 12 km.

JET STREAM, OCEANIC. A fast narrow jet stream of water about 200 miles (500 km) wide which flows from west to east in the equatorial waters of the Indian ocean between the two periods of monsoon. It can reach speeds as high as 100 miles (160 km) per day in April and May but is somewhat weaker during September and October.

JIG BORER. A machine tool for the machining of accurately spaced holes and surfaces of component parts of jigs, fixtures, tools and gauges. It may also be modified to permit its use as an accurate measuring machine.

JIGS AND FIXTURES. Appliances used in engineering production for holding work on a machine tool so that the appropriate machining operations may be carried out quickly and correctly. Generally speaking a jig incorporates means for guiding the cutting tools, whereas a fixture locates and holds the work in position but contains no specific guides.

J–J COUPLING. See: *Coupling.*

JOG. In a dislocation line: the discontinuity or step formed where a dislocation line changes from one glide plane to a parallel glide plane, when the connecting length of dislocation line is about one atomic distance. See also: *Dislocation.*

JOHNSON NOISE. Random fluctuations in a conductor which arise from the thermal agitation of the electrons. It is sometimes called *thermal noise* but is not to be confused with temperature fluctuation. See also: *Electrical noise.*

JOHNSON–RAHBEK EFFECT. The increase in frictional force between two electrodes in contact with a semiconductor, which occurs when a potential difference is applied. It is to be distinguished from the increase in frictional force arising from the attractive force between the electrodes. The effect has been applied in such devices as electromagnetic clutches.

JOLY BLOCK PHOTOMETER. A photometer consisting of two equal paraffin wax blocks separated by a thin opaque sheet. The two light sources under comparison are arranged to illuminate one block each and their distances adjusted until the blocks appear equally bright.

JORDAN LAG. A type of magnetic viscosity in which the angular lag of the induction behind the magnetic field, and also the energy loss per cycle, are independent of frequency.

JOSEPHSON EFFECT. An effect concerning the behaviour of two superconductors separated by such a short gap (say in the form of an oxide layer junction of thickness some 10^{-6} mm) that their wave functions could overlap. Josephson predicted (and the prediction was subsequently confirmed) that under certain conditions the frequencies and phases of the two sets of wave functions would be related and that there would be tunnelling of elec-

tron pairs across the gap; and that there would be interactions between this tunnelling current and the magnetic and electric fields in which the junction was situated. Analogous effects were also expected in superfluid helium and have also been observed.

JOSHI EFFECT. The change in the current passing through an ozonizer tube when the gas or vapour in the tube is irradiated with visible light.

JOULE. The unit of work and energy in MKS and SI units. It is the work done (i.e. the energy transferred) when the point of application of 1 N is displaced a distance of 1 m in the direction of the force, and is equal to 10^7 erg. See also: *Ampere. Volt. Watt. Appendix II.*

JOULE CYCLE. A reversible sequence of operations of a heat engine in which air is used as the working fluid. It consists of adiabatic compression, heating at constant pressure, adiabatic expansion, and cooling at constant pressure to the initial state.

JOULE EFFECT. (1) The heating effect of an electric current flowing through a resistance. It is also known as *Joule heat.* (2) The initial slight increase in length of a ferromagnetic rod subjected to a gradually increasing longitudinal magnetic field. This is also known as the *Joule magnetostriction effect.* (3) The change in temperature of a compressed gas when it undergoes adiabatic throttled expansion, usually known as the *Joule–Thomson effect.*

JOULE EQUIVALENT: MECHANICAL EQUIVALENT OF HEAT. The mechanical energy equivalent to unit quantity of heat. It is about 4.185×10^7 ergs/cal. Where the quantity of heat is expressed in terms of energy (as in MKS and SI units) the need for the concept no longer exists.

JOULE EXPERIMENT. An experiment designed to detect intermolecular attraction in a gas by the decrease in temperature observed when the gas is allowed to expand into a vacuum. The decrease arises from the loss in kinetic energy and gain in potential energy of the individual molecules.

JOULE HEAT. See: *Joule effect.*

JOULE, INTERNATIONAL. The work done per second when a current of one international ampere is passed through a conductor whose resistance is one international ohm. It has now been superseded by the joule.

JOULE LAW OF ELECTRIC HEATING. States that the heat produced by an electric current, I, flowing through a resistance, R, for a time, t, is proportional to I^2Rt.

JOULE LAW OF ENERGY CONTENT. For a

perfect gas: states that the internal energy of a given mass of a perfect gas is a function of temperature only. This is also known as *Mayer's hypothesis*.

JOULE MAGNETOSTRICTION EFFECT. See: *Joule effect*.

JOULE–THOMSON COEFFICIENT. For the Joule–Thomson effect: the ratio of the change in temperature to the change in pressure.

JOULE–THOMSON EFFECT. See: *Joule effect*.

JOULE–THOMSON INVERSION TEMPERATURE. The temperature at which the Joule–Thomson coefficient changes its sign for a given gas.

JOULE–THOMSON VALVE. A throttling device used for the expansion of the working fluid in refrigerators employing the Joule–Thomson effect.

JOVIGNOT TEST. A test carried out to determine the ductility of a metal sheet, in which a circular plate is clamped at the edges and subjected to fluid pressure on one side. Measurements are made of the fractional increase in surface area necessary to produce fracture.

J/PSI PARTICLE. Also known as "gipsy" particle. A charmed quark, bound in orbit to its anti-quark. It has sometimes been referred to as "charmonium". See: *Quarks*.

JULIAN CALENDAR. The calendar used for reckoning years and months for civil purposes, based on a tropical year of 365·25 days. It was instituted by Julius Caesar in 45 B.C. and is still the basis of our calendar. As modified by a decree of Pope Gregory XIII in 1582, it is known as the *Gregorian calendar*. By this reform certain leap years are omitted to bring the length of the year nearer to the true astronomical value. This calendar was adopted at widely differing times by different countries, e.g. 1600 in Scotland, 1752 in England and not at all in Czarist Russia.

JULIAN DATE. The date expressed in terms of the number of days since an arbitrary zero date, 1st January 4713 B.C. Julian days are used to express the times of most astronomical observations and are reckoned from noon, a portion of a day being expressed as a decimal to the required degree of precision. The name Julian is derived from that of Julius Scaliger, the father of J. J. Scaliger who introduced the system in 1582. There is no connection with the Julian calendar. (*Note:* On 1st January 1971 the Julian date was 2439951.)

JUNCTION, HYBRID. A type of waveguide circuit with four branches which, when the branches are properly terminated, has the property of enabling energy to be transferred from any one branch into only two of the remaining three.

JUNCTION, P–N. See: *Semiconductor junctions*.

JUNO. See: *Asteroid*.

JUPITER. The largest planet in the solar system and fifth in order of distance from the Sun. Its diameter is about 11 times that of the Earth, its mass about 318 times, and its period of rotation is about 10 h. It has 12 satellites, of which four (*Io, Europa, Ganymede* and *Callisto*) are named, the rest being referred to by number. It has recently been found to possess a long magnetic tail extending for more than $6·5 \times 10^8$ km (4×10^8 miles or 4·3 A.U.).

k

KAON. See: *Meson.*

KAPITZA BALANCE. A magnetic balance designed for the measurement of susceptibilities in large magnetic fields when the force is large and applied only momentarily. It is not suitable for absolute measurements.

KAPITZA LIQUEFIER. (1) A hydrogen liquefier, employing Joule–Thomson expansion after precooling with liquid nitrogen. (2) A helium liquefier employing, after precooling with liquid nitrogen, a gas-lubricated single-cylinder reciprocating engine.

KAPITZA RESISTANCE. An effect involving a wire immersed in liquid ^{4}He, and heated by a passing current, whereby the wire continues to remain appreciably hotter than the surrounding liquid after the current is cut off. It is a quantum mechanical effect not yet fully understood.

KAPTEYN SELECTED AREAS. Areas selected as the basis of a system of sampling in the study of faint stars owing to the prohibitive cost, both in time and money, of making a complete catalogue.

KAPTEYN UNIVERSE. An early model of the Galactic System which has now been superseded.

KÁRMÁN BOUNDARY LAYER THEOREM. States that the rate of change of momentum across an area between a boundary layer of a fluid and the surface is equal to the sum of the difference of the pressures across the area, and the skin friction.

KÁRMÁN SIMILARITY THEORY. Of turbulent flow in fluids: states that the local flow pattern is statistically similar at or near all points in the fluid, only the time and length scales being different.

KARMAN–TSIEN METHOD. A method of approximating to the equations for compressible fluid motion in two dimensions, which leads to a simple rule for the estimation of compressibility effects in subsonic flow.

KÁRMÁN VORTEX STREET. A regular arrangement of vortices situated alternately in two parallel rows as are street lamps. Such a vortex street resembles a double trail of vortices in the wake of a bluff cylinder in a certain range of Reynolds number, and was proposed by von Kármán as a mechanism for the drag of a solid body in the flow of a fluid of small viscosity.

KATABATIC WIND. A wind caused by the downward motion of the air, e.g. by cold air flowing into a valley from high ground. Such winds are common in desert ravines and may reach very high speeds.

KATATHERMOMETER. A form of mercury thermometer for measuring the rate of cooling of the body, used, for example, in mining to estimate the degree of comfort experienced. It is used to find the time taken for the mercury to drop from $38\,^\circ$C to $35\,^\circ$C.

KATHAROMETER. An instrument for determining the composition of a mixture of known gases by the measurement of thermal conductivity as revealed by the cooling effect of the gas as it flows past a heated wire or thermistor.

K-BETA FILTER: BETA FILTER. A filter used in X-ray diffraction studies to remove essentially all the $K\beta$ radiation from a beam of characteristic K X-rays. It usually takes the form of a foil of an element which has an absorption edge between the $K\alpha$ and $K\beta$ wavelengths.

K-ELECTRON CAPTURE: K-CAPTURE. See: *Orbital electron capture.*

KELVIN. The unit of thermodynamic temperature, formerly called the *degree Kelvin.* It is $1/273{\cdot}16$ of the thermodynamic temperature of the triple point of water. See also: *SI units. Temperature scale, Kelvin: Temperature scale, thermodynamic. Appendix II.*

KELVIN CLAMP. See: *Kinematic design of instruments.*

KELVIN CONTRACTION: KELVIN-HELMHOLTZ CONTRACTION. The contraction of a star as a consequence of the radiation of thermal energy.

KELVIN DOUBLE BRIDGE. See: *Bridge, electrical.*

KELVIN THEOREM. For fluid flow: states that,

in a region of flow in which the entropy is constant, the circulation around a closed path moving with the fluid also remains constant. In other words, vorticity cannot be created or annihilated inside a homentropic region of fluid flow.

KENNELLY–HEAVISIDE LAYER: HEAVISIDE LAYER: E-LAYER. The lower of the two main strata or "layers" in the ionized atmosphere, the other being the Appleton or F-layer. It occurs at a height of about 100 km.

KENOTRON. A high-voltage (above 10 kV) thermionic diode rectifier.

KEPLER LAWS. Three laws which initiated the modern mathematical treatment of planetary motions. They may be stated as follows: (a) Every planet moves in an ellipse of which one focus is the Sun. (b) The radius vector from the Sun to a planet sweeps out equal areas in equal times. (c) The squares of the periodic times which the various planets take to describe their respective orbits are proportional to the cubes of their mean distances from the Sun, the mean distance being taken as the length of the semi-major axis.

KERNEL. A known function of (usually) two variables which occurs in an integral equation or integral transform. Its use is exemplified in the following entries from the field of reactor physics. See also: *Integral transform.*

KERNEL, DIFFUSION. In nuclear reactor theory: a function which, for a finite homogeneous medium, relates the thermal flux density at a given position to the strength of a source of thermal neutrons at another place. A more accurate description is given by the *transport kernel*, which is formed from the diffusion kernel by the addition of a "single collision" kernel.

KERNEL, DISPLACEMENT. In nuclear reactor theory: a function which relates, for an infinite, homogeneous and isotropic medium, the flux density or slowing down density of neutrons at a given point to the strength of a point source of neutrons at another point, and is dependent only on the distance between the two points.

KERNEL, SLOWING-DOWN. In nuclear reactor theory: a function which, for a homogeneous medium, gives the probability per unit volume that a neutron will go from one specified position to another while slowing down through a specified range of energy.

KERNEL, TRANSPORT. See: *Kernel, diffusion.*

KERR CELL. A fast optical shutter employing the Kerr effect. It consists essentially of a cell containing an optically transparent isotropic material (solid or liquid) and a pair of plates for applying a large electric field across it. Between crossed nicol prisms (or crossed polaroids) no light passes through the cell until the field is turned on. Such cells have been employed in television equipment, in the determination of the speed of light and in measuring the decay times of fluorescent phenomena.

KERR EFFECT. (1) Electro-optical: the occurrence of birefringence in a transparent isotropic medium when it is placed in an electric field. The medium then behaves like a uniaxial crystal with its optic axis lying in the direction of the field. (2) Magneto-optical: the occurrence of elliptical polarization in plane polarized light when it is reflected from the pole of a magnet. See also: *Pockels effect.*

KERSTEN THEORY. A theory, now superseded, which explained the existence of non-zero coercive forces in ideal bulk ferromagnetic material as arising from the existence of non-magnetic inclusions located at the corners of a simple cubic lattice.

KETTELER DISPERSION FORMULA. See: *Sellmeier dispersion formula.*

K-FACTOR. See: *Gamma-ray constant, specific: Gamma-ray emission, specific.*

KICKSORTER. See: *Pulse height analyser.*

KIKUCHI LINES. Parallel black (intensified) and white (absorbed) lines observed in electron diffraction photographs from single crystals in addition to Laue spots. The lines are similar in origin to those observed in the Borrmann and Kossel effects.

KILLER. In luminescence: an impurity in a phosphor which inhibits luminescence, and which presumably competes as an electron trap with the luminescent centres.

KILOGRAM. The unit of mass in the MKS and International (SI) systems. It is the mass of a standard platinum–10% iridium alloy in the form of a cylinder which is kept at the Bureau International des Poids et Mésures in Sèvres. See: *SI units. Appendix II.*

KILOGRAMME DES ARCHIVES. The original prototype kilogramme, intended to represent the mass of a cubic decimetre of pure water at its temperature of maximum density (4°C), of which the standard kilogramme is an exact copy. The volume of 1 kg of water at 4°C is, however, 1·000028 dm³, so that the standard kilogramme is 28 mg heavier than was intended. The litre, which was formerly defined as the volume of 1 kg of pure water at 4°C, is on this basis equal to 1·000028 dm³. It has, however, been redefined by the CGPM (Conférence Génerale des Poids et Mésures) as 1·000 dm³ exactly, with the recommendation that neither the word "litre" nor its symbol "l" should be used to express results of high precision. See also: *Litre.*

KILOGRAM FORCE: KILOGRAM WEIGHT. The force with which a mass of 1 kg is attracted by the Earth. It is equal at Sèvres to 9·806 65 N. See also: *Units, gravitational. Appendix II.*

KINEMATIC DESIGN OF INSTRUMENTS. A technique or techniques for the maintenance of correct positional relationships between component parts. A classic example is the *Kelvin clamp* for the location of an instrument, having three ball feet resting on a plane surface which carried a "hole and a slot", now usually a trihedral hollow and a vee-groove.

KINEMATIC (OR KINETIC) POTENTIAL: LAGRANGIAN FUNCTION. The difference between the kinetic energy and potential energy of a dynamic system. This function enables the equations of motion of classical mechanics and Hamilton's principle to be written in a simple form. See: *Hamilton principle.*

KINEMATICS. The study of the motion of particles and bodies without reference to the forces associated with that motion. Its three basic parameters are displacement, rate of change of displacement, and rate of change of velocity (i.e. acceleration). See also: *Dynamics. Mechanics. Statics.*

KINEMATIC THEORY OF ELECTRON AND X-RAY DIFFRACTION. See: *Dynamical theory of electron and X-ray diffraction.*

KINEMATIC VISCOSITY. The coefficient of viscosity of a fluid divided by its density. It measures the kinematic effect of the viscosity. Typical values at 20°C are: for water 0·01 cm² s⁻¹, and for glycerine 6·8 cm² s⁻¹. The CGS unit of kinematic viscosity is the *stokes*, which is exactly equal to 10^{-4} m² s⁻¹, the (unnamed) SI unit being m² s⁻¹

KINETIC ENERGY. Energy stored in a system by virtue of the velocities of various moving masses within the system. The kinetic energy of a body of mass m and velocity v is $\frac{1}{2}mv^2$. See also: *Equipartition of energy.*

KINETIC HEAD: VELOCITY HEAD. For a perfect fluid in steady flow: one-half the ratio of the square of the flow velocity to the gravitational acceleration. It is the height of a column of fluid giving a hydrostatic pressure of $\frac{1}{2}\varrho v^2$, where ϱ is the density and v the flow velocity. Pressure head + Velocity (or Kinetic) head + Elevation head = constant, according to the Bernoulli equation. See also: *Pressure head. Elevation head.*

KINETIC MOMENTUM. Of a charged particle in an electromagnetic field: the vector $\mathbf{p} - (e/c)\,\mathbf{A}$, where $\mathbf{A}$ is the field, $\mathbf{p}$ the momentum, e the charge, and c the speed of light.

KINETIC THEORY OF GASES. The theory which relates the macroscopic properties of a gas to the motion of its individual molecules. It assumes that heat can be identified with molecular motion, that the molecules can be regarded as elastic spheres, that the interaction of gas molecules with each other and with the walls of the container may be treated according to the laws of classical mechanics, and that the methods of statistical analysis may be used.

KINETIC THEORY OF LIQUIDS. The theory which relates the macroscopic properties of a liquid to the motion of its individual molecules. Its most direct confirmation is found in the Brownian motion; but the theory has to take account of the different type of motion in a liquid from that in a gas. Thus, the mean free paths of the molecules are effectively zero and the molecules themselves are in continual interaction with each other.

KINK BAND. See: *Deformation band.*

KINKING OF METALS. Any process leading to the localized bending of the crystal lattice under the influence of external stress. See also: *Deformation band.*

KIRCHHOFF LAWS OF ELECTRICAL NETWORKS. For electrical networks carrying steady currents: state (a) that the sum of all currents flowing towards a node of the network is zero, a node denoting a common junction of several conductors; and (b) that the sum of the voltages encountered in traversing the network along a closed loop is zero.

KIRCHHOFF RADIATION LAW. States that the ratio of the emissive power to the absorptive power for thermal radiation of a given wavelength is the same for all bodies at the same temperature and is equal to the emissive power of a black body at that temperature.

KIRCHHOFF VAPOUR PRESSURE FORMULA. For the variation of vapour pressure with temperature: may be stated as $\ln p = A - (B/T) - C \ln T$, where p is the vapour pressure, T the temperature, and A, B and C are constants. It is valid only over a limited range of temperature.

KIRKENDALL EFFECT. The shift in the interface between two metals or alloys, bonded together by, say, pressure welding or electrodeposition, when they are annealed to allow the two to diffuse into one another. It is sometimes known in the French literature as the *Smigelskas effect*.

KIRKWOOD FORMULA. An equation relating the dielectric constant ε of a polar liquid to the polarizability α and dipole moment μ of its molecules. It may be written: $(\varepsilon - 1)(2\varepsilon + 1)/3\varepsilon = 4\pi N(\alpha + \mu\bar\mu/3kT)$, where N is the number of molecules per unit volume, $\bar\mu$ is the total moment of a finite macroscopic sphere polarized by a single fixed molecule when it is immersed in a medium of its own dielectric constant, k is Boltzmann's constant, and T is the temperature.

KLEIN–GORDON EQUATION. A relativistic form of the Schrödinger equation in nuclear quantum theory. It may be written as

$$\nabla^2\psi + \frac{1}{\hbar^2 c^2}[(E - V)^2 - (mc^2)^2]\,\psi = 4\pi\varrho,$$

where ψ is the Schrödinger wave function, E is the total energy of the particle, V is the potential energy, m is the rest mass, c is the speed of light, and ϱ is proportional to the density of the nucleons. See also: *Schrödinger equation*.

KLEIN–NISHINA FORMULA. An expression for the total or differential cross-section of an unbound electron for the Compton scattering of a photon, according to Dirac's electron theory.

KLYSTRON. A velocity-modulated electron beam tube for the generation of very-high-frequency oscillations at high power. It consists of an electron gun, drift tunnel (with usually not less than two resonant cavities—the *buncher* and the *catcher*), and a beam collector. See also: *Electron phase focusing*.

KNIGHT SHIFT. The fractional increase in the magnetic resonance frequency of an atomic nucleus in a metal relative to that of the same nucleus in a non-metallic compound in the same external magnetic field. It is a function of the spin paramagnetism of the conduction electrons and their magnetic coupling to the nuclei.

KNOCK-ON PARTICLE. A particle which is displaced or recoils after a collision with an energetic particle moving through matter. A knocked-on atom may possess sufficient energy to displace other atoms.

KNOT. A speed of one nautical mile per hour, now defined as 1·852 km per hour exactly. See: *Mile*.

KNUDSEN ABSOLUTE MANOMETER. A vacuum gauge for the absolute measurement of low pressure, involving the measurement of the repulsion exerted between two closely-spaced metal surfaces when a small temperature difference is maintained between them. See also: *Pressure gauge, Knudsen*.

KNUDSEN COSINE RULE. For the distribution of the direction of reflection of gas molecules at an irregular solid surface: states that $ds = \dfrac{d\omega}{\pi}\cos\theta$, where ds is the probability that a molecule will leave the surface within the solid angle $d\omega$ forming an angle θ with the normal to the surface.

KNUDSEN FLOW. The flow of gas through a long tube at pressures such that the mean free path of the gas molecules is much greater than the radius of the tube.

KNUDSEN NUMBER. A parameter λ/L, where λ is the mean free path of the molecules and L is a characteristic length, which is important in the case of fluid flow of low molecular density.

KOHLRAUSCH LAWS OF ELECTROLYTIC CONDUCTION. (1) Every ion at infinite dilution contributes a definite amount towards the equivalent conductance of an electrolyte, irrespective of the nature of the other ions with which it is associated. (2) The equivalent conductance of a strong electrolyte in very dilute solution, plotted against the square root of the concentration, yields a straight line.

KOPP LAW. States that the specific heat of a solid element is the same whether it is free or part of a solid compound. Alternatively: the molar heat of a solid compound is equal to the sum of the atomic heats of its constituents.

KOPP–NEUMANN LAW. States that, for compounds of the same type (e.g. Al_2O_3, Cr_2O_3, ...), the specific heat is inversely proportional to the molecular weight.

KOSSEL EFFECT. The production of a series of cones of reflected X-rays when characteristic X-rays are generated within a single crystal (e.g. by an electron or X-ray beam). These cones, recorded on a flat film, manifest themselves as ellipses, parabolae and hyperbolae, which are known as *Kossel lines*. See also: *Borrmann effect. Kikuchi lines*.

KRAMERS THEOREM. States that the lowest

189

energy level in a paramagnetic material is degenerate if the magnetic ions have an odd number of electrons, and that the degeneracy is at least two-fold.

K-STAR. See: *Stars, spectral classes of*.

KUNDT LAW OF ABNORMAL DISPERSION. When the refractive index of a solution increases, e.g. because of change of composition, its optical absorption bands are displaced towards the red. This law does not always hold in practice.

KUNDT TUBE. For the measurement of the speed of sound in gases: a tube containing a light powder which reveals the positions of the nodes when stationary sound waves of known frequency are set up in the tube, thus giving the wavelength and hence the speed.

KURIE PLOT: FERMI PLOT. Of a β-particle spectrum: a graph in which a suitable function of the observed intensity is plotted against the particle energy, the function being chosen so that the graph is a straight line for allowed β-transitions. It is used in determining the character of the β-transition and the maximum energy.

KUTTA–JOUKOWSKI HYPOTHESIS. In the theory of the uniform flow of a fluid past two-dimensional aerofoils: provides a criterion for specifying the circulation round a closed contour at a large distance from the aerofoil, which must be satisfied to permit a satisfactory analytical study to be made. The hypothesis states that, in the irrotational flow of a uniform stream of inviscid fluid past an aerofoil with a cusped or wedge-shaped trailing edge, the circulation is chosen so that the fluid speed at the trailing edge is finite.

KYMOGRAPHY. A method of recording in a single radiograph the excursions of a moving organ (such as the heart or stomach) in the body.

1

LABELLED COMPOUNDS. See: *Isotopically labelled compounds: Labelled compounds.*

LABORATORY SYSTEM. See: *Coordinates, laboratory system of.*

LADDER NETWORK. A series of identical symmetrical four-terminal networks connected together to form a line with continuously repeated impedance sections.

LAEVOGYRIC: LAEVOROTATORY. Refers to an optically active substance that rotates the plane of polarization of a transmitted light beam in an anti-clockwise direction, the observation being made looking through the substance towards the light source. See also: *Dextrogyric: Dextrorotatory.*

LAGGING. (1) The use of materials of low thermal conductivity to hinder the passage of heat from hot bodies (e.g. a steam boiler) or to cold ones (e.g. in refrigerating plant). (2) Short for lagging material.

LAGGING CURRENT. The steady-state current in an a.c. circuit, the maximum of which, owing to inductive reactance, lags behind the maximum of the applied voltage.

LAGRANGE BRACKET. The expression

$$\sum_{r=1}^{n} \left(\frac{\partial q_r}{\partial u} \frac{\partial p_r}{\partial v} - \frac{\partial p_r}{\partial u} \frac{\partial q_r}{\partial v} \right)$$

where $q_1 \ldots q_n$ and $p_1 \ldots p_n$ are functions of two variables u and v, and possibly other variables. It plays a big part in the theory of partial differential equations and, in particular, in analytical dynamics.

LAGRANGE EQUATIONS. A set of equations of motion for a dynamical system, which may be regarded as a formulation of Newton's laws of motion in a general frame of reference. See also: *Fluid flow, equations of motion for.*

LAGRANGIAN FUNCTION: KINEMATIC (OR KINETIC) POTENTIAL. The difference between the kinetic energy and potential energy of a dynamic system. This function enables the equations of motion of classical mechanics and Hamilton's principle to be written in a simple form. See also: *Hamilton principle.*

LAGRANGIAN POINTS. Five points in the orbital plane of two massive particles in orbit about a common centre of gravity, at any of which a third particle of negligible mass can remain in equilibrium. They represent particular solutions of the three-body problem. See also: *Trojans.*

LAGUERRE EQUATION. The differential equation $xy'' + (1 - x) y' + vy = 0$, where v is a constant but not necessarily an integer. Its solutions are the *Laguerre polynomials.*

LAMBDA LEAK. A leak of liquid helium II through small holes impassable for normal liquids. It is also known as a *superleak.*

LAMBDA PARTICLE. See: *Hyperon.*

LAMBDA POINT. (1) The temperature at which the transition between the two phases of liquid helium takes place at the saturated vapour pressure of helium. It is about 2·19 K. (2) The temperature at which various second order transitions take place in other substances, e.g. the temperature at which the specific heat of a substance reaches a sharp maximum.

LAMBERT–BOUGUER LAW OF ABSORPTION. See: *Bouguer–Lambert law of absorption: Bouguer–Beer law.*

LAMBERT COSINE LAW. See: *Light emission, cosine law of.*

LAMBERT PROJECTION. See: *Orthomorphic projection: Conformal projection.*

LAMB–RETHERFORD SHIFT. The difference between the positions of atomic energy levels as calculated by the Dirac theory and by quantum electrodynamics. It arises from the neglect by Dirac of the interaction of electrons with the radiation field.

LAMINAR FLOW CONTROL. Boundary layer control, especially at high speeds. See also: *Boundary layer control. Fluid flow, laminar.*

LAMP. Originally a source of light in which liquid

fuel and a wick were used. It now denotes any artificial source of light, whether for home and street lighting or for scientific purposes.

LAMP, ARC. A lamp which consists essentially of electrodes between which an electric arc is maintained.

LAMP, ATOMIC BEAM. A lamp in which the light is produced by exciting the atoms in an atomic beam, thus producing light which covers a small spectral range.

LAMP, CADMIUM VAPOUR. A cadmium vapour discharge lamp giving several spectral lines of which the red line was formerly used as a wavelength standard. See also: *Lamp, isotope. Lamp, Michelson.*

LAMP, CARCEL. The former French standard lamp. It burned colza oil and used a wick.

LAMP, CERAMIC. Essentially consists of an electroluminescent panel in which a specially prepared zinc sulphide phosphor is sandwiched between two conducting sheets (one of which is transparent). This panel emits light when subjected to the effect of an a.c. electric field. See also: *Electroluminescent panel.*

LAMP, COMPARISON. A lamp having constant but not necessarily known luminous intensity, with which a standard lamp and a light source under test are successively compared by means of a photometer.

LAMP, DAYLIGHT. A lamp giving light with the same spectral quality as sunlight from a moderately overcast sky, as far as is possible.

LAMP, DISCHARGE. A lamp which consists essentially of a tube filled with gas or vapour, containing two electrodes between which a discharge passes.

LAMP, ELECTRIC. A generic term including filament or incandescent lamps on the one hand and discharge lamps on the other.

LAMP, ELECTRODELESS. A lamp in which a ring discharge is formed by the action of an intense high-frequency magnetic field in which the lamp is placed.

LAMP, FLUORESCENT. A mercury vapour discharge lamp producing ultraviolet light which is converted to visible light by the excitation of a layer of fluorescent salt deposited on the inner surface of the tube.

LAMP, HEFNER. The former German standard lamp. It burned amyl acetate and used a wick. It is also known as the *Hefner–Alteneck* lamp.

LAMP, ISOTOPE. A vapour lamp containing vapour of a single isotope, and hence producing light of high spectral purity. The best known of such lamps are the *mercury isotope lamp*, employing ^{198}Hg; the *cadmium isotope lamp*, employing ^{112}Cd, ^{114}Cd or ^{116}Cd; and the *krypton isotope lamp*, employing ^{86}Kr, which has replaced the cadmium vapour lamp as a wavelength standard. See also: *Length, standards of.*

LAMP, MERCURY VAPOUR. A discharge lamp containing mercury vapour. It yields a blue-green light, rich in ultraviolet and near infrared, the precise character of which depends on the pressure at which the mercury vapour is maintained. See also: *Lamp, isotope.*

LAMP, MICHELSON. A special form of cadmium vapour lamp used by Michelson in his investigations of the emitted spectrum, which led to the adoption of the red line of cadmium as a primary standard.

LAMP, QUARTZ. See: *Lamp, ultraviolet.*

LAMP, STANDARD. (1) A standard light source in terms of which the luminous intensity of another light source can be expressed. Originally such standards consisted of some form of luminous flame burning under specified conditions, but these have now been replaced by a radiation standard, depending on the light emitted by a black body under specified conditions. See also: *Candela. Lamp, Carcel. Lamp, Hefner. Lamp, Vernon Harcourt.* (2) A *secondary standard*, consisting of a gas-filled tungsten filament lamp, which is calibrated against the standard black-body radiator. See: *Light, primary standard of.*

LAMP, TUNGSTEN–HALOGEN. A lamp containing a halogen which combines with the tungsten evaporated from the filament, the tungsten then being released and deposited back on to the filament, and so on indefinitely. The life of the lamp is much greater than that of a normal tungsten lamp and also the lamp has a higher efficiency. The halogens normally used are iodine and bromine.

LAMP, ULTRAVIOLET. Although conventional light sources may emit useful amounts of ultraviolet radiation the term is usually restricted to specially designed sources. Among common types are the hydrogen discharge lamp, the mercury arc and mercury discharge lamps, the fluorescent lamp, the enclosed carbon arc, the xenon discharge lamp, open arcs with impregnated carbon or metal electrodes, and the cadmium arc or spark source. The visible radiation is sometimes absorbed in an outer envelope which is transparent to the required region of the ultraviolet spectrum, as in the *quartz lamp*.

LAMP, VERNON HARCOURT. The former

British standard lamp. It burned pentane and did not use a wick.

LANDAU DAMPING. The damping of a space charge oscillation by a stream of particles moving at a velocity which is slightly less than the phase velocity of the associated wave. The axial velocity of the particles is thereby increased slightly at the expense of the amplitude of the oscillation.

LANDAU FLUCTUATIONS. The fluctuations in the observed rate of the energy loss of fast particles when this loss is measured by the ionization produced in "thin" detectors.

LANDÉ G-FACTOR: LANDÉ SPLITTING FACTOR. A factor of proportionality, g, introduced by Landé to describe the anomalous Zeeman effect for which the change in electronic energy level induced by a magnetic field (the "splitting") is given by $g\beta HM$, where β is the Bohr magneton ($eh/4\pi mc$, where e is the electronic charge, h is Planck's constant, m is the electron rest mass and c is the speed of light), H is the magnetic field and M the magnetic quantum number. The value of g depends on the orbital angular momentum and the spin angular momentum quantum numbers, and ranges from 1, for pure orbital momentum, to 2, for pure spin momentum.

LANDÉ INTERVAL RULE. For Russell–Saunders coupling: states that the interval between the level of a given component J and that of $J - 1$ is proportional to J.

LANE LAW. States that if a star contracts, its internal temperature rises.

LANGMUIR LAW. States that, in a thermionic diode with space charge limited current, the anode current density is proportional to the $^3/_2$ power of the anode–cathode voltage.

LANTHANIDE CONTRACTION. The decrease in the value of the atomic radius which occurs in the rare earth elements that follow lanthanum in the Periodic Table.

LANTHANIDE ELEMENTS. The rare earth elements following lanthanum in the Periodic Table, with atomic numbers 58 to 71. They may be regarded as a transition group within a transition group and, accordingly, have very similar chemical properties. See also: *Rare earths. Appendix I.*

LAPLACE AZIMUTH STATIONS. Triangulation stations set up to control the measurement of azimuth on the surface of the Earth.

LAPLACE EQUATION. The basic equation of potential theory:

$$\frac{\partial^2 \phi}{\partial x^2} + \frac{\partial^2 \phi}{\partial y^2} + \frac{\partial^2 \phi}{\partial z^2} = 0,$$

or, in terms of the Laplace operator, $\nabla^2 \phi = 0$.

LAPLACE OPERATOR. The vector operator div grad, denoted by ∇^2. In cartesian coordinates it is
$$\frac{\partial^2}{\partial x^2} + \frac{\partial^2}{\partial y^2} + \frac{\partial^2}{\partial z^2}.$$

LAPLACE–STIELTJES TRANSFORM. Of a function $g(t)$: the function $G(s) = \int_0^\infty e^{-st} \, dg(t)$, where s is a complex number.

LAPLACE TRANSFORM. Of a function $f(x)$ defined for all real positive values of x: the function
$$F(s) = \int_0^\infty f(x) \, e^{-sx} \, dx.$$
It is a special case of the Fourier transform.

LAPLACIAN. (1) The Laplace operator. (2) In nuclear reactor theory: the negative of the geometrical buckling. See also: *Buckling.*

LAPORTE SELECTION RULE. For atomic spectra: states that for electric dipole radiation, terms for which the sum of the azimuthal quantum numbers of the individual electrons are even combine only with those terms for which the sum is odd, and vice versa.

LAPSE RATE. The rate of decrease of temperature with height in the troposphere and mesosphere. When the temperature increases with height the lapse rate is said to be negative and we speak of a *temperature inversion.* The *dry adiabatic lapse rate* is the rate at which an element of unsaturated air ascending (descending) adiabatically is cooled (warmed). The *saturated adiabatic lapse rate* is the corresponding rate for saturated air. Values of the lapse rate are in the region of $1\,°C/100$

LARGE-ANGLE BOUNDARY. See: *Grain boundary, large-angle.*

LARMOR PRECESSION. The precessional motion of the orbit of a charged particle which is subjected to a magnetic field. The precession occurs about the direction of the field. For an electron revolving about a nucleus the angular velocity of the Larmor precession is given by $eH/2mc$, where e is the charge, m is the mass, c the speed of light and H the magnetic field strength.

LARMOR PRECESSION FREQUENCY. The frequency of the Larmor precession, also known as the *gyromagnetic frequency.* See also: *Larmor precession.*

LASER. An acronym for "*Light amplification by stimulated emission of radiation*". A laser uses the maser technique in the optical region, and has been termed an *optical maser*. However, sub-millimetre lasers are now also in use. The emphasis in the optical region, in contrast with that in the microwave region, has been in the development of the laser oscillator, which provides a source of light that is completely phase-coherent; and an intense beam may be provided by the use of resonance techniques. The laser effect has been observed in solids, liquids, and gases. See also: *Maser. Optical pumping.*

LASER APPLICATIONS. Since a laser beam can be made monochromatic, is phase coherent and can be collimated and focused to a degree limited only by diffraction, lasers have found many different applications, for example (1) *Energy concentration:* e.g. in welding, melting, drilling, surgery. (2) *Spectral uses:* e.g. in holography, absorption or transmission spectroscopy. (3) *Determination of distance:* e.g. in velocity measurements, surveying, navigation. (4) *Communications:* e.g. the transmission of sound and pictures by a modulated laser beam. (5) *Computer or data processing:* e.g. processing itself, the performance of linear or non-linear operations, large display presentation.

LASER CASCADE. A cascade of laser lines obtained in gases when, in a fluorescence spectrum, a given electron produces several photons related to each other by energy jumps between common energy levels.

LASER, INFRARED. A term sometimes used for a maser operating in the infrared region of the spectrum (up to about 750 μm).

LASER PROTECTION. Protection against the harmful effects of laser beams. Of particular concern are the production of damage to the eye, sometimes leading to permanent blindness; to the skin, where serious burns may occur; and to other parts of the body, where tumours may be formed.

LATENT HEAT. The energy associated with the change of physical state of any substance at constant temperature and pressure under equilibrium conditions. It includes heats of fusion, vaporization, sublimation and transition (e.g. from one crystal form to another), and the heats of their reverse processes. It does not in general include heats of absorption, adsorption, desorption, neutralization, solution, dilution, mixing, expansion or compression, dissociation or combination, and chemical reactions.

LATENT IMAGE. The invisible image formed on a photographic emulsion on its exposure to light or other radiation, which may be developed by suitable chemical processes. Its precise nature is still open to discussion but it is generally agreed that it consists of small specks of silver on or in the grains of the photographic emulsion, which have been produced by photolysis of the silver halide present.

LATIN SQUARE. An arrangement of n different letters in a square of n rows and n columns such that each letter occurs once and only once in each row and column. Latin squares are of importance in the statistical design of experiments, enabling, for instance, the separate testing of factors which may influence the outcome of an experiment.

LATITUDE. That coordinate of a point on the surface of a sphere or spheroid which specifies the angular elevation above the equator of a radius or normal passing through the point. See the following terms for particular examples.

LATITUDE, ASTRONOMICAL. Of a point on the Earth's surface: the angle between the axis of rotation of the Earth and the plane tangent to the geoid at the point of interest.

LATITUDE, CELESTIAL: LATITUDE, ECLIPTIC. Of a celestial body: its latitude on the celestial sphere referred to the ecliptic as equator and to the corresponding poles. See also: *Declination.*

LATITUDE, GALACTIC. Of a celestial body: its latitude on the celestial sphere referred to the galactic poles and galactic equator.

LATITUDE, GEOCENTRIC. Of a point on the Earth's surface: the angle between the equatorial plane and the radius of the Earth which passes through the point in question.

LATITUDE, GEODETIC. Of a point on the Earth's surface: the angle between the axis of rotation of the Earth and the plane tangent to the spheroid at the point in question.

LATITUDE, GEOGRAPHICAL: LATITUDE, TERRESTRIAL. Of a point on the Earth's surface: the angle between the equatorial plane and the normal to the surface of the Earth at the point in question.

LATITUDE, GEOMAGNETIC. Latitude defined in the same way as geographical latitude but with respect to the geomagnetic axis of the Earth.

LATITUDE, PHOTOGRAPHIC. The range of exposure permissible or the range of density usefully obtainable in a photographic emulsion. A measure of this is the quantity D_s/γ, where D_s is the saturation density produced by development of all the grains and γ is the slope of the straight line portion of the characteristic curve.

LATITUDE, VARIATION OF. A variation of a few tenths of a second of arc about a mean value, arising from the fact that the Earth's axis of rotation is not quite fixed with respect to the crust.

LATOUR CIRCUIT: GREINACHEP CIRCUIT. A constant potential voltage-doubling circuit in which two capacitors connected in series are each charged in alternate half-cycles.

LATTICE. (1) Of a crystal: a term loosely used for space lattice. See also: *Space lattice: Bravais lattice.* (2) Of a nuclear reactor: an array of nuclear fuel and other materials arranged according to a regular pattern.

LATTICE CONDUCTION OF HEAT. One of the two modes of heat conduction in a metal, the other being electron conduction. Lattice conduction arises from the motion of atoms vibrating about their equilibrium positions and the heat may be considered to be transmitted by the motion of vibrational quanta or phonons. See also: *Electron conduction of heat.*

LATTICE CONSTANTS: LATTICE DIMENSIONS: LATTICE PARAMETERS. Terms loosely used to denote unit cell dimensions. Tney specify the size and shape of the unit cell of a crystal structure in terms of the cell edges and their angles of intersection.

LATTICE ENERGY. Of an ionic crystal: the energy per ion pair necessary to separate the crystal into individual ions at infinite distance at 0 K.

LATTICE ROTATION. In the plastic deformation of a metal: the progressive change of orientation of a crystallite, relative to the direction of the applied force.

LAUE EQUATIONS. A set of three equations governing the diffraction of X-rays which must be satisfied to permit reinforcement of the contributions scattered from atoms at successive equivalent points of a crystal along each of its coordinate axes. This set of equations represents an alternative to the Bragg law as a method of expressing the conditions for selective reflection. The Laue equations also hold for the diffraction of electrons, neutrons, etc.

LAUE METHOD: LAUE PHOTOGRAPHY. The examination of the diffracted beams which are produced for any arbitrary setting of a stationary crystal from a beam of "white" radiation (i.e. a beam containing a wide continuous range of wavelengths). Each set of crystal planes selects the wavelength that will satisfy the Laue equations and produces a diffraction "spot" on a photographic film.

LAURENT SERIES. A generalization of the Taylor series which makes it possible to develop a function of the complex variable about a singular point.

LAVAL NOZZLE. A nozzle which first converges and then diverges. When used to produce a steady stream of gas at a supersonic speed the flow is subsonic in the converging part, supersonic in the diverging part and sonic at the throat (where the area of cross-section is a minimum).

LAVES PHASES. Intermetallic phases the structures of which are associated with relations between the atomic size ratios, as was first pointed out by Laves.

LAYER LATTICE. A term used in crystallography instead of the more correct term *layer structure* to describe an atomic arrangement in which the atoms are largely concentrated in a set of parallel planes, with the intervening regions relatively vacant. Typical examples are graphite and the clays.

LAYER LINE. One of a series of horizontal lines obtained on a cylindrical film when a single crystal is rotated in a beam of monochromatic X-rays about a crystallographic axis which coincides with the axis of the cylinder.

LAYER STRUCTURE. See: *Layer lattice.*

LCAO. An abbreviation for "linear combination of atomic orbitals". See also: *Molecular orbital.*

LEAD EQUIVALENT. In radiation protection: the thickness of metallic lead which affords the same protection as a material of interest under the same conditions of irradiation.

LEADING CURRENT. In an a.c. circuit with capacitive (or negative) total reactance: the current which reaches its maximum value at an earlier instant in the cycle than the impressed voltage.

LEAKAGE CURRENT. The electrical current which results from charge leaking along the surface or through the body of an insulator. It is usually negligible except for a dirty, moist or cracked insulator.

LEAK, VIRTUAL. The apparent leak in a vacuum system arising when the vapours condensed in a freezing trap escape. Such an escape occurs when the temperature in the trap is not low enough to ensure that the vapour pressure at that temperature is less than that in the system.

LEAST ACTION, PRINCIPLE OF. States that a conservative dynamical system in passing from one configuration to another does so in such a way that the action of the system remains stationary. This stationary value could be a maximum but is usually found to be a minimum. See also: *Action. Fermat law: Fermat principle. Hamilton principle.*

LEAST CONSTRAINT: LEAST CURVATURE.
A general principle, comparable to the principle of least action, by means of which the path traced out by any dynamical system when acted upon by a given set of external forces may be determined. It states that, of all paths consistent with the constraints, the actual path is that for which the constraint or curvature is least. See also: *Constraint.*

LEAST SQUARES, METHOD OF.
A method for obtaining the most likely value from a set of observations on the assumption that the sum of the squares of the deviations of the observed values from the most likely one is a minimum.

LEAST TIME, PRINCIPLE OF.
See: *Fermat law: Fermat principle.*

LEAST WORK, PRINCIPLE OF.
States that the deflections of individual parts of a structure subject to applied stresses are such that the load will be carried with a minimum storage of energy in the elastic members.

LE CHATELIER PRINCIPLE.
States that, if a system in equilibrium is subjected to a change of conditions which tends to modify the equilibrium, the system will alter in such a way as to counteract the imposed change. The principle is of wide application, one example of which is Lenz's law.

LEED.
An acronym for "*Low energy electron diffraction*". See also: *Electron diffraction, low energy.*

LEESON DISK.
A modification of a Bunsen disk in which the translucent spot is star-shaped to provide a very fine line of demarcation. See also: *Bunsen photometer.*

LEGENDRE EQUATION.
The second order differential equation

$$(1 - z^2)\frac{d^2u}{dz^2} - 2z\frac{du}{dz} + n(n-1)u = 0,$$

where z is a complex variable and n is a constant. The solutions are known as *Legendre functions* if n is non-integral and *Legendre polynomials* for integral values of n.

LEGENDRE EQUATION, ASSOCIATED.
The second order differential equation

$$(1 - z^2)\frac{d^2u}{dz^2} - 2z\frac{du}{dz} + \left\{n(n+1) - \frac{m^2}{1-z^2}\right\}u = 0,$$

where z is a complex variable and n and m are constants. The solutions are known as *Legendre associated functions* if n and m are non-integral and

Legendre associated polynomials for integral values of n and m.

LEIBNITZ THEOREM.
For the nth derivative of a product: states that, if u and v are functions of x,

$$\frac{d^n}{dx^n}(uv) = \frac{d^nu}{dx^n}v + {}^nC_1\frac{d^{n-1}u}{dx^{n-1}}\frac{dv}{dx} + \cdots$$
$$+ {}^nC_r\frac{d^{n-r}u}{dx^{n-r}}\frac{d^rv}{dx^r}\cdots u\frac{d^nv}{dx^n}.$$

LEIDENFROST PHENOMENON.
For a liquid dropped on a hot surface: the existence of a critical temperature above which the liquid does not wet the surface but is insulated from it by a layer of vapour which retards evaporation. Thus drops of water on a hot plate form into globules (the *spheroidal state*) which dance about for some time without sensible evaporation. The phenomenon is a special case of nucleate boiling. See also: *Heat transfer, boiling.*

L-ELECTRON CAPTURE: L-CAPTURE.
See: *Orbital electron capture.*

LENGTH STANDARDS FOR ENGINEERING.
(1) *End standards*, i.e. secondary standards whose form is such that their definitive length is the overall length from end to end. (2) *Line standards*, which consist of scales ruled on metal bars or glass. See also: *Length, standards of.*

LENGTH, STANDARDS OF.
Formerly signified the Imperial standard yard held in London and the international prototype metre held at Sèvres. These are *line standards* whose value is defined by the distance between two lines on a stable metal. The internationally agreed standard of length is now, however, based on the radiation emitted by the krypton isotope ^{86}Kr, the metre being defined as the length equal to $1\,650\,763 \cdot 73$ wavelengths in vacuum of the radiation (orange in colour) corresponding to the transition between the levels $2p_{10}$ and $5d_5$ of that isotope, the radiation being emitted in the absence of any perturbing effects which cause a wavelength change, such as Doppler, pressure, or Stark effects. In addition the inch $\left(\text{being } \frac{1}{36} \text{ yd}\right)$ is legally $2 \cdot 54 \times 10^{-2}$ m. See also: *Metre. Yard.*

LENGTH, UNITS OF.
Two units are in common use: the *yard*, the unit of the British system; and the *metre*, the unit of the Metric system. The Imperial standard yard has now been superseded by international agreement whereby one *inch* $\left(\text{being } \frac{1}{36} \text{yd}\right)$ is equal to $2 \cdot 54 \times 10^{-2}$ m exactly. See also: *Metre. Toise. Yard. Appendix II.*

LENS. (1) Optical: a portion of a homogeneous refracting medium bounded by two spherical surfaces and with an axis of symmetry—the *principal axis*—which passes through the centre of curvature of the two surfaces. A beam of parallel light will converge or diverge in passing through the lens according as the main surface is convex or concave. (2) Magnetic: an arrangement of electric or magnetic fields designed to focus an electron beam. See: *Electron lens.* (3) Dielectric: a lens of dielectric material used for refraction at radio frequency. (4) Of the eye: a fibrous lens situated between the aqueous and vitreous humours and behind the iris. It is often known as the *crystalline lens.*

LENS, ACHROMATIC. A lens designed to minimize chromatic aberration. In its simplest form it is composed of two lenses, one convergent and the other divergent, made of glasses having different dispersive powers, the ratio of their focal lengths being equal to the ratio of their dispersive powers.

LENS, ANAMORPHOTIC. A lens containing a cylindrical component, used in photography for producing distorted images.

LENS, APLANATIC. A lens in whose construction use is made of aplanatic points, whereby it possesses the property of giving a sharp image for rays making large angles with the axis.

LENS, APOCHROMATIC. A compound lens that is sensibly free from chromatic errors, from spherical aberration for two wavelengths, and from central coma for one wavelength.

LENS, ASPHERICAL. A lens whose surface is made not quite spherical in order to reduce aberrations. The process of treating a spherical surface (e.g. by polishing) so as to remove aberrations is known as *figuring* and a lens so treated is known as a *figured lens.*

LENS, BILLET SPLIT. A lens which has been cut across a diameter and the two halves separated slightly in a direction perpendicular to the cut and to the lens axis. Such a system produces two coherent images of, say, a slit, which can produce interference fringes. Such a lens is also known as a *half lens.*

LENS, CARDINAL POINTS OF. Six points of a thick lens which are used in determining image positions. They consist of two *focal points*, two *principal points* and two *nodal points* defined as follows. The first focal point is the point on the axis at which a point object will form an image at infinity. The second focal point is the point at which a parallel beam of light parallel to the axis will form an image. The two principal points are the inter-sections with the axis of the two *principal planes* (conjugate planes of unit positive lateral magnification). The two nodal points are conjugate points on the axis having unit positive angular magnification.

LENS, COMPOUND. A lens consisting of a combination of several simple lenses, usually made from different types of glass. Such lenses are employed to reduce the various aberrations exhibited by simple lenses.

LENS, CONTACT. A lens worn underneath the eyelid and in contact with part of the eyeball. It is usually made of plastic.

LENS, CURVATURE OF FIELD OF. For an astigmatic lens system: the curvature of the spherical surface which forms the locus of the circle of least confusion.

LENS, DELAY. A lens designed for use with low microwave frequencies in which, instead of using a dielectric, as with high frequencies (which would require a prohibitive size), an "artificial" dielectric is employed, consisting of regular arrays of conductors which, on account of their polarization, behave like a crystalline dielectric.

LENS, DOUBLET. One consisting of two components, often cemented together.

LENS, FOCAL LENGTH OF. For a thin lens the focal length is the distance from the lens at which a parallel beam of light is brought to focus. It is given by $\frac{1}{f} = \frac{1}{u} + \frac{1}{v}$, where f is the focal length, and u and v are the axial distances from the lens of the object and image respectively. For a lens made of material of refractive index n, the focal length is given by $\frac{1}{f} = (n - 1)\left(\frac{1}{r_1} + \frac{1}{r_2}\right)$, where r_1 and r_2 are the radii of curvature of the surfaces. For a thick lens no such simple relationships exist. See also: *Lens, cardinal points of. Lens, sign convention for.*

LENS, FOCAL POINT AND PLANE OF. The focal point is the point on the axis at which parallel rays of light are brought to a focus. The focal plane is the plane perpendicular to the axis at this point. See also: *Lens, cardinal points of.*

LENS, FRESNEL. A compound stepped lens of annular refracting prisms designed for lighthouses and intended to be free from spherical aberration. It has also been developed for use in signal lenses.

LENS, GAS. A device which produces and maintains a flow of gas in which appropriate gradients of refractive index permit the focusing of light. Such lenses are characterized by extremely low reflection and scattering, and their development has been stimulated by advances in laser technology.

LENS, HALF. See: *Lens, billet split.*

LENS, NODAL POINTS OF. See: *Lens, cardinal points of.*

LENS, POWER OF. The reciprocal of the focal length. It is usually expressed in metres^{-1}, i.e. *dioptres.*

LENS, PRINCIPAL PLANES AND POINTS OF. See: *Lens, cardinal points of.*

LENS, RECTILINEAR. One which is free from distortion, reproducing straight lines as such, regardless of their orientation.

LENS, SIGN CONVENTION FOR. All real distances are taken as positive and all virtual distances as negative. Thus a converging lens has a positive focal length and a diverging lens a negative focal length.

LENS, TORIC. A lens one surface of which is a portion of the surface of a torus (a surface generated by a circle rotating about an axis in its own plane, which does not pass through the centre). Such lenses are in wide use as spectacle lenses.

LENS, VARIFOCAL. A photographic lens system consisting of a converging system followed by a diverging system. The equivalent focal length can be altered by altering the distance between the two systems.

LENS, WIDE ANGLE. A composite photographic lens with a wide angle of view, the focal length of the lens being less than the diagonal of the film or plate used.

LENS, ZOOM. A lens whose components are adjustable so as to permit the angle of view (and thus the magnification) to be changed without changing the focus.

LENZ LAW. States that the direction of a current induced in a circuit as a result of a change in the linkage between that circuit and a magnetic field is such as to oppose the change. The law may be regarded as an example of the Le Chatelier principle.

LEPTONS. Elementary particles which take part essentially only in weak or electromagnetic interactions. They consist of electrons, muons, and the electron and muon neutrinos, together with their anti-particles. It has been suggested that matter might be made up only of leptons and quarks. See also: *Neutral currents. Quarks.*

LEPTON NUMBER. The number of leptons minus the number of anti-leptons in a system. It is believed to be conserved in any conceivable process.

LET. See: *Linear energy transfer: LET.*

LEWIS–RAYLEIGH AFTERGLOW. See: *Active nitrogen.*

LIAPUNOV FUNCTION. One of a number of functions used in the investigation of the stability of isolated equilibrium points of the system of real first order ordinary differential equations given by

$$\frac{dx_i}{dt} = fi(x_1, x_2, ..., x_n, t), \text{ where } i = 1, 2, 3, ..., n.$$

Such a system has reference to many problems of dynamic analysis and automatic control theory.

LIBRA, FIRST POINT OF. See: *Equinox, autumnal.*

LIBRATION. Of a celestial body: an oscillation of the body associated, for example, with a variation in its rate of revolution. In the case of the Moon one result is that 59% of the Moon's surface is visible at one time or another.

LICHTENBERG FIGURE. The pattern of the discharge on the surface of a sheet of dielectric when the plate is subjected to a corona discharge by being placed between two appropriate high-voltage electrodes. The pattern exists only as a latent image until it is "developed", for example by dusting the surface with dielectric powder. Also, if the dielectric plate is a photographic plate or film this image may be developed photographically in the usual way.

LIESEGANG RINGS. Stratified precipitates formed in gels, under certain conditions, by allowing one reactant to diffuse into another. They were first seen as concentric rings round a crystal of silver nitrate placed on a glass slide covered with a dilute solution of potassium dichromate in gelatin, but may also occur as stratified bands in a tube containing an electrolyte which diffuses downwards through a gelatin solution.

LIFT. On a body in a flowing fluid: the component of the force exerted by the fluid on the body at right angles to the direction of flow. See also: *Drag.*

LIGAND. A chemical group which has one or more pairs of electrons available for the formation of bonds in complex compounds (coordination compounds). It may be an ion (e.g. CN^-) or a neutral molecule (e.g. NH_3), and may be organic or inorganic. Ligands are designated as *monodentate*, *bidentate*, *tridentate*, and so on, according to the number of linkages attached to a central atom, i.e. according to the number of pairs of electrons taking part.

LIGHT. Visible electromagnetic radiation of wavelength about 4000 to 7500 Å, or 0·4 to 0·75 μm, together with the invisible radiation whose wavelengths lie immediately on either side of this range. See also: *Light, theories of.*

LIGHT, ABSORPTION COEFFICIENT FOR. The fraction of a beam of light which is removed by absorption on passing through unit thickness of a material, measured for very small thicknesses. It is the coefficient α in the expression $e^{-\alpha x}$ for the fraction of light transmitted through thickness x. See also: *Absorption coefficient for radiation. Bouguer–Lambert law of absorption: Bouguer–Beer law. Light, attenuation coefficient for. Optical density. Optical extinction coefficient.*

LIGHT ADAPTATION. See: *Adaptation (vision)*

LIGHT, ATTENUATION COEFFICIENT FOR. The fraction of a beam of light which is removed by scatter and absorption in passing through unit thickness of a material, measured for very small thicknesses. It is the sum of the absorption and scattering coefficients and is one minus the transmission coefficient. See also: *Attenuation coefficient for radiation.*

LIGHT DISTRIBUTION CURVE. A curve indicating the distribution of light emitted by a source. It may be in polar or cartesian coordinates and may show the luminous intensity in a horizontal or vertical plane, or show the variation with the angle of emission. A *solid of light distribution* is a surface such that the radius vector from origin to surface in any direction is proportional to the luminous intensity of the source in same direction.

LIGHT DISTRIBUTION SOLID. See: *Light distribution curve.*

LIGHT EMISSION, COSINE LAW OF. For a perfectly radiating or diffusing surface: states that the luminance in a given direction is proportional to the cosine of the angle between that direction and the normal to the surface. It is sometimes known as *Lambert's cosine law.*

LIGHT, EXTINCTION COEFFICIENT FOR. The common logarithm analogue of the absorption coefficient. It is the coefficient k in the expression 10^{-kx} for the fraction of light transmitted through thickness x. See also: *Bouguer–Lambert law of absorption: Bouguer–Beer law. Light, absorption coefficient for. Optical density.*

LIGHT-GAS GUN. For the study of aerodynamic ballistics, aerophysics, and impact: a gun employing a light propellant gas, such as helium or hydrogen, to reduce the mass of the propellant and permit the achievement of high speeds (over 8 km/s—the re-entry speed for an orbiting vehicle).

LIGHTNING. A large-scale spark discharge which is associated with charged water droplets or ice particles in a thunder cloud. The discharge may be entirely within the cloud, and is then known as sheet lightning; or it may reach the ground, when it is known as *forked lightning*. Lightning flashes also occur which penetrate the cloud but do not reach the ground. The main charge within the cloud is negative and the difference in potential between this and the ground is about 10^8 V. Lightning flashes to the ground contain several successive *strokes*, i.e. partial discharges, which may be broadly classified as *leader* (downward) and *return* (upward) strokes. See also: *Stepped leader stroke.*

LIGHTNING, BALL. A slowly moving ball of fire (up to about a foot in diameter) associated with a thunderstorm, which is said to vanish with a loud report. Accounts of such phenomena are numerous but their reliability has not yet been firmly established.

LIGHT OUTPUT RATIO. (1) Of a light source: the luminous flux per unit energy consumption, expressed in lumens per watt, also known as the *efficiency* of the source. (2) Of a light fitting: the ratio of the total luminous flux emitted from the fitting to that emitted by a reference source.

LIGHT, POLARIZATION OF. See: *Polarization.*

LIGHT, PRIMARY SOURCE OF. A body or object emitting light by virtue of a transformation of energy into radiant energy within the body or object itself.

LIGHT, PRIMARY STANDARD OF. A standard light source which is accurately reproducible from a specification. The present international standard is a full radiator at the temperature of solidification of platinum. See also: *Candela. Lamp, standard.*

LIGHT, QUANTITY OF: LUMINOUS ENERGY. The time integral of luminous flux.

LIGHT, SCATTERING COEFFICIENT FOR. The fraction of a beam of light which is removed by scattering in passing through unit thickness of a material, measured for very small thicknesses. See also: *Light, attenuation coefficient for.* Various entries under *Scattering.*

LIGHT, SCATTERING OF, BY LIGHT. A phenomenon predicted by quantum theory as a consequence of coupling between the electron field of

Dirac and the electromagnetic field of Maxwell.

LIGHT, SECONDARY SOURCE OF. A body or object transmitting or reflecting light falling on it from any other source whether primary or secondary.

LIGHT SOURCE, EFFICIENCY OF. See: *Light output ratio*.

LIGHT STIMULUS. Light of such a wavelength as to stimulate a visual phenomenon. A physical measure of the stimulus is provided by the energy, dominant wavelength and spectral purity of the light.

LIGHT, THEORIES OF. (1) The wave theory: the classical theory which considers electromagnetic radiation to be a wave motion, and which explains such properties as interference and diffraction. (2) The quantum theory: the theory which considers electromagnetic radiation to consist of particles of energy (quanta), called photons, and which explains many of the interactions of such radiation with matter, such as the photoelectric effect and the Compton effect. The two theories are to be regarded as complementary rather than contradictory and represent two aspects of electromagnetic radiation. The energy of a photon is equal to the frequency of the corresponding waves multiplied by Planck's constant. See also: *Corpuscular theory of light. Quantum theory*.

LIGHT, TOTAL REFLECTION OF. The complete reflection of a beam of light at the interface between two media of refractive index n_1 and n_2 respectively (where the light is travelling in the medium of refractive index n_1) when the angle of incidence exceeds the critical angle, given by $\sin^{-1} n_2/n_1$, and n_1 is greater than n_2. Total reflection also occurs with sound, radio and X-rays and may be described as *internal* or *external* according to circumstances. For example, for X-rays falling upon a solid, the total reflection is external, since the refractive index is smaller in the solid than in air. For light totally reflected at the inside surface of a prism, for example, it is internal. In the latter case there is some evidence that the light actually penetrates the interface, or is propagated along it, before "reflection". See also: *Critical angle*.

LIGHT, TRANSMISSION COEFFICIENT FOR. The fraction of a beam of light which is transmitted through unit thickness of material, measured for very small thicknesses. It is one minus the attenuation coefficient.

LIGHT YEAR. The distance travelled by light, *in vacuo*, in one astronomical year. It is equal to about $9\cdot46 \times 10^{12}$ km, or $5\cdot9 \times 10^{12}$ miles.

LIMB. The edge or rim of a celestial body having a visible disk, particularly that of the Sun and Moon.

LIMIT GAUGE. A gauge applied to mechanical components to verify that the component size lies between the permitted high and low limits.

LIMNOLOGY, PHYSICAL. The application of the principles of physics to the examination of the properties of and the processes taking place within lakes, and at their boundaries. Topics studied include the physical properties of the water, its fluid dynamics and variations in density, the production of waves in lakes and electromagnetic radiation exchange processes.

LINDE LIQUEFACTION PROCESS. A method of gas liquefaction using the *Joule–Thomson effect*. After expansion the gas is returned to the compressor through a counter-current heat exchanger in which the incoming gas is cooled.

LINDEMANN ELECTROMETER. A form of quadrant electrometer which is insensitive to changes in level and which employs a needle as vane, whose movement is observed through a microscope for measurement.

LINDEMANN GLASS. A lithium–beryllium borate glass containing only elements of low atomic number and so having a low absorption coefficient for X-rays. It is used, in the form of capillary tubes, for holding powders in X-ray diffraction analysis and, like beryllium, for the windows of X-ray tubes.

LINDEMANN MELTING-POINT FORMULA. Associates melting with the point at which, under the influence of heat, the amplitude of oscillation of atoms in a crystal is so great that neighbouring atoms come into contact. It may be written as $T = \text{constant} \times v^2 M V^{2/3}$, where T is the melting point in K, v is the Debye characteristic frequency, M is the atomic weight and V the atomic volume. The formula is reasonably accurate for a wide variety of substances. See also: *Melting. Simon melting equation*.

LINEAGE STRUCTURE. A structure occurring in large metallic grains arising from misalignment within the grain. It often appears as a columnar structure within the grain.

LINEAR ABSORPTION COEFFICIENT. See: *Absorption coefficient for radiation*.

LINEAR ACCELERATOR. A straight-line accelerator for charged particles in which a number of electrodes are so arranged that, when a potential difference is applied at the proper radio frequency, the particles passing through them receive successive increments of energy. In the *travelling wave* linear accelerator, the particles are accelerated by the electric component of a travelling wave field set up in a wave guide.

LINEAR AMPLIFIER. An amplifier in which the output amplitude is directly proportional to the input amplitude. The term generally refers to a pulse amplifier. See also: *Pulse amplifier.*

LINEAR ENERGY TRANSFER: LET. The average energy locally imparted to a medium per unit path by a charged ionizing particle of specified energy. It may be considered as a measure of the stopping power of that medium for that particle. The concept is used mainly in radiation protection and radiation chemistry.

LINEARLY-POLARIZED RADIATION: PLANE-POLARIZED RADIATION. Electromagnetic radiation so polarized that the vibrations lie in one single plane, the *plane of vibration*. The electric vector lies in this plane, which is sometimes called the *plane of polarization*, although the term is also applied to the plane at right angles to this (i.e. the plane containing the magnetic vector), which may lead to confusion.

LINEAR MOTOR. An electromagnetic machine of the induction type, designed to produce force or motion in a straight line. It has been compared in principle to a conventional rotary machine which has been cut along one side and unrolled. Linear motors may be used for propulsion (the "rotor" being fitted to the track), in stationary machines such as pumps, compressors and impact machines, or in *tubular* or *axial flux* motors.

LINEAR PROGRAMMING. Refers to the problem of optimizing a linear function subject to linear constraints, as in operational research. For example, that of finding the non-negative values of x_1, x_2, ..., x_i, ..., x_n for which $\sum_{i=1}^{n} a_1 x_1$ is a maximum, subject to m linear constraints.

LINE BROADENING. (1) In X-ray diffraction: the increase of line width in an X-ray powder photograph arising from small crystallite size, from crystallite strain, or both. The effect forms the basis for the measurement of both these phenomena. (2) Of spectral lines: the broadening of spectral lines for instrumental or other reasons, such as *self-absorption, radiation broadening, pressure broadening, Doppler broadening, resonance broadening* and *Stark broadening*, all of which are separately defined.

LINE FREQUENCY. Of a television system: the number of lines scanned per second by the cathode-ray beam. It is a product of the number of lines in the picture and the number of times per second that the picture is scanned.

LINE OF APSIDES. See: *Apsis: Apse.*

LINE OF FORCE. An imaginary line in an electric or magnetic field, the tangent to which, at a given point, represents the direction of the field at that point.

LINE OF MAGNETIC INDUCTION. One of a series of lines which are thought of as passing from a magnetized body into the air at a north pole, entering the body again at the south pole, and returning through the body to the north pole to form a closed loop. This concept forms the basis of several definitions in the field of magnetism.

LINE SQUALL. A squall advancing on a wide front, caused by the replacement of a warmer by a colder body of air. It may extend for some hundreds of miles and its passage is marked by a rapid change in wind direction, a rapid rise in barometric pressure, a rapid fall in temperature, and violent changes in weather.

LINE STANDARDS. See: *Length, standards of.*

LINE VOLTAGE. The voltage between two line conductors of a polyphase a.c. system.

LINKAGE MECHANISM. A restricted form of mechanism for the transmission of motion. It consists of a series of rigid members joined together with constraints so that motion can be both amplified and redirected.

LIQUEFACTION OF GASES. See: *Claude–Heylandt liquefaction process. Claude liquefaction process. Collins machine. Kapitza liquefier. Linde liquefaction process. Philips liquefier. Simon liquefier. Turbine liquefier.*

LIQUID. Describes that state or phase in which matter ultimately assumes the shape of its containing vessel, under gravitational forces, up to a definite level—the liquid surface. A liquid is essentially an incompressible fluid, with strength but little or no rigidity.

LIQUID CRYSTALS. Liquids which show double refraction and exhibit interference phenomena in polarized light. These liquid phases are known as *mesomorphic phases* and are described as *smectic* or *nematic* according as the molecules are arranged in sheets, or are more or less parallel without being so arranged. Liquid phases in which the molecules are spiral are known as *cholesteric* and may be regarded as a sub-group of the nematic group of phases. Liquid crystals are widely used in alpha numeric displays in clocks, calculators and scientific instruments.

LIQUID DEGENERACY. The process by which a liquid, cooled below a certain temperature, loses its entropy of liquid disorder without going into the solid state.

LIQUID JUNCTION POTENTIAL: DIFFUSION POTENTIAL. The potential difference set up across the boundary between electrolytes of different composition. It arises from differences in the rates of diffusion of oppositely charged ions across the boundary.

LIQUID–LIQUID EXTRACTION: SOLVENT EXTRACTION. A process for the separation of the components of a liquid mixture by bringing it into intimate contact with an immiscible or partially miscible liquid (the *solvent*) in which one or more components of the mixture are soluble.

LIQUID STRUCTURE. The arrangement of atoms or molecules in a liquid. In general the structure is characterized by the presence of short-range order and the absence of long-range order. See also: *Radial density function.*

LIQUIDUS. A line in an equilibrium diagram indicating the temperatures at which solidification begins or melting is completed in alloys of different composition. Thus, all phases above the liquidus curve are liquid. See also: *Solidus.*

LISSAJOUS FIGURES. Patterns which arise from the combination of two simple harmonic motions at right angles to each other. They can be produced in a cathode-ray oscilloscope, for example, by supplying harmonically related voltages to the deflection plates; and may be used for comparing the frequencies and phases of the two motions.

LITRE. The unit of volume in MKS and SI units. It is equal by definition to 1 dm³. It was formerly defined as the volume of 1 kg of pure water at 4°C which led to a value of the litre of 1·000 028 dm³. This was rescinded in 1964 by the Conférence Génerale des Poids et Mésures, which defined the litre as 10^{-3} m³ (1 dm³) exactly, with the recommendation that neither the word litre nor its symbol "l" should be used to express results of high precision, to avoid confusion. See also: *Kilogramme des archives. Appendix II.*

LITTORAL CURRENT. The current along the seashore (or littoral zone). It causes intense marine abrasion. See also: *Mass transport by waves.*

LITTROW MIRROR. A plane mirror mounted nearly normal to the dispersed beam emerging from the prism in a prism spectrometer. For one particular wavelength the beam is reflected back upon itself, and the spectrum may be scanned by rotating the mirror instead of the prism. A spectrograph in which a Littrow mirror is used together with an autocollimating device is known as a *Littrow spectrograph.*

LITZ WIRE. Fine stranded wire in which each strand is individually insulated. It is used for high-frequency currents since it has a reduced skin effect.

LOAD. An abbreviation for *L*aser *o*pto-*a*coustic *d*etection. See: *Opto-acoustic spectroscopy.*

LOBE. That portion of the overall radiation pattern of an aerial which is contained within a region bounded by directions of minimum intensity. The term is sometimes used to describe the radiation within this region.

LOCALIZED ELECTRON THEORY. See: *Collective electron theory. Ferromagnetism, Heisenberg theory of.*

LOBE SWITCHING: BEAM SWITCHING. In radar: a method of determining the direction of a target by successive comparisons of the signals corresponding to two or more beam directions differing slightly from the target direction.

LODESTONE. The first known magnet, consisting of the mineral magnetite, Fe_3O_4.

LOGARITHM. Of a number: the index to which a given base (usually 10 or e) must be raised to produce the number. Such an index is known as the logarithm of the number to the base 10, or e, or whatever base is chosen. The whole number preceding the decimal point of the logarithm is the *characteristic*, the decimal fraction being the *mantissa*. Logarithms to the base 10 are known as *common logarithms* (written log x). Those to the base e are *natural logarithms* (written ln x). See also: *Exponential function.*

LOGARITHMIC DECREMENT. (1) For the decay of an oscillatory motion whose amplitude is decreasing exponentially: the decrease per swing in the natural logarithm of the amplitude. (2) For the elastic scattering of neutrons by nuclei of which the kinetic energy is negligible compared with that of the neutrons: the decrease per collision in the natural logarithm of the neutron energy.

LOGARITHMIC SERIES. The series

$\ln(1+x) = x - \dfrac{x^2}{2} + \dfrac{x^3}{3} - \dfrac{x^4}{4} + \cdots$, where x lies between 1 and -1.

LOGICAL OPERATION. A term used in computer technology to signify a systematic self-consistent process used in the derivation of output information from input information. See also: *Computer, digital. Logic circuit.*

LOGIC CIRCUIT. A circuit producing discrete output signals from discrete input signals, these signals being connected by specific relationships, the most elementary of which are the Boolean "logical" relationships AND, OR and NOT. A collection of logic circuits is known as a *logic network*.

LONE-PAIR ELECTRONS. Two electrons of opposite spin, and occupying an s, p, or $s-p$ hybrid orbital of the valency shell of an atom, which play no part in the formation of a covalent or ionic bond.

LONGITUDE. That coordinate of a point on the surface of a sphere or spheroid which specifies the azimuth of the point in the equatorial plane. See the following terms for particular examples.

LONGITUDE, ASTRONOMICAL. Of a point on the Earth's surface: the angle between the meridian planes which are parallel to the geoidal verticals at the point in question and at Greenwich.

LONGITUDE, CELESTIAL: LONGITUDE, ECLIPTIC. Of a celestial body: its longitude on the celestial sphere if the ecliptic is taken as the equator. It is measured eastwards from the first point of Aries. See also: *Oblique ascension. Right ascension.*

LONGITUDE, GALACTIC. Of a celestial body: its longitude on the celestial sphere referred to the galactic poles and galactic equator.

LONGITUDE, GEODETIC. Of a point on the Earth's surface: the angle between two planes, each containing the minor axis of the spheroid, the one containing the point on the spheroid corresponding to the point of interest, and the other being parallel to the vertical at some defined point (the origin of the local survey) or inclined to that vertical at a defined angle. To bring this longitude into Greenwich terms there must be added to it the astronomical longitude of the origin.

LONGITUDE, GEOGRAPHICAL: LONGITUDE, TERRESTRIAL. Of a point on the Earth's surface: the azimuthal angle between the meridian through the point and the meridian of Greenwich, measured from 0° to 180° east or west of Greenwich.

LONGITUDINAL WAVE. See: *Compressional wave.*

LOOMING. See: *Mirage.*

LOOP, IN AN ELECTRICAL CIRCUIT. A closed path around which actual or hypothetical currents can flow.

LORENTZ CONTRACTION. A hypothesis put forward by Fitzgerald to account for the null result of the Michelson–Morley experiment, which states that a body moving with velocity v is contracted in the direction of motion by the factor $\sqrt{(1 - v^2/c^2)}$, where c is the speed of light. It is also known as the *Lorentz–Fitzgerald* contraction and the *Fitzgerald–Lorentz* contraction. See also: *Relativity theories.*

LORENTZ FACTOR (CRYSTAL ANALYSIS). A factor which occurs in expressions for the intensity of reflection of X-rays or neutrons by crystal planes. It takes account of the effect on the intensity of the orientation of a crystal plane or the length of time (e.g. during the rotation of a single crystal) that the plane is in a reflecting position.

LORENTZ FORCE. The force acting on a charge and current in a magnetic field.

LORENTZ INVARIANT. A physical quantity $f(\mathbf{r}, t)$ in relativity theory which remains unaltered under a Lorentz transformation.

LORENTZ–LORENZ FORMULA. A formula relating the refractive index, n, and the density, ϱ, of a gas. For a given wavelength and state of aggregation it may be written as $\dfrac{n^2 - 1}{n^2 + 2} = \varrho \times$ constant. For small changes in n it leads to the Gladstone–Dale law. It can also be applied to the variation of dielectric constant with density, when it leads to the Clausius–Mosotti equation.

LORENTZ TRANSFORMATION. This gives the relationship, according to the special theory of relativity, between the description of an event in one frame of reference and that of the same event in a second frame of reference which is moving with a uniform velocity with respect to the first. See also: *Space–time.*

LORENTZ UNIT. A frequency unit in terms of which Zeeman splitting may be expressed.

LORENZ NUMBER: LORENZ CONSTANT. The ratio of the thermal conductivity of a metal to the product of electrical conductivity and absolute temperature. It is approximately constant for many metals. See also: *Wiedemann–Franz law.*

LOSCHMIDT NUMBER. The number of molecules per cubic centimetre of an ideal gas at 0°C and normal atmospheric pressure. It is equal to about 2.687×10^{-19} cm^{-3}. The term is also used, particularly in Germany, to denote *Avogadro's constant*.

LOSS ANGLE. In a dielectric subjected to alternating electric stress: the angle by which the angle of lead of the current over the voltage is less than 90°.

LO SURDO TUBE. A discharge tube with an extended cathode dark space used for observing the Stark effect in spectral lines emitted from that space.

LOVE WAVE. In an elastic medium: a wave which propagates along a stratum which is bounded on each side by a medium which has elastic properties which differ from those of the stratum. The particles of the medium are displaced in a direction parallel to that of the stratum.

LOVIBOND TINTOMETER. A colorimeter in which the colour is expressed in terms of standardized red, yellow and blue filters.

L–S COUPLING. See: *Coupling.*

LUBRICATION. The application or introduction of interfacial films to reduce the friction or wear between rubbing surfaces.

LUBRICATION, BOUNDARY. Lubrication under conditions (e.g. at high pressures and low speeds) where a film of liquid lubricant cannot prevent the moving surface from touching. Boundary lubrication consists in covering the surfaces with a closely adherent film of, at most, a few molecules in thickness. Fatty acids are effective boundary lubricants at low temperatures (below about 200°C), but at higher temperatures *extreme pressure additives* are employed. These may be active sulphur or chlorine, or lamellar solids such as graphite, molybdenum disulphide, or boron nitride.

LUBRICATION, FLUID: LUBRICATION HYDRODYNAMIC. Lubrication in which the rubbing surfaces are completely separated by a fluid lubricant or viscous oil film which is induced and sustained by the relative motion of the surfaces.

LUBRICATION, GAS. Fluid lubrication in which the lubricant is a gas.

LUBRICATION, WEEPING. Lubrication by liquid squeezed out of a soft, permeable bearing impregnated with the liquid, when a load is applied.

LÜDER BANDS: STRETCHER STRAINS. Surface markings observed on mild steel and certain non-ferrous metals when strained beyond the elastic limit.

LUMEN. The SI unit of luminous flux. See also: *Luminous flux. Appendix II.*

LUMINAIRE. A term now adopted internationally to denote lighting fittings.

LUMINANCE. At a point of a surface: the luminous intensity emitted per unit area. The SI unit is the candela per square metre, formerly known as a *nit*. Other units are the *stilb* (1 cd/cm^2), and the *foot-lambert* (3·426 cd/m^2), formerly known as the *equivalent foot-candle*. See also: *Illuminance: Illumination. Appendix II.*

LUMINESCENCE. A general term denoting the emission of light as a result of causes other than high temperature, e.g. fluorescence. Such emission may result from bombardment by cathode rays (*cathodoluminescence*), radiations from radioactive materials (*radioluminescence*), or ions (*ionoluminescence*), the effect of an electric field (*electroluminescence*), optical excitation (*photoluminescence*), irradiation by ultrasonic waves (*sonoluminescence*), or the effect of grinding or pulverizing (*triboluminescence*). See also: *Anti-Stokes emission. Phosphor.*

LUMINOSITY. (1) Of a star: a measure of the total radiant energy output per second. See also under *Stellar magnitude*. (2) In light: the luminous intensity.

LUMINOUS EFFICIENCY. Of a radiation: the quotient of the luminous flux by the corresponding radiant flux.

LUMINOUS EMITTANCE. At a point of a surface: the luminous flux emitted per unit area at the point of interest.

LUMINOUS ENERGY: LIGHT, QUANTITY OF. The time integral of luminous flux.

LUMINOUS FLUX. The rate of flow of luminous energy. It is the quantity characteristic of a radiant flux which expresses its capacity to produce visual sensation. The SI unit is the *lumen*, i.e. the flux emitted within unit solid angle of one steradian by a point source having a uniform intensity of one candela. See also: *Illuminance: Illumination. Luminance. Radiant flux. Appendix II.*

LUMINOUS INTENSITY. In a given direction: the luminous flux emitted per unit solid angle in that direction. More precisely, it is the quotient of the luminous flux emitted by a source, or by an element of a source, in an infinitesimal cone containing the given direction, by the solid angle of that cone. The unit of luminous intensity is the *candela*. See also: *Illuminance: Illumination. Luminance. Appendix II.*

LUMMER–BRODHUN CUBE. An optical device,

employing two right-angled glass prisms, for assessing the brightness of a light source relative to that of a standard source. The prisms are so made and arranged that the light from one source passes to the eye through the centres of the prisms and is seen as a disk, while the light from the other source is peripherally reflected to the eye and is seen as a ring adjacent to and outside the disk. When the two sources have the same brightness the line of demarcation between the disk and the ring disappears.

LUMMER–GEHRKE PLATE. A type of interferometer in which parallel beams of light produced by multiple reflection within a thick plate of glass or quartz give high order interference. The resolving power obtainable is of the order of 10^6.

LUMPED CIRCUIT. A circuit in which the energy storage and energy loss are essentially concentrated in relatively small regions possessing inductance, capacitance or resistance.

LUSEC. A unit of inleakage into a gas system, equal to 1 litre of gas per second at 1 μm mercury pressure. (*Note:* 1 mm mercury pressure has now been defined as 133·322 N/m^2 exactly.)

LUX. The SI unit of illuminance, equal to 1 lumen/m^2. See also: *Illuminance: Illumination. Appendix II.*

LYMAN GHOSTS. See: *Spectral ghosts.*

LYMAN SERIES. A series in the line spectrum of the hydrogen atom. See also: *Hydrogen atom, line spectrum of.*

LYOLYSIS: SOLVOLYSIS. A general term descriptive of an ionic reaction between a solvent and a solute. It includes such reactions as hydrolysis, sut not solvation.

LYOPHILIC. Describes a colloid which is readily dispersed in a suitable medium and may be redispersed after coagulation.

LYOPHOBIC. Describes a colloid which is dispersed only with difficulty, yielding an unstable solution which cannot be re-formed after coagulation.

m

MACH CONE. The wave front of a Mach wave.

MACHINE. An assemblage of moving parts capable of transmitting power. If the main purpose of the assemblage is to produce a desired motion it is called a *mechanism*, so that a machine may itself comprise an assembly of mechanisms.

MACHINE TOOL. A power-driven machine for changing the shape or size of a workpiece by the displacement or removal of material.

MACH NUMBER. The ratio of the speed of an object to the speed of sound in the undisturbed medium in which the object is moving. See also: *Dynamical similarity principle*.

MACH WAVE. The shock wave set up by an object moving with a Mach number greater than unity.

MACLAURIN THEOREM. A special case of Taylor's theorem which states that, if $f(x)$ and all of its derivatives remain finite at $x = 0$, then

$$f(x) = f(0) + f'(0)\, x + f''(0)\, \frac{x^2}{2!} + \cdots$$
$$+ f^{(n-1)}(0)\, \frac{x^{n-1}}{(n-1)!} + R_n$$

where R_n is the *remainder* after n terms. When R_n converges as n increases the result is the *Maclaurin series* for $f(x)$ at $x = 0$.

MACROMOLECULE. See: *Polymer*.

MACROSCOPIC SYMMETRY ELEMENT. See: *Crystal symmetry, space group in*.

MADELUNG ENERGY. The sum of the electrostatic interactions for an ionic crystal in the expression for the Coulomb energy of the crystal. The electrostatic energy per ion pair is $\alpha_M e^2 / v_0$, where α_M is the *Madelung constant*, e the electronic charge, and v_0 the inter-ionic distance. This constant is a pure number, determined only by the crystal structure.

MAGELLANIC CLOUDS. Two irregular galaxies in the southern hemisphere, which appear to the naked eye like detached portions of the Milky Way. They are, in fact, the two nearest stellar systems outside the Milky Way.

MAGIC NUMBERS IN NUCLEI. The numbers 2, 8, 20, 28, 50, 82 and 126. Nuclides possessing these numbers of neutrons or of protons have exceptional stability. The name arose before the advent of the nuclear shell model. See also: *Nuclear models*.

MAGMA. The molten material from which igneous rocks crystallize. At the Earth's surface it appears as volcanic lava.

MAGNET. Any body which has the power of attracting iron.

MAGNETIC AMPLIFIER. A device in which the non-linear properties of a ferromagnetic material are used to amplify a small signal into one of greater power. Magnetic amplifiers control the alternating current flowing through a circuit containing an inductance by controlled saturation of the inductance. They are analogous to *dielectric amplifiers* in which the non-linear properties of a ferroelectric material are used in the same way.

MAGNETIC AXIS: GEOMAGNETIC AXIS. Of the Earth: the axis of the dipole whose field represents the major portion of the Earth's magnetic field. It makes an angle of about 12° with the axis of rotation of the Earth.

MAGNETIC AXIS POLES. See: *Magnetic poles*.

MAGNETIC BALANCE. A balance for the measurement of the small forces involved in the determination of para- and diamagnetic susceptibility. The choice of balance is decided by the accuracy required, the order of magnitude of the susceptibility, the temperature range and the nature and amount of the material concerned. See also: *Curie–Cheveneau balance. Faraday balance. Gouy balance. Kapitza balance. Quincke balance. Rankine balance. Sucksmith ring balance*.

MAGNETIC BAY. A disturbance in which the horizontal intensity of the Earth's magnetic field deviates slightly from normal, but returns later to its undisturbed value, the deviation of the magnetic curve resembling the outline of a coastal bay.

MAGNETIC BUBBLES. Cylindrical domains of magnetization which occur in thin uniform crystal platelets of certain materials, e.g. rare earth ortho-ferrites and some garnets. These bubbles may be produced at will and are moveable under certain conditions. They can form the basis of devices for logic, memory, switching and counting circuits. See also: *Domain (ferromagnetic)*.

MAGNETIC CIRCUIT. A device which includes either a permanent magnet or a current-carrying coil of wire which provides a well-defined path (or paths) for the magnetic flux. It is in some ways analogous to a simple electrical circuit.

MAGNETIC CONDUCTANCE. Of a magnetic circuit: the reciprocal of the magnetic reluctance, i.e. $\mu A/l$, where l is the length of the circuit, A its cross-sectional area, and μ the permeability. Also known as *permeance*.

MAGNETIC COOLING. (1) Of a paramagnetic salt: the most common method of producing temperatures below 1 K. It involves the isothermal magnetization of the salt, followed by adiabatic demagnetization, and temperatures as low as about 10^{-3} K may be obtained in this way. (2) Of a substance with nuclear moments: a method involving the isothermal magnetic alignment of the nuclear spins at temperatures of about 10^{-2} K, followed by adiabatic demagnetization. The production of temperatures as low as about 10^{-6} K should be feasible.

MAGNETIC CROCHET. A displacement of small amplitude in the continuous record of the horizontal component, declination or variation, or vertical component of the Earth's magnetic field.

MAGNETIC DATING. The dating of archaeological objects by comparing the direction of magnetization in ferrous objects *in situ* with the present direction of the geomagnetic field. See also: *Magnetization, thermoremanent*.

MAGNETIC DAYS, QUIET AND DISTURBED. See: *Disturbed days: Active days. Quiet days: Calm days*.

MAGNETIC DECLINATION: MAGNETIC VARIATION. At a point on the Earth's surface: the bearing of magnetic north, east or west of the celestial pole, i.e. the angle between magnetic north and geographic north.

MAGNETIC DEVIATION. The angle between the magnetic north and the compass direction, the latter being affected by the presence of local ferro-magnetic material, as on board ship. The deviation will, in general, change with the direction of the ship's head.

MAGNETIC DIP: MAGNETIC INCLINATION. At a point on the Earth's surface: the angle between the horizon and the direction of the Earth's magnetic field.

MAGNETIC DIP POLES. See: *Magnetic poles*.

MAGNETIC DOUBLE REFRACTION: VOIGT EFFECT. The double refraction exhibited when light is passed through a vapour in a direction perpendicular to a strong magnetic field.

MAGNETIC ELEMENTS. In geomagnetism: the numerical characteristics of a magnetic field encountered at a given time or place. They are, with their usual symbols in parenthesis: magnetic declination (D), magnetic inclination (I), the intensity of the magnetic field, alsoc alled total intensity (F), the horizontal component of F (H), the north component of $H(X)$, the east component of $H(Y)$, and the vertical component of F (Z).

MAGNETIC ENERGY: INTRINSIC INDUCTION. Defined as $\mathbf{B} - \mathbf{H}$, where $\mathbf{B}$ is the magnetic induction in a magnetic field $\mathbf{H}$. It is also called *ferric induction*. The CGS unit is the gauss-oersted and the SI unit the tesla-ampere per metre ($= 40\ \pi$ or about 125.6 gauss-ocrstcd). See: *Appendix II*.

MAGNETIC EQUATOR. The line on the Earth's surface for which the vertical component of the magnetic field is zero.

MAGNETIC FIELD. (1) The region in the neighbourhood of a magnetized body in which magnetic forces can be detected. (2) Used interchangeably to refer to magnetic induction or magnetic intensity.

MAGNETIC FIELD STRENGTH. An alternative term for *Magnetic intensity*.

MAGNETIC FLUX. Through a closed figure (e.g. a circular or rectangular loop): the product of the area of the figure and the average component of the magnetic induction normal to the area, i.e. the surface integral of the magnetic induction normal to the surface. The SI unit is the weber; and the CGS unit, the maxwell, is equal to 10^{-8} Wb. See: *Maxwell. Weber. Appendix II*.

MAGNETIC FLUX DENSITY: MAGNETIC INDUCTION. May be defined as the magnetic flux per unit area at right angles to the flux, or as the product of the magnetic intensity and permeability. The SI unit is the tesla (equal to 1 Wb/m²); and the CGS unit, the gauss, i.e. 1 maxwell(or line)/cm², is equal to 10^{-4} tesla. See: *Appendix II*.

MAGNETIC HARDNESS. Of a ferromagnetic material: a qualitative term expressing the size of the magnetic field required to produce saturation. The greater this field the harder the material.

MAGNETIC HYSTERESIS. In a ferromagnetic material in a magnetic field whose strength is varied: the lagging behind of the magnetic flux in relation to the magnetic field. If the magnetic field strength is plotted against the magnetic flux density, for values of field strength between zero and some given value, and back again to zero, a closed loop is obtained, the *hysteresis loop*, the area of which represents the energy expended in taking the material through the cycle. This energy is known as the *hysteresis loss*, and appears as heat. See also: *Hysteresis.*

MAGNETIC IMAGE. The magnetic analogue of the electrical image. See also: *Electrical image.*

MAGNETIC INCLINATION. An alternative term for *magnetic dip.*

MAGNETIC INDUCTION. An alternative term for *Magnetic flux density.*

MAGNETIC INTENSITY: MAGNETIC FIELD STRENGTH. At a point in a magnetic field: the magnitude of the force experienced by a unit pole situated at that point. The CGS unit, corresponding to a force of 1 dyne, is the *oersted*. The SI unit, corresponding to a force of 1 newton, is the *ampere per metre,* 1 oersted being equal to 79·6 A/m. See: *Appendix II.*

MAGNETIC LATITUDE See: *Latitude, geomagnetic.*

MAGNETIC LEAKAGE FACTOR. The ratio of the total flux which must be produced in a magnetic circuit or system to the useful flux in the circuit or system.

MAGNETIC LENS. A lens for focusing an electron beam, which may depend either on coils (*electromagnetic lens*) or permanent magnets (*magnetostatic lens*). See also: *Electron lens. Electron microscope, transmission.*

MAGNETIC MERIDIAN. Of a point on the Earth's surface: the great circle passing through that point and the magnetic axis poles.

MAGNETIC MOMENT. (1) Of a magnet: the torque exerted on the magnet in a magnetic field of unit strength, when the axis of the magnet is at right angles to the field. (2) Of a flat coil of wire carrying a current: the torque exerted on the coil by a magnetic field of unit flux density, when the plane of the coil is parallel to the direction of the field. See also: *Dipole moment.*

MAGNETIC MOMENT, ATOMIC. The magnetic moment arising from the magnetic moment of the nucleus, and the (relatively large) magnetic moment of the electrons. See also: *Bohr magneton.*

MAGNETIC MOMENT, ELECTRON. The magnetic moment associated with the electron spin on the one hand and the orbital motion on the other. The moment of a single electron spin is one Bohr magneton. See also: *Bohr magneton.*

MAGNETIC MOMENT, NEUTRON. The magnetic moment associated with the neutron spin. It is about −1·9 nuclear magnetons.

MAGNETIC MOMENT, NUCLEAR. The magnetic moment associated with nuclear spin. Nuclei with zero spin have zero magnetic moment, other nuclei have magnetic moments between about −2 and +6 nuclear magnetons.

MAGNETIC MOMENT, PROTON. The magnetic moment associated with the proton spin. It is about 2·79 nuclear magnetons.

MAGNETIC POLE. A convenient fiction for describing certain magnetic phenomena. It denotes the points of a magnet from which the magnetic field appears to diverge or towards which it appears to converge. The *pole strength* of a magnet is the magnetic moment divided by the distance between the poles. A *unit pole* is an isolated magnetic pole of such a pole strength that, when placed at unit distance from a similar pole, it would experience unit repulsive force. The CGS unit is the oersted and the SI unit is expressed in amperes per metre. Although such a pole does not exist, an approximation to it may be achieved by using a long magnet of small diameter. See also: *Magnetic intensity: Magnetic field strength. Maxwell. Weber.*

MAGNETIC POLES. Of the Earth: (1) *Magnetic axis poles:* the intersections of the magnetic axis (geomagnetic axis) with the Earth's surface, i.e. the geomagnetic poles. The north pole is at about 79°N, 70°W; and the south pole is at about 79°S, 110°E. (2) *Magnetic dip poles:* the points on the Earth's surface where the magnetic field is directed vertically downward or upward. The north pole is at about 75°N, 101°W; and the south pole is at about 67°S, 143°E. See also: *Dipole, centred,*

MAGNETIC POTENTIAL. (1) At a point: the work done in bringing a unit north pole from infinity to that point. (2) In a magnetic circuit: the magnetomotive force.

MAGNETIC RECORDING. The recording of signals as magnetic variations in a magnetic medium using recording transducers. The signals are reproduced via the same transducers and the magnetic medium may be in the form of tapes, strips, disks, wires, drums, or loops.

MAGNETIC RELUCTANCE. Of a magnetic circuit: the analogue of electrical resistance. It is given by $l/A\mu$, where l is the length of the circuit, A its cross-sectional area, and μ the permeability. It is the reciprocal of the magnetic conductance.

MAGNETIC RESONANCE. The absorption of microwaves or radio waves by atoms or nuclei respectively in the presence of a static magnetic field. See also: *Electron paramagnetic resonance (EPR): Electron spin resonance (ESR). Nuclear magnetic resonance.*

MAGNETIC ROTATION. See: *Faraday effect.*

MAGNETIC SATURATION. The condition, approached only as the temperature approaches absolute zero, in which all magnetic spins are aligned in the same direction. In practice magnetic saturation usually signifies the condition in which the intensity of magnetization no longer continues to increase with increasing field strength. This is sometimes referred to as *technical saturation.*

MAGNETIC SCREENING: MAGNETIC SHIELDING. The screening from a magnetic field which exists at the interior of an enclosure of high permeability material (e.g. a hollow ferromagnetic cylinder). The lines of induction tend to be strongly refracted as they pass into this material, leaving a relatively field-free space within.

MAGNETIC SHELL. A hypothetical double layer of magnetic "charge", which is the magnetic analogue of the electrical double layer. A magnetic shell is said to be uniform if its dipole moment per unit area is the same at all points on the surface.

MAGNETIC SIGMA−T (σ − T) CURVES. See: *Magnetothermal analysis.*

MAGNETIC STORM. A worldwide magnetic disturbance, accompanied by an enhanced display of the aurora, and associated with sunspot activity.

MAGNETIC STRUCTURE. Of a crystalline material: the structure related to the magnetic unit cell, which is not the same as the chemical unit cell in antiferromagnetic and ferrimagnetic crystals, owing to the existence of antiparallel spins in chemically identical atoms. The magnetic structure may be studied very effectively by neutron diffraction.

MAGNETIC SUSCEPTIBILITY. See: *Susceptibility.*

MAGNETIC TAGGING. The use of Curie temperature as a means of identifying materials.

MAGNETIC VARIATION. See: *Magnetic declination: Magnetic variation.*

MAGNETIC VARIATION, LUNAR. The change in magnetic variation arising from the effect on the ionosphere of lunar tides in the atmosphere.

MAGNETIC VARIATION, SECULAR. The change in magnetic variation with time. Such a change may be slow, arising from sources within the Earth, or quick, arising from changes in the upper atmosphere.

MAGNETIC VARIATION, SOLAR DAILY. The daily change in magnetic variation arising from the effects of solar radiation on the upper atmosphere.

MAGNETIC VARIOMETER. Any instrument that indicates or monitors departures of a magnetic element from some arbitrary reference value. See also: *Magnetograph.*

MAGNETIC VISCOSITY. The existence of a delay in the change of magnetic induction in a ferromagnetic material when the applied magnetic field undergoes a sudden change, which is considerably longer than is to be expected from the consideration of eddy currents alone.

MAGNETISM. The science dealing with the study of magnetic forces and fields. It includes ferromagnetism, ferrimagnetism, antiferromagnetism, paramagnetism, and diamagnetism.

MAGNETIZATION BY ROTATION: BARNETT EFFECT. The magnetization of an initially unmagnetized specimen by the rotation of the specimen in the absence of any external magnetic field. The discovery of the effect by Barnett, in 1914, was the first successful experiment in the field of gyromagnetic phenomena. The inverse effect, the slight rotation of a suspended iron cylinder when suddenly magnetized, was discovered at about the same time by Einstein and de Haas. See also: *Einstein–de Haas effect: Rotation by magnetization.*

MAGNETIZATION CURVES. Curves relating the magnetic flux density in a magnetic specimen to the magnetic field strength, i.e. **BH** *curves.* Such curves may be curves of simple magnetization of a demagnetized material, known as *initial, virgin* or *normal* magnetization curves; or related to a specimen in which hysteresis is suppressed by the superimposition of a demagnetizing field and known as *ideal* or *anhysteric* magnetization curves.

MAGNETIZATION, ENERGY OF. Of a ferromagnetic material: the average work per unit volume required to magnetize the material.

MAGNETIZATION, INTENSITY OF. For a uniformly magnetized body: the magnetic moment of the body per unit volume, or alternatively the magnetic pole strength per unit area. See also: *Magnetic intensity: Magnetic field strength.*

MAGNETIZATION, REMANENT. The magnetization left in a specimen after the removal of a magnetic field.

MAGNETIZATION, SPONTANEOUS. The persistence of magnetization below the Curie point. The spontaneous magnetization of individual domains of a ferromagnetic material is sometimes called intrinsic magnetization.

MAGNETIZATION, THERMOREMANENT. The remanent magnetization acquired when a substance is cooled in a magnetic field. This can be very different, for weak magnetization, from ordinary isothermal remanent magnetism, the difference being particularly noteworthy in baked clays and volcanic rocks, where it forms the basis of geomagnetic and archaeological applications. See also: *Archaeomagnetism.*

MAGNETOACOUSTICS. The study of the interaction between magnetic fields and ultrasonic waves. Such a study can yield information about the electronic structure of a metal since, under appropriate conditions, ultrasonic energy may be absorbed by conduction electrons and returned as thermal vibrations to the lattice.

MAGNETOCALORIC EFFECT. The reversible heating and cooling of a specimen by changes of magnetization.

MAGNETOCHEMISTRY. The study of the magnetic changes associated with chemical reactions.

MAGNETOELASTIC EFFECTS. The effects of stress and strain upon the magnetic properties of a ferromagnetic material.

MAGNETOGRAPH. An apparatus for recording fluctuations in the Earth's magnetic field. It commonly comprises three variometers, each of which responds to a different magnetic element and gives a continuous record (a *magnetogram*) of the magnitudes of these elements.

MAGNETOHYDRODYNAMIC GENERATION OF ELECTRICITY. The generation of electricity by the movement of a stream of highly ionized gas across a magnetic field. See also: *Direct conversion of heat to electricity.*

MAGNETOHYDRODYNAMICS (MHD). The study of electromagnetic phenomena in electrically conducting fluids. The fluid may be a molten metal or ionized gas (plasma).

MAGNETOHYDRODYNAMIC WAVES. Material waves in an electrically conducting fluid in the presence of a magnetic field. They are of importance in plasma physics and astrophysics. See also: *Alfvén wave.*

MAGNETOMETER. An instrument for measuring the intensity of a magnetic field. This may be the Earth's field or a field external to a magnetized body, when the magnetization of the body is measured in terms of the external field it produces. See also *Fluxgate magnetometer. Proton magnetometer.*

MAGNETOMOTIVE FORCE (M.M.F.). The analogue of electromotive force. It is the line integral of the magnetic field along any closed path in the field. See also: *Ampere-turn.*

MAGNETOPAUSE. The outer boundary of the geomagnetic field, some 10 Earth-radii from the Earth's centre.

MAGNETOPHOTOPHORESIS: EHRENHAFT EFFECT. The (usually) helical movement of fine dust particles in a gas along the lines of force of a magnetic field during their irradiation by light. It is one type of *photophoresis*. See also: *Photophoresis.*

MAGNETORESISTANCE: MAGNETORESISTIVITY. The change of electrical resistivity brought about by an external magnetic field. It occurs in all metals but is most pronounced in the ferromagnetic metals and their alloys.

MAGNETOSPHERE. The region between the upper atmosphere and the magnetopause. It is the outer region of the ionosphere.

MAGNETOSTATICS. The study of magnetic phenomena which can be regarded as not changing with time.

MAGNETOSTRICTION. The change in the physical dimensions of a ferromagnetic substance when it is magnetized or when its magnetization is changed. When a change in magnetization is brought about by elastic tension, the effect is usually described in terms of a change in elastic constants and is referred to as the *delta E effect.*

MAGNETOSTRICTION EFFECT, JOULE. The initial slight increase in length of a ferromagnetic rod subjected to a gradually increasing longitudinal magnetic field.

MAGNETOSTRICTION EFFECT, WIEDEMANN. The twisting of a ferromagnetic rod under the simultaneous action of circular and longitudinal magnetic fields.

MAGNETOSTRICTION OSCILLATOR. The oscillation of a rod of ferromagnetic material when placed in a coil carrying an alternating current of appropriate frequency. The effect may be used to provide a feedback system in an electronic oscillator, or for the production of ultrasonic waves.

MAGNETOTHERMAL ANALYSIS. The study of the magnetic behaviour of materials at various temperatures to obtain information concerning their physical state. A well-known example is the study of the change of the saturation magnetization with temperature to determine the constitution of alloy systems. The curves employed are often referred to as σ–T curves.

MAGNET, PERMANENT. One whose magnetic field is maintained at the same strength for a considerable period of time. The material of which the magnet is made must have a high coercive force and high residual flux density. See also: *Hard magnetic material.*

MAGNETRON. The name originally given to a diode with cylindrical outer anode and inner cathode, in an axial magnetic field. It was found to generate high-frequency oscillations and was developed, in the form of the *cavity magnetron* (in which the anode takes the form of a ring of resonant cavities), as a power valve in pulse-operated microwave radar equipment.

MAGNETRON CUT-OFF. The value of the anode voltage for a given magnetic field, or of the magnetic field for a given anode voltage, below which electrons are unable to reach the anode in a magnetron.

MAGNET, SUPERCONDUCTING. See: *Superconductivity.*

MAGNIFICATION, EMPTY. See: *Magnification, useful.*

MAGNIFICATION, OPTICAL. The ratio of the apparent size of a final image formed by an optical system to the actual size of the object. Where the linear dimensions are involved this is known as the *linear magnification.* Where the angles subtended at the eye are involved it is known as the *angular magnification* or the *magnifying power* of the system.

MAGNIFICATION, USEFUL. Magnification up to the point at which no further resolution of detail occurs. Beyond this point the magnification is termed *empty.*

MAGNON. A quantum of magnetic energy, analogous to the photon and phonon. It is invoked in considering the excitation or de-excitation of waves of magnetic spin by neutrons.

MAGNUS EFFECT. The development of a circulation around a rotating cylinder which is held transversely in a stream of fluid, whereby a lift force is generated, equal to $\varrho v \gamma$, where ϱ is the density of the fluid, v the relative flow velocity, and γ the circulation. The effect formed the basis of the *Flettner rotor*, whereby the lift force on a rotating cylinder was to be used to propel ships by the action of the wind on vertical rotating cylinders. It also accounts for the "slice" of a golf ball, and similar phenomena.

MAIN CONE (COSMIC RAYS). See: *Cosmic rays, allowed and forbidden regions for.*

MALLEABILITY. The property of a material of being able to be beaten or rolled into a thin sheet without splitting.

MANGIN MIRROR. A lens–mirror combination giving an intense parallel beam of light from a small source placed at the focus. It has been widely used in searchlights.

MANOMETER. An instrument for measuring low fluid pressures or pressure differences. A variety of types exists, the simplest being a U-tube containing mercury or some other liquid of low vapour pressure, in which one limb of the tube is connected to the vessel within which the pressure is to be measured and the other is closed or is left open to the atmosphere.

MANOSTAT. A device for maintaining constant gas pressure in an enclosure, usually involving the use of a pressure-sensitive valve which admits gas to or exhausts gas from the enclosure in question.

MANTISSA. See: *Logarithm.*

MANY-BODY PROBLEM. The problem of determining the results of interactions between a number of free particles which are subjected to known forces (for example, attractive or repulsive forces which are inversely proportional to the square of the distance), and for which the initial positions and velocities are specified. For more than two bodies there is, in general, no rigorous solution. Examples of the many-body problem are the calculation of the motions of the planets, the treatment of many-electron systems in atomic physics, and of nucleon interactions in the atomic nucleus.

MARS. The fourth planet in the solar system in order of distance from the Sun. Its diameter is about half that of the Earth, its mass about one-tenth, and its period of rotation about $24\frac{1}{2}$ h. It has two satellites, Phobos and Demos. Satellite exploration has shown that the atmosphere of Mars consists in the main (about 95%) of carbon dioxide, the oxygen content being about 0·15%. The composition of the surface is broadly similar to that of the Earth's crust and of the surface of the Moon. There is no evidence of life.

MARTENSITIC TRANSFORMATION. A change of crystal structure of the type occurring in the formation of martensite, i.e. one which is diffusionless, in which each atom tends to retain the same neighbours throughout the change. The new structure is thus obtained from the old by a process of deformation. See also: *Metallic transformations.*

MARX CIRCUIT. A circuit for the production of high-voltage pulses (i.e. an *impulse generator*), in which capacitors can be charged in parallel and discharged in series.

MASCON. One of a number of areas of concentration of mass on the Moon (from *mass concentration*). Mascons reveal themselves by anomalous gravitational effects but their origin remains uncertain.

MASER. Formerly an acronym for "*M*icrowave *a*mplification by *s*timulated *e*mission of *r*adiation", it has now come to stand for Molecular rather than "Microwave" amplification since the techniques have been extended to the infrared and optical regions. The stimulated emission is provided by exciting the atoms in an atomic or molecular system to a higher state by the absorption of radiation of a suitable frequency, the atoms or molecules then reverting to their former states and emitting photons whose energy is the difference between the states. The stimulated emission, which is phase-coherent, is used in amplifier and oscillator techniques. See also: *Laser. Optical pumping.*

MASS. The quantity of matter in a body. The unit in the MKS and International (SI) systems is the *kilogram,* in the CGS system is the *gram,* and in the Imperial system is the *pound.* See also: *Electromagnetic mass. Kilogram. Weight. Appendix II.*

MASS ABSORPTION COEFFICIENT. See: *Absorption coefficient for radiation.*

MASS ACTION, LAW OF. States that, in a homogeneous reaction, the rate of a chemical change is at any instant proportional to the product of the effective concentrations of the reacting substances at that instant.

MASS, ACTIVE. The molecular concentration, generally expressed in moles per unit volume.

MASS COEFFICIENT OF REACTIVITY. In a nuclear chain-reacting medium: the partial derivative of reactivity with respect to the mass of a given substance in a specified location.

MASS DECREMENT. Of a nuclide: the difference between the atomic weight and the mass number. See also: *Packing fraction.*

MASS DEFECT. Of a nucleus: the difference between the mass of the nucleus and the sum of the masses of its constituent nucleons. See also: *Packing fraction.*

MASS, DETERMINATION OF. See: *Weighing.*

MASS DOUBLETS. In mass spectroscopy: doublets arising from ions or atoms with almost identical mass numbers.

MASS, EFFECTIVE. A value of the mass which is consistent with the behaviour of an electron in a solid or a nucleon in a nucleus, but which differs from that normally associated with the particle in question.

MASS–ENERGY EQUIVALENCE. The concept, expressed by Einstein in the relation $E = mc^2$, where E is the energy, m the mass and c the speed of light. This concept follows from the theory of relativity and has been strikingly illustrated by the release of energy in nuclear fission. See also: *Conservation of mass.*

MASS NUMBER. Of an atomic nucleus: the number of nucleons in that nucleus.

MASS, RELATIVISTIC. The total mass of a moving particle. It is equal to $m_0 \left(1 - \dfrac{v^2}{c^2} \right)^{-\frac{1}{2}}$, where m_0 is the rest mass, v is the speed of the particle and c is that of light.

MASS SPECTROGRAPHY: MASS SPECTROMETRY. The identification of an atom, molecule or compound by virtue of its atomic or molecular weight. Positive ions of the material concerned are separated by the use of electric or magnetic fields, or both. In the *mass spectrograph* the ions are recorded on a photographic plate according to their mass/charge ratios; and in the *mass spectrometer* it is usual to detect the particles electrically.

MASS TRANSFER COEFFICIENT. Another term for diffusion coefficient or diffusivity. See also: *Fick diffusion laws.*

MASS TRANSPORT BY WAVES. In a liquid containing progressive waves: the translation of the liquid in the direction in which the waves advance. The motion of translation decreases rapidly with increasing depth.

MASS UNIT, CHEMICAL. See: *Atomic mass unit, unified.*

MATCHED TERMINATION: MATCHED LOAD For a waveguide: a termination producing no reflected wave at any transverse section of the wave guide.

MATERIAL PARTICLE. An atomic or subatomic particle which, although it exhibits wave properties, also possesses rest mass. See also: *Matter. Wave mechanics.*

MATHEMATICAL SERIES. A set of (possibly complex) numbers a_1, a_2, a_3 arranged in order, e.g. an arithmetic or geometric progression. The problem usually considered is that of finding the sum of such a series to n terms, i.e. the sum of the first n terms.

MATHEMATICS. The construction and operation of deductive systems, and the comparative study of such systems.

MATHIEU EQUATIONS. (1) The equation
$$\frac{d^2y}{dx^2} + (\lambda + \gamma \cos 2x)\, y = 0.$$

(2) The equation, known as the *modified Mathieu equation*,
$$\frac{d^2y}{dx^2} - (\lambda + \gamma \cosh 2x)\, y = 0.$$
The solutions of these equations are known as *Mathieu functions*.

MATRIX. (1) In mathematics: a system of elements of real or complex numbers, associated with a set of simultaneous equations, which are arranged in a rectangular formation. For example, the set of linear equations $\sum_{j=1}^{j=n} a_i j \cdot x_i = y_i$, may be written as $A \cdot x = y$, where A is the matrix of the coefficients. Matrices may be treated as single entities and manipulated by well-defined rules, the rules of *matrix algebra*. Unlike a determinant a matrix, being a set of coefficients, is not characterized by a numerical value, in the ordinary sense of the term. Examples of the applications of matrices are to be found in the transformation of axes, in electrical theory, and in the mathematical expression of quantum mechanics. A matrix having m rows and n columns is said to be of *order* or *dimension* $(m \times n)$. (2) In colour television: an array of coefficients symbolic of a colour coordinate transformation. The performance of such a transformation by electrical or optical methods is sometimes referred to as *matrixing*.

MATRIX MECHANICS. A form of quantum mechanics, initiated by Heisenberg, which sets out to work with directly observable quantities. It was developed prior to and independently of wave mechanics, but the matrix elements, as was shown by Schrödinger, may be defined in terms of normalized wave functions. See also: *Quantum mechanics. Schrödinger equation. Wave mechanics.*

MATRIX MECHANICS, COMMUTATION RELATIONS IN. A set of relations which must be satisfied if two matrices are to commute, i.e. if, for matrices A and B, AB is to be equal to BA.

MATRIX, UNIT. A matrix whose diagonal elements are unity, and whose non-diagonal elements are zero.

MATTER. An aggregation of material particles, i.e. particles that possess rest mass.

MATTHIESSEN RULE. An approximate rule stating that the electrical resistance of a metal is the sum of the resistance due to the scattering of electrons by the thermal vibration of the atoms (the *ideal resistance*) and that due to scattering by defects of various kinds (the *residual resistance*). See also: *Resistance, electrical.*

MAXIMUM. Of a function: the point at which the value of the function is greater than that at all points in the immediate neighbourhood.

MAXWELL. The CGS unit of magnetic flux. It is equal to 10^{-8} Wb. By definition 4π maxwells of flux emanate from a unit magnetic pole. See also: *Magnetic flux. Appendix II.*

MAXWELL BRIDGE. See: *Bridge, electrical.*

MAXWELL COLOUR TRIANGLE. A scheme for representing mixtures of colours, according to which the three primary colours, red, blue and green, are placed at the corners of an equilateral triangle. On the sides facing these corners are placed the corresponding complementary colours, and any colour is represented by the coordinates of a point inside the triangle.

MAXWELL DEMON. An imaginary being, conjured up by Maxwell, able to operate a trap door in a partition between two chambers containing an isolated mass of gas initially at the same temperature, so as to let only fast molecules into one chamber and slow ones into the other.

MAXWELL DISTRIBUTION LAW: MAXWELL–BOLTZMANN DISTRIBUTION LAW. For a perfect gas: a law giving the average number of molecules having speeds within well-defined limits. The fraction of molecules of mass m having speeds between c and $c + dc$ is given by
$$4\pi \left(\frac{m}{2\pi kt}\right)^{3/2} c^2 \exp\left(-\frac{mc^2}{2kt}\right),$$
where k is Boltzmann's constant and T the absolute temperature.

MAXWELL EQUATIONS. A set of equations summarizing the classical theory of the electromagnetic field. They may be stated in both rationalized MKSA and SI units as follows:
$$\text{Curl } \mathbf{H} = \frac{\partial \mathbf{D}}{\partial t} + \mathbf{J}, \quad \text{Curl } \mathbf{E} = -\frac{\partial \mathbf{B}}{\partial t}$$
and
$$\text{Div } \mathbf{D} = \varrho, \quad \text{Div } \mathbf{B} = 0.$$
The second pair are implicit in the first but are normally stated explicitly. In these equations $\mathbf{H}$ is the magnetic field strength (or intensity), $\mathbf{D}$ is the displacement (or electric induction), $\mathbf{J}$ is the current density, $\mathbf{E}$ is the electric field strength (or intensity), $\mathbf{B}$ is the magnetic induction (or flux density), ϱ is the volume density of electric charge, and t is the time. All the components of $\mathbf{E}$ and $\mathbf{H}$ satisfy the

equation of wave motion $\nabla^2\phi = \mu\varepsilon\dfrac{\partial^2\phi}{\partial t^2}$, showing that the field vectors can propagate as waves of velocity $v = 1/\sqrt{\mu\varepsilon}$, where μ is the permeability and ε the permittivity. This, when first stated, provided a basis for the electromagnetic theory of light and opened up the entirely new field of electromagnetic radiation.

MAXWELLIAN VIEWING SYSTEM. For optical instruments: an arrangement in which the field of view is observed by placing the eye at the focus of a lens, instead of using an eyepiece.

MAXWELL LAW OF VISCOSITY. States that the viscosity of a gas is independent of the gas pressure. It does not hold at very high or very low pressures.

MAXWELL MODULUS. The factor by which the actual thermal conductivity of a gas is greater than that predicted from simple kinetic theory. It varies from 1·7 to 2·4, being greater for monatomic than for polyatomic gases.

MAXWELL RELATIONS. Four relations which apply to the equilibrium state of a system. They are:

$$\left(\frac{\partial T}{\partial V}\right)_S = -\left(\frac{\partial P}{\partial S}\right)_V \qquad \left(\frac{\partial P}{\partial T}\right)_V = \left(\frac{\partial S}{\partial V}\right)_T$$

$$\left(\frac{\partial T}{\partial P}\right)_S = \left(\frac{\partial V}{\partial S}\right)_P \qquad \left(\frac{\partial V}{\partial T}\right)_P = -\left(\frac{\partial S}{\partial P}\right)_T$$

where T is the absolute temperature, V the volume, P the pressure, and S the entropy.

MAXWELL–WAGNER EFFECT. For a heterogeneous mixture subjected to an electric field: an initial accumulation of charge at the boundaries between the various particles or layers of different dielectric constant and conductivity, which continues for a finite time until equilibrium is reached. The effect is of importance in studying biological systems.

MAYER HYPOTHESIS. See: *Joule law of energy content.*

M-CENTRE. See: *Colour centres.*

MEAN: AVERAGE. One of a small number of specific parameters used in summarizing statistical data, the most useful measure of which is the *arithmetic mean*, defined, for a series of values $x_1, x_2, ..., x_n$, as $(x_1 + x_2 + \cdots + x_n)/n$. In the case of a frequency distribution with f_i members in a class interval, each class having a mean x_i ($i = 1, 2, ..., n$), then the arithmetic mean is given by $(f_1 x_1 + f_2 x_2 + \cdots + f_n x_n)/N$, where $N = f_1 + f_2 + \cdots + f_n$. This also defines the *weighted mean* of the quantities x_1, the effect of the x_1 being proportional to the weights f_i, which then allow for their relative reliability or importance. See also: *Moving average.*

MEANDER. Of a river: the curved path of a river across flat country. It consists of wide loops which tend to be accentuated by scouring at the concave bend and deposition at the convex bend. See also: *Coriolis effect.*

MEAN FREE PATH. Of a specified particle in a given medium: the average distance travelled by the particle before a specified type of interaction occurs. It may be specified for all types of interaction, when it is known as the *total mean free path*, or for particular types, such as *collision, scattering, capture,* or *ionization.* The particle may, for example, be a molecule in a gas, an electron in a metal, or a neutron in a nuclear reactor.

MEAN LIFE: AVERAGE LIFE. The average lifetime for an atomic or nuclear system in a specified state. For an exponentially decaying system it is the average time for the number of atoms or nuclei in a specified state to decrease by a factor of e (2·718 ...). This is $1/\ln 2$ times the half-life and for a radioactive nuclide it is the reciprocal of the disintegration constant. See also: *Radioactive decay constant: Radioactive disintegration constant.*

MEAN SUN. See: *Solar day.*

MECHANICAL EQUIVALENT OF HEAT: JOULE EQUIVALENT. The mechanical energy equivalent to unit quantity of heat. It is about $4·185 \times 10^7$ ergs/cal. Where the quantity of heat is expressed in terms of energy (e.g. in SI units), the need for the concept no longer exists.

MECHANICAL EQUIVALENT OF LIGHT. The ratio of the radiant flux in watts to the luminous flux in lumens at the wavelength of maximum visibility. It is about $1·5 \times 10^{-3}$ W/lumen.

MECHANICAL IMPEDANCE. The ratio of the (complex) force acting on a given area of an acoustic medium or mechanical device to the resulting linear velocity in the direction of the force. The units are the *mechanical ohm* (dyne s cm^{-1}) or the *MKS (and SI) mechanical ohm* (N s m^{-1}). See also: *Acoustic impedance. Impedance.*

MECHANICS. That branch of applied mathematics which deals with the motions of bodies, the forces by which these motions are produced, and the balance of forces in a body at rest. That part of the subject dealing with forces which mutually balance, and therefore cause no motion, is called *statics;*

that part dealing with forces that do cause motion is called *dynamics*; and that part dealing with motion alone, taking no account of the forces causing the motion, is called *kinematics*.

MECHANICS, CELESTIAL. The theory of the motion of celestial bodies in their mutual gravitational fields.

MECHANISM. See: *Machine*.

MECHANOCALORIC EFFECT. The increase in temperature of helium II in a tube when some of the helium II flows out through a porous plug. Conversely, the cooling of helium II when some of it flows into a tube, closed by a porous plug, which is plunged into the liquid. The temperature change in each case is about 0·01 deg.

MEDIAN. Of a statistical distribution: that value of the variate above and below which equal numbers of items lie. The data can be divided into four equal parts by using three values of the variate of which the middle value is the median. The other two are called the *upper quartile* and the *lower quartile* respectively.

MEDICAL PHYSICS. The study of human medical problems by the application of physical knowledge and methods. See also: *Biophysics, Health physics. Hospital physics. Ultrasonics, medical applications of*.

MEDIUM. A region of space in which physical phenomena take place, or through which energy can be transmitted. It may be a material medium (i.e. one containing matter) or a non-material medium (e.g. empty space).

MEGGER. A trade name for an instrument manufactured by Evershed and Vignoles for the measurement of very high electrical resistance.

MEIOSIS. The process of the division of a biological cell.

MEISSNER EFFECT. Concerns the behaviour of a superconductor as a perfect diamagnetic in a magnetic field. When a superconductor is cooled in a magnetic field from above the critical temperature to below that temperature the magnetic flux is expelled from the material with a sudden change in the field strength outside.

MEKOMETER. An electronic device for the measurement of distances by means of a light beam modulated at microwave frequency. It is based on the Pockels effect and employs gigahertz frequencies (much higher than those employed by the Geodimeter and Tellurometer) and is therefore suitable for the highly accurate measurement of short distances. It can achieve an accuracy of 1 part in 10^6.

MELLONI THERMOPILE. An early form of thermopile consisting of a block of antimony and bismuth.

MELTING. The process of passing from the solid to the liquid state. It is believed that the long-range order, characteristic of a crystalline solid, is replaced in the liquid state by some form of short-range order. In the *lattice theory* of melting a relation is sought between the solid and liquid atomic or molecular sites. Melting is sometimes known as *fusion*. See also: *Freezing*.

MELTING CURVE. A curve showing the variation of melting point with pressure.

MELTING POINT. The temperature at which the solid and liquid phases of a substance are in equilibrium at atmospheric pressure. See also: *Lindemann melting-point formula. Simon melting equation*.

MELTING POINT, CONGRUENT. Describes the melting point of a substance which melts at constant temperature to form a liquid of the same composition.

MELTING POINT, INCONGRUENT. Describes the "melting" point of a substance which is transformed at constant temperature into a liquid of different composition and another solid phase.

MEMBRANE. An essentially two-dimensional structure which lies at the boundary between two separate three-dimensional phases, is physically or chemically distinct from both of them, and resists to differing extents the passage of the various molecular species constituting the two bulk phases.

MEMBRANE POTENTIAL. See: *Donnan potential*.

MENISCUS. The curved surface of a liquid in a tube or other container. The curvature is due to capillary action, the sense of the curvature depending on the angle of contact between the liquid surface and the material of the container.

MENSURATION. That branch of mathematics which deals with the calculation of lengths, areas and volumes, associated with geometrical figures, from data which can be obtained by direct measurement.

MERCALLI SCALE. A scale of earthquake intensity. See also: *Earthquake intensity scale*.

MERCATOR PROJECTION. See: *Orthomorphic projection: Conformal projection*.

MERCURY. The nearest planet to the Sun. Its diameter is just over a third (0·38) of that of the Earth, its mass is about one-twentieth (0·055), and its period of rotation 58d. 16h.

MERIDIAN. (1) In astronomy: the great circle on the celestial sphere which passes through its poles and through the zenith of the observer. (2) Of a point on the Earth's surface: the plane including the Earth's axis and the point in question. It is also called a *meridian of longitude*. See also: *Prime vertical.*

MERIDIAN CIRCLE: TRANSIT CIRCLE. An instrument for the precise determination of the right ascension and declination of a celestial object. It is the fundamental instrument of positional astronomy and consists ideally of a rigid refracting telescope mounted with its optical axis accurately constrained in the meridian plane. The telescope may be rotated about an east–west line in this plane and its setting accurately determined. The position of the object under observation is determined from a recording of the time and zenith distance at which it crosses the meridian. In the *mirror transit circle* the telescope is replaced by a rotatable mirror and the object under observation is viewed by one or other of two horizontally mounted telescopes.

MERIDIAN PLANE. (1) Of the Earth or the celestial sphere: any plane passing through the poles. (2) Of an optical system: any plane containing the optic axis.

MERIDIAN TRANSIT. The passage of a celestial body across the meridian of an observer. Measurements of the times at which meridian transits occur are commonly used to determine longitude.

MERIDIONAL FOCUS. See: *Astigmatic foci.*

MERSENNE LAW. See: *String, vibration of.*

MERTON NUT. A nut whose threads are made of a substance sufficiently elastic (e.g. cork) to take up the errors of a screw. No threads are, or can be, cut: they are formed by compressing the material into the screw. The use of this elastic nut permits the cutting of screws which are free from periodic errors. Other errors are averaged out.

MERZ SLIT. A variable width bilateral slit for spectrographs, in which the centre of the slit remains in a fixed position.

MESNAGER TEST. See: *Impact tests.*

MESOCOLLOID. A colloid composed of particles whose dimensions lie between about 250 and 2500 Å (0·025–0·25 μm). See also: *Eucolloids. Hemicolloid.*

MESOMORPHIC PHASES. See: *Liquid crystals.*

MESON. One of a series of particles of mass intermediate between that of an electron and that of a proton. The mesons, in ascending order of mass, are shown in the following table, the muon being included for convenience although, unlike the pions and kaons (which are hadrons), it is now known to be a lepton.

Particle	Mass (MeV)	Charge and symbol		Mean Life (sec)
Muon	~ 106	Positive ($\mu+$)		$2\cdot2 \times 10^{-6}$
		Negative ($\mu-$)		$2\cdot2 \times 10^{-6}$
Pion	~ 139	Positive ($\pi+$)		$2\cdot56 \times 10^{-8}$
		Negative ($\pi-$)		$2\cdot56 \times 10^{-8}$
		Neutral ($\pi°$)		10^{-16}
Kaon	~ 496	Positive ($K+$)		$1\cdot23 \times 10^{-8}$
		Negative ($K-$)		$1\cdot23 \times 10^{-8}$
		Neutral ($K°$ and anti-particle $\overline{K}°$)	$K_1°$: $0\cdot9 \times 10^{-10}$ $K_2°$: 6×10^{-8}	

(*Note* 1: the neutral kaon behaves as a mixture of $K°$ and $K°$, $K_1°$ as a symmetric mixture, and $K_2°$ as an antisymmetric mixture of wave functions). (*Note* 2: the mass of the electron is $9\cdot1095 \times 10^{-31}$ kg = $0\cdot511$ MeV. See: *Appendix III*).

MESON FIELD. See: *Meson theory of nuclear forces.*

MESONIC ATOM: MESIC ATOM. (1) Refers strictly to an atom in which one of the orbital electrons is replaced by a negative muon or pion; but the term is often extended to include atoms in which an orbital electron is replaced by a negative kaon, a negative sigma particle, or an anti-proton. (2) A transient system consisting of an atom in which the nucleus is replaced by a positive meson. It is also known as *pionium*. The above atoms are collectively known as *exotic atoms*.

MESON–NUCLEON COUPLING CONSTANT. A number which, in the meson theory of nuclear forces, is a measure of the strength of interaction between the meson and nucleon involved.

MESONS, POLARIZATION OF. See: *Polarization.*

MESON THEORY OF NUCLEAR FORCES. A theory which explains the short-range (about 10^{-13} cm) nucleon–nucleon forces in the nucleus in terms of the exchange of a particle between them.

The particle, originally postulated by Yukawa, was subsequently discovered and called a *meson*. Various mesons are now known, but that principally concerned in nucleon–nucleon forces is the π-meson, sometimes termed the *nuclear force meson*. These mesons are the free quanta of the *meson field* as photons are the free quanta of the electromagnetic field. The meson theory of nuclear forces aims to relate the properties of such mesons to the characteristics of the nuclear forces involved. See also: *Nuclear forces*.

MESON, W. See: *Boson, intermediate*.

MESOPAUSE. See: *Mesosphere*.

MESOSPHERE. That layer of the atmosphere whose height is between about 55 and 80 km. Its lower boundary is known as the *stratopause* and its upper boundary as the *mesopause*. See also: *Atmosphere*.

METACENTRE. Of a floating body displaced from its equilibrium position by rotation: the intersection of the vertical line through the new centre of buoyancy (centre of gravity of the displaced liquid) with the (displaced) line which formerly passed vertically through the centre of gravity of the body and the original centre of buoyancy. For stability the centre of gravity must be below the metacentre. There are, in general, two metacentres according as the rotation is about a horizontal or vertical axis.

META-ELEMENT. Another name for *transition element*.

METAGALAXY. The total recognized assemblage of galaxies. It is essentially the measurable material Universe, and includes whatever there may be in the way of gas, particles, planets, stars and star clusters in the spaces between the galaxies.

METAL. An element characterized by the presence of relatively free electrons, and hence by high thermal and electrical conductivity, opacity and optical reflecting power or "lustre". The distinction between metals and non-metals is, however, not sharp in certain instances.

METALLIC LINE STARS. Stars whose spectra resemble those of class A stars, but have stronger metallic lines. Such stars are usually placed in class A but their metallic lines correspond to about class F. See also: *Stars, spectral classes of*.

METALLIC RADIUS. See: *Atomic radius*.

METALLIC TRANSFORMATIONS. Changes in the configuration of the atoms in a metal by the alteration of the external constraints in such a way that the initial configuration becomes less stable than another configuration. They are usually classified by the growth process involved into *nucleation and growth* transformations, in which diffusion plays the main part, and *martensitic transformations*, in which distortion plays the main part.

METALLOID. An element in whose properties both metallic and non-metallic behaviour can be discerned.

METALLOGRAPHY. That branch of metallurgy concerned with the study of the internal structure of metals and alloys in relation to their properties. See also: *Equilibrium diagram: Phase diagram: Constitution diagram*.

METALLURGY. The study of metals in all their aspects. It covers the practical and theoretical aspects of their extraction, refining, alloying, fabrication and working; and includes the whole field of metallography.

METALLURGY, POWDER. The art and science of the preparation of metal powders and their subsequent fabrication into useful forms.

METALS, ELECTRON THEORY OF. The theory, now generally accepted, that the behaviour of metals is explicable in terms of the existence within them of free electrons. See also: *Band theory of solids. Brillouin zones. Fermi energy: Fermi level. Fermi surface*.

METALS, PAULING THEORY OF. A theory in which familiar chemical ideas, suitably modified, are applied to metals. These include the notions of valency bonds each involving the pairing of two electrons, of bond order (or number), of bond length, and of hydridization.

METALS, REFRACTORY. A term applied arbitrarily to chromium (melting point 1875°C) and to metals with higher melting points. However, titanium (m.p. 1800°C), platinum (m.p. 1773°C) and zirconium (m.p. 1660°C) are sometimes included.

METALS, TRANSITION. See: *Transition elements*.

METAMICT STATE. The amorphous condition brought about in certain originally crystalline minerals containing uranium or thorium as a result of radioactive bombardment.

METAMORPHISM. The recrystallization of rocks within the Earth's crust by heat and stress.

METASTABLE STATE. (1) In a chemical system: a state in which the energy of the system is greater than that of its most stable state, but in which the system has not yet become unstable. (2) Of an atom:

an excited state for which all possible quantum transitions to lower states are forbidden. (3) Of a nucleus: an isomeric state.

METEORITE. A meteor which is not completely burnt out during its passage through the Earth's atmosphere. They occur in four main types, the first (*aerolites*) consisting almost entirely of stone, the second (*siderolites*) containing stone and a fair amount of metal (principally iron and nickel), the third (*siderites*) consisting almost entirely of metal, and the fourth (*tektites*) consisting mostly of siliceous material. The majority of meteorites are in the aerolite group. Meteorites range in size from dust particles (*meteoric dust*) to large specimens weighing up to over 50 tons, or 50000 kg.

METEOROLOGY. The study of the general circulation of the atmosphere and the local variations which are concerned with weather and climate. See also: *Climate*. Entries under *Weather*.

METEOR RADIANT. The point in the atmosphere from which visible tracks of shower meteors appear to diverge. The appearance arises from the parallel motion of the meteors and is the effect of perspective. The position of the radiant is fixed relative to the fixed stars each time the shower reappears, and a shower may therefore be identified by the name of the constellation nearest to it. Typical names of such showers are the *Perseids, Orionids, Geminids*, etc.

METEORS. Stony or metallic bodies travelling in elliptical orbits round the Sun. Those groups moving in apparently unrelated orbits are termed *sporadic meteors*, and those moving in very similar orbits and reappearing at the same time each year form a number of *meteor streams*, and are known as *shower meteors*. When a meteor enters the Earth's atmosphere it produces ionization (*meteor ionization*), which is detectable by radar. It also becomes incandescent, and is seen as a *shooting star*.

METONIC CYCLE. Of the Moon: a period of 19 years (during which there are very nearly exactly 235 lunations or synodic months) after which the phases of the Moon recur on the same days of the year. The rule was propounded by Meton in Athens in 435 B.C., and is in error by only 2 h in the whole 19-year cycle.

METRE. The unit of length in MKS and SI units. It is now defined as the length equal to 1 650 763·73 wavelengths in vacuum of the radiation (orange in colour) corresponding to the transition between the levels $2p_{10}$ and $5d_5$ of the krypton isotope ^{86}Kr, the radiation being emitted in the absence of any perturbing effects which cause a wavelength change, such as Doppler, pressure, or Stark effects. The wavelength of this radiation is about 6058 Å, i.e.

0·6058 μm. The metre was originally conceived as one ten-millionth part of the distance between the pole and the equator on the meridian of Paris and the standard metre, the *metre des archives,* was the length of a bar of platinum, a copy of which became the *international prototype metre,* kept at Sèvres. The new definition is not intended to change the length of the metre but to improve the accuracy with which it can be realized so as to ensure that it shall not change with time. See also: *Length, standards of. Length, units of. SI units. Wavelength standards. Appendix II.*

METRE-CANDLE. A unit of illumination, equal to the lux, i.e. 1 lumen/m². See also: *Illumination.*

METRICAL TENSOR. See: *Riemannian space.*

METROLOGY. The study of the accurate measurement of mass, length and time, and of their direct derivatives such as area, volume, density, angle, velocity and so on. It also includes the measurement of such quantities as temperature, barometric pressure and thermal expansion. *Engineering metrology* is connected with the provision of engineering standards, and *physical metrology* with the science of measurement itself.

METROLOGY, SYSTEMS OF LIMITS AND FITS IN. An internationally agreed system according to which the limits of size of the types of fit between, and the tolerances permitted for, parts of engineering components can be specified.

MEYER HARDNESS NUMBER. For the Brinell test: the load divided by the projected area of the indent. The value of this number when the indenting ball is immersed to its equator is known as the *ultimate ball hardness,* and is a constant for the material under test. The variation of Meyer hardness number with the shape of the impression is a measure of the ability of a material to be cold worked. See also: *Hardness.*

MHO. The unit of conductance. It is the reciprocal of the ohm, and in the International System (SI) is known as the *Siemens.*

MICELLE. A colloidal particle consisting of an aggregation of small molecules or ions.

MICHELSON–MORLEY EXPERIMENT. The best known of a series of experiments devised to test the relationship between the laws of mechanics and those of the electromagnetic field, and in particular to detect the motion of the Earth through the ether. It involved the comparison of the times of travel of two coherent beams of light, one travelling in the direction of translation of the Earth, and the other at right angles to it. No motion was detected, a result which is in agreement with Einstein's theory of special relativity. See also: *Mössbauer effect*.

MICHELSON STELLAR INTERFEROMETER. An interferometer designed to be placed on a telescope and used to measure the angular diameter of a star. See also: *Interferometer, Interferometer, optical, classification of. Interferometer stellar*.

MICROANALYSIS. See: *Electron probe microanalysis. Nuclear microprobe analysis*.

MICROCANONICAL ASSEMBLY. See: *Canonical assembly*.

MICROCIRCUIT. A minute electrical circuit formed on one semiconductor crystal (typically silicon), often known as a "chip". Such chips are used as the basis of microcomputers and microprocessors. See also: *Integrated circuit*.

MICROCLIMATE. The physical properties of the atmosphere, such as temperature, humidity and wind speed, in the first few metres above the Earth's surface. These properties are subject to rapid variation with height and are of considerable importance in plant and animal ecology.

MICRODENSITOMETER: MICROPHOTOMETER. An instrument, which may or may not be of the recording type, for the measurement of the absorption of light in successive small areas of a specimen, by the determination of the optical density. Most microdensitometers are designed for measuring densities on a photographic plate, but may also be designed for direct use on, for example, biological specimens, where such things as the distribution of pigment may be of interest.

MICROFOCUS X-RAY TUBE. An X-ray tube, as its name implies, having a very fine focus. The object of such a tube is to provide a high intensity beam, which is achieved at a small power input, owing to the high specific loading and rapid dissemination of heat.

MICROMANOMETER. A manometer used for the measurement of a very small difference of pressure between near atmospheric pressures. The difference may be balanced by a liquid column or by a diaphragm.

MICROMETER. A device for measuring small dimensions or displacements by the motion of an accurately made screw of uniform pitch. The best known is the engineer's micrometer, designed for the accurate measurement of the dimensions of solid objects. The screw rotates in a nut attached to one end of a C-shaped frame, and measures the gap between its own flat end and the anvil which it approaches.

MICROMETRE. A unit of length equal to 10^{-6} m. It was formerly known as a *micron*.

MICROPHONE. A device for converting sound waves into their corresponding electrical vibrations, i.e. an electro-acoustical transducer. Microphones may employ various methods of energy transformation, e.g. variations in electrical resistance, as in the *carbon microphone*; in electrical capacitance, as in the *condenser microphone*; electromagnetic variations, as in the *moving conductor microphone*, examples of which are the *moving coil* and *ribbon* types, and the *moving armature microphone*; in temperature, as in the *hot wire microphone*; piezo-electric variations, as in the piezo-electric microphone; variations in magnetostriction, as in the *magnetostrictive microphone*; and in electrostriction, as in the *electrostrictive microphone*. In all cases the aim is to ensure that the electrical and acoustic vibrations correspond as accurately as possible, so that the eventual sound reproduction may be faithful.

MICROPHOTOGRAPHY. The production of microscopic images on high-resolution photographic emulsion. It should not be (but often is) confused with photomicrography. See also: *Photomicrography*.

MICROPHOTOMETER. See: *Microdensitometer: Microphotometer*.

MICROPROJECTION. The use of the optical system of a compound microscope to project on to a screen an image of an object situated on the microscope stage.

MICRORADIOGRAPHY. The radiography of thin sections of material in such a way that the resulting image may be enlarged to reveal microstructure. In *contact microradiography* the section is placed in contact with a high-resolution photographic emulsion. In *projection microradiography* direct magnification is obtained by increasing the distance between the section and the film, and using a point source of X-rays. See also: *X-ray microscopy*.

MICROSCOPE. An optical system designed to provide an enlarged image of a small object, together with increased resolution. The most common is the *compound microscope* consisting essentially of a substage condenser, to provide suitable illumination of the (transparent) object; an *objective* lens system, to produce a real image of the object; and an *eyepiece* lens system, to provide a virtual image which is viewed by the eye. The microscope *stage*, which supports the object in such a way that the portion of interest can be readily illuminated, lies between the condenser and the objective. See also: *Dark-field illumination. Immersion lens, optical*.

MICROSCOPE, ATOM-PROBE FIELD-ION. See: *Field-ion microscope, atom-probe*.

MICROSCOPE, BLINK. A microscope used in astronomy, in which two photographic plates of the same region are viewed at the same time, one with each eye. Each plate is obscured in turn in rapid succession, small differences being thereby revealed. An example of its use is in the search for proper motions of faint stars.

MICROSCOPE, FIELD EMISSION. An instrument which produces an image of the surface of a metal tip or of material adsorbed by the tip, by the projection of field emission electrons on to a fluorescent screen. The resolution is some 20 or 30 Å.

MICROSCOPE, FIELD ION. An instrument similar to the field emission microscope in which the image from the surface of a metal tip is produced by gaseous ions which originate close to that surface. By *field evaporation* lower and lower depths of the original tip may be examined. The resolution is good enough to permit the visualization of individual atoms.

MICROSCOPE, INTERFERENCE. A microscope in which details of an object in which there is no appreciable absorption are revealed by the interference between a direct beam and one which passes through the object. As in a conventional interferometer, one light beam is divided so as to traverse two paths. See also: *Microscope, phase-contrast.*

MICROSCOPE, METALLURGICAL. A microscope for the examination of opaque objects by reflected light, in which the incident beam of light impinges on the object along the microscope axis, being reflected by a plane sheet of glass situated in the microscope tube. Provision for phase-contrast, dark-field illumination, polarized light illumination, etc., may be provided as in the conventional microscope.

MICROSCOPE, PHASE-CONTRAST. A microscope which reveals details of an object in which there is no appreciable absorption, by the enhancement of phase difference. It is a form of interference microscope. An annular source of light is produced by an annular diaphragm placed at the front focal plane of the substage condenser, and a phase plate is placed at the first focal plane of the objective. The phase plate carries an annular ring of transparent material which increases the optical path of light passing through both rings by one quarter of a wavelength relative to light which does not pass through the rings. See also: *Microscope, interference.*

MICROSCOPE, POLARIZING. A compound microscope equipped with apparatus for the examination of the optical properties of crystalline and paracrystalline material. The essential features are a polarizer and an analyser (usually Nicol prisms), but various other components (e.g. a Bertrand lens) are often included.

MICROSCOPE, ULTRAVIOLET. A microscope in which the object is illuminated by ultraviolet light mainly to increase the resolution but also, in some cases, to reveal features which have higher absorption or reflection in ultraviolet than in visible light. The optical components are usually constructed of quartz (which is transparent to ultraviolet light) and the image is normally registered photographically.

MICROSCOPIC SYMMETRY ELEMENT. See: *Crystal symmetry, space-group in.*

MICROSEISMS. Small continuous movements of the ground, detectable by sensitive seismographs. Most arise from local sources, but *microseism storms* occur simultaneously over wide areas of the Earth's surface and often accompany hurricanes, cold fronts, and large-scale depressions. They may also be correlated with waves at sea.

MICROSPECTROMETRY. The process of obtaining spectrometric or spectrophotometric information about microscopically sized objects.

MICROTRON. An electron accelerator operating on a modified cyclotron principle.

MICROWAVE REFLECTOMETER. An instrument for the study of travelling wave phenomena in transmission lines. It depends on the measurement of the incident waves and the reflected waves produced by a discontinuity.

MICROWAVES. A term commonly applied to electromagnetic radiation of which the wavelength lies between about 1 mm and 10 cm (i.e. in the GHz range of frequency). The sources of power most commonly used for the generation of microwaves are the klystron and the magnetron, and the most common detector is the crystal rectifier in coaxial or waveguide mounts. Frequency measurements normally involve the use of a cavity resonator. See also: *Sub-millimetre waves.*

MICROWAVE SPECTROSCOPY. The measurement of spectral lines in the wavelength range of about 10 cm to 0·5 mm (i.e. in the frequency range of about 3000 to 600000 MHz). In its simplest form a microwave spectrometer consists of a klystron (to produce highly monochromatic radiation), an absorption cell and a suitable detector (e.g. a silicon–tungsten crystal detector), to measure the change of absorption with frequency. There are two basic kinds of spectroscopy in the microwave region, *electron paramagnetic resonance* and *gaseous microwave spectroscopy*, which are defined separately. See also: *Spectra, radiofrequency.*

MICROWAVE SPECTROSCOPY, GASEOUS.
The measurement of the rotational and other spectra of gaseous molecules in the millimetre wavelength region. See also: *Spectra, radiofrequency.*

MICROWAVE ULTRASONICS. Artificially generated very high frequency vibrations in solids. They are similar in nature to thermal vibrations and are sometimes known as *microwave phonons.* They are, however, unlike thermal phonons, coherent, polarized in a given direction, and have a single frequency.

MIGRATION AREA. For a fission neutron in an infinite homogeneous medium: the sum of the slowing down area from fission energy to thermal energy and the diffusion area from thermal energy to capture. It is one-sixth of the mean square displacement of the neutron from its point of origin to its point of capture.

MIGRATION LENGTH. For a fission neutron in an infinite medium: the square root of the migration area.

MILE. (1) *Statute mile:* equal to 1760 yd. (1·609344 km exactly). (2) *Nautical mile:* equal to $1/60$ of a degree of latitude at the equator, but now defined internationally as 1·852 km exactly. The UK nautical mile (6080 ft) is equal to 1·853 km. See also: *Length, units of. Yard. Appendix II.*

MILKY WAY. See: *Galaxy.*

MILLER INDICES. Indices specifying a crystallographic plane. They are the reciprocals of the intercepts made by the plane on the crystallographic axes, expressed in terms of the axial lengths and as the simplest proportional whole numbers, a prominent crystal face (the *parametral plane*) being defined as the (111) plane. In diffraction work the term is extended to mean the reciprocals of the intercepts made by a plane on the unit cell axes. The indices are then expressed in terms of the edges of the unit cell and not as the simplest proportional whole numbers. Thus (110), (220) and (330) describe three parallel planes. See also: *Crystal direction, indices of.*

MINERALOGY. The study of *minerals* (i.e. inorganic bodies produced in nature). It includes their mode of formation, the methods used in their discovery (i.e. prospecting), their chemical and physical properties, and their crystallography (macroscopic, microscopic, and sub-microscopic).

MINIMUM. Of a function: the point at which the value of the function is smaller than that at all points in the immediate neighbourhood.

MINIMUM ENERGY THEOREM. See: *Variation method.*

MIRAGE. (1) An effect caused by the total reflection of light at the upper surface of shallow layers of hot air in contact with the ground, whereby the reflected sky appears as pools of water on the ground. (2) The reverse effect (known as *looming*) when the layer of air near the ground is colder than that above. An object on the ground then appears to be in the sky. Where the ground layer consists of several layers of varying refractive index, multiple images are seen in the sky, sometimes much elongated, and the effect is known as the *Fata Morgana.*

MIRROR. A highly polished boundary between two optical media at which light is reflected. In the conventional mirror the boundary surface is silvered, a process which enhances the brightness of the reflection. A *half-silvered* mirror will both reflect and transmit light. The commonest forms of mirror are plane, spherical (convex or concave), and paraboloidal.

MIRROR NUCLEI: MIRROR NUCLIDES. Two atomic nuclei with the same number of nucleons, such that each nucleus has the same number of protons as the other has neutrons, e.g. ^8_3Li and ^8_5B. Mirror nuclides with odd mass number, in which the atomic number and neutron number differ by one, e.g. ^3_1H and ^3_2He, or $^{29}_{14}Si$ and $^{29}_{15}P$, are known as *Wigner nuclei.*

MIST. A greyish haze arising from the presence of small water droplets in the atmosphere. See also: *Haze.*

MIXED CRYSTALS. A term formerly used in crystallography to denote what is now termed solid solution. See also: *Solid solution.*

MIXING LENGTH. In the theory of the turbulent flow of a fluid: the distance through which small volumes of the fluid may be transported by turbulent motion before becoming mixed completely with their surroundings.

MKK SYSTEM OF STELLAR CLASSIFICATION. See: *Stars, spectral classes of.*

MKSA UNITS. See: *Units, MKSA.*

MKS UNITS. See: *Units, MKS.*

MOBILITY. Of charge carriers moving in a solid, liquid or gas under the influence of an electric field: the drift velocity per unit electric field. The term may be qualified by other words to denote the nature of the carrier, as in *ionic mobility.* See also: *Drift velocity.*

MOCK SUN: PARHELION. A luminous image of the Sun most frequently seen at an angular distance of about 22° from the Sun. The phenomenon is caused by refraction of the Sun's rays by ice crystals. See also: *Halo.*

MODE. (1) In statistics: that value of the variate which occurs most commonly in a distribution. (2) Of a vibrating body: the pattern of motion of the individual particles of which the body is composed. (3) In a given resonant system: one of the several possible states of oscillation that may be sustained in the system: alternatively one of several methods of exciting the system.

MODE, CRITICAL FREQUENCY OF. In a waveguide: that frequency below which a travelling wave cannot be maintained.

MODE, DEGENERATE. See: *Degeneracy*.

MODE, DOMINANT: MODE, FUNDAMENTAL. Of a waveguide: the mode with the lowest *critical frequency*.

MODE, PI. Of a magnetron: that mode of operation in which there is a phase difference of π between adjacent cavities.

MODERATING RATIO. Of a moderator in a nuclear reactor: the ratio of the slowing down power to the thermal macroscopic absorption cross-section. It expresses the effectiveness of the moderator.

MODERATOR. A material used to reduce neutron energy by scattering without appreciable capture.

MODIFIED X- OR γ-RAY SCATTER. The incoherent component of the scattered radiation when X- or γ-rays interact with extranuclear electrons. Also known as *Compton scatter*. See also: *Compton effect*.

MODULATION. (1) The process by which the essential characteristics of a signal wave (the *modulating wave*) are impressed upon another (the *carrier wave*). Such modulation may be achieved by varying the amplitude, frequency, or phase of the carrier, giving *amplitude modulation, frequency modulation*, or *phase modulation* respectively. (2) The control of electron or ion beams by means of electric or magnetic fields. Such modulation may be achieved by varying the velocity or intensity, or by deflecting the beam, giving *velocity modulation, intensity modulation*, or *deflection modulation* respectively.

MODULUS. Of a complex number: the absolute value of the number. It is the length of a vector representing the number. For a complex number A, equal to $a + ib$, the modulus is $(a^2 + b^2)^{\frac{1}{2}}$ and is written $|A|$.

MODULUS OF RIGIDITY: SHEAR MODULUS. Of an elastic body: the tangential force necessary to produce unit angular deformation. See also: *Elastic modulus*.

MOHOROVIČIĆ DISCONTINUITY. A discontinuity in both seismic velocities (compression and shear) which marks the boundary between the Earth's crust and the outer or upper mantle. See also: *Earth, crust of. Earth, mantle of. Earth, structure of*.

MOHS SCALE OF HARDNESS. A scale in which a number of minerals were grouped by Mohs according to their average hardness, each mineral being capable of scratching any other mineral which was lower in the scale. The scale served as a basis for expressing the hardness of any mineral or crystal and its use has now been extended to cover other materials. The scale is as follows: 1, talc; 2, gypsum; 3, calcite; 4, fluorspar; 5, apatite; 6, felspar; 7, quartz; 8, topaz; 9, corundum; 10, diamond. See also: *Hardness*.

MOIRÉ FRINGES. A series of interference fringes arising from the superposition of line grids, the lines of which are slightly inclined to each other, or are otherwise not in register. Such fringes have been used to measure displacements of one grid against another in metrological problems such as the measurement of errors in screw-cutting lathes, or the checking of the accuracy of ruling of diffraction gratings; and of the distortion of one grid with respect to another in such problems as the measurement of surface strain from the deformation of a model or structure on which a grid has been printed.

MOLAL ELEVATION OF THE BOILING POINT. Of a solvent: that elevation of the boiling point which would be observed if 1 mole of a solute were dissolved in 1000 g, and if the solution behaved ideally. It is independent of the nature of the solute. Also known as the *ebullioscopic constant*.

MOLALITY. The number of moles of a substance per unit mass of a phase or system, e.g. of a solute per kilogram of solvent.

MOLAL SOLUTION. One containing 1 mole of solute per kilogram of solvent.

MOLAR. Refers to 1 mole. For example, the value of a given physical or chemical quantity, per mole of a specified substance, is known as the molar value. Thus we speak of the molar susceptibility, the molar absorption coefficient, and so on.

MOLAR ABSORPTION COEFFICIENT. See: *Absorption coefficient for radiation*.

MOLAR GAS CONSTANT: GAS CONSTANT. The constant of proportionality R in the equation of state for a perfect gas: $pv = RT$, referred to 1 mole of gas. In this equation p is the pressure,

v the volume, and T the absolute temperature. R is equal to 8·314 J/K, or 1·985 cal/K. It may be interpreted as being two-thirds the total kinetic energy due to transitional motion of the molecules of an ideal gas at 1 K.

MOLAR SOLUTION. One containing 1 mole of solute per litre of solvent, i.e. per dm^3.

MOLE. The amount of substance which contains the same number of molecules (or ions, atoms, electrons, etc. as specified) as there are atoms in exactly 12 g of the pure carbon nuclide ^{12}C. Symbol *mol*. Where not specified it is usually assumed that the term refers to molecules. See also: *Amount of substance. SI units. Appendix II.*

MOLECULAR BEAM: ATOMIC BEAM. A narrowly defined stream of neutral molecules (atoms) moving through a highly evacuated enclosure, the distance between the molecules (atoms) both of the stream and of the surrounding space being so large that collisions or intermolecular (interatomic) forces can be neglected. Measurements with molecular and atomic beams have been used, among other things, for studying the kinetic properties of gases, the magnetic properties of atoms, molecules and nuclei, the hyperfine structure of spectra, and the interactions between gases and solids.

MOLECULAR COMPOUND. A compound consisting of two or more molecules held together by weak forces.

MOLECULAR DIAGRAM. A structural formula which also shows such things as bond lengths, bond angles, bond orders, net atomic charges and free valencies.

MOLECULAR EFFUSION: MOLECULAR FLOW. The passage of a gas through a fine tube or orifice under conditions such that the mean free path of its molecules is large compared to the internal dimensions of the tube or orifice.

MOLECULAR ORBITAL. The analogue of an atomic orbital in a molecule. A molecular orbital may be constructed from the orbitals of its component atoms by assuming these orbitals to be centred on the atoms of the molecule and expressing the molecular orbital as a linear combination of the atomic orbitals (*LCAO*). Molecular orbitals are usually considered to correspond to localized bonds but such *localized orbitals* can be transformed into strictly equivalent *non-localized orbitals*, which may be a more convenient basis for discussing properties such as ionization and electronic excitation. See also: *Atomic orbital.*

MOLECULAR POTENTIAL CURVE. A curve showing the potential energy of a molecule as a function of the separation of its component nuclei.

MOLECULAR SHELL. A group of molecular orbitals having the same principal quantum number, and hence corresponding to an electron shell in an atom. See also: *Electron shell.*

MOLECULAR WEIGHT. Of a specified substance: the sum of the atomic weights of the constituent atoms of a molecule of that substance.

MOLECULE. Of a pure substance: may be defined as the smallest association of atoms that can be regarded as a chemical unit of structure. The concept is without significance, however, for ionic crystals, for example, where the molecule so defined is either non-existent or infinite according to the point of view; and is of little significance for the mammoth organic molecules found in nature, where the crystallographic unit is of greater importance.

MOLE FRACTION. Of a constituent in a homogeneous mixture: the ratio of the number of moles of the constituent to the total number of moles present in the mixture.

MOLLIER DIAGRAM. For a gas– or vapour–liquid system: a diagram in which enthalpy is plotted against entropy for constant pressure, temperature or volume, thus giving a series of curves which represent certain thermodynamic properties. See also: *Thermodynamic diagram.*

MOMENTA, GENERALIZED. See: *Coordinates and momenta, generalized.*

MOMENT OF A FORCE. A measure of the turning effect of the force about a point, given by the product of the magnitude of the force and the perpendicular distance of the point from the line of action of the force.

MOMENT OF INERTIA. A measure of the resistance offered by a body to angular acceleration. For a given body, it depends on the axis of rotation chosen; and, if a representative particle of the body has mass m_i and is situated at a perpendicular distance r_i from the axis of rotation, it is given by $\sum_i m_i r_i^2$. Where the moment of inertia is written as Mk^2, where M is the mass of the body, k is termed the *radius of gyration.*

MOMENT OF MOMENTUM. See: *Angular momentum.*

MOMENTUM. (1) *Linear momentum:* the product of the mass of a moving body and its velocity. (2) *Angular momentum:* the product of the moment of inertia of a rotating body and its angular velocity. See also: *Angular momentum. Conservation of angular momentum. Conservation of linear momentum.*

MOMENTUM SPACE: RECIPROCAL SPACE. The space used in the band theory of solids. It is the same as that in which the reciprocal lattice is plotted.

MONAD. See: *Polyad.*

MONOCHROMATIC. Describes light, X-rays, or other radiation which consists of one wavelength only. In practice, the radiation so described usually consists of a narrow band of wavelengths.

MONOCHROMATOR. A device for picking out radiation of a narrow band of wavelengths from a beam in which a wide range of wavelengths is present. For light it may consist of a prism or a diffraction grating; for X-rays it is usually a diffracting crystal, often bent; for neutrons it may be a diffracting crystal or a time-of-flight device.

MONOCLINIC SYSTEM. A system of crystal symmetry containing those crystals which can be referred to two axes, oblique to each other, and a third axis at right angles to their plane, all the axes being unequal. The system is characterized by having one axis of two-fold symmetry, which may have a plane of symmetry perpendicular to it.

MONOLAYER. A film one molecule thick at the interface between two bulk phases. Such films may occur at gas–liquid, liquid–liquid, solid–liquid or gas–solid interfaces. They are also known as *monomolecular films.*

MONOLITHIC CIRCUIT. An integrated circuit which uses either thin film or silicon chip techniques but not a mixture of both.

MONOMER. A single molecule or a substance consisting of single molecules. Substances consisting of associated molecules are known as *polymers,* special cases being *dimers* (two molecules), *trimers* (three molecules) etc. See also: *Polymer.*

MONOPOLE. See: *Multipole moments.*

MONOTROPY. The existence of more than one structural or crystalline form of a substance, only one of which is stable. See also: *Allotropy. Enantiotropy. Polymorphism.*

MONOVARIANT SYSTEM. A system of several phases in equilibrium, having only one degree of freedom.

MONSOON. A seasonal wind which blows with regularity and constancy during one part of the year, and which is absent or blows from another dierction during the remainder of the year. The most striking monsoons occur in Asia although similar wind changes do occur in Australia, Africa and the U.S.A. The term has also been applied to the wind régime above about 20 km, where there is a regular reversal of wind about the equinoxes in temperate latitudes.

MONTE CARLO METHOD. A method of solving numerically problems arising in mathematics, physics and other sciences, by constructing for each problem a random process whose parameters are equal to the required quantities and on which observations can be made by ordinary computational means. From these observations, made on the random process, an estimate is made of the required parameters. In the *probabilistic* type of problem the actual random processes are simulated by suitably chosen random numbers and, from a sufficiently large number of observations using these numbers, the required information is obtained. Examples are investigations in neutron physics, problems of queueing and congestion, and the behaviour of epidemics. The *deterministic* type of problem is usually concerned with mathematical processes (e.g. the solution of differential equations, the evaluation of definite integrals, and the inversion of matrices) which can be stated analytically but not always solved. A probabilistic problem is then sought which has the same analytic expression as the one under investigation.

MONTH. (1) *Anomalistic:* the interval between successive passages of the Moon through perigee (i.e. through the point of its orbit which is nearest to the centre of the Earth). Its length is just over 27·5 mean solar days. (2) *Nodical:* the interval between successive passages of the Moon through a given node (i.e. a point at which the plane of its orbit passes through the plane of the ecliptic). Its length is slightly more than 27·2 mean solar days. This month is also called the *Draconitic month.* (3) *Sidereal:* the interval between successive passages of the Moon through the same point in its orbit relative to the fixed stars. Its length is just over 27·3 mean solar days. (4) *Synodical:* the interval between successive conjunctions (or oppositions) of the Moon and the Earth. Its length is about 29·5 days. It is the original (*lunar*) month used by the ancient astronomers. (5) *Tropical:* the interval between successive passages of the Moon across the same meridian of longitude. Its length is very slightly less than that of the sidereal month (27·3216 mean solar days as against 27·3217).

MOON. (1) The Earth's satellite. It has a diameter of about 3476 km (2160 miles) and revolves about the Earth in a variable orbit at a mean distance of about 380 000 km (239 000 miles) in a period of about 29·5 days, always presenting the same face to the Earth. Its mass is about 0·0123 times that of the Earth (i.e. about one-eightieth). No atmosphere has been detected and the composition of the surface is broadly similar to that of the Earth's crust and the surface of Mars. (2) By extension, any satellite of a planet.

MORPHOLOGY. The study of the shape or "habit" of minerals and crystals.

MORPHOTROPIC SERIES. A series of crystalline substances having similar crystalline structures (axial ratios and interfacial angles) which change progressively as one element or radical is replaced by another. Such a series is exemplified by the sulphates of Li, Na, K, Rb, Cs, Ag, Tl and NH_4. A series of this kind is said to exhibit the property of *morphotropy*.

MORSE CURVE. A curve in which the potential energy of a diatomic molecule is plotted against the internuclear distance. The potential energy E_r, for an internuclear distance of r, is given by the *Morse potential energy equation:*

$$E_r = D\{1 - \exp[-a(r - r_0)^2]\},$$

where D is the dissociation energy of the molecule, r_0 is the equilibrium separation of atoms, and a is a constant.

MORSE RULE. An empirical relationship between the equilibrium internuclear distance, r_e, of a diatomic molecule and the equilibrium vibrational frequency, w_e. It states that $w_e r_e^3 = 3000 \pm 120\,cm^{-1}$ See also: *Birge–Mecke rule. Clark rule.*

MOSAIC SCREEN. (1) A screen used for colour photography by the additive process. It consists of a minute pattern of red, green and blue filter elements. (2) The storage element in an electronic storage tube, or a television camera tube working on the storage principle.

MOSAIC STRUCTURE. A type of imperfection exhibited by most crystals, the crystal containing *mosaic blocks* consisting of microcrystals misoriented relative to each other by a few minutes of arc, and being joined by dislocation arrays. Such a crystal is known as a *mosaic crystal.*

MOSELEY LAW. States that the frequencies of characteristic X-rays are related to the atomic number of the target element by the relation: $v^{\frac{1}{2}} = K(Z - \sigma)$, where v is the frequency, Z is the atomic number, and K and σ are constants. The relation applies to all the X-ray series (K, L, M, etc.) with appropriate values of K and σ. If $\sqrt{k/R}$, where k is the wave number (the reciprocal of the wavelength) and R is the Rydberg constant for an atom of infinite mass, is plotted against the atomic number, a series of straight lines is obtained, one for each characteristic wavelength, which converge to the origin. This is known as a *Moseley diagram.*

MÖSSBAUER EFFECT. The emission of γ-rays from the nuclei of certain crystals in such a way that the recoil is taken up by the whole crystal, the line width of such rays is then very small since it is determined by the lifetime of the corresponding nuclear state alone. The effect may be considered as a particular aspect of nuclear resonance absorption. It is most easily observed in the 14·4 keV radiation from ^{57}Fe, for which 63% of the γ-ray photons are recoil-less at room temperature. The effect has been used, among other things, to investigate the strength of atomic binding, for investigations in solid state physics, and to check "ether drift" theories.

MOTION, NEWTON LAWS OF. State: (1) That every body continues in a state of rest or uniform motion unless acted upon by a force. (2) That the rate of change of momentum of a body in a given direction is proportional to the resultant force applied to it in that direction. (3) That to every action there is an equal and opposite reaction.

MOTT SCATTERING FORMULA. (1) A formula giving the differential cross-section for the mutual scattering of two identical charged particles at non-relativistic speeds. (2) A formula for the scattering of a fast electron by an idealized nucleus, obtained by the solution of the Dirac equation.

MOUNTAIN BUILDING. See: *Diastrophism. Earth, thermal contraction of.*

MOVING AVERAGE. In a series of numbers, each of which is correlated with neighbouring numbers, the moving average is a weighted mean of each value taking into account the preceding and succeeding values. This has the effect of smoothing out the data so as to eliminate irregularities. See also: *Mean: Average.*

MOVING-BOUNDARY METHOD. A method for the determination of the transport number of an ion in solution, based upon the determination of the speed at which the boundary or boundaries between appropriate electrolytes in a tube, say, move when an electric current is passed through the tube.

MOVING-COIL INSTRUMENT. A sensitive instrument for the measurement of electric current, in which the rotation of a current-carrying coil is observed in the field of a permanent magnet.

MOVING-FIELD THERAPY. See: *Radiotherapy, cross-fire technique in.*

MOVING-IRON INSTRUMENT. An instrument for the measurement of electric current in which the deflection of a soft-iron armature or other magnetic material is observed in the field of a current-carrying coil. The deflection is in the same sense whatever the direction of the current and thus such instruments are suitable for alternating currents.

M REGIONS. On the Sun: restricted magnetically active regions which are responsible for geomagnetic disturbances by their emission of radiation, but not for magnetic storms.

M-STAR. See: *Stars, spectral classes of.*

MULTIGROUP THEORY. A theory of neutron transport in which it is assumed that the neutron population is divided into a number of groups each of which has a constant neutron energy. See also: *Neutron energy group.*

MULTINOMIAL. See: *Polynomial.*

MULTIPACTOR EFFECT. A type of resonant discharge which occurs at high frequencies in a tenuous gas and leads to electrical breakdown in the gas.

MULTIPLE-BEAM INTERFERENCE. The interference of light beams reflected by a surface to reveal small variations in the flatness of the surface, e.g. contours, cleavage steps, microscopic pits.

MULTIPLET, SPECTRAL. A spectral line which is split as a result of energy-level splitting. See also: *Multiplicity.*

MULTIPLICATION FACTOR: MULTIPLICATION CONSTANT. In a neutron fission process: the ratio of the total number of neutrons produced during a given time interval (excluding neutrons produced by sources whose strengths are not a function of fission rate) to the total number of neutrons lost by absorption and leakage during the same time interval. For an infinite medium or infinite repeating nuclear reactor lattice it is known as the *infinite multiplication factor*. For a finite medium or lattice it is referred to as the *effective multiplication factor*. See also: *Neutron multiplication. Nuclear reactor, four-factor formula for.*

MULTIPLICATIVE ARRAY. See: *Acoustic array, multiplicative.*

MULTIPLICITY. In atomic spectra: The number $2S + 1$ where S is the total spin angular momentum quantum number. It represents the number of ways of vectorially coupling the orbital angular momentum vector **L** with the spin angular momentum vector **S** of an atom, i.e. the number of relatively closely spaced energy levels or terms in an atom which result from the coupling process. Evenness or oddness of multiplicity refers to the evenness or oddness of $2S + 1$. An even number of electrons gives rise to odd multiplicity and an odd number to even multiplicity, hence the alternation law. See also: *Alternation law of multiplicities.*

MULTIPOLE MOMENTS. The strengths of the equivalent multipoles (electric charge or *monopole*, electric or magnetic *dipoles*, *quadrupoles*, *octupoles*, etc.) which represent the effects, at points outside the system, of the electric and magnetic fields produced by a system of charges and currents, such as those arising from the electrons in an atom or from the protons in a nucleus; and which are assumed to be located at the centre of the system. The resultant field may be represented by an infinite series of terms, but where the series converges quickly only the first few of these need be considered. In atomic and nuclear physics the highest term of importance is the quadrupole term.

MULTI-SHOCK COMPRESSION. The deceleration to subsonic of the supersonic airflow in the intake of a supersonic aircraft engine or in the diffuser of a supersonic wind tunnel by a series of oblique shocks.

MUON. See: *Meson.*

MUONIUM. An atom in which the proton of a hydrogen atom is replaced by a positive muon.

MUONIUM. See: *Mesic atom.*

MUSICAL SCALE. An ordered stepwise series, arranged in order of frequency, of the sounds found in a musical composition or in the music of a people or period. The pattern of intervals (i.e. frequency ratios between the various notes within an octave) repeats every octave. A large number of scales are or have been in use, since any combination of intervals is theoretically possible, although certain intervals (e.g. the octave and the perfect fifth) are common in most.

MUTAROTATION. A change in optical rotary power observed when certain optically active compounds are dissolved. An example is afforded by the solution of either pure α- or β-glucose in water when an equilibrium mixture of α- and β-glucose results.

MYOELECTRIC CONTROL. The control of artificial or paralysed limbs, or other parts of the body which have become immobile, by signals emanating from the brain and central nervous system.

MYRIOTIC FIELD. A quantized field for which there are creation and annihilation operators satisfying specific commutation rules, but for which there is no vacuum state. See also: *Amyriotic field.*

n

NABLA. See: *Differential vector operator*.

NADIR. That point on the celestial sphere which is vertically below the observer, i.e. the point diametrically opposite to the zenith.

NARROW BEAM. In beam attenuation measurements for ionizing radiation: a beam in which none of the scattered radiation reaches the detector. See also: *Broad beam*.

NAVIER–STOKES EQUATIONS FOR FLUID FLOW. See: *Fluid flow, equations of motion for*.

NAVIGATION. The science of finding the position of a vehicle on the surface of the Earth (on land, sea, or in the air) and directing its course. The position may be found by observations of the Sun, Moon and selected fixed stars (*astronomical navigation*), by reference to radio beacons (*radio navigation*) by reference to a grid based on pulse-modulated signals (*radar navigation*), by use of the radio Doppler effect to measure ground speed and direction (*Doppler navigation*), or by visual means. Allowances then have to be made for tidal currents, wind speed, and so on, according to the medium in or on which the vehicle is moving. In *inertial guidance* inertial properties are used to measure the magnitude and direction of motion, employing accelerometers and gyroscopes. See also: *Inertial guidance. Omega. Sextant*.

N-CENTRE. See: *Colour centre*.

NEAR POINT. Of the eye: the nearest point of clear vision when the eye is relaxed. It is also known as *punctum proximum*.

NEBULA. A term formerly applied to all celestial objects which presented a hazy appearance in the telescope, but now restricted to the *gaseous nebulae*, generally referred to as *diffuse bright nebulae*. These consist of *emission nebulae*, in which the interstellar material is excited to emit light by a neighbouring hot star, and *reflection nebulae*, which are believed to be rendered visible by scattering the light from a nearby super giant star. The best known emission nebula is the nebula in Orion, and the best known reflection nebula is that in Pleiades. See also: *Stars, spectral classes of*.

NEBULA, DARK. A nebular appearance occasioned by a cloud of dust or gas which obscures the light from the stars beyond it. Two well-known examples are the *Coalsack nebula* and the *Great Rift* of the Milky Way, the latter being not one dark nebula but a conglomeration of superposed dark nebulae. See also: *Coalsack nebula. Great rift*.

NEBULA, EXTRAGALACTIC. A galaxy outside our own Galaxy. It is not strictly a nebula at all and the term is being superseded by the word "galaxy". Such a galaxy may be *elliptic* (a symmetrical galaxy, more or less flattened, but having no apparent structure), *spiral* (the largest class of galaxy, having an apparently spiral structure), or *irregular* (the smallest class of galaxy, having no symmetry or structure). One of the best known spiral galaxies is the "nebula" in Andromeda.

NEBULA, PLANETARY. A small regular emission nebula appearing as a greenish disk whose colour resembles that of Uranus or Neptune. Such a nebula derives its light from a hot central star and some hundreds are known, all in the region of the Galaxy. The best known example is the Ring Nebula in Lyra.

NEBULA, VARIABLE. An emission nebula associated with a variable star.

NÉEL TEMPERATURE: NEEL POINT. The temperature at which the susceptibility of an antiferromagnetic material has a maximum value. Only below this temperature does the ordered arrangement of magnetic moments exist which is characteristic of antiferromagnetism. It is sometimes termed the *antiferromagnetic Curie point*. See also: *Antiferromagnetism*.

NEGATIVE CRYSTAL. See: *Optic sign*.

NEGATIVE GLOW. In a glow discharge tube: a luminous region adjacent to the Faraday dark space on the cathode side. See also: *Discharge, glow*.

NEGATIVE PROTON. A synonym for anti-proton. See also: *Anti-particle*.

NEGATIVE THERMODYNAMIC TEMPERA-TURE. A "temperature" of a nuclear spin system in a magnetic field at which a small increase in energy produces a decrease in entropy.

NEGATRON: NEGATON. A negatively charged electron. See also: *Electron.*

NEMATIC PHASES. See: *Liquid crystals.*

NEPER. A logarithmic unit of attenuation used in electrical systems. If I_0 is the initial value of an electrical quantity (e.g. power, current, voltage) and I is the final value, the expression $I = I_0 e^{-N}$ signifies an attenuation of N nepers. The concept has been extended to include scalar quantities in mechanics and acoustics. (*Note:* 1 neper = 8·686 decibels.)

NEPHELOMETRY. The measurement of the cloudiness of a medium containing a suspension of small particles. Where the instruments employed determine the decrease in intensity of a transmitted light beam the process is termed *turbidimetry.* Where the scattered light is measured the term is *Tyndallimetry.*

NEPHOSCOPE. A meteorological instrument for the determination of the direction of motion and relative speed of clouds passing directly overhead.

NEPTUNE. The eighth planet in the solar system in order of distance from the Sun. Its diameter is about $3\frac{1}{2}$ times that of the Earth, its mass about 17 times, and its period of rotation is about $15\frac{3}{4}$ h. It has two satellites, Triton and Nereid, whose masses are about 1/40 and 1/100 000 that of the Earth respectively.

NEPTUNIUM SERIES. The radioactive family of elements beginning with ^{237}Np and ending with ^{209}Bi (ordinary bismuth). Also known as the *$4n + 1$ series* (since the atomic number of each member of the series can be expressed in this way). ^{237}Np gives its name to the series since it is the most stable of the $4n + 1$ type nuclides above ^{209}Bi.

NERNST EFFECT. The development of a transverse electric potential across a conductor through which heat is flowing, when a magnetic field is applied at right angles to the direction of flow. The potential gradient, magnetic field and direction of heat flow are mutually perpendicular. The effect is analogous to the Hall effect.

NERNST EQUATION. An equation for electrode potential involving the osmotic pressure of the ions and the solution pressure of the electrode.

NERNST GLOWER. A form of electric lamp in which the passage of a current through a rod of material, composed of a mixture of zirconia and rare earth oxides, causes the rod to glow. The main present-day use of the Nernst glower is as a labora-tory source of infrared radiation. Under appropriate conditions its emission is approximately that of a black body. See also: *Globar.*

NERNST HEAT THEOREM. States that, at absolute zero, the entropy difference disappears between all those states of a system which are in internal thermodynamic equilibrium. It leads to a denial of the possibility of attaining absolute zero. It is also known as the third law of thermodynamics. See also: *Thermodynamics, laws of.*

NERNST HIGH-FREQUENCY CAPACITANCE BRIDGE. See: *Bridge, electrical.*

NERNST–LINDEMANN THEORY OF SPECIFIC HEATS. A modification of the Einstein theory in which the single frequency of atomic vibration is replaced by two frequencies, one of which is twice the other. See also: *Debye theory of specific heats. Einstein theory of specific heats. Specific heat theories.*

NERVE. A structure in the animal body whose special function is the transmission of information by electrical impulses. It consists of a bundle of nerve fibres enclosed in a sheath and in each fibre is a conducting element (the *axon*), which is a threadlike protoplasmic outgrowth. See also: *Action potential. Resting potential.*

NETWORK ANALYSIS: ELECTRICAL CIRCUIT ANALYSIS. The use of mathematical techniques to predict the response of an electrical system to a given stimulus. By analogy the behaviour of non-electrical systems (e.g. acoustic, hydraulic, magnetic, mechanical) can also be treated by electrical circuit analysis. See also: *Electrical network.*

NEUMANN FUNCTION. A Bessel function of the second kind. See also: *Bessel equation.*

NEUTRAL CURRENTS. "Currents" observed in weak interactions between leptons and hadrons in which the charge on the lepton does not change. The term "current" arises from the concept of the lepton-hadron interaction as an interaction between two currents, the leptonic (weak) current and the hadronic (electromagnetic) current. Only the electro-magnetic current had been regarded as a *"charge non-changing"* (or *neutral*) current, the leptonic current being regarded as a *"charge changing"* (or *charged*) current, until the discovery of neutral leptonic currents. The discovery supports the view that weak and electromagnetic interactions are different aspects of the same law of force. See also: *Field theory, unified. Gauge theories.*

NEUTRAL FILTER, OPTICAL. A filter which absorbs light, within a specified range, to the same extent (or very nearly so) for all wavelengths. In a *neutral wedge* the faces of the filter subtend a small angle and the optical density varies linearly with the distance along the wedge so formed.

NEUTRAL POINT. (1) The star point of a three-phase machine, transformer or star-connected set of impedances. (2) The mid-point of a three-wire d.c. system. (3) That point, in each cross-section of a twisted thick beam, at which the elastic strain is zero. See also: *Centre of compression and twist.*

NEUTRAL SURFACE. In elastic bending: a surface of zero stress.

NEUTRETTO. The neutrino associated with the muon.

NEUTRINO. A particle with no charge and essentially zero rest mass, and with spin $\frac{1}{2}$. Two kinds of neutrino are known, one associated with the emission of electrons and the other with that of μ-mesons. In each case there exists an anti-neutrino.

NEUTRON. An elementary particle with no charge and a rest mass of $1\cdot67495 \times 10^{-27}$ kg. It decays to a proton, with a mean life of about 1000 s. It is one of the constituents of the atomic nucleus (together with the proton); and, according to meson theory, is dissociated therein at times into a proton and a negative meson cloud. See also under *Neutrons.*

NEUTRON AGE. See: *Fermi age. Fermi age equation. Fermi age theory.*

NEUTRON CAPTURE. The capture of a neutron by an atomic nucleus. The most common form is *radiative*, i.e. capture followed by the emission of γ-radiation, but α-particles, protons or neutrons may also be emitted. See also: *Capture.*

NEUTRON DIFFRACTION. Diffraction of neutrons analogous to that of X-rays, which arises from the wave aspect of their nature. Diffraction of neutrons by atoms, like that of X-rays or electrons, permits the study of solid state structure with important differences since neutrons are mainly scattered by nuclei and X-rays by electrons. See also: *Electron diffraction. X-ray diffraction.*

NEUTRON DIFFRACTOMETER. See: *Diffractometer.*

NEUTRON DIFFUSION. The migration of neutrons from regions of high neutron density to those of low neutron density, in a medium in which neutron capture is small compared to neutron scattering. An approximate theory for such diffusion is based on the assumption that, in a homogeneous medium, the neutron current density is proportional to the gradient of the neutron flux density. See also: *Neutrons, diffusion area for. Neutrons, diffusion length for. Transport theory.*

NEUTRON DISPERSION, ANOMALOUS. A phenomenon analogous to anomalous X-ray dispersion, in which a nuclear resonance replaces the absorption edge. However, only a few isotopes (^{6}Li, ^{113}Cd, ^{149}Sm, ^{157}Gd) possess resonances in the wavelength range of thermal neutrons (1–2 Å). See also: *X-ray dispersion, anomalous.*

NEUTRON ENERGY GROUP. One of a set of groups consisting of neutrons having energies within arbitrarily chosen intervals. Each group may be assigned effective values for the characteristics of the neutrons within the group, and such groups form the basis of the multigroup theory of neutron transport. See also: *Multigroup theory.*

NEUTRON FLUX. The particle flux density for neutrons. The correct term is *neutron flux density* but the term neutron flux is commonly used. See also: *Particle flux density.*

NEUTRON FLUX DENSITY, ASYMPTOTIC. The neutron flux density at a point far from boundaries, localized sources, and localized absorbers.

NEUTRON GENERATOR. Usually denotes a device for the production of neutrons by the bombardment of a suitable target by projectiles from a particle accelerator or by X-rays generated by such projectiles. Nuclear reactors on the one hand and isotopic sources on the other are therefore excluded. Typical reactions employed are the reaction between deuterium and tritium, which yields an α-particle and a neutron; and that between X-rays and beryllium-9, which yields beryllium-8 and a neutron.

NEUTRON HARDENING. The spectral hardening of neutrons. See also: *Spectral hardening.*

NEUTRON LIFETIME. The mean lifetime between production and absorption of a neutron in a given medium. See also: *Mean life: Average life.*

NEUTRON MONOCHROMATOR. See: *Monochromator.*

NEUTRON MULTIPLICATION. The production of neutrons in a medium containing fissionable material, as a consequence of a neutron chain reaction. See also: *Chain reaction, neutron. Multiplication factor: Multiplication constant.*

NEUTRON (NUMBER) DENSITY. The number of free neutrons per unit volume. Partial densities may be defined for neutrons characterized by such parameters as energy or direction.

NEUTRON OPTICS. The study of those aspects of neutron behaviour which are associated with wave properties, e.g. diffraction, scattering, refraction, reflection, polarization.

NEUTRON POLARIZATION. See: *Polarization.*

NEUTRON RADIOGRAPHY. Radiography by means of neutrons obtained from nuclear reactors, particle accelerators or radioactive sources. Neutron radiography is complementary to X-ray and γ-ray radiography, since neutron absorption varies from element to element in a fashion quite different from that experienced with X- or γ-rays. Thermal neutrons are most commonly used but cold, epithermal and fast neutrons have also been employed. See also: *Radiograph*.

NEUTRON SCATTERING. See: various entries under *Scattering*.

NEUTRON SCATTERING FACTOR. See: *Atomic scattering factor*.

NEUTRONS, COLD. Neutrons with an average kinetic energy less than that of thermal neutrons. They may be produced by passing a beam of thermal neutrons through a solid (e.g. graphite or beryllium) having high scattering power and low absorption.

NEUTRONS, DELAYED. Neutrons emitted by excited fission products formed by beta decay. (The neutron emission itself is prompt, so that the observed delay is due to the preceding β-emission or emissions.)

NEUTRONS, DIFFUSION AREA FOR. One-sixth of the mean square displacement of a thermal neutron in an infinite homogeneous medium from appearance to disappearance.

NEUTRONS, DIFFUSION LENGTH FOR. The square root of the diffusion area.

NEUTRONS, EPICADMIUM. Neutrons of kinetic energy greater than the effective cadmium cut-off.

NEUTRONS, EPITHERMAL. Neutrons of kinetic energy greater than that of thermal agitation. The term is often restricted to energies just above thermal, i.e. energies comparable with those of chemical bonds.

NEUTRONS, FAST. Neutrons of kinetic energy greater than some specified high value. In reactor physics this is frequently taken as 0·1 MeV, but other values may be used in other contexts (e.g. dosimetry). (*Note:* 1 eV ≈ $1·6021 \times 10^{-19}$ J.)

NEUTRONS, FISSION. Neutrons originating in the fission process which have retained their original energy.

NEUTRONS, INTERMEDIATE. Neutrons of kinetic energy between the energies of slow and fast neutrons.

NEUTRONS, MIGRATION AREA FOR. See: *Migration area*.

NEUTRONS, MODERATION OF. See: *Moderating ratio. Moderator*.

NEUTRON SPECTROMETER. A device for isolating neutrons of some prescribed energy from a continuous spectrum of neutrons, for spectroscopic measurements. See also: *Monochromator*.

NEUTRON SPECTROSCOPY. The study of the variation of the number of neutrons with neutron energy under specified conditions.

NEUTRONS, PROMPT. Neutrons accompanying nuclear fission without measurable delay.

NEUTRONS, PULSED. Short bursts of neutrons, usually intense, produced by a specially designed neutron generator or nuclear reactor, or by the use of a chopper.

NEUTRONS, RESONANCE. Neutrons having kinetic energies in the resonance energy range.

NEUTRONS, RESONANCE ABSORPTION OF. Neutron absorption in the resonance energy range.

NEUTRONS, RESONANCE ESCAPE PROBABILITY FOR. The probability that a neutron, when slowing down in an infinite medium, will traverse all, or some specified portion of, the range of resonance energies, without being absorbed. See also: *Escape probability*.

NEUTRONS, SLOW. Neutrons of kinetic energy less than some specified low value. In reactor physics this is frequently taken as 1 eV, in dosimetry the effective cadmium cut-off is used. (*Note:* 1 eV ≈ $1·6022 \times 10^{-19}$ J.)

NEUTRONS, SLOWING DOWN POWER FOR. A measure of the ability of a medium to slow down neutrons. It is the product of the average logarithmic energy decrement and the macroscopic scattering cross-section.

NEUTRONS, SUBCADMIUM. Neutrons of kinetic energy less than that of the effective cadmium cut-off.

NEUTRON STAR. A star composed of neutrons, whose density is comparable with that of an atomic nucleus, and whose existence represents the final stage of the life cycle of an appropriately massive star. It is generally accepted that a pulsar is a rapidly rotating neutron star. See also: *Black hole. Nuclear matter. Pulsar. Stellar evolution. Ylem*.

NEUTRONS, THERMAL. Neutrons in thermal equilibrium with the medium in which they exist.

NEUTRONS, VIRGIN. Neutrons from any source, before they make a collision.

NEUTRON TEMPERATURE. The temperature assigned to a population of neutrons which can be described by a Maxwellian distribution.

NEUTRON YIELD PER ABSORPTION:
η **FACTOR.** The average number of primary fission neutrons (including delayed neutrons) emitted per neutron absorbed by a fissionable nuclide or by a nuclear fuel, as specified. It is a function of the

energy of the absorbed neutrons, and is less than the neutron yield per fission (v). See also: *Nuclear reactor, four-factor formula for.*

NEUTRON YIELD PER FISSION: v FACTOR. The average number of primary fission neutrons (including delayed neutrons) emitted per fission. It is a function of the energy of the absorbed neutrons. See also: *Nuclear reactor, four-factor formula for.*

NEWTON. The unit of force in MKS and SI units. It is the force which gives to 1 kg an acceleration of 1 m/sec², and is equal to 10^5 dyne. It is also equal, within about 2% ,to the gravitational pull on 100 g mass. See: *Appendix II.*

NEWTON LAW OF COOLING. States that the rate at which a body loses heat is proportional to the temperature difference between the body and the surrounding air. The law is true for quite large temperature differences, if the air is flowing past the body in a forced draught.

NEWTON LAW OF GRAVITATION. States that every particle of matter in the Universe attracts every other particle with a force acting along the line joining the particles. For two particles of masses m_1 and m_2, separated by a distance d, the force between the two particles, known as the *gravitational force,* is given by Gm_1m_2/d^2, where G is the *gravitational constant.* Its value is about $6·67 \times 10^{-8}$ dyn cm²/g² ($6·67 \times 10^{-11}$ N m²/kg²). See: *Appendix III.*

NEWTON LAWS OF MOTION. State: (1) That every body continues in a state of rest or uniform motion unless acted upon by a force. (2) That the rate of change of momentum of a body in a given direction is proportional to the resultant force applied to it in that direction. (3) That to every action there is an equal and opposite reaction.

NEWTON RINGS. Interference fringes, in the form of concentric rings, formed when a slightly convex lens is placed on a plane sheet of glass and the point of contact viewed by reflected light. Such rings may also occur whenever optical contact between two surfaces is not quite achieved.

NICOL PRISM. See: *Polarizing prism.*

NILE (NUCLEAR REACTOR TECHNOLOGY). A unit of reactivity numerically equal to 0·01. The smaller unit, the milli-nile, is more convenient for small changes in reactivity, and is the same as a *p.c.m.* See: *Nuclear reactor, reactivity of.* See also: *p.c.m.*

NIPKOW DISK. A mechanical scanning device used in early television. It consisted of a flat circular disk provided with a spiral of small holes. As the disk rotated across a field of view, the whole field was scanned upon each rotation.

NIT. A unit of luminance equal to 1 cd/m². See also: *Luminance.*

NOBLE GAS: INERT GAS: RARE GAS. Any element of Group 0 of the periodic system, namely helium, neon, argon, krypton, xenon and radon. The electron shells are completely filled and hence these gases are practically devoid of chemical properties, but a number of crystalline compounds have been prepared with fluorine, oxygen and certain metals.

NOBLE METAL. Metals which are not readily attacked by chemical reagents and are resistant to atmospheric oxidation. They comprise mercury, silver, gold, platinum, palladium, ruthenium, rhodium, iridium and osmium.

NOCTILUCENT CLOUDS. Tenuous luminous clouds, at a height (about 80 km) far above that of ordinary clouds, that are visible only after sunset in a clear summer sky, when the Sun descends beyond about 6° below the horizon. They glow with a pearly silvery light, often showing a bluish tinge.

NODE. (1) In an atomic orbital: a point at which the value of the wave function is zero. (2) In a system of stationary waves: points of minimum displacement, spaced half a wavelength apart. Points of maximum displacement, midway between the nodes, are termed *anti-nodes.* (3) In an electrical circuit: a terminal of any branch of a network, or a common terminal of two or more branches. (4) Of the orbit of a planet or satellite: the two points, diametrically opposite to each other, at which the orbit cuts the plane of the ecliptic. For an artificial satellite, however, the plane of reference is usually that of the Earth's equator. The node at which a planet's satellite passes from south to north is termed the *ascending node*, the passage from north to south being made at the *descending node.*

NODICAL MONTH. The interval between successive passages of the Moon through a given node (i.e. a point at which the plane of its orbit passes through the plane of the ecliptic). Its length is slightly more than 27·2 mean solar days. This month is also called the *Draconitic month.* See also: *Month.*

NOISE. Any undesired sound. By extension, noise is any unwanted disturbance, such as undesirable electrical fluctuations, spurious signals on a cathode ray tube, and so on. See also: *Electrical noise. Johnson noise. Shot noise: Shot effect. Temperature fluctuation noise.*

NOISE, AERODYNAMIC. Sound generated by a fluctuating aerodynamic flow. It is a by-product of the instability of the flow and is to be distinguished from sound produced by the vibration of solids. An example of aerodynamic noise is afforded by the Aeolian tone. See also: *Aeolian tone.*

NOISE FROM DEFORMED METALS. See: *Acoustic emission from materials.*

NOMOGRAM. A chart consisting of a set of (usually) parallel lines on which are drawn scales corresponding to the variables in a given equation. The chart is designed so that sets of corresponding values for these variables lie on straight lines intersecting all the scales.

NON-ATTACHING GAS. A gas in which electron attachment does not occur.

NON-BONDING ELECTRONS. Electrons, usually in the outer valence shell of an atom, which do not participate in forming bonds with other atoms.

NON-BONDING ORBITALS. Those orbitals of an atom which are essentially unaffected when a bond is formed with another atom.

NON-CROSSING RULE. States that the potential energy curves of diatomic molecules which are of the same electronic species never cross.

NON-DESTRUCTIVE TESTING. The testing of materials and components in such a way that, in a particular context, the material or component being tested will not be damaged to such an extent, or in such a fashion, as to render it incapable of being used for the purpose for which it was originally intended. The techniques employed involve radiological, elastic, electrical, magnetic, optical, thermal, mechanical, atomic, nuclear, chemical and penetrant methods.

NON-DIAGRAM LINES. X-ray emission lines which arise from doubly- or multiply-ionized atoms by a single electron jump, or by a two-electron jump with one emitted quantum, and which do not therefore fit on the Moseley diagram.

NON-EUCLIDEAN GEOMETRY. A type of geometry which postulates, among other things, that more than one line can be drawn through a given point parallel to a given line (*hyperbolic geometry*) or that there are no parallel lines and therefore all lines in a plane will intersect (*elliptic geometry*). A convenient model of elliptic geometry is provided by the surface of a sphere but no such simple model exists for hyperbolic geometry. Both types of geometry are special cases of *projective geometry*, which has to do with any space (a *projective space*)

which avoids the special situations encountered in Euclidean geometry as a result of parallelism, i.e. situations involving the meeting of lines, planes, etc., at infinity.

NON-METAL. An element not possessing free electrons, lacking lustre and ductility, and having low thermal and electrical conductivity. See also: *Metal. Metalloid.*

NON-NEWTONIAN LIQUIDS. Liquids which show a decrease in viscosity as their rate of flow or velocity gradient increases. Such liquids are said to exhibit *anomalous viscosity.*

NON-POLAR MOLECULES. Molecules having no resultant dipole moment.

NON-STATIONARY DATA. See: *Random data.*

NORMAL DISTRIBUTION: GAUSSIAN DISTRIBUTION. The most important theoretical distribution in statistics. It is that which gives, among other things, the distribution of errors of observation. It is described by the equation

$$p(x) = \frac{1}{\sigma \sqrt{(2\pi)}} \exp -\left(\frac{(x-\mu)^2}{2\sigma^2}\right),$$

where $p(x)\,dx$ gives the probability that x lies between x and $x + dx$, μ is the mean of the data and σ the standard deviation. The probability that a randomly selected value of x shall have a value greater than X is given by

$$\int_X^\infty p(x)\,dx = \frac{1}{\sqrt{(2\pi)}} \int_u^\infty \exp\left(-\frac{1}{2}u^2\right)\,du$$

where $u = (X-\mu)/\sigma$. See also: *Error function. Frequency distribution.*

NORMALIZING HEAT TREATMENT. The elimination of internal stresses in steel and the refinement of its crystal grains by heating above a specified temperature and subsequent cooling in air.

NORMAL SURFACE. One of a number of geometrical constructions for the determination of propagation velocities and vibration directions in doubly-refracting crystals. It shows the distribution of phase velocities. See also: *Optical indicatrix: Fletcher indicatrix. Ray ellipsoid: Fresnel ellipsoid. Ray surface.*

NORMAL TEMPERATURE AND PRESSURE (N.T.P.). Refers to a temperature of 0°C and a pressure of 760 mm of mercury. (*Note:* a pressure of 1 mm mercury is equal to 133·322 N/m².) See also: *Atmosphere, standard.*

NORMAL VIBRATION. Of a molecule: an internal molecular oscillation in which all the atoms execute simple harmonic motion, are moving in phase, and have the same frequency. The number of distinct normal vibrations for a molecule with N atoms is $3N - 6$, or $3N - 5$ for a linear molecule.

NORTH POLAR DISTANCE. Of a celestial body: is given by $90° - \delta$, where δ is the declination (the latitude of the body on the celestial sphere), and is counted as positive for north declinations and negative for south.

NOVA. A star which, after existing as a faint star, "erupts" and becomes extremely bright, perhaps by a factor of many thousands, and then relapses into relative faintness. It is a radio star, the radio intensity long outlasting the optical. No wholly satisfactory explanation is as yet available. See also: *Radio stars. Stars, spectral classes of. Supernova.*

N-STAR. See: *Stars, spectral classes of.*

N.T.P. See: *Normal temperature and pressure (N.T.P.).*

NUCLEAR ASTROPHYSICS. The study of the contribution of nuclear reactions to the power output and total energy of individual stars.

NUCLEAR BATTERY. A small isotope-powered thermoelectric battery which uses heat from Pu-238. It may be employed as a heart pacemaker.

NUCLEAR CHARGE. The charge on an atomic nucleus. It is equal to the product of the atomic number and the charge on the electron.

NUCLEAR CHARGE, EFFECTIVE. The apparent charge of a nucleus, which is less than that appropriate to the atomic number on account of screening by the space charge of the inner electrons.

NUCLEAR CHEMISTRY. That part of chemistry which deals with the study of nuclei and nuclear reactions using chemical methods. The term is also used to cover all the chemical aspects of nuclear science.

NUCLEAR DISINTEGRATION. The transformation of a nucleus, possibly a compound nucleus, involving a splitting into more nuclei or the emission of particles. If the disintegration is spontaneous it is said to be *radioactive*. See also: *Radioactive decay: Radioactive disintegration. Radioactive decay constant: Radioactive disintegration constant.*

NUCLEAR EMULSION. A photographic emulsion capable of recording the passage through it (the *tracks*) of individual charged particles.

NUCLEAR EMULSION, GRAIN DENSITY OF. The number of grains per unit length along the track of a particle in a nuclear emulsion.

NUCLEAR ENERGY. Energy released in nuclear reactions or transitions, often called *atomic energy*.

NUCLEAR ENERGY LEVEL. One of the energy values at which a nucleus can exist for an appreciable time ($>10^{-22}$ s). The lowest level (the *ground state*) has a sharply defined value, but the others have a spread of values—the level *width*, which is inversely proportional to the mean life of the level. *Partial widths* may be assigned (e.g. neutron width, gamma width) which are proportional to the respective transition probabilities. See also: *Energy level width: Energy level breadth.*

NUCLEAR EXPLOSIVES. Bombs, missiles or other devices in which explosive power is derived from nuclear energy, i.e. from nuclear fission or nuclear fusion. See also: *Nuclear fission. Thermonuclear reaction.*

NUCLEAR FERROMAGNETISM. Magnetism associated with dipole or exchange coupling between the spin of nucleons (spin–spin coupling). The spins tend to become parallel as in normal ferromagnetism, but the effect is much smaller. See also: *Nuclear paramagnetism.*

NUCLEAR FISSION. The division of a heavy atomic nucleus into two parts with masses of equal order of magnitude, usually accompanied by the emission of neutrons, gamma-radiation and, infrequently, small charged nuclear fragments. See also: *Fissile. Fissionable. Nucleon fission, ternary.* Entries under *Fission.*

NUCLEAR FISSION, SPONTANEOUS. Nuclear fission which occurs without the need for the addition of particles or energy to the nucleus.

NUCLEAR FISSION, TERNARY. Nuclear fission in which three fragments of comparable mass are produced. It is extremely rare and the evidence for it is scanty.

NUCLEAR FISSION, THERMAL. Nuclear fission caused by thermal neutrons.

NUCLEAR FORCES. The forces acting between nucleons at close quarters, but excluding electromagnetic interactions. Such forces hold the nucleons together in atomic nuclei and are believed to involve exchange forces. While a wholly satisfactory theory of nucleon–nucleon forces is not yet available, the meson theory of Yukawa has proved very successful. See also: *Charge independence. Exchange. Meson theory of nuclear forces.*

NUCLEAR FUSION. See: *Thermonuclear reaction.*

NUCLEAR INDUCTION. Another name for nuclear magnetic resonance.

NUCLEAR MAGNETIC RESONANCE (NMR).
The resonant absorption of electromagnetic energy by a system of atomic nuclei situated in a magnetic field. It is a branch of radiofrequency spectroscopy. The frequency of the magnetic resonance coincides with the frequency of the Larmor precession of the nuclei in the magnetic field and is proportional to the strength of the field. Typical values of this frequency lie in the range 1–100 MHz. The phenomenon is confined to nuclei having a magnetic moment (i.e. with non-zero spin) and the resonances of such nuclei may be detected as they occur in ordinary bulk matter. Nuclear magnetic resonance may be used for the measurement of nuclear magnetic moments; for the precise measurement of magnetic fields; for obtaining information regarding the environment of nuclei in molecules, liquids and solids; and in chemical analysis. See also: *Spectra, radiofrequency.*

NUCLEAR MAGNETON. The unit of nuclear magnetic moment, given by $eh/4\pi mc$, where e is the electronic charge, h is Planck's constant, m is the rest mass of a proton, and c is the speed of light. Its value is about $5\cdot05 \times 10^{-24}$ erg/gauss (or $5\cdot05 \times 10^{-27}$ J/T), i.e. about 1/1840 that of the Bohr magneton. See also: *Bohr magneton.*

NUCLEAR MATTER. The matter of which the atomic nucleus is composed. The term may also refer to matter of which the density is comparable with that of the nucleus, such as is believed to occur in some stars. A further use of the term in the legislation of some countries is to denote fissionable, radioactive, and similar materials. See also: *Neutron star. Ylem.*

NUCLEAR MICROPROBE ANALYSIS. A technique, analogous to electron-probe microanalysis, for the identification and estimation of the component elements of a selected micro-volume. Nuclear particles or X-rays are used, produced by nuclear reactions, and a spatial resolution of 4 μm is comparatively easy to achieve. See also: *Electron-probe microanalysis.*

NUCLEAR MODELS. Descriptions of atomic nuclei based on a variety of simplifying assumptions. Many models have been proposed, all of which describe some phenomena but fail to account for others. The best known are (a) the *liquid drop model*, an early model now mainly of historical interest, in which nucleon motion was treated like that of molecules in a liquid; (b) the *shell, quasi-atomic, independent particle* or *Hartree–Fock model*, in which the motions of individual nucleons are assumed to be essentially uncorrelated with each other but occupy energy "shells" analogous to the electron shells in an atom, and which explains the observed "magic" numbers as shell closures; (c) the *unified model*, which is a synthesis of the liquid drop and shell models; (d) the *Fermi gas model*, which is equivalent to the quasi-atomic model but assumes so many particles that surface effects are of secondary importance; and (e) the generalized *Hartree Fock model*, in which an "effective" internucleon potential is introduced to make some correlation between the motions of individual nucleons. See also: *Magic numbers in nuclei. One-particle model. Optical model of particle scattering.*

NUCLEAR MOMENTS. The angular momentum, nuclear magnetic dipole moment, nuclear electric quadrupole moment, and nuclear magnetic octupole moment. See also: *Angular momentum in atoms and nuclei. Dipole moment. Multipole moments.*

NUCLEAR PARAMAGNETISM. The magnetism arising from nuclei with non-zero spin and hence having a magnetic dipole moment. An assembly of such nuclei is paramagnetic but the paramagnetism is so small that it can be observed directly only in diamagnetic substances. See also: *Nuclear ferromagnetism.*

NUCLEAR PHYSICS. The study of the physical properties of atomic nuclei and their mathematical treatment.

NUCLEAR POISON. A substance which, because of its high neutron absorption, reduces the reactivity of a nuclear reactor.

NUCLEAR POLARIZATION. See: *Polarization.*

NUCLEAR POTENTIAL. The potential energy of some specified particle (e.g. proton, neutron) as a function of its distance from the atomic nucleus.

NUCLEAR POTENTIAL WELL. A pictorial description of the interaction between a nucleus and a nucleon (or a small group of nucleons as in a deuteron or α-particle). The potential is constant and negative up to a certain distance from the nucleus, beyond which it is zero—hence the concept of a well.

NUCLEAR QUADRUPOLE RESONANCE. The resonance interaction between a nuclear electric quadrupole moment and a molecular electric field gradient. Transitions between the associated energy levels give rise to *nuclear quadrupole resonance spectra* with frequencies 1–1000 MHz. See also: *Spectra, radiofrequency.*

NUCLEAR RADIUS. The distance from the centre of an atomic nucleus at which the density of nuclear matter drops sharply. It is roughly equal to $r_0 A^{1/3}$, where A is the atomic weight of the nucleus and r_0 is a "constant" ranging from $1\cdot4 \times 10^{-13}$ cm for the lightest nuclei to $1\cdot2 \times 10^{-13}$ cm for the heaviest.

NUCLEAR REACTION. Any reaction involving a change in the balance of the constituent nucleons of an atomic nucleus or in the energy state of the nucleus. If the Q-value is negative (i.e. if there is a net loss of kinetic energy) the reaction is *endoergic*, and if the Q-value is positive it is *exoergic*.

NUCLEAR REACTIONS, NOTATION FOR. An abbreviated notation indicating the nature of the incident and resultant particles in nuclear reactions. Such a reaction may be described, for example, as a (n, p) reaction, an (α, d) reaction, a (γ, n) reaction and so on, where n, p, α, d and γ refer to a neutron, a proton, an α-particle, a deuteron and a γ-ray photon respectively. The notation is also used to describe a reaction in more detail. Thus, the reaction $^9\text{Be} + \alpha \rightarrow {}^{12}\text{C} + n$, may be written $^9\text{Be}(\alpha, n) {}^{12}\text{C}$

NUCLEAR REACTOR. An assembly in which a self-sustaining neutron chain reaction can be maintained and controlled. See also: *Chain reaction. Chain reaction, neutron. Nuclear reactor types.*

NUCLEAR REACTOR BLANKET. A region of fertile material placed around or within a nuclear reactor core for the purpose of conversion. By extension the term may be used when the purpose is not conversion but the transformation of non-fertile material. See also: *Nuclear reactor, conversion in.*

NUCLEAR REACTOR, BREEDING IN. Denotes conversion when the conversion ratio is greater than unity. See also: entries under *Breeding.*

NUCLEAR REACTOR CONTROL. The intentional variation of the reactivity of a nuclear reactor to maintain a desired state of operation. It may be achieved by one or more of the following: *absorption control*, in which the neutron absorption is varied by adjusting the properties, position or quantity of neutron-absorbing material (excluding the fuel, moderator and reflector material); *configuration control*, in which the configuration of the fuel, reflector, coolant or moderator is varied; *fluid poison control*, in which the position or quantity of a fluid nuclear poison is adjusted; *fuel control*, in which adjustment is made of the properties, position, or quantity of fuel; *moderator control*, in which adjustment is made of the properties, position or quantity of the moderator; *reflector control*, in which adjustment is made of the properties, position or quantity of the reflector; and *spectral shift control*, in which the neutron spectrum is intentionally changed (this is a special type of moderator control).

NUCLEAR REACTOR, CONTROL MEMBER OR CONTROL ELEMENT FOR. An adjustable part of a nuclear reactor used for reactor control. Such a control member often takes the form of a *rod*.

NUCLEAR REACTOR, CONVERSION IN. The nuclear transformation of a fertile substance into a fissile substance. The *conversion ratio* (or *conversion factor*) is the ratio of the number of fissile nuclei produced by conversion to the number of fertile nuclei destroyed.

NUCLEAR REACTOR, CORE OF. That region of a nuclear reactor which contains the fuel elements and in which a chain reaction can take place. It is normally bounded by the reflector or blanket.

NUCLEAR REACTOR, FLATTENING OF. The achievement of an approximately uniform flux density in the core.

NUCLEAR REACTOR, FOUR-FACTOR FORMULA FOR. A formula used to calculate the infinite multiplication factor, k_∞, of a given thermal reactor. It is $k_\infty = \eta \varepsilon p f$, where η is the neutron yield per absorption, ε is the fast fission factor, p is the resonance escape probability, and f is the thermal utilization factor. See also: *Fast fission factor. Multiplication factor: Multiplication constant. Neutron yield per absorption: η factor. Neutrons, resonance escape probability for.*

NUCLEAR REACTOR FUEL. Material containing fissionable nuclides which, when placed in a nuclear reactor, enables a self-sustaining neutron chain reaction to be achieved. When the fuel is sealed in a *can*, or protected by the application of an external layer of material (*cladding*), it is known as a *fuel element*. A group of such elements which remains intact during the charging and discharging of a reactor with fuel is termed a *fuel assembly*.

NUCLEAR REACTOR FUEL, ENRICHED. Fuel containing uranium which has been enriched in one or more of its fissile isotopes, or to which chemically different fissile nuclides (e.g. plutonium) have been added. See also: *Fuel spike: Fuel seed.*

NUCLEAR REACTOR LATTICE. An array of fuel and other materials arranged according to a regular pattern.

NUCLEAR REACTOR, NEUTRON CYCLE IN. The series of events experienced by neutrons, beginning with fission and continuing until the neutrons have leaked out of the reactor or been absorbed.

NUCLEAR REACTOR, NEUTRON ECONOMY IN. The detailed account of neutrons produced and lost. The term is also used qualitatively to refer to the extent to which neutrons are used in desired ways instead of being lost by leakage or useless absorption.

NUCLEAR REACTOR OSCILLATOR. A device which produces periodic variations of reactivity by the oscillatory movement of a sample. It is used for measuring reactor properties, or nuclear cross-sections of the sample. The device was formerly known as a pile oscillator.

NUCLEAR REACTOR PERIOD. See: *Nuclear reactor time constant.*

NUCLEAR REACTOR POWER COEFFICIENT. The rate of change of the reactivity with respect to the thermal power of the reactor. In a homogeneous reactor it is the same as the temperature coefficient; but in a heterogeneous reactor the power and temperature coefficients are different, owing to the different temperatures of the various components.

NUCLEAR REACTOR POWER DENSITY. The power generated per unit volume of a nuclear reactor core.

NUCLEAR REACTOR, REACTIVITY OF. A parameter giving the deviation from criticality of a neutron chain-reacting medium (and hence of a nuclear reactor), positive values of which correspond to a supercritical state and negative values to a subcritical state. Quantitatively the reactivity, ϱ, is given by $\varrho = 1 - 1/k_{eff}$, where k_{eff} is the effective multiplication factor. It is expressed in different countries in terms of many different units, such as *dollar, cent, inhour, nile* and *p.c.m.*

NUCLEAR REACTOR, REFLECTOR FOR. Part of a reactor placed adjacent to the core so as to scatter some of the escaping neutrons back into it.

NUCLEAR REACTOR, SEED OR SPIKE IN. See: *Fuel spike: Fuel seed.*

NUCLEAR REACTOR SHIELD. (1) *Biological shield:* material which reduces ionizing radiation in a given region to biologically permissible levels. (2) *Thermal shield:* material which prevents the biological shield from becoming overheated, by absorbing β-, γ- and X-rays the heat from which could otherwise damage that shield.

NUCLEAR REACTOR, SPECIFIC POWER OF The power generated per unit mass of fuel in the reactor core.

NUCLEAR REACTOR TEMPERATURE COEFFICIENT. The rate of change of the reactivity with respect to the temperature of the reactor. The coefficient may also refer to the temperature of some specified location or component of the reactor. See also: *Nuclear reactor power coefficient.*

NUCLEAR REACTOR, THERMAL COLUMN OF. A large body of moderator, adjacent to or inside a reactor, used to provide thermal neutrons for experiment or measurement.

NUCLEAR REACTOR TIME CONSTANT. The time required for the neutron flux density in the reactor to rise or fall by a factor *e*. It is also known as the reactor *period.*

NUCLEAR REACTOR TYPES. May be specified in terms of one or more of the following: nuclear design, engineering design, and purpose. (1) *Nuclear design:* the reactor may be described as *heterogene-ous*, in which the fuel and moderator are separate; or *homogeneous,* in which the fuel and moderator present an effectively homogeneous medium to the neutrons. It may be a *bare* reactor, having no reflector; or a *breeder,* one which produces more fissile material than it consumes. Or it may be described in terms of the fuel (*natural uranium, enriched uranium,* or *plutonium*), the nature of the neutrons mainly responsible for fission, e.g. *fast, intermediate, thermal*; or the control system, e.g. *spectral shift.* (2) *Engineering design:* the reactor may be described in terms of the coolant, e.g. *water-cooled* (perhaps *pressurized* or *boiling*), *heavy-water-cooled, gas-cooled,* liquid-metal (e.g. sodium) cooled; or of the moderator, e.g. *water-moderated, heavy-water-moderated, graphite-moderated.* It may be described in terms of the form or arrangement of the fuel, e.g. a *ceramic* reactor, one whose fuel is a ceramic, being an oxide, carbide or nitride of a fissile metal; a *pebble bed* reactor, in which some or all of the fuel (and sometimes the fissile material and moderator) is in the form of a stationary bed of small balls or "pebbles"; or a *fluidized bed* reactor, in which fine fuel particles are maintained in a state of non-circulating suspension during reactor operation by the flow of the fluid coolant. Or it may be described in terms of its mobility. Thus it may be *transportable,* capable of being moved when not critical and possibly partly dismantled; *mobile,* designed to be mobile during operation; or *packaged,* a compact power reactor specially designed to simplify shipping and assembly. Well-known examples in the category of engineering design are the *pool* or *swimming pool* reactor, the fuel elements of which are immersed in a pool of water which serves as moderator, coolant, and biological shield; and the *integral* reactor, in which the heat exchanger between the primary and secondary coolant circuits is contained within the reactor vessel. (3) *Purpose:* this is often self-evident, as in an *experimental* reactor, *breeder* reactor, *research* reactor, *training* reactor, *power* reactor (including *propulsion*), and *production* reactor (usually for plutonium production). Not quite so obvious is a *zero-energy* or *zero-power* reactor, which is designed to be used at a power so low that no cooling system is needed. One reactor which does not quite conform to the above scheme is the *Magnox* reactor which is, in fact, a first-generation British power reactor taking its name from the "magnox" (an alloy of aluminium and magnesium) used as the canning material for the fuel elements.

NUCLEAR REACTOR VESSEL. The principal vessel surrounding the reactor core.

NUCLEAR RELAXATION. (1) *Nuclear spin lattice relaxation:* the process through which the nuclear spin system achieves thermal equilibrium with a crystal lattice. (2) *Nuclear spin–spin relax-*

ation: the tendency of an assembly of nuclei, initially precessing in phase about a uniform magnetic field, to lose phase coherence. See also: *Relaxation phenomenon.*

NUCLEAR RESONANCE ENERGY. The kinetic energy of an incident particle (expressed in the laboratory system) that excites an energy level in a compound nucleus.

NUCLEAR STATISTICS. See: *Bose–Einstein statistics. Fermi–Dirac statistics.*

NUCLEAR TRANSITION. Of a nuclide: a change in nuclear configuration. It may involve transformation to a different nuclide or a change in energy level accompanied by the emission of γ-rays. The latter is known as a *radiative transition* and has been termed *electric* or *magnetic* according as electric or magnetic multipoles are involved. See also: *Transition, allowed. Transition, forbidden.*

NUCLEATE BOILING. See: *Heat transfer, boiling.*

NUCLEATION AND GROWTH. In recrystallization: a type of transition in which domains of new crystallographic orientation nucleate and grow at the expense of the parent crystals without accompanying changes in composition. See also: *Metallic transformations.*

NUCLEIC ACIDS. Polymer molecules which, like the proteins, are examples of large molecules found in biological cells. They are of two kinds, deoxyribonucleic acid (DNA), which stores the genetic information in chromosomes, and ribonucleic acid (RNA). Their structures have been deduced from X-ray diffraction observations.

NUCLEON. A proton or a neutron, especially when regarded as a constituent of the atomic nucleus. See also: *Nucleor.*

NUCLEOPHIL. A reagent which, in a reaction involving an aromatic compound, preferentially approaches positions at which there is the smallest concentration of π electrons. See also: *Electrophil.*

NUCLEOR. The core of a nucleon, on the theory that a nucleon is composed of a nucleor surrounded by a pion cloud.

NUCLEOTIDE. Any one of four bases (adenine, thymine, guanine and cytosine) of which desoxyribonucleic acid (DNA) may be composed. See also: *Genetic code.*

NUCLEUS. (1) Of an atom: the positively charged central portion of the atom, of extremely small radius (about 10^{-13} cm) but associated with almost the whole atomic mass. It is composed of protons and neutrons. (2) Of a comet: See: *Comet.* (3) Of a biological cell: a central corpuscle, lying in the protoplasm and contained within a thin membrane. It may be globular, elongated, or flattened in shape.

NUCLIDE. A species of atom characterized by its mass number, atomic number, and nuclear energy state.

NUCLIDES, CHART OF. A partly graphical representation of the relationships between the nuclides and of the properties of the nuclides. It is sometimes known as an *isotope chart.*

NUMBERS, THEORY OF. The study of numbers as such. Examples are: the subdivision and distribution of prime numbers: the study of equations that must be satisfied in integers only (*diophantics*); the decomposition and factoring of numbers; *Fibonacci sequences* (in which each term after the second is the sum of the two preceding terms); and many others. Euclid's proof of the existence of an infinitude of primes is usually taken as the natural starting point of the theory of numbers. Although academic in origin the theory of numbers has found many applications in physics, e.g. in Einstein–Bose and Fermi–Dirac statistics. See also: *Fermat last theorem.*

NUMERICAL ANALYSIS. That branch of mathematical analysis which deals with the conversion of mathematical processes into operations with numbers. Its aim, typically, is to provide a method of computation that will give, in the most economical manner, numerical answers to a physical problem expressed in mathematical terms.

NUMERICALLY-CONTROLLED MACHINE TOOLS. Machine tools in which information that automatically controls the position of the cutting tool at each stage of the machining operation is fed to the machine in numerical form. The information is usually conveyed in the same way that numbers are fed into a digital computer, i.e. by punched holes in paper tape, or magnetic signals on magnetic tape; but it can also be fed in manually.

NUSSELT NUMBER. A dimensionless quantity appearing in the semi-empirical formula for the loss of heat by convection from solid surfaces. It expresses the increase of heat transfer due to the motion of a fluid.

NUTATION. The oscillation of the axis of a rotating body about its mean position as in a spinning top or, more particularly, the axis of the Earth. See also: *Earth, motions of.*

NYQUIST FORMULA. A formula for applying the theory of the Brownian motion to the problem of the "noise" in electrical networks that arises from the random movement of electrons. See also: *Brownian motion.*

NYQUIST THEOREM. For a feedback system: a theorem concerning the relationships between the parameters and dynamic behaviour of the system, originally propounded in relation to feedback amplifiers. It offers a method of testing for stability and assessing the degree of damping.

O

OBJECTIVE. That part of an optical system which lies nearest to the object viewed, and on which the efficiency of the system primarily depends. It may be in the form of a mirror or a lens, depending on the instrument, and is designed to provide corrections for the aberrations appropriate to the instrument concerned, e.g. chromatic and spherical aberrations, coma, astigmatism.

OBJECT SPACE. Of an optical system: the space which light rays traverse before passing through the refracting system. See also: *Image space.*

OBLIQUE ASCENSION. Of a celestial body: the longitude of the body on the celestial sphere, measured eastwards from the *first point of Aries* (i.e. the point at which the ecliptic crosses the celestial equator at the vernal equinox) along the ecliptic. It is also known as the celestial or ecliptic longitude. See also: *Longitude, celestial: Longitude ecliptic. Right ascension.*

OBSERVATORY. (1) *Astronomical:* an installation for observing and recording the positions, motions, brightness, etc., of celestial bodies. (2) *Magnetic:* an installation for measuring the elements of the Earth's magnetic field and monitoring their changes.

OCCLUSION. (1) The retention of a gas or liquid by a solid mass or on the surface of solid particles. (2) The meteorological condition arising when a cold front overtakes a warm front. See also: *Front (meteorology).*

OCCULTATION. The interposition of a celestial body between an observer and another celestial body, such as the hiding of a star or planet by the Moon (*lunar occultation*), or the satellites of a planet by the planet itself.

OCCUPATION NUMBER, MEAN. Of a quantum state: the mean number of particles in that state.

OCEANOGRAPHY. The study of the hydrodynamics of the ocean, and the evolution and structure of the ocean basin.

OCTET RULE. States that atoms combine to form molecules in such a way as to give each atom an outer shell of eight electrons, i.e. an *octet*. The octet rule is only valid for non-transition elements, and not always then, but is useful in rationalizing the Periodic Table.

OCTUPOLE. One of the multipoles which represent the effects, at points outside the system, of the electric and magnetic fields produced by a system of charges and currents such as those arising from the electrons in an atom or from the protons in a nucleus; and which are assumed to be located at the centre of the system. In its simplest form an electric octupole may be considered as an array of eight equal charges positioned at the corners of a parallelepiped, alternate charges being positive and negative. See also: *Multipole moments.*

OCTUPOLE MOMENT. See: *Multipole moments.*

OERSTED. The CGS unit of magnetic field strength. It is equal to 79·6 A/m. See also: *Magnetic intensity: Magnetic field strength. Appendix II.*

OHM. The MKSA and SI unit of electrical resistance. It is defined as the resistance between two points of a conductor when a constant difference of potential of 1 V, applied between these two points, produces in the conductor a current of 1 A, the conductor not being the source of any electromotive force. See also: *Ampere. Appendix II.*

OHM, ACOUSTIC. See: *Acoustic impedance.*

OHM, INTERNATIONAL. The former standard of electrical resistance, defined as the resistance offered to an unvarying electric current by a column of mercury at the temperature of melting ice, 14·4521 g in mass, of a constant cross-sectional area, and of a length of 106·300 cm. One international ohm is equal to 1·00049 absolute ohms (abohms). See also: *Abohm.*

OHM LAW. Is usually written as $I = E/R$, where I is the steady current in a metallic circuit, E is the constant electromotive force operating in the circuit, and R is the electrical resistance of the circuit.

OHM, MECHANICAL. See: *Mechanical impedance.*

OHMMETER. An instrument that gives the electrical resistance of a component by a direct reading.

OIL-DROP EXPERIMENT. The experiment, carried out by Millikan, by which the charge on the electron was determined for the first time. It was based on the measurement of the rate of fall of a charged oil drop in an electric field.

OKLO PHENOMENON. Refers to the operation of a natural nuclear reactor in a uranium deposit near Oklo, in Gabon (Western Africa), more than 1700 million years ago. The reactor, sometimes referred to as a *fossil* reactor, is believed to have operated for at least 100 000 years.

OLBERS PARADOX. The fact that the sky is dark at night. If the Universe is uniformly populated with stars which have always been shining, the night sky should be as bright as day. The paradox is resolved by the consideration (a) that the stars have not been shining for an infinite time, and (b) that, owing to the red shift of the light from the distant galaxies, its energy is reduced. These two factors are sufficient to account for the observed value of the light intensity of the night sky.

OMEGA. A form of accurate radar navigation system employing very low frequencies whose corresponding wave lengths are comparable with the height of the ionosphere, enabling operation over distances of hundreds of miles, since the space between sea and ionosphere behaves as a waveguide. As in earlier pulse-modulated systems a reference-grid is provided for the navigator but in the case of Omega a world-wide chain of sites is in operation instead of the usual chain of three. See also: *Navigation*.

OMEGA PARTICLE. A negative hyperon, usually denoted by Ω^-, which was predicted by considerations of particle symmetry before its actual discovery. Its anti-particle has now also been discovered. See also: *Hyperon. Particle symmetry.*

ONE-GROUP THEORY. A theory of neutron transport in which it is assumed that all the neutrons belong to the same energy group. See also: *Multigroup theory. Neutron energy group.*

ONE-PARTICLE MODEL. Of the atomic nucleus: a model in which the nuclear properties are attributed to the effect of one nucleon. See also: *Schmidt lines.*

ONSAGER FORMULA. An equation relating the dielectric constant of a liquid or gas to the high-frequency dielectric constant and the permanent dipole moment of the molecules.

ONSAGER RECIPROCITY THEOREM. A theorem concerning the behaviour of unbalanced systems. It includes, but is not limited to, the behaviour of thermodynamic irreversible systems.

OPACITY. The reciprocal of the transmission factor.

OPALESCENCE. (1) The interference colours observed with certain minerals such as opal, which arise from the existence of very thin surface films. (2) The iridescent appearance of a solution arising from the reflection of light from suspended particles.

OPEN CIRCUIT. One which does not provide a complete path for an electric current.

OPEN CLUSTER: GALACTIC CLUSTER. See: *Star cluster.*

OPEN CYCLE. Of a heat engine: a cycle of operations or in which the heat transfer fluid is used only once and then replaced by fresh fluid, instead of being recirculated.

OPERA GLASS. Binoculars in which two Galilean telescopes are employed, with correspondingly small magnifications. No real image is formed so that the image is not inverted and no prism system is required to produce an erect image. See also: *Binoculars.*

OPERATOR. A symbol representing the performance of an operation such as addition, extraction of a square root, differentiation and so on. Examples are $\dfrac{d}{dx}$ (the differential operator) and $\dfrac{\partial}{\partial r}$ or ∇ (the differential vector operator). See also: *Differential vector operator.*

OPHTHALMOSCOPE. An optical instrument for the examination of those parts of the eye (e.g. the retina) lying behind the crystalline lens.

OPPENHEIMER–PHILLIPS PROCESS. The capture of the neutron from a deuteron by an atomic nucleus without the deuteron entering that nucleus. It is a particular case of stripping.

OPPOSITION. Of a celestial body: the instant when the body is in line with the Earth and the Sun.

OPTIC–ACOUSTIC EFFECT: TYNDALL–RÖNTGEN EFFECT. The production of periodic pressure fluctuations and sound emission from a radiation-absorbing gas or vapour exposed to periodically interrupted thermal radiation. It arises from the successive heating and cooling of the gas or vapour.

OPTICAL ABSORPTION COEFFICIENT. See: *Light, absorption coefficient for.*

OPTICAL ACTIVITY. The property of rotating the plane of polarization of light. It may arise from the asymmetric arrangement of molecules in a crystal or of atoms in a molecule.

OPTICAL COMMUNICATION SYSTEMS. Systems by which signals are transmitted over short and long distances by optical means rather than electrical or electromagnetic. The most promising systems employ glass fibres to act as optical waveguides and the development of such systems has led to rapid developments in the field of fibre optics. See also: *Fibre optics.*

OPTICAL CONSTANTS. The refractive index, n, and the absorption coefficient, α, which together determine the complex refractive index $n - i\alpha$ of an absorbing medium.

OPTICAL DATA PROCESSING. The processing of data of electrical origin by optical means in circumstances where the ability to process in two dimensions, and the capability of achieving high packing densities or high speed of access are advantageous. It does not replace electronic processing for normal use. An important application is to the processing of radar and sonar signals. See also: *Optical information processing*

OPTICAL DENSITY. For light transmitted by an absorbing medium the optical density is defined as $\log_{10} I_0/I$, where I_0 is the incident and I the transmitted intensity. It is synonymous with the optical *extinction.* For partly scattering media such as photographic negatives terms such as *specular* or *diffuse density* are used to explain how I and I_0 are measured. See also: *Density, diffuse. Density, specular. Photographic density.*

OPTICAL DEPTH: OPTICAL DISTANCE. See: *Optical thickness.*

OPTICAL DIFFRACTOMETER. A device designed for observing the optical diffraction patterns of "masks" representing portions of crystal structures, used in the study of complicated structures. The device produces an optical image of the reciprocal lattice of the crystal structure under consideration.

OPTICAL EXTINCTION COEFFICIENT. The extinction per unit path. See also: *Light, extinction coefficient for.*

OPTICAL FLAT. A surface used in engineering metrology for the measurement of the flatness of lapped surfaces, etc. It consists of a plate of glass, fused silica or quartz with one or both surfaces worked and polished flat to within about 10^{-6} cm. The parallelism between the faces, where two faces are involved, is of the same order of accuracy. See also: *Flatness.*

OPTICAL GLASS. Glass which exhibits complete homogeneity throughout its mass.

OPTICAL INDICATRIX: FLETCHER INDICATRIX. A geometrical construction for the determination of propagation velocities and vibration directions in a doubly-refracting crystal. It takes the form of an ellipsoid, the semi-axes of which are proportional to the principal indices of refraction, and the directions of which are parallel to the corresponding vibration directions. The required velocities and vibration directions can be obtained from the axes of the ellipse passing through the centre of the ellipsoid and at right angles to the direction of the wave normal. See also: *Normal surface. Ray ellipsoid: Fresnel ellipsoid. Ray surface.*

OPTICAL INFORMATION PROCESSING. The function performed by any system which processes data in a parallel rather than a serial fashion, and which uses light at some stage as a carrier of the data. The initial data may be in the form of light or a transformation into light may take place later. A simple example of a parallel system is a photographic transparency. Two of these systems may be multiplied by superposing them and illuminating them by a suitably collimated beam, the transmitted intensity distribution then containing the required data for a set of parallel products. See also: *Data processing. Optical data processing.*

OPTICAL LENGTH. The length of path traversed by a ray of light. See also: *Fermat law: Fermat principle.*

OPTICAL LEVER. A device for the measurement of small relative displacements of two objects, or of two parts of the same object, by means of the angular displacement of a beam of light.

OPTICAL MICROMETER. A device in which a subdivided linear or circular scale may be viewed through an optical instrument such as a telescope, microscope, or projector.

OPTICAL MODEL OF PARTICLE SCATTERING. An optical analogy designed to describe the interaction of incident particles with nuclei. The nucleus is considered as a semitransparent sphere with a refractive index and absorption coefficient. The refracted and incident waves then interfere to produce the phenomena observed experimentally. The usual (real) nuclear potential well is replaced by a complex one. Alternative names: *Complex potential model. Cloudy crystal ball model.* See also: *Acoustic model of particle scattering.*

OPTICAL PUMPING. The process of using optical radiation to excite phase-coherent radiation by the maser action. See also: *Laser. Maser.*

OPTICAL ROTATION. The rotation of the plane of polarization of a beam of light by an optically active substance. See also: *Optical activity. Polarization, rotatory.*

OPTICAL TEST PLATE. A polished master gauge which is used to determine the accuracy of an optically polished surface by measurement of the interference fringes between the test plate and the surface.

OPTICAL THICKNESS. Of a given thickness of a transparent medium: the distance that light would travel in a vacuum in the same time that it takes to travel through the given thickness of the medium. It is the geometrical thickness of the medium multiplied by the refractive index, and is also known as *optical depth* and *optical distance.*

OPTICAL WEDGE. A wedge-shaped neutral density filter.

OPTIC AXIS. One of the directions of propagation, in an optically anisotropic crystal, in which no double refraction occurs. See also: *Biaxial crystal. Uniaxial crystal.*

OPTICS. Originally denoted the scientific study of the behaviour of visible light. The term now embraces the study of almost the whole of the electromagnetic spectrum in one aspect or another, and even includes the behaviour of electrons and neutrons in an optical context (e.g. electron optics, the interference of neutron waves).

OPTIC SIGN. Of a doubly-refracting crystal: a uniaxial crystal is *negative* if the ordinary refractive index is greater than the extraordinary, and *positive* if the reverse is true. A biaxial crystal is *negative* if the acute bisectrix coincides with the direction of vibration for the smallest value of the refractive index and *positive* if it coincides with that of the greatest.

OPTICS, NON-LINEAR. The study of the phenomena associated with the non-linear terms in the mathematical expression for the electric polarization of a medium through which light waves are passing. Incoherent optical sources do not normally produce any significant non-linear effects and the subject is mostly concerned with high-power lasers.

OPTO-ACOUSTIC EFFECT. The degradation of light pulses to heat, when the light is absorbed by a gas, with the resultant emission of sound. The sound arises from the alternating compression and rarefaction of the gas, the pitch being determined by the pulse frequency. Characteristic peaks of intensity are found, corresponding to the presence of various molecules. The effect occurs at all optical wavelengths.

OPTO-ACOUSTIC SPECTROSCOPY. The use of the opto-acoustic effect for investigating molecular structure and detecting small quantities of pollutant gases. The technique has recently been extended to cover the optical absorption and thermal properties of small solid samples. Another development is the use of a laser beam as a source of radiation and an opto-acoustic cell to monitor the degradation of light pulses to heat, in a technique known as LOAD (*L*aser-*o*pto-*a*coustic *d*etection).

OPTOMETER. A generic term for any instrument designed to measure the refractive state of the human eye.

ORBIT. (1) In astronomy: the path of a celestial body which is moving about another under the influence of gravitational attraction. (2) In atomic and nuclear physics: the path of an electron moving around the nucleus of an atom or of one particle around another. See also: *Bohr orbit. Sommerfeld orbit.*

ORBITAL. See: *Atomic orbital. Molecular orbital. Node.*

ORBITAL ANGULAR MOMENTUM. The angular momentum of a particle or system of particles that revolves in an orbit (or behaves as though it did). It is expressed in units of $\hbar$, i.e. $h/2\pi$.

ORBITAL, ANTI-BONDING. A molecular orbital in which the addition of an electron reduces the bonding. The electrons in such an orbital are termed *anti-bonding electrons.*

ORBITAL, BONDING. A molecular orbital in which the addition of an electron increases the bonding. The electrons in such an orbital are termed *bonding electrons.*

ORBITAL ELECTRON CAPTURE. The radioactive decay process in which an orbital electron is captured by the nucleus, with the emission of a neutrino and the production of X-rays characteristic of the daughter atom, and sometimes Auger electrons. It decreases the atomic number by one but does not change the mass number. *K-electron capture, L-electron capture*, etc., are terms used where the shell occupied by the captured electron is known.

ORBITALS, LOCALIZED AND NON-LOCALIZED. See: *Atomic orbital. Molecular orbital.*

ORDER. In interference phenomena: See: *Interference, order of. Spectrum, order of.*

ORDER–DISORDER TRANSFORMATION. (1) In a solid solution: the transformation (often effected by an increase in temperature) from an *ordered* phase, i.e. one in which atoms of the various components take up preferred sites in the structure, to a *disordered* phase, i.e. one in which the atoms are distributed at random over these sites. In the ordered phase the preferred arrangement of atoms defines another lattice, a *superlattice*, upon the original one. See also: *Superlattice*. (2) In a solid dielectric: a reduction in the static dielectric constant with increasing temperature, corresponding, in the case of dipolar materials, to a change from a state of long-range orientational *order* to one of orientational *disorder*.

ORDINARY RAY. See: *Double refraction: Birefringence.*

ORDINARY REFRACTIVE INDEX. See: *Double refraction: Birefringence.*

ORIENTATION OF CRYSTALLITES. See: *Preferred orientation.*

ORIFICE METER. A fluid meter in which the rate of flow is determined from the pressure difference created by a restriction in a closed conduit.

ORIGIN. (1) Of a graphical plot referred to two or more variables: the point at which the values of the variables are zero. (2) Of a plane projection of the Earth's surface: the point from which the co-ordinates of points on the projection are measured.

OROGENESIS. See: *Diastrophism. Earth, thermal contraction of.*

ORTHICON. A type of television camera tube in which a low-velocity electron beam scans a photo-active mosaic. In the *image orthicon* the beam scans a storage target on which has been focused an electron image produced by a photo-emitting surface. See also: *Camera tube.*

ORTHOGRAPHIC PROJECTION. A plane projection used for displaying the positions of the poles of a crystal. These poles are projected from the surface of the reference sphere on to the equatorial plane by dropping perpendiculars to that plane from the individual poles. See also: *Gnomonic projection. Stereographic projection.*

ORTHOHELIUM. Helium in which the wave function is antisymmetric, i.e. changes sign when its two electrons are interchanged. See also: *Parhelium.*

ORTHOHYDROGEN. Molecular hydrogen in which the two nuclear spins in each molecule are parallel. A similar definition obtains for *ortho-deuterium.* See also: *Parahydrogen.*

ORTHOMORPHIC PROJECTION: CONFORMAL PROJECTION. Of the Earth's surface: one which gives no local distortion, the scale being the same in all directions at a given point, but varying from point to point. One of the best known is the *Mercator projection* for a wide extent of latitude, in which the spheroid is projected on to a cylinder whose axis usually coincides with that of the Earth, although it can be oblique. Another is the *Lambert projection*, which may be visualized as projection on to a cone whose axis also coincides with that of the Earth.

ORTHOPOSITRONIUM. Positronium in which the positron and electron spins are parallel. It decays into three photons with a mean life of about 10^{-7} s. See also: *Parapositronium.*

ORTHORHOMBIC SYSTEM. A system of crystal symmetry containing those crystals which can be referred to three unequal axes at right angles. The system is characterized by having axes of two-fold symmetry (rotation or inversion) in three mutually perpendicular directions.

ORTHOTOMIC SYSTEM. An optical system containing only rays which may be cut perpendicularly by a suitably constructed surface.

ORTON CONES. See: *Pyrometric cones.*

OSCILLATING CRYSTAL METHOD. In crystal structure analysis: a modification of the rotating crystal method in which the movement of the crystal is restricted to an oscillation of a few degrees to simplify correlation between the observed reflections and the various sets of crystal planes. See also: *Rotating crystal method.*

OSCILLATION. (1) A periodic energy variation in a mechanical, electrical or atomic system. (2) One complete period of an oscillation.

OSCILLATION, DAMPED. An oscillation whose amplitude decays exponentially with time.

OSCILLATION, FORCED. An oscillation set up by the application of a periodic excitation and having the same period as that excitation. See also: *Forced vibrations.*

OSCILLATION, FREE. An oscillation initiated by an excitation which is wholly or partly transient. Its frequency is determined by the parameters of the oscillating system itself.

OSCILLATION, RELAXATION. A self-maintained oscillation whose waveform shows very rapid changes of slope or height at certain points in the cycle.

OSCILLATIONS, LINEAR AND NON-LINEAR. Oscillations of a physical system whose governing differential equations are linear or non-linear respectively. In general, linear behaviour is characteristic of small amplitudes, and is replaced by non-linear behaviour at large amplitudes.

OSCILLATION, SYMPATHETIC. A forced oscillation.

OSCILLATION, TRANSIENT. An oscillation which is set up by a sudden disturbance, and which is quickly damped out.

OSCILLATOR. (1) An oscillating electronic circuit designed to generate undamped signals of a desired frequency and waveform. (2) An electrical circuit designed to produce an oscillating current when its equilibrium is disturbed. See also: *Barkhausen–Kurz oscillator. Colpitts oscillator. Dynatron. Hartley oscillator. Hartmann oscillator. Klystron. Magnetron.*

OSCILLOGRAPH. An instrument which produces a graph representing the variation with time of the instantaneous magnitude of a physical quantity.

OSCILLOSCOPE. An instrument which displays the variation with time of the instantaneous magnitude of a physical quantity. One of the most common types is the cathode-ray oscilloscope, in which a cathode-ray tube is used.

OSEEN APPROXIMATION. A modification of the Navier–Stokes equations for fluid flow past a small object, based on the assumption that the perturbation in the fluid velocity due to the presence of the object is small. See also: *Fluid flow, equations of motion for.*

OSMOMETER. An instrument for the measurement of osmotic pressure.

OSMOSIS. The spontaneous diffusion of a chemical species through a semi-permeable membrane from a region of higher concentration to one of lower.

OSMOSIS, ELECTRICAL. The tendency of a liquid to pass through a porous diaphragm towards the cathode, when a current is passed through the liquid.

OSMOSIS, ISOTOPIC. The separation of isotopes of the same element by selective diffusion through a porous barrier, as in the separation of $^3_2$He and $^4_2$He in the liquid state. See also: *Isotope separation.*

OSMOTIC COEFFICIENT. A factor introduced into the equations for the chemical potential of ideal solutions to permit their use for non-ideal solutions.

OSMOTIC PRESSURE. The pressure developed in the process of osmosis. It may be expressed as the pressure which must be applied to a solution in order just to prevent osmosis into that solution.

O-STAR. See: *Stars, spectral classes of.*

OSTWALD DILUTION LAW. For the ionization of a weak electrolyte: an application of the law of mass action, yielding the expression $[\alpha^2/(1 - \alpha)] V = K$, where α is the degree of ionization, V the volume containing one mole of electrolyte, and K the ionization constant. See also: *Ionization constant.*

OTTO CYCLE. The thermodynamic cycle upon which the operation of spark ignition engines is based. It consists of adiabatic compression, heat addition at constant volume, adiabatic expansion, and heat rejection at constant volume.

OUTER EFFECT. In X-ray diffraction: the effects of neighbouring atoms and molecules as opposed to those occurring inside a particular atom or molecule (the *inner effect*). It is of particular consequence in diffraction by liquids.

OVERPOTENTIAL. A measure of the electrolytic polarization occurring at an electrode when a current passes through the electrode–solution interface. It is given by the difference between the electrode potential at a given current density and the reversible equilibrium potential of the same electrode.

OVERTONE. A partial whose frequency is higher than, but not necessarily an integral multiple of, the fundamental frequency.

OXIDATION. The addition of oxygen to a substance. More generally, the loss of electrons from an atom. See also: *Reduction.*

OXIDATION POTENTIAL. The potential drop involved in the oxidation of a neutral atom to a cation, of an anion to a neutral atom, or of an ion to a more highly charged state.

OXIMETRY. A technique for estimating the percentage of blood haemoglobin saturated with oxygen.

OXYGEN TENSION. The partial pressure of oxygen in a solution, especially with reference to the blood.

OZONIZER. A machine for producing ozone for industrial use, for example for water treatment or sewage waste purification. It commonly involves the passage of a corona discharge through either air or oxygen in conditions which preclude the formation of arc or spark discharges.

p

PACKING FRACTION. (1) Of a nuclide: the mass decrement per nucleon expressed as a fraction. (2) Of a nucleus: the mass defect per nucleon expressed as a fraction. When expressed in terms of energy (or actual mass) the two are equivalent for a given nuclide and its nucleus except for the energies (or masses) of the atomic electrons. The packing fraction is positive for light and heavy nuclides but negative for most other nuclides. See also: *Mass decrement. Mass defect.*

PAIR-DISTRIBUTION FUNCTION. For a system of particles: the probability of finding two particles at a specified distance apart.

PAIR PRODUCTION: PAIR CREATION. The formation of a particle and its anti-particle through the interaction of a photon or fast particle with the field of an atomic nucleus or other particle, or through the de-excitation of an excited atomic nucleus (known as *internal pair production*). The best-known example is the production of an electron–positron pair by a photon. See also: *Annihilation. Electron–positron pair.*

PAIR SPECTROMETER. For γ-rays of energy above about 3 MeV: a spectrometer in which the energy of the γ-rays is obtained from the measurement of the energy of the electron–positron pairs produced by these rays.

PALAEOGEOPHYSICS. The extension of geophysical methods to events occurring throughout geological time. It includes the study of the Earth's magnetic field from rocks; climatology from rock types, fossil flora and fauna; and, on a more speculative basis, the Earth's thermal history, geoelectricity, seismology, rotation and gravitational field.

PALAEOMAGNETISM. The permanent magnetism of rocks in remote times as determined by geological observations on the direction of magnetization. Such observations indicate the position of the geomagnetic pole at the time of the rocks' formation. See also: *Archaeomagnetism.*

PALAEOTEMPERATURES. Temperatures assigned to various geological periods and ages, as determined, for example, by a study of the distribution of flora and fauna as revealed by geological evidence and, for marine temperatures, by the palaeothermometer.

PALAEOTHERMOMETER. A sensitive mass spectrometer for measuring marine temperatures in remote times by analysing the calcium carbonate from shells deposited at the relevant times. The abundance ratio of ^{18}O and ^{16}O in carbon dioxide obtained from the carbonate is determined, and since it is temperature-dependent leads to an estimate of the temperature required. Among other things the results support the theory that the extinction of the great dinosaurs was due to a general decline in temperature over the whole Earth.

PALAEO WIND DIRECTIONS. Wind directions in remote times which are determined from geological evidence such as the distribution of volcanic ash, and the orientation of sand dunes. Studies of such evidence show that the prevailing winds in many parts of the world, but not all, were very different from what they are at present.

PALLAS. See: *Asteroid.*

PANRADIOMETER. An instrument for recording or measuring radiant heat irrespective of wavelength, in which the radiation is absorbed by a black body, of which the temperature is measured.

PARABOLA. See: *Conic: Conic section.*

PARABOLOID. See: *Quadric.*

PARACHOR. A quantity which may be regarded as the molecular volume of a liquid when its surface tension is unity. It is given by $M\sigma^{1/4}/(\varrho_l - \varrho_g)$, where M is the molecular weight, σ the surface tension, and ϱ_l and ϱ_g are the liquid and gas densities respectively.

PARAFOVEA. See: *Fovea.*

PARAHYDROGEN. Molecular hydrogen in which the two nuclear spins are antiparallel. A similar definition obtains for *paradeuterium.* See also: *Orthohydrogen.*

PARALLACTIC ANGLE. Of a celestial body: the angle between the great circle passing through the positions, on the celestial sphere, of the celestial body and the celestial pole, and the great circle passing through the position of that body and the zenith. It is also known as the *angle of situation*.

PARALLACTIC ELLIPSE. The small ellipse which a star appears to describe annually on the celestial sphere. It is a consequence of the orbital motion of the Earth about the Sun.

PARALLAX. (1) General: the apparent change in the position of an object seen against a reference background when the viewpoint is changed. (2) Of a star: the angle subtended at the star by two ends of a base-line of known length.

PARALLAX, ANNUAL. See: *Parallax, stellar.*

PARALLAX, CHROMATIC. Parallax arising in an optical system not corrected for chromatic aberration, arising from the dependence of the position of the focal plane on colour.

PARALLAX, GEOCENTRIC. Of a body in the solar system: the angular difference in its direction as seen by a real observer on the Earth's surface and an imaginary one at the Earth's centre. For a body in the zenith it is thus zero. For a body on the horizon it is known as the *horizontal parallax*.

PARALLAX, HORIZONTAL. See: *Parallax, geocentric.*

PARALLAX, SECULAR. The apparent angular displacement impressed upon a star by the motion of the Sun and planets as a whole through space. See also: *Secular variation.*

PARALLAX, SOLAR. The angle subtended at the Sun by the Earth's equatorial radius, at the Sun's mean distance from the Earth.

PARALLAX, SPECTROSCOPIC. The estimation of the distance of a remote star by combining the spectroscopically determined absolute magnitude with the observed apparent magnitude.

PARALLAX, STELLAR. The angle subtended at a star by the mean radius of the Earth's orbit round the Sun, i.e. by an astronomical unit. It is also known as the *annual parallax* of the star.

PARALYSIS CIRCUIT. For a counting system recording pulses: a circuit which renders the system inoperative for a predetermined time after a recorded pulse and thus allows pulses to be distinguished which would otherwise have overlapped.

PARAMAGNETIC RELAXATION. The relaxation occurring in a paramagnetic material, when a magnetic field is suddenly applied, or when the magnitude of the field is suddenly changed.

PARAMAGNETIC RESONANCE. See: *Electron paramagnetic resonance (EPR): Electron spin resonance (ESR).*

PARAMAGNETISM. The property shown by many substances of becoming magnetized in the direction of the applied field, but not retaining this directional magnetization when the field is removed. Such a substance may be regarded as an assembly of magnetic dipoles which are normally directed at random. It has a positive magnetic susceptibility (*paramagnetic susceptibility*), and is said to be *paramagnetic*. All ferromagnetic and antiferromagnetic materials become paramagnetic at temperatures above the Curie point or Néel point respectively. See also: *Antiferromagnetism. Diamagnetism. Ferromagnetism.*

PARAMETRAL PLANE. In crystallography: a prominent crystal plane used in the specification of Miller indices, for which these indices are, by definition, (111). See also: *Miller indices.*

PARAMETRIC AMPLIFIER: REACTANCE AMPLIFIER. An amplifier for which the equations describing its operation contain one or more time-dependent reactance parameters. The operation of the amplifier depends on the fact that a time-varying reactance can exhibit negative resistance under some conditions and can act as a frequency converter under others.

PARAPOSITRONIUM. Positronium in which the positron and electron spins are anti-parallel. It decays into two photons with a mean life of about 10^{-10} s. See also: *Orthopositronium.*

PARASITIC CAPTURE. Of neutrons: neutron absorption in a reactor which does not result in fission or any other desired process.

PARASITIC E.M.F. An unwanted signal in a measuring apparatus, arising from spurious causes.

PARAXIAL RAYS. Rays lying close to the principal axis of an optical system, at such an angle θ that $\sin \theta = \theta$ to a good approximation.

PARHELION: MOCK SUN. A luminous image of the Sun most frequently seen at an angular distance of about 22° from the Sun. The phenomenon is caused by refraction of the Sun's rays by ice crystals. See also: *Halo.*

PARHELIUM. Helium in which the wave function is symmetric, i.e. remains unchanged when its two electrons are interchanged. The ground state of helium is a parhelium state. See also: *Ortho-helium.*

PARITY. A symmetry property of a wave function. The parity is said to be even (or +) if the wave function is unchanged by an inversion (reversing the signs of all the coordinates) and odd (or −) if the sign of the wave function is thereby changed.

PARSEC. A unit of astronomical distance. It is the distance at which the mean radius of the Earth's orbit would subtend an angle of 1″ if viewed normally. It is thus the distance at which an object would have a parallax of 1″ if the astronomical unit were used as a base line, hence its name (*par*-allax of one *sec*-ond). (Note: 1 parsec = 206 265 astronomical units $\approx$ 3·263 light years $\approx$ 3·0857 $\times$ 10^{13} km). See also: *Astronomical unit. Light year.*

PARTIAL. Any pure tone component of a complex tone. It may be higher or lower than the fundamental frequency and may or may not be an integral multiple or submultiple of that frequency. See also: *Harmonic. Overtone. Subharmonic.*

PARTIAL MOLAR QUANTITY. Of each component in a multi-component system, for a given extensive quantity: the increase in the magnitude of the quantity which would result from the addition to the system of one mole of the component. See also: *Extensive variable.*

PARTIAL PRESSURE. Of each gas in a mixture of gases: that pressure which the gas would exert if all the other gases had been removed. See also: *Dalton law of partial pressures.*

PARTICLE. A general term denoting a small entity of matter, whether it be solid, liquid, or gas. Unless otherwise qualified (either explicitly or implicitly) the term, especially in such expressions as *particle technology,* applies to such things as disperse systems of particles and bulk collections in which statistical rather than individual properties are significant. Smokes, fumes, fogs, and mists come into these categories. See also: *Elementary particles. Material particle.*

PARTICLE ACCELERATOR. Any machine designed to impart large kinetic energy to charged particles such as electrons, protons and atomic nuclei. See also: *Betatron. Cockcroft–Walton apparatus. Cyclotron. Electrostatic generator. Linear accelerator. Microtron. Synchro-cyclotron. Synchrotron. Van de Graaff accelerator.*

PARTICLE FLUENCE. See: *Particle flux density.*

PARTICLE FLUX DENSITY. Of atomic or nuclear particles at a given point in space: the number of particles incident per unit time on an imaginary sphere of unit cross-sectional area centred at that point. It is identical with the product of the particle density (number per unit volume) and the average speed of the particles. Particle flux density is commonly, but incorrectly, termed *particle flux.* The time integral of particle flux density is sometimes known as *particle fluence.*

PARTICLE ORBIT. The closed or nearly closed path of a charged particle in an accelerator which employs a guiding magnetic field, e.g. a cyclotron. Coils which cause the radii of the orbits to decrease or increase are known as *particle–orbit contractors* or *expanders* respectively.

PARTICLE SYMMETRY. Concerns the classification of the large number of subatomic particles discovered in recent years in terms of symmetry relationships between them. The most successful way of doing this is by the use of a continuous symmetry group, known as $SU(3)$, by which the Ω-particle was first predicted. The concept of quarks and anti-quarks also arose from the SU(3) symmetry group. See also: *Particle symmetry group.*

PARTICLE SYMMETRY GROUP. A group which shows a symmetrical grouping of particles into electric charge multiplets (doublets, triplets, octuplets, etc.) when the difference between the charge on a particle and the average charge of the multiplet to which it belongs is plotted against the hypercharge. A number of such groups have been found, the most striking being the SU(3) group, which comprises a decuplet in the form of an inverted triangle at the apex of which is the Ω-particle, a particle which was indeed predicted from the consideration of this symmetry group.

PARTITION COEFFICIENT: DISTRIBUTION COEFFICIENT. The ratio of the equilibrium concentrations of a given substance dissolved in two specified immiscible solvents.

PARTITION FUNCTION. For a particular method of averaging a thermodynamic property of an assembly over all possible states: an expression giving the distribution of molecules in different energy states in the assembly. It may be written as

$$Z = \sum q_r \, e^{-\varepsilon_r / kT},$$

where Z is the partition function, q_r is the statistical weight of the rth energy state ε_r, k is the Boltzmann constant, T the absolute temperature, and the summation is taken over all the energy states of the system. The energy levels ε_r may be those attributed to rotation, translation, vibration, etc., and, where they may be treated as independent, the complete partition function of the assembly may be written as $Z = Z_{rot} \times Z_{trans} \times Z_{vib} \times \cdots$, where Z_{rot}, Z_{trans}, Z_{vib}, etc., are the *rotational, translational, vibrational*, etc., partition functions.

PASCAL (PA). The unit of pressure in the International System of Units (SI). It is a pressure of 1 newton per square meter (N/m^2), and is equal to 10^{-2} millibar. See: *Appendix II.*

PASCHEN–BACK EFFECT. The change in the Zeeman pattern as the external magnetic field strength reaches a critical value. At very high values, however, the normal Zeeman pattern is restored.

PASCHEN CIRCLE. A name sometimes given to a spectrographic mounting which employs only a small part of a Rowland circle.

PASCHEN LAW. States that the breakdown potential of a gas between parallel plate electrodes is a function of the product of the gas pressure and the electrode separation.

PASCHEN–RUNGE MOUNTING. A method of mounting a concave diffraction grating in which the slit, grating, and part of the Rowland circle are fixed in position.

PASCHEN SERIES. A series in the line spectrum of the hydrogen atom. See also: *Hydrogen atom, line spectrum of.*

PASS-BAND. The frequency range within which a band-pass filter transmits without appreciable attenuation. See also: *Band-pass filter. Stop band.*

PASSIVE NETWORK. A network containing capacitance, inductance, or resistance, but no internal voltage or current sources.

PASSIVITY. (1) Of a metal or alloy: the ability of a metal or alloy to behave like a more noble metal (e.g. in its resistance to corrosion) in certain environments. It may arise from anodic polarization, or from a protective layer arising as a result of immersion in certain solutions or exposure to air. (2) In instrument design: the design of instruments in such a way that the measuring devices incorporated in the instruments do not interfere with the phenomena being measured.

PASTEURIZATION. Partial sterilization of a liquid by a moderate increase of temperature for a limited time. Thus the micro-organisms in milk are reduced considerably in about 30 min at 62–65°C.

PATCH EFFECT. The fluctuation in the response of a photocathode when a narrow pencil of light of constant intensity and composition is passed over its surface. The emission appears to be concentrated at a number of discrete patches, depending on the size and position of active crystals on that surface.

PATTERSON FUNCTION. In crystal structure analysis: a Fourier summation, based on the fact that the square of the structure factor of a crystal plane is a measure of the reflecting power, which can be transformed to give a vectorial representation of the interatomic distances in the crystal. The problem of determining the phase of the reflections is thereby overcome at the expense of obtaining interatomic distances rather than atomic positions. See also: *Fourier synthesis.*

PATTERSON MAP. A two-dimensional Fourier projection in which the distance between the origin and any peak represents the vectorial distance between two atoms in a crystal. The ease of interpretation depends on the ability to identify some of the peaks.

PAUCIMOLECULAR FILM. A film of adsorbed gas, several molecules in thickness, which is formed on the surface of a solid.

PAULI EXCLUSION PRINCIPLE. A principle enunciated by Pauli which states that no two identical fermions (i.e. particles of spin $\frac{1}{2}$) in any system may occupy states which have the same set of quantum numbers. This principle holds, therefore, for electrons, protons and neutrons, but not for photons or π-mesons.

PAULI SPIN MATRIX. One of a set of matrices introduced by Pauli in connection with electron spin in non-relativistic wave mechanics.

P CYGNI STAR. A star of spectral class B with a surface temperature of about 27 000 K, a radius 8 times that of the Sun, and a magnitude about -4. See also: *Stars, spectral classes of.*

P.C.M. (POUR CENT MILLE). A unit of reactivity (of a nuclear reactor) numerically equal to 10^{-5}, and hence the same as a *milli-nile*. See also: *Nile. Nuclear reactor, reactivity of.*

PECLET NUMBER. A dimensionless number giving the ratio of convected to conducted heat in a flowing heat-transfer medium.

PECULIAR STARS. Stars whose spectra do not fit into any of the standard classifications. See: *Stars, spectral classes of.*

PEIERLS STRESS: PEIERLS FORCE. The resolved shear stress required to move a dislocation through a perfect lattice.

PELLIN–BROCA PRISM. A constant-deviation

prism consisting of two 30° prisms connected by a 45° total-reflecting prism.

PELORUS. A dummy compass, complete with graduated scale and sight(s), which is used for taking relative bearings. It is commonly used to facilitate the adjustment of ships' compasses.

PELTIER COEFFICIENT. The amount of heat generated per second by the Peltier effect when unit current is passed through a junction of two dissimilar metals.

PELTIER COOLING: THERMOELECTRIC COOLING. Cooling on a useful scale by use of the Peltier effect.

PELTIER EFFECT. The liberation of heat at one junction and the absorption of heat at the other when a current is passed round a circuit consisting of two different metals joined at two points. It is the opposite of the Seebeck effect. See also: *Thermoelectric effects.*

PENCIL OF LIGHT. A narrow beam of light, having a small or zero angle of convergence or divergence.

PENDULUM. A body suspended so as to be free to swing or oscillate. See also: following articles.

PENDULUM, COMPOUND. A rigid body oscillating under the action of gravity about an axis (usually horizontal and often a knife edge) which does not pass through its centre of mass.

PENDULUM, CONICAL. A pendulum consisting of a body suspended by a "weightless" thread, which moves in a horizontal circle, so that the thread describes a cone.

PENDULUM, CYCLOIDAL. A simple pendulum in which the swing of the suspending thread is limited by two solid metal cheeks of cycloidal form, rendering the time of oscillation independent of the amplitude.

PENDULUM, DOUBLE. Two rigid bodies smoothly jointed together, one of which is also smoothly supported at a fixed point, so that the assembly can swing about that point.

PENDULUM, EQUIVALENT LENGTH OF. For a given pendulum: the length of the simple pendulum which has the same period.

PENDULUM, FOUCAULT. A simple pendulum in which the suspension is a long wire. The plane of the swing remains constant but, owing to the rotation of the Earth, appears to rotate. The time of a complete rotation is $24/\sin \lambda$ hours, at latitude λ.

PENDULUM, HERBERT. See: *Hardness.*

PENDULUM, KATER. A reversible compound pendulum used for the accurate determination of the acceleration due to gravity. It consists essentially of a bar carrying two knife-edges, set facing each other, one on each side of the centre of mass, from which the pendulum is suspended in turn.

PENDULUM, REVERSIBLE. A compound pendulum used for the accurate determination of the acceleration due to gravity, from the measurement of the periods corresponding to two different suspension lengths.

PENDULUM, SIMPLE. A pendulum consisting of a body suspended by a "weightless" thread, which moves in a vertical plane. The period of oscillation for small amplitudes of swing is $2\pi \sqrt{(l/g)}$, where l is the length of the thread and g is the acceleration due to gravity.

PENDULUM, TORSIONAL. A pendulum which consists essentially of a body suspended by a wire, the body being free to oscillate backwards and forwards about the axis of the wire. The body executes simple harmonic vibrations with a constant period which is independent of amplitude.

PENETRAMETER. An assembly of pieces of matter having different opacities to X-rays, neutrons etc. (e.g. a step wedge), whose main uses are in judging the quality of a radiograph and in the radiographic calibration of an X-ray or other generator.

PENETRATION DEPTH. See: *Skin effect. Superconductivity, London equation for.*

PENNING GAUGE. See: *Ionization gauge.*

PENTAD. See: *Polyad.*

PENTODE. A thermionic valve or tube having five electrodes: a cathode, a control grid, a screen (or auxiliary grid) maintained at approximately the anode potential, a suppressor grid maintained at approximately the cathode potential, and a cathode. It has characteristics similar to those of a screen-grid valve or tube, but secondary emission effects are suppressed.

PENUMBRA. The outer, partially dark, portion of the shadow of an object cast by a source of finite size, e.g. the shadow of the Earth or Moon cast by the Sun, as in an eclipse. The inner, completely dark, part of the shadow is known as the *umbra.*

PERFECT CRYSTAL. A crystal in which the regular periodicity of the structure is strictly maintained throughout. Imperfections such as thermal agitation, which do not impair the periodicity, are ignored.

PERFECT GAS LAWS. See: *Gas, perfect: Gas, ideal.*

PERFECT NUMBER. A number which is equal to the sum of all its factors (including unity). If the sum is less than the number, the latter is said to be *deficient*; and if the sum is greater, the number is said to be *abundant*.

PERIASTRON. Of the orbit of one star about another in a binary system: the point in that orbit which is nearest to the other star. See also: *Apastron.*

PERIGEE. Of the orbit of a body about the Earth: the point in that orbit which is nearest to the Earth. The term is sometimes used when speaking of the Sun's apparent orbit about the Earth. Similar terms (e.g. *Perijove, Perisaturnium*) are used to refer to orbits about other planets. See also: *Apogee.*

PERIHELION. Of the orbit of a body about the Sun: the point in that orbit which is nearest to the Sun. See also: *Aphelion.*

PERIHELION, ADVANCE OF. The change in the position of the perihelion of a planet with time. The excess of the observed value for Mercury (43″ in a rotation of 574″ of arc per century) over that predicted by classical mechanics can be accounted for by the general theory of relativity.

PERIOD. Of an oscillation: the time for one complete cycle to take place.

PERIODIC TABLE OF THE ELEMENTS: PERIODIC SYSTEM. A table, originally due to Mendeléef and Meyer, in which the elements are arranged in a systematic grouping wherein elements having like properties occur in related positions, in horizontal or vertical sequence. The arrangement is based on the atomic numbers of the elements, the properties varying periodically with those numbers. There are nine groups and seven periods, the last element in each period (i.e. that in the ninth group of the period) being a inert gas corresponding to closed shells of electrons). For the first six periods these elements are He, Ne, Ar, Kr, Xe, and Rn. The seven periods begin at H, Li, Na, K, Rb, Cs and Fr respectively, but some gaps occur. See: *Appendix I.*

PERISCOPE. An optical instrument in which an arrangement of reflecting surfaces allows viewing in a direction displaced from the direct line of vision. Its most frequent use is to provide a view over a parapet or other obstacle, as from a military tank or submarine, or for the remote observation of radioactive material.

PERITECTIC. In a binary alloy system: the structure found in certain binary alloys in which a secondary crystal formed during cooling envelopes the primary crystal and prevents its contact with the melt.

PERMALLOYS. Alloys of iron and nickel, with or without additions of other elements, which have unusually high magnetic permeability.

PERMANENT DIPOLE. See: *Dipole, electric.*

PERMEABILITY. Of a given medium: the magnetic flux density produced in the medium per unit magnetic field strength. The SI and MKSA unit is henry per metre (H/m). The (dimensionless) CGS unit is gauss per oersted. The permeability of free space in CGS units is unity, and in SI units is $4\pi \times 10^{-7}$ H/m.

PERMEABILITY, DIFFERENTIAL. The slope of the magnetization curve.

PERMEABILITY, IDEAL. The permeability obtained from an ideal or anhysteric magnetization curve.

PERMEABILITY, INCREMENTAL. The permeability attributable to an alternating magnetic field superimposed on a steady field. Thus if the alternating field has amplitude H_a, and causes an alternating induction B_a, the incremental permeability is B_a/H_a.

PERMEABILITY, INITIAL. The slope of the initial magnetization curve at the origin.

PERMEABILITY, NORMAL. The permeability of a material in the cyclic state.

PERMEABILITY, RELATIVE. Of a given medium: the ratio of the magnetic flux density produced in the medium to that which would be produced in a vacuum by the same magnetizing force.

PERMEABILITY, REVERSIBLE. The limit approached by the incremental permeability as the amplitude of the alternating magnetic field approaches zero. It can also be defined as $\Delta B/\Delta H$, where B is the magnetic flux density and H the magnetic field strength.

PERMEAMETER. An instrument for measuring permeability.

PERMEANCE. The reciprocal of magnetic reluctance, i.e. the magnetic conductance.

PERMITTIVITY: ABSOLUTE PERMITTIVITY. Of a medium: the electric flux density or displacement produced in a medium by unit electric force. The term is often used to mean *relative permittivity*. See also: *Dielectric, refractive index of. Permittivity, relative.*

PERMITTIVITY, INCREMENTAL. Of a medium: the ratio $\Delta D/\Delta E$, where ΔD is the change in displacement produced by a change in field strength ΔE.

PERMITTIVITY OF FREE SPACE. The value of the permittivity in a vacuum. In SI and MKSA units it is about 8.854×10^{-12} F/m, and in CGS (e.s.u.) units it is unity (and dimensionless).

PERMITTIVITY, RELATIVE. The ratio of the electric flux density produced in a medium to that which would be produced in a vacuum by the same electric force. It is the same as the dielectric constant. See also: *Dielectric constant: Specific inductive capacity: Relative permittivity.*

PERMUTATION GROUP. A type of group suitable for describing symmetry properties, and therefore of application in crystallography and quantum-mechanical theory. See also: *Group.*

PERMUTATION (MATHEMATICS). The assignment of a group of objects into two or more mutually exclusive sets, having regard to their order. The number of ways (or permutations) of selecting r objects from a set of n objects in this way is thus greater than the number of combinations, and is in fact given by $n!/(n-r)!$ See also: *Combination (mathematics).*

PERPETUAL MOTION. Motion which, once started, continues indefinitely. Two kinds are distinguished. The *first* refers to the continual operation of a device which creates its own energy and thus violates the first law of thermodynamics. The *second* refers to a device which can convert heat completely into work, thus violating the second law of thermodynamics.

PERSONAL EQUATION. The correction which is to be applied to a measurement on account of the systematic error of the observer's readings. For a given observer and instrument it is usually constant.

PERSORPTION. The absorption of a gas by a solid in such a complete manner as to approximate to a solid solution.

PERSPECTIVE. The deliberate distortion (e.g. by making parallel lines meet at suitable points) of a two-dimensional reproduction of a three-dimensional object, undertaken to give the representation an appearance of reality.

PERTURBATION METHOD: PERTURBATION THEORY. The approximate solution of the equations for a complicated system by first solving the equations for a similar system chosen because the solution is relatively easy, and then considering the effect of small changes (i.e. perturbations) on this solution. It may be applied to problems in both classical and quantum mechanics, e.g. to the motion of a planet round the Sun (in which the influence of other planets is at first neglected) or to the solution of the Schrödinger equations (in which the inter-electronic energy terms are treated as perturbations of an artificial atom whose electrons interact with the nucleus but not with each other).

PERVEANCE. The constant A in the Child–Langmuir equation: $I = AV^{3/2}$, where I is current density and V is potential. See also: *Child–Langmuir equation.*

PETROGRAPHY. That part of petrology which is concerned with the geological mode of occurrence, mineralogical composition, chemical constitution, and physical properties of rocks.

PETROGENESIS. That part of petrology which is concerned with the more theoretical aspects of the subject, such as the interpretation of the factual findings of petrography.

PETROLOGY. That branch of geological science which deals with the study of rocks as the materials of construction of the Earth's crust.

PFAFFIAN DIFFERENTIAL FORM. The expression

$$dW = \sum_{i=1}^{n} X_i \, dx_i,$$

where X_i is a function of the independent variable and dW is a total differential.

PFUND SERIES. A series in the line spectrum of the hydrogen atom. See also: *Hydrogen atom, line spectrum of.*

PH. The quantitative expression of the degree of acidity or basicity of a liquid. It was originally the common logarithm of the reciprocal of the hydrogen ion concentration in an aqueous solution. In a neutral solution this is 7, corresponding to a hydrogen ion concentration of 10^{-7}. However, the hydrogen ion concentration is now often replaced by the hydrogen ion activity. See also: *Hydrogen electrode.*

PHANTOM MATERIAL. In radiation-dose measurements: a solid or liquid whose absorbing and scattering properties for a given radiation are similar to those of a given biological material, such as part or all of the human body. A body of such material, which may be constructed to simulate some special shape (e.g. part of the human body) or which is large enough to provide full back-scatter, is known as a *phantom.*

PHASE. (1) Of alternating current, voltage, or other sinusoidal quantity: the fractional part of a period through which the quantity has advanced from an arbitrary origin, usually taken to be its last previous passage through zero in the negative-to-positive direction. This phase is often expressed as the *phase angle*. (2) In chemistry and metallurgy: any homogeneous and physically distinguishable portion of a system having definite boundaries. It is characterized by its chemical composition, state of aggregation, and, particularly in metallurgy, by its crystal structure. (3) Of a planet or satellite: a term descriptive of the changing shape of the visible illuminated surface of the planet or satellite arising from changes in the relative positions of the Earth, Sun, and the object in question. Thus in the case of the Moon, the phases, starting from *new moon*, are *crescent, first quarter, gibbous, full moon, gibbous* again, *third quarter*, and back to *new moon*. The phase may be expressed as a decimal fraction of the total surface which is illuminated.

PHASE-CHANGE COEFFICIENT: PHASE CONSTANT. For an electrical transmission line: the imaginary part of the propagation coefficient. Also known as the *wavelength constant*.

PHASE CONTRAST. See: *Microscope, phase-contrast*.

PHASE DIAGRAM: EQUILIBRIUM DIAGRAM: CONSTITUTION DIAGRAM. A temperature-composition map of the solid and liquid phases of an alloy system. It shows the ranges of composition and temperature over which the various phases exist in equilibrium with each other, and the phase transformations which occur on heating or cooling. The construction of such diagrams is vital to the metallographic study of structures in an alloy system. Binary systems are represented by two-dimensional diagrams, and ternary usually by sections of a three-dimensional figure.

PHASE-INTEGRAL METHODS. Approximate analytical techniques for the solution of equations which are not soluble in terms of the various transcendental functions. Many problems concerning the propagation of waves in inhomogeneous media may be solved by such techniques.

PHASE MODULATION. The process of imparting information to a carrier wave by causing the frequency of the carrier to vary in accordance with an input signal See also: *Modulation*.

PHASE PLATE. A quarter-wave plate used in phase-contrast microscopy. See also: *Microscope, phase-contrast*.

PHASE RULE: GIBBS PHASE RULE. A mathematical expression showing the conditions of equilibrium between various phases in a system. It may be stated as $P + F = C + 2$, where P is the number of phases, F the number of degrees of freedom, and C the number of independent components (independent variable constituents in terms of which the composition of each phase can be expressed mathematically).

PHASE SEQUENCE. For alternating current: the order in which the various phases of a polyphase system attain a maximum potential of a given sign.

PHASE SHIFT, ELECTRICAL. A change made in the phase angle of an alternating current so as to improve the power factor.

PHASE TRANSFORMATION. See: *Transformation*.

PHASE VELOCITY. The velocity of propagation of any one phase state, such as the point of zero instantaneous field, in a steady train of sinusoidal waves. It is also known as the wave velocity. For a dispersive medium it differs from the group velocity. See also: *Characteristic velocity of an electromagnetic wave front. Group velocity*.

PHASITRON. A frequency-modulator tube used to produce direct phase modulation.

PHENOMENOLOGICAL THEORY. A theory which expresses mathematically the results of observed phenomena without paying detailed attention to their fundamental significance.

PHILIPS LIQUEFIER. An air liquefier based on the Stirling cycle with helium or hydrogen as the working fluid. The compression and expansion steps are carried out in a common cylinder using two pistons driven independently from a common crankshaft.

PHON. The unit of equivalent loudness of sound. The equivalent loudness of a given sound is equal to the sound pressure level of a standard pure tone of frequency 1000 Hz which is judged to be equal in loudness to the sound in question. In phons, it is the number of decibels by which the pressure level of this matching tone exceeds that of a reference tone of the same 1000 Hz frequency, the reference level being 2×10^{-4} dyn/cm^2 or 2×10^{-5} N/m^2.

PHONIC WHEEL. (1) A device, controlled by a tuning fork, which enables the speed of rotation of a motor to be kept constant. (2) An elementary synchronous motor driven from a valve oscillator.

PHONON. A quantum of sound, analogous to the photon. The term is used chiefly in connection with atomic vibrations in crystals, where the thermal motions of atoms may be regarded as a set of phonons travelling through the lattice.

PHONON AVALANCHE. In a paramagnetic crystal at a low temperature: an interaction between phonons and magnetic ions which results in the cumulative production of resonant phonons. See also: *Phonon bottleneck*.

251

PHONON BOTTLENECK. In a paramagnetic crystal at a low temperature: an interaction between phonons and magnetic ions which results in the escape of non-equilibrium phonons being impeded. See also: *Phonon avalanche.*

PHOSPHOR. Any substance which exhibits luminescence. The use of the term has been extended to cover the emission of light outside the visible range, a well-known example of which is the *E phosphor* whose emission band lies partly within the erythemal ultraviolet region. Most phosphors require the presence of a small quantity of a specific activator.

PHOSPHORESCENCE. Luminescence which persists for more than 10^{-8} s after the excitation is removed (and may well last for hours or days), in contrast to fluorescence, which stops within about 10^{-8} s. Phosphorescence and luminescence together make up the *after-glow.* See also: *After-glow. Anti-Stokes emission. Fluorescence. Luminescence.*

PHOT. A unit of illumination equal to 1 lumen/cm^2. See also: *Illumination.*

PHOTOCATALYSIS. The acceleration or retardation of the rate of a chemical reaction by light. See also: *Photosensitization.*

PHOTOCATHODE. The photoemissive cathode in a photocell.

PHOTOCELL: PHOTOELECTRIC CELL: PHOTOEMISSIVE CELL. A device for converting radiant energy (commonly light) into the kinetic energy of free electrons. It consists essentially of a photoemissive cathode in the form of a slotted cylinder (or its equivalent) surrounding a central wire electrode which collects the electrons. See also: *Photovoltaic cell. Quantum efficiency: Quantum yield.*

PHOTOCELL, DARK CURRENT IN. The current passed by a photocell in the absence of any light stimulus.

PHOTOCELL, RECTIFIER. See: *Photovoltaic cell.*

PHOTOCHEMICAL CELL. See: *Photovoltaic cell.*

PHOTOCHEMICAL EQUIVALENCE. According to Einstein's law of photochemical equivalence there is a simple integral relationship between the primary photochemical yield and the number of light quanta absorbed. The yield may be greater than unity if secondary processes arise, or less when the recombination of products occurs.

PHOTOCHEMICAL PROCESS. A chemical process brought about by light in the wavelength range of about 1000–10000 Å (i.e. $0 \cdot 1$–$1 \cdot 0$ μm). It differs from a radiation–chemical process in that each photon is absorbed by a single molecule, whereas in a radiation–chemical process the highly energetic radiation dissipates its energy over a long path, affecting a great number of molecules.

PHOTOCHEMICAL YIELD. The number of molecules chemically changed in a photochemical reaction per quantum of incident light of a given wavelength. It is also known as the *quantum efficiency.* See also: *Photochemical equivalence.*

PHOTOCHEMISTRY. The study of chemical reactions induced by light.

PHOTOCHROMISM: PHOTOTROPY. A reversible colour change induced in a substance by exposure to violet or ultraviolet light. A similar change induced by infrared radiation is known as *thermochromism.*

PHOTOCONDUCTION. Electrical conduction produced in a solid by the influence of electromagnetic radiation in the range from ultraviolet to infrared, inclusive.

PHOTODISINTEGRATION. A nuclear disintegration induced by photons.

PHOTODISSOCIATION. The dissociation of a chemical compound into simpler molecules, or of a molecule into its component atoms, by the action of ultraviolet and visible radiation. See also: *Grotthus–Draper law. Photochemical equivalence.*

PHOTOELASTICITY. The changes produced in the optical properties (notably the production of temporary double refraction) in normally isotropic transparent materials, when the materials are subjected to mechanical stress. It forms the basis of a technique for analysing the stress distribution in a body under a complex system of loading by passing polarized light through a model of the body and observing the birefringent stress patterns.

PHOTOELECTRIC ABSORPTION. The complete absorption of a photon by a nucleus with the emission of an orbital electron.

PHOTOELECTRIC EFFECT: PHOTOEMISSIVE EFFECT. The emission of electrons from the surface of a solid when it is bombarded by visible or ultraviolet light. See also: *Einstein equation for photoelectric emission.*

PHOTOELECTRIC YIELD The quantum efficiency of a photocathode.

PHOTOELECTRON. An electron emitted as a result of the photoelectric effect.

PHOTOFERROELECTRIC EFFECT. The formation of a lasting optical image on the surface of a special ceramic to which a suitable voltage is applied. It has been applied to an image-storage device which eliminates the need for photo-conductive layers. The ceramic used is a transparent lead lanthanum zirconate titanate.

PHOTOFINISH CAMERA. A camera used to record the finishing positions of contestants in a race. Commonly the film is moved behind a narrow slit in the focal plane to compensate for image movement. Sharp images of the contestants are then shown in the order in which they crossed the finishing line.

PHOTOFISSION. Nuclear fission induced by photons.

PHOTOFLUOROGRAPHY. Another name for fluorography, i.e. the photography of a fluoroscopic image.

PHOTOGRAMMETRY. The scientific measurement of photographs. One of its most important applications is to the making of maps from surveys carried out by aerial photography.

PHOTOGRAPHIC DENSITY. A measure of the opacity of a photographic image. For a transparency it is expressed as the *transmission density*, given by $\log(L_i/L_t)$, where L_i is the intensity of the incident light and L_t that of the transmitted light. For an opaque object, such as a paper print, it is expressed as the *reflection density*, given by $\log(1/R)$, where R is the ratio of the light reflected by the base (e.g. paper) to that reflected by the image. See also: *Callier coefficient. Density, diffuse. Density, specular.*

PHOTOGRAPHIC DEVELOPMENT. The conversion of a latent image into a visible image by treating a photographic emulsion with a suitable chemical solution.

PHOTOGRAPHIC EMULSION. A suspension of photosensitive material, commonly silver halide grains, in a medium such as gelatine. See also: *Graininess. Granularity. Resolution.*

PHOTOGRAPHIC EMULSION, CHARACTERISTIC CURVE OF. A curve showing the variation in photographic density with the logarithm of exposure, under specified conditions of processing. The slope of the approximately straight portion between the toe and the shoulder is a measure of the contrast. This slope is often known as the *gamma* of the emulsion.

PHOTOGRAPHIC EXPOSURE. The time integral of the illumination (i.e. of the luminous flux per unit area).

PHOTOGRAPHIC FOG. The darkening of a developed photographic emulsion or a portion of such an emulsion, when the portion affected had not been exposed to light or other radiation.

PHOTOGRAPHIC INERTIA. Of a photographic emulsion: the exposure corresponding to zero density when the linear portion of the characteristic curve is produced backwards.

PHOTOGRAPHIC INTENSIFICATION. The chemical treatment of a processed emulsion to increase the contrast or density or both.

PHOTOGRAPHIC REDUCTION. The chemical treatment of a processed emulsion to reduce the density. For certain methods of reduction the contrast may also be changed.

PHOTOGRAPHIC SENSITOMETRY. The measurement of the response, or sensitivity, to light of photographic materials. It involves a study of the treatment received by such materials (e.g. exposure and development) and the resultant blackening. See also: *Photographic density. Photographic exposure. Photographic speed.*

PHOTOGRAPHIC SPEED. A measure of the sensitivity of a photographic material. The higher the speed, the lower is the exposure necessary to produce a given density under the same conditions. Various criteria have been used for describing the speed, of which the best known are the *Hurter and Driffield*, which depends on the inertia; the *Scheiner*, which employs the threshold; the *DIN* (Deutsche Industrie Norm), which is related to a specified density; and the *ASA* (American Standards Association), which was formerly related to a specified fractional gradient of the characteristic curve, but is now also related to a specified density.

PHOTOGRAPHIC ZENITH TUBE (PZT). A photographic telescope with the optical axis pointed to the zenith. It is used for the exact determination of the time at which a star crosses the zenith, and for the measurement of latitude variation.

PHOTOGRAPHY. The production of more or less permanent images by exposing a suitable material to light or other radiation, followed by the appropriate treatment (usually chemical) of that material.

PHOTOGRAPHY, COLOUR. Photographic reproduction in colour, based upon the Helmholtz theory of colour vision. All processes involve the taking of photographs through red, blue, and green colour filters. In *additive processes* a mosaic screen (a minute pattern of red, green, and blue filter elements) is used and the primary colours are

253

printed; and in *subtractive processes* (now more usual) an *integral tripack* (three layers of emulsion sensitive to blue, green, and red light respectively, coated one above the other) is used, and the complementary colours are printed.

PHOTOGRAPHY, RACE-FINISH. The use of a special camera to record the finishing positions of contestants in a race. Commonly the film is moved behind a narrow slit in the focal plane, to compensate for image movement. Sharp images of the contestants are then shown in the order in which they crossed the finishing line.

PHOTOGRAPHY, SPARK. High-speed photography employing a short period of illumination afforded by a spark discharge. The actinic time of the spark may be as short as 10^{-6} s.

PHOTOGRAPHY, STREAK. A form of one-dimensional high-speed photography used, for example, in measurements on explosions, in plasma physics research, and in the study of electrical discharges. The field of view is restricted by a slit and the image moved progressively along the film or converted to an electron image in an image converter.

PHOTOGYRATION. The motion of ultra-microscopic particles (10^{-5} to 10^{-7} cm) suspended in a gas and subjected to intense illumination.

PHOTOIONIZATION. Ionization produced by photons of light, X-rays, or γ-rays.

PHOTOLOFTING. A process used in the shipbuilding and aircraft industries to enable dies and metal sheets to be cut accurately to plan. Negatives from photographs of the plans are projected onto sensitized sheets of metal. Development then leaves the lines of the full scale plan on the metal parts. See also: *Photomechanical process. Phototemplate.*

PHOTOLYSIS. The decomposition or dissociation of a molecule by light.

PHOTOMAGNETIC EFFECT. Photodisintegration ascribable to the magnetic vector of the photon.

PHOTOMECHANICAL PROCESS. Any technique employing a photographic process at some stage in the production of a forme (type matter assembled and locked in a frame) for mechanical printing. See also: *Photolofting. Phototemplate.*

PHOTOMESON. A meson produced by the interaction of a photon with an atomic nucleus.

PHOTOMETER. An instrument for the measurement of light, usually the luminous flux or the luminous intensity. It is mainly used for comparing two sources, typically an unknown with a standard. See also: *Spectrophotometer.*

PHOTOMETER, FLICKER. See: *Photometer, visual.*

PHOTOMETER HEAD. That portion of a visual photometer in which photometric comparison is effected, or that part of a physical photometer which contains the light-sensitive element.

PHOTOMETER, LUMINOSITY. See: *Photometer, visual.*

PHOTOMETER, PHYSICAL. A photometer in which measurement is made of some physical effect rather than by a visual method.

PHOTOMETER, VISUAL. A photometer in which an equality setting between an unknown source and a comparison source is made visually. It may involve matching luminosity (a *luminosity photometer*), matching contrast (a *contrast photometer*), or adjustment to minimize the flicker on a screen illuminated alternately and in rapid succession by the two sources (*flicker photometer*).

PHOTOMETRIC INTEGRATOR: INTEGRATING PHOTOMETER. An instrument which enables the luminous flux emitted by a source to be determined by a single measurement. The source is placed inside a large sphere, cube, or globe, the interior surfaces of which are painted matt white. The light is thus diffused throughout the sphere and some is allowed to escape through an aperture on to a photometer, usually photoelectric.

PHOTOMICROGRAPHY. The production of macroscopic images of microscopic objects, by an optical system of which a microscope forms a part. It should not be (but often is) confused with microphotography. See also: *Microphotography.*

PHOTOMULTIPLIER: ELECTRON MULTIPLIER. A sensitive detector of light in which the initial electron current, derived from photoelectric emission, is amplified by a series of stages of multiplicative secondary electron emission.

PHOTON. A quantum of electromagnetic radiation. Its energy is $h\nu$, where h is Planck's constant and ν is the frequency; and it has a spin of 1.

PHOTONEUTRON. A neutron produced in a photonuclear reaction.

PHOTONS, POLARIZATION OF. See: *Polarization.*

PHOTONUCLEAR REACTION. A nuclear reaction initiated by a photon.

PHOTOPHORESIS. The migration of small particles (10^{-5} to 10^{-1} cm), suspended in a gas, owing to irradiation by light. In the simple case the direction of movement is helical, about an axis that coincides with the direction of the incident light, but in the presence of a magnetic or electric field the axis coincides with the direction of the field. See also: *Electrophotophoresis. Magnetophotophoresis: Ehrenhaft effect.*

PHOTOREACTIVATION: PHOTORESTORATION. The recovery of a biological material from the effects of ultraviolet radiation, which is effected by further irradiation to light of somewhat longer wavelength (i.e. longer wavelength ultraviolet or short wavelength visible light).

PHOTOSCULPTURE. The use of enlarged photographs of a subject, taken at various angles, as a guide to the sculpturing of the subject.

PHOTOSENSITIZATION. The addition of a light-sensitive but non-reacting substance to induce or catalyse a photochemical reaction which would not otherwise take place. See also: *Photocatalysis.*

PHOTOSPHERE. Of a star: that region of the star which emits the main part of the flux of radiant energy escaping to the outside and being visible to the eye. See also: *Solar atmosphere.*

PHOTOSYNTHESIS. The process by which organisms containing chlorophyll utilize the energy of visible or near infrared light in the assimilation of carbon compounds.

PHOTOTEMPLATE. An exact full-scale pattern, obtained photographically, used as a guide in cutting out or machining some part of a manufactured article or mechanical structure. The process used involves the projection of a photographic negative on to a sensitized metal sheet, development of which produces an outline of the pattern on the metal. See also: *Photolofting. Photomechanical process.*

PHOTOTROPY. See: *Photochromism: Phototropy.*

PHOTOVOLTAIC CELL. Also known as a *photochemical cell.* (1) An electrolytic cell which sets up an electromotive force when exposed to radiation, commonly light. (2) A cell which generates a potential in the barrier layer of an electrode consisting of two types of material, when radiation is incident upon it. Such cells are in common use as radiation detectors. They are also known as

barrier-layer photocells and *rectifier photocells.* See also: *Barrier layer.*

PHYSICAL MEDICINE. The application of physical methods to the diagnosis and treatment of illness, and to the rehabilitation of patients by the development of artificial limbs, electrically operated "muscles", tools, and so on. The subject includes the use of all types of radiation.

PHYSICAL OPTICS. The treatment of optical phenomena from the point of view of the wave theory of light. See also: *Geometrical optics.*

PHYSICS. The study of the properties of matter and energy, and of their interactions. It involves making observations of, and experiments and measurements on, the phenomena encountered in that study, together with the theoretical treatment of those phenomena, preferably in terms of elementary entities, variously regarded as submicroscopic particles or manifestations of wave motion according to the context. Much of engineering science owes its origin to physics, which is also the basis of a large part of the sciences of chemistry, metallurgy, crystallography, astronomy, and electronics, to name a few. Physics is also of increasing importance in biology and medicine.

PI (π). The ratio of the circumference of a circle to its diameter. It is related to e, the base of the system of natural logarithms, by the expression $e^{i\pi} + 1 = 0$.

PICKERING SERIES: PICKERING–FOWLER SERIES. A series giving the frequencies in the line spectrum of singly ionized helium.

PICTET METHOD. A method of gas liquefaction using compression and expansion.

PIEZOELECTRIC AXES. The directions in a piezoelectric crystal along which tension or compression develops electrical polarization. See also: *Piezoelectric effect.*

PIEZOELECTRIC EFFECT. The production of electric charges of opposite sign at opposite ends of certain crystals when such crystals are compressed or extended in particular directions. Conversely, the production of a strain in certain directions when an electric field is applied. Substances which have a centre of symmetry do not show the piezoelectric effect but all crystals show a second-order effect, *electrostriction* or the *electrostrictive effect*, in which a strain appears when an electric field is applied. See also: *Piezomagnetic effect.*

PIEZOELECTRICITY. Electricity associated with the piezoelectric effect.

PIEZOELECTRIC OSCILLATOR. A crystal,

caused to oscillate at its mechanical resonant frequency by the application of an alternating potential difference, and used as a standard of frequency, or as a frequency stabilizer.

PIEZOMAGNETIC EFFECT. An inverse magnetostriction effect whereby the magnetic susceptibility and remanent (permanent) magnetization are affected by the application of mechanical stress. See also: *Piezoelectric effect.*

PILE. The former name for a nuclear reactor.

PILE OSCILLATOR. See: *Nuclear reactor oscillator.*

PILOT BALLOON. A small balloon, tracked by a theodolite, used to determine the strength and direction of the wind.

PI MESON: PION. The meson that is principally concerned in nucleon–nucleon forces. See also: *Meson. Meson theory of nuclear forces.*

PINACOID. In crystallography: an open crystal form which consists solely of a pair of parallel faces. Such a form may result by repetition of a plane about a centre of symmetry.

PINCH EFFECT. In a liquid conductor, tubular solid conductor, or a plasma column carrying a large current: a constriction in the conductor or plasma due to the interaction of the current with its own magnetic field.

PINHOLE CAMERA. A photographic camera in which the lens is replaced by a small hole. The optimum diameter of the hole is given by $1 \cdot 9 \sqrt{a\lambda}$, where a is the distance of the image screen from the hole and λ is the wavelength of the light.

PION. See: *Pi meson: Pion.*

PIONIUM. See: *Mesonic atom: Mesic atom.*

PIPE DIFFUSION. Atomic migration along dislocation lines in metals.

PIPETTE. A glass tube which is designed to deliver a measured amount of liquid. It may carry calibrated graduations, or may incorporate a device which delivers the required amount automatically.

PISTON ATTENUATOR. In a waveguide: a variable length of cut-off waveguide in which one of the coupling devices is carried on a sliding member like a piston. See also: *Electrical attenuator.*

PISTONPHONE. A mechanical device for generating a desired sound pressure, used for determining the electroacoustic sensitivity of microphones.

PITCH. (1) Of a sound: a subjective property of a simple or complex tone by which the ear is enabled to allocate the position of the tone on a frequency scale. (2) Of an aerofoil cascade: see: *Aerofoils, cascade of.* (3) Of a propeller: the distance through which the propeller advances along its axis during one revolution. Where no thrust is given the distance is known as the *experimental mean pitch*, the actual distance under conditions of thrust being the *effective pitch*. (4) Of a screw: the axial distance between adjacent turns of a single thread.

PITCHING. Of a ship at sea or an aircraft in the air: the movement of the ship or aircraft about the lateral axis. See also: *Rolling. Yawing.*

PI THEOREM. The general theorem of dimensional analysis. It states that: provided an equation connecting the measures of certain physical quantities is true, irrespective of the sizes of the fundamental units, it can always be written in an equivalent form as a relation between non-dimensional combinations of the physical quantities.

PITOT TUBE. Consists essentially of a cylindrical tube open at one end and connected at the other to a suitable device for measuring pressure. Such a tube, with its open end facing into a moving fluid, registers a pressure greater than the static pressure and may be combined (typically in an aeroplane) with a tube which measures that pressure, in a device (a *pitot-static tube*) for determining the speed of movement relative to the fluid. Values measured in this way are accurate only at speeds well below sub-sonic.

PLAIN GAUGE. A limit gauge which serves to check a single dimension, such as diameter, length, or height.

PLANCK CONSTANT. A universal constant (the *elementary quantum of action*) by which the frequency of a quantum of energy is to be multiplied to give the quantum energy. Its value is $6 \cdot 6262 \times 10^{-27}$ erg s, or $6 \cdot 6262 \times 10^{-34}$ Joule s. Its symbol is h, and $h/2\pi$ is often written as $\hbar$. See: *Appendix III.*

PLANCK RADIATION FORMULA. For the intensity of radiation emitted by a black body in the wavelength band between λ and $\lambda + d\lambda$: may be written as

$$E_\lambda \, d\lambda = \frac{hc^3}{\lambda^5} \cdot \frac{d\lambda}{\exp(hc/k\lambda T) - 1},$$

where $E_\lambda \, d\lambda$ is the required intensity, h is the Planck constant, c the speed of light, λ the wavelength, k the Boltzmann constant, and T the absolute temperature. It may be written in terms of wavenumber instead of wavelength, or energy density instead of radiation intensity. In the expression

256

given above, hc^3 is known as the *first radiation constant* and hc/k as the *second radiation constant*.

PLANE, COMPLEX. A plane in which may be represented functions of complex numbers or complex variables, typified by the *Argand diagram*, which is defined separately.

PLANE OF VIBRATION. See: *Plane-polarized radiation: Linearly-polarized radiation*.

PLANE OF WEAKNESS, GEOLOGICAL. A fracture system near the surface of the Earth below which active seismic phenomena take place.

PLANE-POLARIZED RADIATION: LINEARLY-POLARIZED RADIATION. Electromagnetic radiation so polarized that the vibrations lie in one single plane, the *plane of vibration*. The electric vector lies in this plane, which is sometimes called the *plane of polarization*, although the term is also applied to the plane at right angles to this (i.e. the plane containing the magnetic vector), which may lead to confusion.

PLANET. A massive non-luminous body revolving about a star, visible only by the light it reflects from that star. From Earth eight other planets are visible. See also: *Solar system*.

PLANE-TABLE SURVEYING. A method of surveying in which the field work and the plotting are executed simultaneously.

PLANETARIUM. A building in which spectators can see a realistic reproduction of the whole night sky. A complex optical projection instrument, situated at the centre of a hemispherical dome, projects images of the Sun, Moon, planets, and stars on the inner surface of the dome, and the motions of the various celestial objects are presented by appropriate movements of one or more of the projectors of which the instrument is composed.

PLANETARY ABERRATION. See: *Aberration of light*.

PLANETARY RADAR. That branch of radar astronomy which is concerned with such information as the positions, velocities, and physical states of planets.

PLANET, INFERIOR. A planet whose orbit lies inside that of the Earth, viz. Mercury or Venus.

PLANET, MAJOR. A planet whose mass is greater than that of the Earth, viz. Jupiter, Saturn, Uranus, or Neptune. Such a planet is sometimes described as a *giant planet*.

PLANET, MINOR. (1) A planet whose mass is equal to or less than that of the Earth, viz. Mercury, Venus, Earth, Mars, and Pluto. Such a planet is also known as a *terrestrial planet*. (2) An *asteroid*.

PLANET, STATIONARY POINTS OF. Those two points in a planet's orbit where it appears to have no motion as viewed from the Earth. The apparent motion of the planet changes from direct to retrograde at one of these points and from retrograde to direct at the other.

PLANET, SUPERIOR. A planet whose orbit lies outside that of the Earth, viz., Mars, Jupiter, Saturn, Uranus, Neptune, and Pluto.

PLANIMETER. A mechanical integrator for measuring the area of a plane figure.

PLAN POSITION INDICATOR (PPI). An intensity-modulated radar display indicating, as on a map, the relative positions of echo-producing objects.

PLASMA. A highly ionized gas in which the ionization is so great that the dynamical behaviour of the gas is dominated by the interaction of free ions and electrons. It is thus an electrically neutral assembly of electrons and positive ions. Such a plasma has sometimes been regarded as a fourth state of matter.

PLASMA, COLD. A plasma in which the positive ions remain near room temperature, although the electrons reach much higher temperatures.

PLASMA FREQUENCY. The natural frequency of oscillation of a plasma, arising from the collective motion of the electrons. It is proportional to the square root of the electron density.

PLASMA, SOLID STATE. A system of mobile charges which is contained by a fixed matrix of atoms and responds collectively to external stimuli. The number density of the positive and negative mobile charges need not be the same, as in gaseous plasma, since excess charge may be balanced by charges in the matrix. When the number densities, however, do happen to be equal the plasma is termed a *compensated plasma*. Typical solid state plasmas are the one-component plasma of electrons moving in the field of fixed positive ions in a metal, and the two-component plasma of electrons and positive holes in a semiconductor.

PLASMA, THERMAL. A plasma which is at or near to thermal equilibrium so that a single temperature will suffice to specify most of its properties.

PLASMA TORCH. A device used to heat gases electrically to very high temperatures. The hot gases can be used for such purposes as chemical synthesis and metal fabrication.

PLASTIC DEFORMATION. Deformation which occurs at a level of stress above the elastic limit, and is therefore permanent.

PLASTOMETER. An instrument for determining plasticity, typically from the measurement of the rate of approach, for a given load, of two plates between which the substance under investigation is placed.

PLATE TECTONICS. The explanation of continental drift in terms of large-scale movements of sections of the outer shell of the Earth, known as *plates,* which move relative to each other. It is believed that there are six such plates, five of which carry continental crust. See also: *Continental drift. Tectonics.*

PLATINUM METALS. A well-defined group of metals in the Periodic System of the elements, characterized by high melting point and resistance to chemical attack. The group comprises platinum, ruthenium, rhodium, osmium, iridium, and palladium. The platinum metals are widely used as catalysts in chemical technology.

PLEOCHROIC HALO. A small coloured ring or group of concentric rings found in certain minerals (e.g. mica) which contain radioactive inclusions. The colour is due to damage caused by α-particles, which produce maximum ionization at the end of their range, which is 10–50 μm in most minerals.

PLEOCHROISM. See: *Dichroism.*

PLETHYSMOGRAPHY. The measurement of volume changes at the extremity of a part of the body caused by changes in the rate of blood flow.

PLUTO. The outermost planet in the solar system. Its diameter is about half that of the Earth, its mass much less than that of the Earth, and its period of rotation about 6 days. Its temperature is about 60 K.

PNEUMATIC GAUGING. A technique by which a dimensional change is converted into a change in flow rate in a pneumatic circuit. Typically the measurement involves the position of a barrier situated in front of an air outlet, the displacement of the barrier changing the flow through the outlet, which may be accurately measured.

POCKELS EFFECT. The alteration in the refractive properties in a piezoelectric crystal by the application of a strong electric field. The effect is used in the Mekometer to produce a light beam modulated at microwave frequency. See also: *Kerr effect. Mekometer.*

POGGENDORF COMPENSATION METHOD. A method of determining an e.m.f. by balancing it against the voltage drop along a uniform resistance wire in a potentiometer.

POIKILOTHERMIC. Having or concerning a temperature which varies with the surroundings. It describes, for example, a cold-blooded animal.

POINT BRILLIANCE. The illumination produced by a point source of light at the pupil of the eye. It is usually measured in microlux (i.e. microlumen/m^2).

POINT DEFECT. A crystal defect which is located at a point in the crystal structure. It may arise from a missing atom which leaves behind a *vacant lattice site*; or it may be an intruder atom which takes up a position between the normal atomic positions, and is known as an *interstitial atom*; or it may be an impurity atom, which may occur as an interstitial atom or may be substituted for a normal atom. A *vacancy-interstitial pair*, produced when an atom is displaced (leaving a vacancy) to take up an interstitial position, is known as a *Frenkel defect.* A *Schottky defect* is formed either as a vacancy when an atom is displaced and moves to the surface, or as an interstitial when an atom moves in from the surface. In this context "surface" is meant to include grain boundaries or dislocations.

POINT-GROUP. One of the thirty-two groups of macroscopic symmetry elements which correspond to the thirty-two crystal classes. See also: *Crystal class.*

POINT PROJECTION MICROSCOPE. See: *Electron microscope, point projection.*

POISE. The unit of dynamic viscosity in the CGS system, defined as the tangential force per unit area (dyn/cm^2) required to maintain unit difference in relative velocity (1 cm/s) between two parallel plates separated by 1 cm of fluid. It is equal to 1 dyn-s/cm^2 or 10^{-1} Ns/m^2. The SI unit of viscosity is Ns/m^2 or pascal second. See also: *Kinematic viscosity. Viscosity, coefficient of. Appendix II.*

POISEUILLE FLOW. Laminar viscous flow through a long pipe of circular cross-section. See also: *Poiseuille equation.*

POISEUILLE EQUATION. For liquid flow through a capillary: states that

$$V = \frac{\pi r^4 p}{8 \eta l},$$

where V is the volume of liquid flowing per second through a capillary of length l and radius r under the influence of a pressure difference p, and η is the coefficient of viscosity.

POISON. (1) In a nuclear reactor: a substance which, because of its high neutron absorption, reduces the reactivity of the reactor. (2) In a phosphor: a material that reduces the sensitivity of the phosphor. (3) In a cathode: a material which reduces the emission from the cathode.

POISSON DISTRIBUTION. A frequency distribution that applies when the probability of an event happening is very small. The frequency distribution for 0, 1, 2, 3, ..., occurrences is given by the successive terms of the *Poisson series*

$$Ne^{-m}\left(1, m, \frac{m^2}{2!}, \frac{m^3}{3!}, \cdots\right),$$

where N is the total number of observations and m is the mean number of occurrences.

POISSON EQUATION. A partial differential equation which describes the electric field or potential produced by a given distribution of charge density.

POISSON RATIO. For an elastic body subjected to elongation: the ratio of the lateral strain (contraction) to the longitudinal strain (extension). See also: *Elastic modulus.*

POLAR AXIS. Of a crystal: an axis of crystal symmetry which has neither a plane of symmetry nor an axis of even symmetry at right angles to it. The two ends of such an axis are dissimilar and the crystal may show such properties as pyroelectricity.

POLAR BOND. An alternative name for an ionic bond.

POLARIMETER. An instrument for measuring optical rotation, usually by the use of a compensator.

POLARIMETRY. The measurement of optical rotation.

POLARIZATION. A term used in a variety of contexts to denote the production of directional dependence in some physical property which is normally independent of direction or dependent in a different sense. The use of the term is illustrated in the following examples. (1) *Polarization of a dielectric:* (a) the separation of electrical charges of opposite sign under the influence of an electric field, so as to produce electric dipoles; (b) a vector quantity expressing the dipole moment per unit volume. (2) *Polarization of an electrolyte:* the production in an electrolytic cell of an e.m.f. acting in opposition to the e.m.f. of the cell itself. (3) *Polarization of ions in a crystal:* the distortion of the electronic structure of the ions in an electric field. It is to be distinguished from the displacement in opposite directions of the ions themselves. The effect is sometimes known as the *polarization of atom cores.* (4) *Polarization of submicroscopic particles with non-zero spin:* the tendency of the spins to be aligned in a particular direction, often as the result of some process such as scattering. The particles may be *atomic nuclei, electrons, neutrons,* protons, mesons, etc., including *photons.* Polarization of the latter is, however, usually described as (5) *Polarization of electromagnetic radiation:* the partial or complete suppression of the electromagnetic vibrations in certain directions.

POLARIZATION, CIRCULAR. Of a beam of electromagnetic radiation: the type of polarization produced when two mutually perpendicular, plane-polarized components, of equal amplitude, differ in phase by $90° \pm n\pi$ $(n = 0, 1, 2)$. Where the two components are not of equal amplitude, the resultant polarization is *elliptical.* See also: *Fresnel rhomb.*

POLARIZATION CURRENT. See: *Current.*

POLARIZATION, DIRECTION OF. For electromagnetic radiation: the direction of the electric or magnetic vector as specified.

POLARIZATION, ELLIPTICAL. See: *Polarization, circular.*

POLARIZATION FACTOR (X-RAY ANALYSIS). An angle-dependent factor which occurs in the expression for the intensity of X-rays reflected by atoms. It arises because the intensity of scattering by an individual electron depends on the direction of the electric field in the incident radiation. Since neutrons are scattered by the nuclei no such factor occurs for neutron diffraction.

POLARIZATION, PLANE OF. See: *Plane-polarized radiation: Linearly-polarized radiation.*

POLARIZATION, ROTATORY. The rotation of the plane of polarization of a beam of light by anisotropic refraction. See also: *Optical rotation.*

POLARIZATION VECTORS. In the theory of the electromagnetic field: two vectors, **P** and **M**, given by $\mathbf{P} = \mathbf{D} - \varepsilon_0\mathbf{E}$ and $\mathbf{M} = \mathbf{B}/\mu_0 - \mathbf{H}$, where **D** is the displacement, **E** the electric field strength, **B** the magnetic induction, **H** the magnetic field strength, and ε_0 and μ_0 are the permittivity and permeability respectively in empty space. See also: *Electromagnetic field.*

POLARIZER. A device which produces plane-polarized light from a beam of unpolarized light passing through it. It may be a polarizing prism or a polaroid sheet. See also: *Analyser (polarized light). Polarizing prism. Polaroid.*

POLARIZING ANGLE. Of a beam of electromagnetic waves: the angle at which a beam of such waves, reflected by a refracting medium, is completely plane polarized. See also: *Brewster law.*

POLARIZING PRISM. A prism of a doubly refracting crystal which produces plane-polarized light from a beam of unpolarized light passing through it. The most common of such prisms are those of *Nicol, Rochon, de Sénarmont,* and *Wollaston,* which are made from calcite or quartz.

POLAR LIGHTS. See: *Aurorae.*

POLAR MOLECULES. Molecules which possess a permanent dipole moment. See also: *Dipole, electric.*

POLAROGRAPHY. The measurement of the relation between current and potential difference in a solution from the current response of a dropping-mercury electrode and its interpretation in terms of the nature and concentration of the materials involved. It is a special case of voltammetry. See also: *Voltameter: Coulometer. Voltammetry.*

POLAROID. The trade name of a synthetic polarizing filter consisting of preferentially oriented sub-microscopic polarizing crystals supported on a plastic base. Although not as efficient as a Nicol prism, polaroid filters can be produced in large sheets and are thus specially suitable for observation with the naked eye.

POLAR SOLUTION. One in which the molecules of the solvent are strongly polar.

POLE. (1) A magnetic pole. (2) An electrode in an electrical cell. (3) In a crystal projection: the points at which normals to the crystal planes cut the reference sphere, at the centre of which the crystal is assumed to be located.

POLE FIGURE. A stereographic projection showing the contours of pole density for a specified set of crystal planes. Pole figures are used to convey an accurate description of preferred orientation in polycrystalline materials.

POLES, AXIS. (1) *Celestial:* the two points at which the Earth's axis of rotation, produced indefinitely, cuts the celestial sphere. (2) *Terrestrial or geophysical:* The two points at which the Earth's axis of rotation cuts the Earth's surface. (3) *Magnetic:* See: *Geomagnetic poles.*

POLYAD. A generic name for groups of spectral terms of the same multiplicity which arise from the addition of an *s, p,* or *d* electron to the electron configuration of an ion. The polyads resulting from such additions are known respectively as *monads, triads,* and *pentads.*

POLYGONIZATION. The re-arrangement of edge dislocations in the form of a wall, which takes place when a plastically bent crystal is heated to a sufficiently high temperature. Such a wall (a *polygon wall, tilt boundary,* or *bend plane*) constitutes a small-angle boundary (known as a *sub-boundary*) between two sub-grains which have lost their original elastic curvature. See also: *Grain boundary.*

POLYGON WALL. See: *Grain boundary. Polygonization.*

POLYMER. A substance consisting of associated molecules, and therefore containing large numbers of atoms, in which molecular weight and molecular length become extensive quantities. A *high polymer* or *macromolecule* usually denotes a substance having an average molecular weight greater than 5000. Such polymers are used widely as plastics, elastomers, adhesives, etc. See also: *Monomer.*

POLYMERIZATION. The process by which molecules of one or more kinds undergo chemical reactions to yield products of high molecular weight as a result of the formation of repeating units held together by primary valence forces.

POLYMORPHISM. The existence of different structural or crystalline forms of the same chemical compound. It is sometimes known as *allomorphism.* The choice of the form taken depends largely on temperature and pressure. See also: *Allotropy.*

POLYNOMIAL. A rational integral function, also known as a *multinomial,* in n variables, of the form

$$c_1 x_1{}^{\alpha_1} x_2{}^{\beta_1} \ldots x_n{}^{\nu_1} + c_2 x_1{}^{\alpha_2} x_2{}^{\beta_2} + \ldots x_n{}^{\nu_2} + \ldots$$
$$+ c_k x_1{}^{\alpha_k} x_2{}^{\beta_k} \ldots x_n{}^{\nu_k},$$

where, for any one term c_i is the coefficient, α_i is the degree with respect to x_1, β_i with respect to x_2, etc., and the total degree of the term is $\alpha_i + \beta_i + \ldots + \nu_i$. The commonest case is the nth degree polynomial in one variable, which may be written as $a_0 x^n + a_1 x^{n-1} + \ldots + a_{n-1} x + a_n$. Polynomials frequently occur in the solutions of differential equations. If they have two, three, four, etc., variables, they are said to be binary, ternary, quaternary, etc., forms. If their degree is 1, 2, 3, etc., they are said to be linear, quadratic, cubic, etc. forms. See also: *Algebraic equation. Hermite equation.*

POLYPHASE SYSTEM, ELECTRICAL. A system supplied from a polyphase generator in which e.m.f.s of differing time-phase, but of the same frequency, are generated in several phases. These separate voltage sources are usually interconnected in special ways. The commonest system is the three-phase system. See also: *Delta connection.*

POLYSACCHARIDE. An organic compound whose molecules consist of two or more molecules of a sugar, held together by a glycoside linkage.

POLYTROPIC CHANGE OR PROCESS. Refers to any process involving the compression or expansion of a constant mass of gas. All such changes are reversible and can be represented on a pressure–volume graph by a family of curves $PV^n =$ constant, where P is the pressure, V the volume, and n may take any positive value between zero and infinity. When $n = 0$ the change is *isobaric*, when $n = 1$ it is *isothermal*, when $n = \gamma$ (the ratio of the specific heat at constant pressure to that at constant volume) it is *adiabatic*, and when $n = \infty$ it is *isometric*.

POLYTYPISM. Of a crystal: the existence of different forms of crystal structure that are so closely related to each other, being composed of identical layers which differ only in their stacking together, that the use of the term polymorphism would be inappropriate.

POMERANCHUK THEOREM. States that the cross-sections of particles and anti-particles of the same type incident on the same target particle should approach the same, constant, value as the incident energy becomes very large.

POROSITY. (1) The property of a solid of containing pores, i.e. minute channels, and open or closed spaces. (2) The proportion of the total volume occupied by such pores.

POSITION ANGLE. A measure of the orientation of one point on the celestial sphere with respect to another.

POSITION LINE. A line on a navigational chart, obtained from observations of terrestrial or celestial objects, or from radio or radar aids, on which the position of the observer is computed to lie at the time of the observation.

POSITION OF LEAST CONFUSION. See: *Astigmatism.*

POSITIVE COLUMN. In a glow discharge tube or arc discharge: an extensive region consisting of a fully ionized plasma, which fills much of the distance between anode and cathode. In the glow discharge it lies between the anode dark space and the Faraday dark space. See also: *Discharge, glow.*

POSITIVE CRYSTAL. See: *Optic sign.*

POSITIVE RAYS: CANAL RAYS. Positively charged particles, consisting of atoms or molecules of gaseous matter, which are produced in an electric discharge at low pressure. They are impelled towards the cathode and will pass through it, if it is perforated, in the form of a beam. The separation of the positive ions in such a beam forms the basis of the mass spectrograph and mass spectrometer. See also: *Mass spectrography: Mass spectrometry.*

POSITRONIUM. An unstable union of a positron and an electron, which gives a hydrogen-like atom whose proton has been replaced by a positron. See also: *Orthopositronium. Parapositronium.*

POSITRON: POSITON. A positively charged electron. See also: *Electron.*

POST-ACCELERATION. In an electron beam tube: the acceleration of the electrons after deflection.

POTENTIAL. At a point in a field of force: a term used in a variety of contexts to denote the work required to move a unit of an appropriate kind (e.g. a unit of electrical charge) from some reference position (commonly infinity) to the point in question. It may be positive or negative according to circumstances. Examples are: *electrical potential, magnetic potential, nuclear potential,* and *gravitational potential,* which are separately defined. If the state of a substance is involved rather than its position, a similar concept holds for *thermodynamic potential,* which is also separately defined. The term potential is sometimes used in the sense of potential difference. See also: *Chemical potential.*

POTENTIAL BARRIER. The region of high potential energy through which a charged particle must pass on leaving or entering an atomic nucleus. It is also known as a *nuclear barrier* or a *Gamow barrier.* See also: *Coulomb barrier.*

POTENTIAL, BIOLOGICAL. An electrical potential difference existing between the inside of a living cell and the external medium in which it exists.

POTENTIAL, COEFFICIENTS OF. The coefficients which specify the potentials of a system of conductors in terms of their charges.

POTENTIAL DEPRESSION. A region of low potential in a field of force. See also: *Potential well.*

POTENTIAL DIFFERENCE. A general term signifying a difference in potential energy, or of voltage, which exists between two points in a circuit or between two electrically charged bodies. See also: *Electromotive force (e.m.f.).*

POTENTIAL DIVIDER: VOLTAGE DIVIDER. A high resistance, provided with a fixed or adjustable tapping, used to provide a voltage between the tapping and one end terminal of the resistance, which is a known fraction of the applied voltage.

POTENTIAL ENERGY. Energy stored in a body or system by virtue of the position of the body or the configuration of the system. It is the work done in bringing the body from some reference position to the position of interest, or in bringing the system from some reference configuration to the configuration of interest; and is expressed or implied with respect to that reference position or configuration.

POTENTIAL ENERGY CURVES AND SURFACES. Curves and surfaces depicting the potential energy of two or more atoms as a function of coordinates describing their relative positions. See also: *Morse curve. Non-crossing rule.*

POTENTIAL FUNCTION. (1) A function which satisfies Laplace's equation $\nabla^2 \psi = 0$. (2) A scalar function which describes the forces acting on any particle of a conservative system. It satisfies the equation $\mathbf{F} = -\nabla V$, where $\mathbf{F}$ is the resultant force vector and V the potential function.

POTENTIAL HEAD. See: *Elevation head.*

POTENTIAL, SCALAR. A scalar function of position ϕ, such that the field vector $\mathbf{F}$ is given by $\mathbf{F} = \nabla \phi$, where $\mathbf{F}$ is an *irrotational* vector, i.e. $\nabla \times \mathbf{F}$ is zero. $\mathbf{F}$ can represent the gravitational, electrostatic, or magnetostatic field vectors, or the velocity vector of a fluid, etc. In these cases ϕ would represent the *gravitational potential, electrostatic potential, magnetostatic potential,* or *velocity potential* respectively. See also: *Vector potential.*

POTENTIAL SCATTERING. Of a particle by an atomic nucleus: scattering in which the incident particle is considered to be reflected at the surface of the nucleus as though the latter were a hard sphere.

POTENTIAL TRANSFORMER: VOLTAGE TRANSFORMER. A transformer whose primary winding is connected to the main circuit and whose secondary winding to an instrument, e.g. a voltmeter. It permits instruments to be isolated from a high voltage supply and is used to extend their range.

POTENTIAL VECTOR. Of a vector field: the vector $\mathbf{A}$, of zero-divergence, in the expression $\mathbf{F} = \nabla \phi + \text{curl } \mathbf{A}$ for a vector field $\mathbf{F}$ which is finite, uniform and continuous, where ϕ is the scalar potential.

POTENTIAL WELL. A pictorial description of the region within a potential barrier in which a particle is contained unless it can "jump" the barrier or "tunnel" through it. See also: *Nuclear potential well. Potential barrier. Tunnelling: Tunnel effect.*

POTENTIOMETER. (1) An instrument for measuring an unknown potential difference by balancing it wholly or partially by a known potential produced by the flow of current in a resistance network of known value. (2) A name sometimes used for a potential divider.

POTENTIOSTAT. A device for the automatic maintenance of the potential of an electrode in an electrolyte while conditions in the electrolyte or at the surface of the electrode are changing.

POTTER–BUCKY DIAPHRAGM. A device incorporating a moving grid which, while collimating an X-ray beam, eliminates grid shadows from a radiograph.

POUND. The unit of mass in the Imperial system. It is the mass of a platinum cylinder kept by the Board of Trade in London, and is, by definition, equal to 0·453 592 37 kg exactly. See: *Appendix II.*

POUNDAL. That force which will produce an acceleration of 1 ft/s² on a mass of 1 lb. It is equal to 0·138 255 N. See: *Appendix II.*

POUND CALORIE. See: *Centigrade heat unit.*

POUND WEIGHT. The gravitational force acting on a mass of 1 lb. It is equal to g poundals, where g is the acceleration due to gravity at the locality under consideration.

POWDER METHOD. In crystal structure analysis: the examination of polycrystalline material by X-ray, electron, or neutron diffraction. The diffracted beams may be detected, and their positions and intensities measured, either photographically (as in an X-ray *powder camera*) or by means of a *diffractometer.*

POWDER PATTERN. (1) The pattern recorded by an X-ray powder camera. (2) A Bitter figure, or magnetic powder pattern.

POWER. The rate at which work is performed. The SI unit is 1 joule per second (1 W), the CGS unit is 1 erg per second (10^{-7} W), and the British unit is the horsepower (745·700 W). The metric horse power is equal to 735·5 W. See: *Appendix II.*

POWER FACTOR. In an a.c. network: the factor by which the product of the voltage and current must be multiplied to give the mean power. It is given by $\cos \theta$, where θ is the phase difference between the voltage and current.

POWER LOSS. Power dissipated as heat in a dielectric on account of its finite conductivity.

POWER SOURCES, ISOTOPIC. Radioactive isotopes which may be used to produce electrical or mechanical energy. The three main ways are: (a) in a direct-charge cell, charged particles being collected at one electrode and the isotope source being used at the other; (b) in an ionization cell, made for example by depositing a thin layer of isotope on a p–n semiconductor junction; (c) as a source of heat, which may be used to drive a heat engine, e.g. a thermoelectric or thermionic converter, or a Stirling or similar engine. Power densities range from 170 W/g for ^{228}Th to 0·35 W/g for ^{3}H.

See also: *SNAP* (*Systems for nuclear auxiliary power*).

POYNTING THEOREM. States that the rate of flow of electromagnetic energy through a surface is equal to the *Poynting vector*, i.e. to the vector product of the electric and magnetic intensities.

PPI. See: *Plan position indicator* (*PPI*).

PRANDTL–MEYER EXPANSION. An expansion wave produced at a corner by a homentropic two-dimensional supersonic flow.

PRANDTL NUMBER. A dimensionless number given by the ratio of the kinematic viscosity to the diffusivity. If expresses the ratio of the diffusivity of momentum to that of temperature through a fluid. See also: *Dynamical similarity principle*.

PRE-AMPLIFIER: HEAD AMPLIFIER. A unit containing the input stage of an amplifying system and designed to be mounted integrally with, or close to, the detector.

PRECESSION. The motion of the axis of a rotating body about a cone when a torque is applied to it so as to tend to change the direction of the axis. The motion of the axis at any instant is at right angles to the direction of the torque.

PRECESSION OF THE EQUINOXES. The westward motion of the equinoxes through the stars in which they precede the constellations in their apparent motion towards the Sun. The effect is due to the precession of the Earth's axis which in turn is due to the gravitational attraction of the Sun and Moon on the Earth. When the effect was discovered in 120 B.C. the vernal equinox was in Aries. It is now in Pisces. See also: *Earth, motions of*.

PRECIPITABLE WATER. The total mass of water vapour in a vertical column of atmosphere of unit cross-section. It is often expressed in centimetres of water.

PRECIPITATION. (1) In chemistry: the formation of a relatively insoluble solid by a reaction which takes place in solution. (2) In meteorology: moisture falling on the Earth's surface from clouds. In may be in the form of rain, hail, or snow. (3) In metallurgy: the decomposition of a solid solution into more stable phases of different composition.

PRECIPITATION HARDENING. The hardening of a material (usually metallic) resulting from the precipitation of a new phase from solid solution. See also: *Age hardening: Precipitation hardening*.

PRECISION. Of an instrument or measuring device designed to measure a particular quantity: the smallest measurable difference in the value of that quantity.

PREDISSOCIATION. In spectroscopy: the decomposition of an excited molecule into un-ionized fragments without emitting radiation.

PREFERRED ORIENTATION. Of crystallites: the arrangement of the crystallites in a polycrystalline material in such a way that one, two, or three of their crystallographic directions each tend to be parallel to identifiable directions related to the shape or macrostructure of the material. Where only one direction is involved we have a fibre structure; where two are involved we have parallel planes in the material, as in a layer lattice; and where three directions are involved we have an approximation to a single crystal. See also: *Fibre diagram. Layer lattice. Pole figure*.

PREIONIZATION. In spectroscopy: the decomposition of an excited molecule into ionized atoms without emitting radiation.

PRESSURE. A stress applied uniformly in all directions. The term is, however, sometimes loosely applied to a directional stress. The SI unit of pressure is 1 newton per square metre (N/m^2), also known as the pascal. The CGS unit is the dyn/cm^2 (equals $10^{-1} N/m^2$) and pressure in the British system is expressed in pounds or tons per square inch. (One pound—i.e. one pound-force—per square inch is equal to $6·895 kN/m^2$, and one ton—i.e. one ton-force—per square inch is equal to $15·4 MN/m^2$). Other units in common use are the *atmosphere* ($101 325 N/m^2$ or $1013·250$ millibar), the *torr* and the conventional *millimetre of mercury* (both equal to $133·322 N/m^2$), and the *bar* (equals $10^6 dyn/cm^2$ or $10^5 N/m^2$).

PRESSURE ACCUMULATOR. A hydraulic device in which a suitable fluid is compressed and stored for future use.

PRESSURE, BACKING. The pressure above which certain types of vacuum pump cannot operate. An auxiliary pump, known as a *backing pump*, must then be used to reduce the pressure sufficiently.

PRESSURE BALANCE. An instrument for the measurement of high pressures in which the fluid pressure acting on a vertical piston of known effective area is balanced by a load derived from accurately calibrated weights.

PRESSURE BROADENING. Of spectral lines: broadening arising from the perturbing effects of other atoms on a radiating atom. The effect increases with pressure and very high pressures are often employed in its investigation. Pressure broadening is always accompanied by a displacement of the maximum of a line and the asymmetry of the line profile. See also: *Spectral lines, broadening of.*

PRESSURE, CENTRE OF. (1) Of an aerofoil: that point at which the resultant of the aerodynamic forces on the aerofoil cuts the chord line or some similar reference line. Its distance behind the leading edge is usually given as a fraction of the chord length. (2) Of a surface immersed in a fluid: that point at which the resultant pressure on the surface may be taken to act.

PRESSURE, DYNAMIC. The impact pressure associated with a moving fluid. The sum of this and the static pressure, i.e. the pressure which would exist in the absence of motion, is known as the *total head.* The dynamic pressure is also known as the *dynamic head.*

PRESSURE GAUGE. An instrument for the measurement of pressure. It may be a *primary gauge*, in which the pressure is balanced against a known force, or a *secondary gauge*, in which some physical property is measured which varies with pressure, and which requires calibration against a primary gauge. Some well-known types of pressure gauge are described below.

PRESSURE GAUGE, BOURDON. One which consists essentially of a flattened tube, closed at one end and bent into the arc of a circle, which tends to straighten under internal pressure, and may be made to operate a pointer to record pressure changes.

PRESSURE GAUGE, IONIZATION. For the measurement of low pressures: (a) A thermionic triode connected to the vacuum to be measured, the pressure being obtained from measurement of the grid current. (b) A wire ring anode between two metal plate cathodes, all placed within the vacuum to be measured. The pressure is obtained from measurement of the ionization current produced when the gas molecules are ionized by electrons ejected from the plates and caused to take a spiral path by the application of an outside magnetic field. This type of gauge is the *Penning gauge.*

PRESSURE GAUGE, KNUDSEN. For the measurement of low pressures: a gauge in which is measured the force exerted on a light vane suspended in a gas between two heaters when a small temperature difference is maintained between them, the momenta of the molecules striking the two sides of the vane being thereby unbalanced. See also: *Knudsen absolute manometer.*

PRESSURE GAUGE, McLEOD. A gauge used for measuring low gas pressures by compressing a sample by a known amount until its pressure can be measured by ordinary manometric methods.

PRESSURE GAUGE, PENNING. See: *Pressure gauge, ionization.*

PRESSURE GAUGE, PHILIPS. A cold-cathode ionization gauge.

PRESSURE GAUGE, PIRANI. For the measurement of low gas pressures at which the loss of heat from a heated wire is proportional only to pressure. The rate of cooling of such a wire is measured by observing its change of resistance.

PRESSURE GAUGE, THERMOCOUPLE. For the measurement of low pressures: a gauge which depends on the thermal conductivity of the residual gas, and involves the measurement of the thermal e.m.f. of a heated thermocouple when a state of equilibrium has been reached. Calibration is essential.

PRESSURE HEAD. For a perfect fluid in steady flow: the ratio of the static pressure to the product of gravitational acceleration and density. Pressure head + velocity head + elevation head = constant, according to the Bernoulli equation. See also: *Elevation head. Velocity head: Kinetic head.*

PRESSURE, STATIC. Of a moving fluid: the pressure which would be measured by an infinitesmally small instrument at rest relative to the fluid.

PRESSURE, TOTAL. Of a moving fluid: the pressure which would arise if the fluid were brought to rest isentropically and adiabatically.

PREVOST EXCHANGE THEORY. A theory propounded in about 1800, which stated that a body radiates the same amount of energy at the same temperature whatever the temperature of the surrounding bodies. The fact that the body loses energy to colder surroundings and gains energy from hotter surroundings follows from this theory and tends to the attainment of thermal equilibrium at which the energy emitted and that received are equal.

PRIMARY CELL. An electrochemical cell designed to operate by converting the chemical free energy of a limited amount of material. Such a cell has a limited lifetime and cannot, in general, be recharged. The most common form is the *dry battery* in which the electrolyte is thickened by a colloidal additive so that the cell may be used in any position. See also: *Accumulator.*

PRIMARY COLOURS. The three colours—red, green, and blue. These are to be distinguished from the three *primary pigments*—red, yellow, and blue. See also: *Colour vision, theories of.*

PRIMARY FOCUS. See: *Astigmatic foci.*

PRIMARY IONIZATION. In the path of an ionizing particle: the ionization formed directly in the interactions of the particle with the atoms of matter traversed. See also: *Secondary ionization.*

PRIMARY RADIATION. Radiation emerging directly from a source of the radiation concerned. See also: *Secondary radiation.*

PRIME VERTICAL. That great circle on the celestial sphere which passes through the zenith and the east and west points. Any other great circle through the zenith is known as a *vertical circle* and that passing through the north and south points is the *meridian circle* of the observer.

PRIMITIVE LATTICE. See: *Space lattice: Bravais lattice.*

PRINCIPAL PLANE. Of the ordinary or extraordinary ray in a doubly refracting crystal: the plane containing the ray in question and the optic axis.

PRINCIPAL PLANES. Of an optical system: See: *Lens, cardinal points of.*

PRINCIPAL REFRACTIVE INDICES. Of a doubly refracting crystal: the refractive indices corresponding to the principal vibration axes.

PRINCIPAL VIBRATION AXES. Of a doubly refracting crystal: the directions corresponding to the fastest and slowest vibrations, together with the direction at right angles to these vibrations.

PRINTED CIRCUIT. A form of electrical wiring in which the electrical connections consist of flat metallic strips bonded to an insulating surface. The strips are, typically, formed by printing or photographing an image of the circuit on copper foil and etching away the unwanted metal. The electrical components may be added (often automatically) by soldering.

PRISM. (1) In crystallography: An open crystal form with three or more parallel edges. (2) In geometry: a polyhedron with vertical sides and parallel top and bottom faces or bases. The prism is named *triangular, quadrangular,* etc., according to the shape of the base. (3) For optical use: a prism made from glass or other transparent material, to provide beam deviation, displacement, etc., or dispersion.

PROBABLE ERROR. That deviation from the mean value of a series of observations above and below which half of the observed values lie.

PROBABILITY. The mathematical expectation that a specified event will occur, as measured for example by the ratio of the favourable cases to the whole number of possible cases.

PROBABILITY-DENSITY DISTRIBUTION OR FUNCTION. Of an electron cloud: See: *Electron cloud.*

PROBE. (1) A device for extracting electromagnetic energy from a waveguide or cavity with negligible disturbance of the field. If the coupling is magnetic the device is usually known as a *magnetic probe.* (2) A metal electrode in a discharge tube which collects ions and thereby provides information about the density distribution of ions in the tube. (3) Any device for obtaining information without undue disturbance of the object or medium under examination.

PROCA EQUATIONS. One of four equivalent sets of equations for describing particles of spin unity, developed by analogy with Maxwell's equations.

PROJECTIVE GEOMETRY. See: *Non-Euclidean geometry.*

PROJECTIVE SPACE. See: *Non-Euclidean geometry.*

PRONG. In a cloud or bubble chamber or in a nuclear emulsion. See: *Track of ionizing particle.*

PROPAGATION COEFFICIENT: PROPAGATION CONSTANT. Of an electrical transmission line: (1) For a uniform line of infinite length or terminated by a network simulating an infinite line: the natural logarithm of the vector ratio of the steady state amplitudes of a wave at specified frequency, at points in the direction of propagation which are separated by unit length. (2) For a coil-loaded line: the natural logarithm of the vector ratio of the amplitude of a wave at a specified frequency, at points similarly situated with respect to the loading coils, divided by the length of line separating these points, the line being assumed infinite in length or terminated as in (1).

PROPAGATION VECTOR. Of a sinusoidal wave: a vector, **k**, in the direction of propagation of the wave, with magnitude $|k| = 2\pi/\lambda = \omega/u$, where λ is the wavelength, ω the angular frequency, and u the speed of the wave.

PROPELLANT. See: *Rocket.*

PROPELLER, AIRCRAFT, THEORIES OF. (1) *Blade element theory:* a theory in which each element of the blade is considered as an aerofoil, and the lift and drag forces calculated for it, allowing for the influence of the trailing vortex field. (2) *Momentum theory:* a theory in which the thrust of the propeller is related to the difference between the axial momentum of the air ahead of the propeller and that of the air behind it, i.e. in the slipstream. (Note: these theories are also applicable, *mutatis mutandis,* to ship's propellers.)

PROPELLER, PITCH OF. See: *Pitch.*

PROPELLER, SHIP, THEORIES OF. See: *Propeller, aircraft, theories of.*

PROPER MOTION. Of a star: the angular rate of change in the apparent position of the star on the celestial sphere, usually quoted in seconds of arc per annum. It arises largely from alterations of the direction of the Earth's axis in space. The *reduced proper motion* is the proper motion expressed as the linear distance per second, and the *restricted proper motion* is the angular motion relative to other stars. The latter is so small (rarely more than 2″ per annum) that the constellations have not changed materially in historic time.

PROPER TIME. Of a moving particle: the time measured as it would be by a clock travelling at the speed of the particle. See also: *Relativity theories. Time dilation.*

PROPORTIONAL REGION. Of a counter: the range of operating voltage for the counter in which the gas amplification is greater than unity and is independent of the primary ionization. It follows that, in this region, the pulse size is proportional to the number of ions produced as a result of the initial ionizing event.

PROTECTIVE GEAR, ELECTRICAL. See: *Electrical protective gear or cut-out.*

PROTEINS, STRUCTURE OF. Proteins are materials of high molecular weight formed in living systems by the linear polymerization of α-amino acids. Their structures are very complicated and have only comparatively recently been determined (and then only in a few cases) by advanced X-ray diffraction methods.

PROTON. The nucleus of the hydrogen atom and a constituent of atomic nuclei in general. It is a stable elementary particle with a positive charge of 1.602×10^{-19} coulomb (i.e. equal and opposite to that of the electron) and a rest mass of 1.673×10^{-27} kg.

PROTON AFFINITY. Of a chemical compound: the heat evolved when one mole of the compound combines with a proton.

PROTON MAGNETOMETER. A precise and portable instrument, based on the phenomenon of the free precession of atomic nuclei (specifically protons), which is mainly used for the rapid measurement of the total intensity of the geomagnetic field. It can only function when remote from the gradients associated with iron and steel objects, and is therefore unsuitable for use within the laboratory, although eminently suitable as a field surveying instrument. One interesting application is to the detection of buried archaeological remains.

PROTON MAGNETOMETER, VECTOR. An adaptation of the proton magnetometer, involving the use of Helmholtz coils, which enables measurements to be made of individual components of the geomagnetic field, rather than the total magnetic intensity as a whole. It is not so easily portable as the proton magnetometer itself.

PROTON MICROSCOPE. A microscope analogous to the electron microscope in which electrons are replaced by protons, which offer the possibility of increased resolution owing to their greater mass. See also: *Proton scattering microscopy.*

PROTON NUMBER. Another name for atomic number.

PROTON–PROTON CHAIN. A series of thermonuclear reactions initiated by a reaction between two protons and building up via ^{2}H and ^{3}He to ^{4}He. The energy liberated is thought to be an appreciable source of energy in many stars, particularly the cooler stars. See also: *Carbon cycle. Solar energy.*

PROTON RADIOACTIVITY. The emission of protons from nuclei with a measurable delay. So far only one type has been observed, namely the emission of protons by excited nuclei formed in the process of β-decay. As with delayed neutrons the observed half-life is that of the β-decay of the parent nuclide. Two other types are, however, possible theoretically: potential-barrier delayed proton emission (analogous to α-radioactivity) and two-proton radioactivity.

PROTON RADIOGRAPHY. Radiography by protons obtained from a particle accelerator. The technique does not depend on changes of absorption as in the usual methods of radiography, but on scattering; and changes of image intensity occur only at the edges (both internal and external) of the object under examination. Thus aspects of an object are revealed that are different from those shown by X-rays, neutrons etc. See also: *Radiograph.*

PROTON SCATTERING MICROSCOPY. The imaging of a crystalline structure on, say, a fluorescent screen by the scattering of a proton beam, owing to the phenomenon of *blocking*, i.e. the prevention of a proton beam from leaving a crystal in the direction of a close-packed row or along a densely populated plane of atoms, owing to the obstacle presented by such a row or plane. It is to be distinguished from proton microscopy, the proton microscope being separately defined.

PROTONS, POLARIZATION OF. See: *Polarization.*

PROTON SYNCHROTRON. See: *Synchrotron.*

PROTON TRANSFER REACTION. A chemical reaction which involves the transfer of a proton from one chemical species to another.

PROTRACTOR, BEVEL. Another, and older, name for a *clinometer.*

PROXIMITY EFFECT. The change in current distribution in a conductor, with related changes in resistance and capacitance, due to the field produced by an adjacent conductor.

PSI PARTICLE. See: *J/Psi particle.*

PSYCHROMETER. A hygrometer consisting of a similar pair of thermometers, the bulb of one of which is kept wet, and therefore cooled by evaporation. The difference in temperature between the two thermometers is a measure of the relative humidity of the air. See also: *Hygrometer.*

PSYCHROMETRIC TABLES. Tables from which the relative humidity, and sometimes other quantities such as the dew-point temperature, can be derived from wet- and dry-bulb temperatures.

PTOLEMAIC SYSTEM. The geocentric theory of planetary motion. See also: *Copernican system.*

PULSAR. One of a number of celestial objects which behave as "point" sources of radio waves in which almost all the energy of emission is concentrated into short pulses. For each pulsar the time between pulses is very nearly constant and the pulse width is about 5% of this period. The period ranges from about 33 msec in the fastest pulsars to about 4 s in the slowest. Pulsars are now identified as rapidly rotating neutron stars. At least two (the Crab and Vela pulsars) also "flash" optically. See also: *Neutron star.*

PULSATANCE. Of an alternating quantity: the product $2\pi f$, where f is the frequency of alternation. It is usually denoted by ω.

PULSATING CURRENT. One which is the sum of a direct current and an alternating current.

PULSE. A sudden disturbance propagated as a wave, singly, or in a train, or applied to a mechanical system capable of movement.

PULSE AMPLIFIER. An amplifier designed to amplify, and sometimes to shape, random pulses, e.g. pulses from a radiation detector, or carrier frequency pulses.

PULSE-AMPLITUDE ANALYSER. See: *Pulse-height analyser.*

PULSE-AMPLITUDE DISCRIMINATOR. See: *Pulse-height discriminator.*

PULSE-AMPLITUDE SELECTOR. See: *Pulse-height selector.*

PULSE ANALYSER. An instrument for recording the variation of blood pressure during heart beats by measuring the periodic displacement produced in the walls of veins and arteries.

PULSE CODE MODULATION. Modulation of a pulse train in accordance with a pulse code (usually binary) so as to convey information.

PULSE GENERATOR. A generator of electrical pulses frequently of very short duration, e.g. microseconds. Such generators are in general of two types: those that produce a pulse in response to an initiating signal (known as *trigger circuits*), and those that give out a train of regularly spaced pulses (known as *free-running* generators).

PULSE-HEIGHT ANALYSER. An instrument which records or counts an electrical pulse only if its amplitude falls within specified limits. A *single-channel* instrument, which incorporates a single-pulse height selector, has a preset channel width with a variable threshold to permit the scanning of the pulse amplitude spectrum. A *multi-channel* instrument, or *kicksorter*, permits the whole or part of the pulse-amplitude spectrum to be determined in a single measurement by having a series of pulse-height selectors arranged in cascade or by incorporating circuits which generate a time interval proportional to the pulse amplitude. The instrument is also known as a *pulse-amplitude analyser.*

PULSE-HEIGHT DISCRIMINATOR. A circuit which accepts only those electrical pulses having amplitudes greater than a preset level and produces an output pulse of fixed amplitude (and sometimes of fixed width) for each pulse accepted. The circuit is also known as a *pulse-amplitude discriminator.*

PULSE-HEIGHT SELECTOR. A circuit which permits only those electrical pulses that have amplitudes lying between predetermined levels to be passed to the succeeding circuits. It is normally part of a single-channel pulse-height analyser. The circuit is also known as a *pulse-amplitude selector.*

PULSE-INTERVAL ANALYSER. A device for counting the number of events (e.g. for particle time-of-flight measurements or for observations of radioactive processes) occurring in each of a sequence of uniform time intervals.

PULSE RADIOLYSIS. A technique used in radiochemistry whereby high instantaneous concentrations of chemically reactive species are produced in liquids by a short (2–10 μs) intense pulse of electrons from a linear accelerator. The transient intermediates and the subsequent reactions are commonly identified and followed by absorption spectroscopy.

PULSE-SHAPE DISCRIMINATION. A technique for examining the shape of an electrical pulse in the circuit of a radiation detector so as to yield information as to the nature of the initiating event, such as the type of particle entering the detector or the position at which that event occurred.

PUMP. A machine for raising a fluid from a lower to a higher level, for imparting energy to a fluid (e.g. increasing the pressure exerted by the fluid), for transporting a fluid from one place to another, or for producing a degree of vacuum in a chamber containing a fluid.

PUMP, AIR-LIFT. A pump for raising water from a low level (as in an artesian well) by the use of compressed air.

PUMP, AXIAL. A high-speed pump consisting essentially of a row of centrifugal pumps. Such pumps are used typically to move large quantities of water against small heads of up to about 700 cm.

PUMP, BACKING. See: *Pressure, backing*.

PUMP, CENTRIFUGAL. A pump which operates by the rotary motion of one or more impellers.

PUMP, DIFFUSION. A vacuum pump whose operation depends on the entrainment of air or other gas by an essentially one-way stream of vapour molecules travelling towards the backing pump. The working fluid may be either mercury or oil.

PUMP, FRACTIONATING. An oil diffusion pump in which the products of cracking are removed as they are formed.

PUMP, GETTER-ION. A vacuum pump whose operation depends on the presence of a suitable getter and on the reduction of gas pressure by ionization.

PUMP, JET. A pump in which the passage of high-pressure water through a nozzle creates a suction effect.

PUMP, PULSOMETER. A pump which depends on the pressure of steam forcing water out of one of two chambers alternately.

PUMP, ROTARY. A pump in which an eccentrically mounted cylinder rotates at high speed inside, and in rotating contact with, an outer case fitted with inlet and outlet ports, fluid being drawn in at the first and expelled at the second. Important types are the *Gaede*, *Hyvac*, and *Kinney* vacuum pumps, but the principle is also used for pumping liquids.

PUMP, TÖPLER. A vacuum pump in which a Torricellian vacuum is produced over a column of mercury, and is then opened to the vessel to be pumped, the gas being swept away by a mercury piston, this cycle being repeated continuously.

PUNCTUM PROXIMUM. See: *Near point*.

PUNCTUM REMOTUM. See: *Far point*.

PUPIL. The aperture of the eye.

PUPILOMETER. An instrument for measuring the size and shape of the pupil of the eye and its position relative to the iris.

PURKINJE EFFECT. The shift in the spectral sensitivity of the eye from the yellow–green, at a good level of illumination, towards the blue, as the illumination is reduced.

PYKNOMETER. A vessel used for the measurement of the specific gravity of a liquid. It differs from the specific gravity bottle in that the calibration is along a capillary, or sometimes two. See also: *Specific gravity bottle: Density bottle*.

PYRAMID. A polyhedron with a polygonal base and triangular faces which meet at a common vertex. A *regular* pyramid is one whose base is a regular polygon.

PYRANOMETER. See: *Actinometer*.

PYRGEOMETER. An instrument for measuring the loss of heat by radiation from the Earth.

PYRHELIOMETER. See: *Actinometer*.

PYROELECTRICITY. The development of equal and opposite electrical charges on crystals when their temperature is changed. It occurs only with crystals that do not possess a centre of symmetry, the best-known example being tourmaline.

PYROLYSIS CURVES. See: *Thermogravimetric analysis*.

PYROLYSIS. The decomposition of a substance by heat.

PYROMETER. An instrument for the measurement of high temperature, often at a distance.

PYROMETER, COLOUR. A type of optical pyrometer consisting of a calibrated wedge-shaped filter that transmits only red and green light, the temperature of a radiant source being determined from the position along the wedge at which the source appears to be white.

PYROMETER, OPTICAL. A pyrometer with which the temperature of a radiant source is determined from the colour of the radiation. The colour may be judged by eye or, more usually, matched with a standard radiator, as in the *disappearing filament pyrometer*, in which an image of a heated filament is matched against the radiant source by varying its temperature; or the *Wanner* or *polarization pyrometer*, in which the radiant source and a standard source are viewed side by side through a monochromatic filter and a rotatable Nicol prism so that they may be equated in brightness.

PYROMETER, RADIATION: PYROMETER, FÉRY. A pyrometer which measures the temperature of a radiant body from the intensity of the thermal radiation which it receives from the body, the radiation being focused on a thermopile or photocell.

PYROMETER, THERMOCOUPLE. A pyrometer which measures the temperature directly by means of a thermocouple immersed in a hot liquid (e.g. liquid steel) or a hot gas (e.g. furnace gas).

PYROMETRIC CONES. Small ceramic cones that differ in the temperature at which they soften and bend on heating. They are used in industrial furnaces to indicate, within fairly close limits, the temperature reached at the position where the cones are placed. They are also known as *Seger cones* in Europe and *Orton cones* in the United States.

PYROMETRIC SECTOR. A rotating sector disk used in pyrometry to reduce the apparent brightness of a radiating source. It serves the same function as a neutral absorbing filter but has the same transmission for all wavelengths, which can be determined geometrically.

PYROSOL. A cloudy solution, regarded as colloidal, formed in the electrolysis of fused salts, the free metal constituting the disperse phase.

PYTHAGOREAN SCALE. A musical scale in which the frequency intervals are represented by the ratios of integral powers of 2 and 3.

q

Q: Q-VALUE. Of a given system, process, or phenomenon: a term used to denote a measure of the energy balance. It is often used as a figure of merit. Thus, for a *resonant system* (e.g. an electrical circuit) it is ($2\pi \times$ energy stored)/(energy lost per cycle); for a *system in forced vibration* it is the amplitude magnification factor at resonance, and its reciprocal is a direct measure of the damping capacity; for a given *nuclear disintegration* it is the amount of energy released in that disintegration; and for a *nuclear reaction* it is the difference in total kinetic energy between the particles leaving and entering the reacting system.

Q-SWITCHING: Q-SPOILING. Of a laser: the pumping of the laser into a non-resonant mode followed by a switch into resonance. The technique was developed for use with solid-state lasers to provide very high pulsed output powers, but it can also be applied to gas lasers. Typical Q-switches include the Kerr cell, rotating mirrors, and bleachable absorbers, all of which in one way or another act as optical shutters.

QUADRATIC EQUATION. An equation of the second degree, the most general form of which is $ax^2 + bx + c = 0$, the solution of which is
$$x = [-b \pm \sqrt{(b^2 - 4ac)}]/2a.$$

QUADRATURE. (1) In astronomy: the position of the Moon or a superior planet when the line joining it to the Earth is at right angles to the line joining the Earth to the Sun. (2) In electrical engineering: See: *Reactive component*.

QUADRIC. A surface whose points (x, y, z) satisfy an equation of the second degree, which may be written
$$ax^2 + by^2 + cz^2 + 2fyz + 2gzx + 2hxy + 2ux$$
$$+ 2vy + 2wz + d = 0.$$
The best-known are the *ellipsoid*
$$(x^2/a^2 + y^2/b^2 + z^2/c^2 = 1),$$
the *hyperboloid of one sheet*
$$(x^2/a^2 + y^2/b^2 - z^2/c^2 = 1),$$
and the *hyperboloid of two sheets*
$$(x^2/a^2 - y^2/b^2 - z^2/c^2 = 1),$$

which are termed *central quadrics*, and the *paraboloids*, which are either *elliptic*
$$(x^2/a^2 + y^2/b^2 = 2z)$$
or *hyperbolic*
$$(x^2/a^2 - y^2/b^2 = 2z);$$
together with the *cones*
$$(ax^2 + by^2 + cz^2 = 0).$$

QUADRUPOLE. One of the multipoles which represent the effects, at points outside the system, of the electric and magnetic fields produced by a system of charges and currents such as those arising from the electrons in an atom or from the protons in a nucleus; and which are assumed to be located at the centre of the system. In its simplest form an electric quadrupole may be considered as an array of four equal charges positioned at the corners of a parallelogram, alternate charges being positive and negative; but more generally it is regarded as an ellipsoidal distribution of charge about the centre of the system. See also: *Multipole moments*.

QUADRUPOLE MOMENT. See: *Multipole moments*.

QUADRUPOLE RADIATION. Radiation from an oscillating quadrupole. The quadrupole may be either magnetic or electric.

QUALITY CONTROL. The control of the quality of manufactured products by a combination of inspection, sampling, and statistical analysis.

QUALITY FACTOR. In radiation protection: a factor depending on the biological effects of the ionizing radiation concerned, by which, together with other factors, the absorbed dose is multiplied to give the dose equivalent.

QUANTIZATION. The existence in a system, according to quantum theory, of discrete values of the parameters describing the system, in contrast to the continuous range of values permitted classically. The term is also applied to the mathematical procedures employed in calculating these values.

QUANTOMETER. An instrument for carrying out chemical analyses by spectroscopic methods in a semi-automatic manner, and for presenting the results directly as the percentages of the elements of interest present.

QUANTUM. The minimum amount of a physical quantity (e.g. action, angular momentum, or energy) which can exist, according to the quantum theory, and by integral multiples of which changes in that quantity are effected.

QUANTUM BIOCHEMISTRY. The use of quantum mechanics in the study of the electronic properties of biomolecules. See also: *Quantum chemistry*.

QUANTUM CHEMISTRY. The use of quantum mechanics in the study of the electronic properties of isolated molecules and of the interactions between numbers of molecules.

QUANTUM DEFECT. Of a spectral series: the difference for certain atoms between the total quantum number and the nearest integer.

QUANTUM EFFICIENCY: QUANTUM YIELD. (1) Of a photochemical reaction: the number of molecules which change in the reaction per quantum of incident light of a given wavelength. (2) Of a photocathode: the number of photoelectrons emitted by the photocathode per quantum of incident light of a given wavelength. (3) Of other reactions induced by radiation: the number of resultant events per incident quantum.

QUANTUM ELECTRODYNAMICS. The quantized field theory of the interaction between electrons and the electromagnetic field, based on the quantized form of Maxwell's equations. See also: *Gauge theories*.

QUANTUM ELECTRONIC SYSTEMS. Systems having to do with atomic and molecular transitions and the interactions of atoms and molecules with electromagnetic or vibrational waves. The term includes masers, lasers, and electric and magnetic resonance phenomena.

QUANTUM FIELD THEORY. A field theory in which all the physical observables of a system are represented by appropriate operators which obey certain commutation relations. The quantized field can be considered as an assembly of particles each of which is characterized by its own energy, momentum, charge, etc., the total energy, momentum, etc., of the field being built up additively from the individual contributions of the particles present. Any particle may thus be considered as a "quantum" of a corresponding field.

QUANTUM LIMIT: BOUNDARY WAVELENGTH. In a continuous X-ray spectrum: the shortest wavelength present. The term quantum limit is also used to denote the energy appropriate to that wavelength. See also: *X-rays*.

QUANTUM MECHANICS. A general theory which embraces wave mechanics and matrix mechanics (which are defined separately). In it the common features of wave mechanics and matrix mechanics are extracted, and combined into a system which can be used to provide different forms of the theory for use in different types of problem.

QUANTUM METROLOGY. The use of atomic quantum phenomena in defining and maintaining units of measurement (e.g. length and time) and for precision metrology and measurement in general. The use of such phenomena should enable units of measurement to be reproduced everywhere with high precision, such units being independent of material "standards", which change with time.

QUANTUM NUMBER. One of several integral or half-integral numbers which jointly characterize a state of an atomic system (and sometimes, by analogy, a molecular system), according to the quantum theory. Quantum numbers are directly related to the eigenfunctions associated with Schrödinger's wave functions.

QUANTUM NUMBER, ANGULAR. Defines any angular momentum (orbital, spin, total, inner, rotational) and is expressed in units of $\hbar$ (i.e. $h/2\pi$).

QUANTUM NUMBER, AZIMUTHAL OR ORBITAL ANGULAR MOMENTUM. Defines the orbital angular momentum of a free particle or the rotational angular momentum of a dumb-bell model of a diatomic molecule. It is expressed in units of $\hbar$ (i.e. $h/2\pi$) and its usual symbol is l.

QUANTUM NUMBER, INNER. Defines the total angular momentum of an electron in an atom (symbol j) or for all exterior or valency electrons of an atom (symbol J). It is expressed in units of $\hbar$ (i.e. $h/2\pi$).

QUANTUM NUMBER, INTERNAL. One of a set of numbers (including zero) used to label the hadrons according to their behaviour, the three principal ones being isospin, strangeness and charm. They are conserved in any strong interaction.

QUANTUM NUMBER, MAGNETIC. Determines the component of any kind of angular momentum in a given direction in a magnetic field, e.g. orbital or spin. Its usual symbol is m.

QUANTUM NUMBER, MOLECULAR. The molecular analogue of the orbital angular momentum (or azimuthal) quantum number of an atomic system. Essentially it defines the symmetry of the associated molecular orbital. It is expressed in units of $\hbar$ (i.e. $h/2\pi$) and its usual symbol is λ. Following the atomic notation ($s, p, d, ... l$), the electrons corresponding to molecular quantum numbers 1, 2, 3, ..., are known as σ, π, δ, ..., electrons, respectively. See also: *Electron shell.*

QUANTUM NUMBER, PRINCIPAL OR TOTAL.
Defines the energy levels of an atomic system. It is expressed as a positive integer, the values from 1 to 7 corresponding to the different electron shells from K to Q. The usual symbol is n.

QUANTUM NUMBER, ROTATIONAL. Defines the rotational angular momentum of a molecule. It is expressed in units of $\hbar$ (i.e. $h/2\pi$) and its usual symbol is r.

QUANTUM NUMBER, SPIN. Defines the intrinsic angular momentum of an elementary particle or atomic nucleus. It is expressed in units of $\hbar$ (i.e. $h/2\pi$), and is always integral or half-integral. Its usual symbol is s.

QUANTUM OF ACTION. The Planck constant.

QUANTUM STATISTICS. The statistics of the distribution of particles of a given type among the various possible energy values taking the quantization of the latter into account. Two forms of quantum statistics exist: *Bose–Einstein* and *Fermi–Dirac* statistics, which are defined separately. Both reduce to the classical Maxwell–Boltzmann statistics at sufficiently high temperatures where a large number of energy levels are excited.

QUANTUM THEORY. That part of physics which grew out of Planck's quantum postulate, made originally in an attempt to explain the spectrum of black-body radiation. The application of its principles led to quantum mechanics, quantum electrodynamics, quantum field theory, etc.

QUANTUM VOLTAGE. Of a given quantum of energy: the voltage through which an electron must be accelerated to acquire the energy corresponding to the quantum concerned.

QUARKS. A set of four sub-particles, and their anti-particles, which, as constituents of hadrons, can explain their symmetry properties. Quarks carry a charge of one-third or two-thirds of that of the electron and also differ in their masses and decay properties. The various quarks are known as *"up"*, *"down"*, *"strange"* or *"charm(ed)"*, terms related to the so-called internal quantum numbers of the hadrons. They have been produced in experiments made with high-energy accelerators, but so far (1978) no free quarks have been observed. Quarks obtain their name from a passage in James Joyce's *Finnegan's Wake*. Also the German word *Quark* has the colloquial meaning of "nothing". See also: *Particle symmetry.*

QUARTER-WAVE PLATE. In optics: a thin plate, usually of mica or quartz, cut of such a thickness that a path difference of one quarter of a wavelength is introduced between the ordinary and extraordinary vibrations of a ray of light travelling perpendicularly through it. It is commonly used for measuring the sign of double refraction of uniaxial crystals. See also: *Quartz wedge. Wave plate: Retardation plate.*

QUARTET, SPECTRAL. A spectral line which is split into four as a result of spin–orbit interaction.

QUARTILE. See: *Median.*

QUARTZ OSCILLATOR, CUT OF. See: *Cut.*

QUARTZ WEDGE. A very thin wedge of quartz cut parallel to the optic axis by the use of which a desired thickness of quartz may be superposed on a mineral section in the polarizing microscope. It is used for determining the sign of double refraction of biaxial minerals from their interference figures in convergent light. See also: *Wave plate: Retardation plate.*

QUASAR: QSO (QUASI-STELLAR OBJECT). One of a class of astronomical objects of dominantly star-like appearance with emission-line spectra showing abnormally large redshifts. Some are also radio sources and variations of intensity are observed at both radio and optical wavelengths, with time scales of a few days to several years. The nature of quasars and their relation to other astronomical objects is still uncertain.

QUATERNION. An operator, in a system of vector analysis invented by Hamilton, which changes one vector into another by rotation accompanied by a change of magnitude.

QUENCHING. (1) Of a counter: See: *Counter, quenching of.* (2) Of fluorescence, phosphorescence and luminescence: the reduction in the intensity of the emitted light for any reason, e.g. self-absorption, a reduction in the lifetime of the excited states by internal molecular collisions, or the introduction of external agents which carry away energy by collision. (3) Thermal: the rapid cooling of metals or alloys from a high temperature by plunging into a suitable liquid.

QUICK FREEZING. The preparation of foodstuffs for cold storage by passing them through the temperature zone of maximum ice-crystal formation (between about 0° and −4°C) as rapidly as possible to avoid damaging plant and animal tissues through the formation of large crystals.

QUIET DAYS: CALM DAYS. Five days selected each month as international magnetic "quiet" or "calm" days, on the basis of many magnetic observations. See also: *Disturbed days: Active days.*

QUINCKE BALANCE. A balance for the measurement of the magnetic susceptibility of a liquid from the change of liquid level in a U-tube in a strong magnetic field. The balance may also be used for a gas if the latter is placed in the U-tube above a liquid of known susceptibility.

QUINTET, SPECTRAL. A spectral line which is split into five as a result of spin–orbit interaction.

r

RACEMIC COMPOUNDS. Substances consisting of equal quantities of enantiomorphic optically active compounds (i.e. laevo- and dextro-rotatory compounds), which are themselves optically inactive.

RAD. A unit of absorbed dose, equal to 100 erg/g, i.e. 10^{-2} J/kg. The SI unit is the gray. See: *Gray. Appendix II.*

RADAR. An acronym for *Ra*dio *D*etection *A*nd *R*anging, denoting the use of radio waves (usually pulsed), which are reflected from distant objects (*primary radar*) or automatically re-transmitted by them (*secondary radar*), to obtain information regarding the location of those objects. The original use of radar was to locate the position of aircraft, but its use has been extended, *inter alia*, to facilitate navigation, to investigate cloud formations, and to examine astronomical objects. Radar was formerly known as *radiolocation.*

RADAR ASTRONOMY. The application of radar echo methods to the study of astronomical objects. It is a branch of radioastronomy. See also: *Planetary radar.*

RADAR BEACON. A fixed radio transmitter which, on reception of a radar signal, emits an echo together with a coded identification signal, so enabling an aircraft, for example, to determine its relative position.

RADAR-ECHO CROSS-SECTION: RADAR ECHOING AREA. The size of that area which, located at the position of a given object and situated at right angles to the direction of a specified radar transmitter, would provide at the radar receiving aerial the same power flux as that provided by the object itself, on the assumption that the power incident on the area is re-radiated equally in all directions.

RADAR EQUATION. A mathematical equation which relates the transmitted and received powers and aerial gains of a primary radar to the radar-echo cross-section and distance of an object.

RADIAL DENSITY FUNCTION. The function $\varrho(r)$ in the expression $4\pi r^2 \varrho(r)\, dr$ for the average number of atoms lying at a distance between r and $(r + dr)$ from the centre of a specified atom.

It may be obtained from X-ray diffraction measurements, typically of liquids and gases, and used to determine the distances between nearest neighbours, second and third nearest neighbours, and so on. See also: *Zernicke–Prins formula.*

RADIAN. One of two supplementary SI units of angular measurement, the other being the steradian. It is defined as "the plane angle between two radii of a circle which cut off on the circumference an arc equal in length to the radius". It is thus the angle subtended at the centre of a circle by an arc equal in length to the radius. See: *Appendix II.*

RADIANT FLUX DENSITY. The radiant flux per unit area, also known as *irradiance* or *irradiancy.*

RADIANT FLUX: RADIANCE. The rate of flow of radiant energy, usually expressed in watts. See also: *Luminous flux.*

RADIANT INTENSITY. The power per unit solid angle, usually expressed in watts per steradian.

RADIANT TOTAL ABSORPTANCE (OR ABSORPTIVITY). The ratio of the total radiant flux absorbed by a body to the total incident radiant flux.

RADIANT TOTAL EMITTANCE. See: *Emissive power.*

RADIANT TOTAL REFLECTANCE (OR REFLECTIVITY). The ratio of the total radiant flux reflected by a body to the total incident radiant flux.

RADIANT TOTAL TRANSMITTANCE. The ratio of the total radiant flux transmitted by a body to the total incident flux.

RADIATION. The emission of energy in the form of electromagnetic or acoustic waves, or of ionizing particles. The term may also denote the emitted energy itself.

RADIATION BALANCE: RADIO-BALANCE. See: *Calorimetry, micro-.*

RADIATION BELTS. See: *van Allen belts.*

RADIATION BROADENING. Of spectral lines: broadening due to the diffuseness of energy levels associated with finite lifetimes of excited states. It is the most fundamental cause of line broadening.

See also: *Spectral lines, broadening of.*

RADIATION-CHEMICAL YIELD. The number of atoms, ions, radicals, or molecules, as the case may be, which either arise or are destroyed in a given chemical system by the input to the system of 100 eV of energy in the form of ionizing radiation. (Note: 1 eV $\approx$ 1·6021 $\times$ 10^{-19} J.)

RADIATION CHEMISTRY. That part of chemistry which deals with the chemical effects of ionizing radiation. It is to be distinguished from radiochemistry which is defined separately. See also: *Photochemical process.*

RADIATION CONSTANTS. See: *Planck radiation formula.*

RADIATION, CONTINUOUS. Another term for *white radiation.*

RADIATION, CORPUSCULAR, OR PARTICULATE. Radiation which is normally thought of as consisting of particles rather than waves, e.g. α-rays, β-rays, atomic or molecular rays, even though in some contexts the wave-like nature of the particles is paramount.

RADIATION DAMAGE. A deleterious change in the physical or chemical properties of a material as a result of exposure to ionizing radiation. The term is not normally applied to biological systems.

RADIATION, ENERGY DENSITY OF. The radiant energy per unit volume. For a black body it is proportional to the fourth power of the absolute temperature.

RADIATION, HETEROGENEOUS. Radiation of different types, or of the same type but having a variety of wavelengths or quantum energies.

RADIATION, IONIZING. Any radiation, either electromagnetic or particulate, which is capable of producing ions, directly or indirectly, in its passage through a medium. For some purposes, e.g. in connection with nuclear energy legislation, the term is not held to include visible or ultraviolet radiation.

RADIATION, LAWS OF. See: *Kirchhoff radiation law Plack radiation formula. Rayligh–Jeans law. Stefan–Boltzmann law. Wien radiation laws.*

RADIATION LENGTH. For a fast electron in a given medium: the mean path length in that medium for which the electron will have its energy decreased by a factor e owing to radiative collisions.

RADIATIONLESS ANNIHILATION. A rare phenomenon occurring when a positron collides simultaneously with two electrons. The positron and one electron are annihilated and the energy of annihilation is given to the other electron without the emission of any radiation. See also: *Annihilation radiation.*

RADIATION LOSS. The loss of energy by an energetic charged particle by bremsstrahlung emission.

RADIATION MONITOR. (1) An instrument for the detection and measurement of the level of ionizing radiation, or the quantity of radioactive material. It may give an audible or visible warning when the radiation level or quantity of radioactive material exceeds a prescribed limit. (2) A person who, in the course of his duties, uses an instrument such as those described in (1).

RADIATION PRESSURE. (1) The pressure exerted by electromagnetic radiation on a surface on which it is incident. At normal incidence, this pressure is equal to the energy density of the radiation. Although radiation pressures encountered terrestrially are very feeble, such pressures are of importance in astrophysical phenomena. Thus comet tails are due to radiation pressure from the Sun, and radiation pressure plays a large part in preventing certain stars from collapsing under their own gravitational fields. (2) The pressure exerted by a sound wave at a point on an obstacle, excluding forces due to streaming.

RADIATION PROTECTION: RADIOLOGICAL PROTECTION. (1) All measures associated with the limitation of the harmful effects of ionizing radiation, such as the limitation of exposure to such radiation, or of the incorporation of radionuclides in the body; or the prophylactic limitation of bodily injury arising from either of these. (2) All measures designed to limit radiation damage in materials.

RADIATION QUANTITY. For a specified ionizing radiation: a quantity characteristic of that radiation, e.g. energy or flux density, that may be measured and expressed in a quantitative fashion. The term therefore excludes such subjective concepts as hardness.

RADIATION SOURCE. An apparatus or material emitting or capable of emitting ionizing radiation.

RADIATION, TARGET THEORY OF. A theory explaining the biological effects of ionizing radiation on the basis of a target area (or small sensitive region within the cell) on which one, two, or more "hits" (ionizing events) may be necessary to bring about an effect.

RADIATION TEMPERATURE. Of a radiating body: the temperature of a black body having the same specific emission as the radiating body in a given range of wavelengths. See also: *Astrophysical temperatures.*

RADIATIVE CAPTURE. The capture of a particle by an atomic nucleus followed by the immediate emission of gamma radiation.

RADIATIVE COLLISION. A collision between two charged particles (typically an electron and a nucleus) in which part of the kinetic energy is converted to electromagnetic radiation (bremsstrahlung).

RADIATIVE TRANSITION. A nuclear transition which is accompanied by the emission of γ-rays. See also: *Nuclear transition.*

RADICAL. A group of atoms which occurs in different compounds and which may remain unchanged during chemical reactions.

RADIO. See: *Radiocommunication: Radio.*

RADIOACTIVATION ANALYSIS. A method of chemical analysis based on the identification and measurement of characteristic radionuclides formed by irradiation. It is also known as *activation analysis.*

RADIOACTIVE CHAIN. A radioactive series.

RADIOACTIVE CONTAMINATION. The presence of a radioactive substance in a material or other location where it is undesirable.

RADIOACTIVE DECAY CONSTANT: RADIOACTIVE DISINTEGRATION CONSTANT. For a radionuclide: the probability per unit time of the spontaneous decay of a nucleus of that radionuclide. It is $-(dN/dt)/N$, where N is the number of nuclei existing at time t. It is usually denoted by λ and appears in the equation for the decrease with time of the activity a as an exponential coefficient, thus: $a = a_0 e^{-\lambda t}$, where a_0 is the activity at time $t = 0$. This constant λ is the reciprocal of the mean life.

RADIOACTIVE DECAY: RADIOACTIVE DISINTEGRATION. Spontaneous nuclear disintegration. See also: *Nuclear disintegration. Radioactivity.*

RADIOACTIVE DECAY SCHEME. A diagrammatic representation of the modes of disintegration of radioactive atoms and their products, together with information concerning the associated energy levels. It is customary to indicate these energy levels by horizontal lines, spaced on a vertical energy scale, the atomic number increasing from the left to the right of the diagram.

RADIOACTIVE DISINTEGRATION RATE. The activity of a radioactive substance expressed as the number of disintegrations per unit time. See also: *Activity. Curie. Radioactive decay constant: Radioactive disintegration constant.*

RADIOACTIVE EFFLUENT. Any solid, liquid, or gaseous radioactive waste material. Its disposal must be strictly controlled, and is usually effected by dispersal or storage, according to the level and type of activity involved and having regard to the half-life of the material.

RADIOACTIVE EMANATION: EMANATION. The inert radioactive gases, consisting of seventeen radon isotopes, having mass numbers ranging from 204 to 223. The best known are *radon* itself ($^{222}_{86}$Rn), *thoron* ($^{220}_{86}$Rn), and *actinon* ($^{219}_{86}$Rn), which are decay products of radium, thorium, and actinium respectively. See also: *Emanating power.*

RADIOACTIVE EQUILIBRIUM. In a radioactive series: a quasi equilibrium condition in which the products decay approximately with the period of the parent. This occurs if the half-life of the parent is very much longer than that of any of the products and is known as *secular equilibrium*. If, however, the half-life of the parent is not much longer, the condition is known as *transient equilibrium*.

RADIOACTIVE FAMILY. A radioactive series.

RADIOACTIVE MIGRATION. See: *Aggregate recoil.*

RADIOACTIVE SERIES. A series of nuclides in which each member transforms into the next through radioactive decay (excluding spontaneous fission) until a stable nuclide has been formed. It is also known as a *radioactive chain*, a *radioactive family*, or a *decay chain*. See also: *Actinium series. Daughter product. Decay product. Thorium series. Uranium series.*

RADIOACTIVE SOURCE. Any quantity of radioactive material which is intended for use as a source of ionizing radiation. It is often transportable and often of known strength. See also: *Radiation source.*

RADIOACTIVITY. Spontaneous nuclear disintegration with the emission of particles or γ-rays, or of X-rays following orbital electron capture. The term includes spontaneous nuclear fission.

RADIOACTIVITY, ARTIFICIAL. A term sometimes used to mean induced radioactivity.

RADIOACTIVITY, INDUCED. Radioactivity induced by irradiation, e.g. by neutrons.

RADIOACTIVITY, NATURAL. Radioactivity in naturally occurring nuclides.

RADIOACTIVITY STANDARD. A radionuclide (usually incorporated in, or combined with, another material) the activity or quantity of which is accurately known. Such standards are commonly certified by a recognized standardizing organization, being determined by such an organization at a specified time and with a specified accuracy. The earliest forms of such standards were of *radium* and its products, before the advent of nuclear technologies.

RADIO ALTIMETER. A device which displays continuously the height of an aircraft above the ground or sea. It depends upon the measurement of the time taken for a radio signal to travel to and from the aircraft, the signal being reflected by ground or sea as the case may be.

RADIOASTRONOMY. The study of celestial bodies (both visible and invisible) by the observation and measurement of their radio emissions, or of their radio echoes (*radar astronomy*). The radio wavelengths involved range from about $100 \mu m$ to 20 m. See also: *Pulsar. Quasar. Radar astronomy. VLBI.*

RADIOAUTOGRAPH. A term sometimes used instead of *autoradiograph*.

RADIOBIOLOGY. That branch of biology which deals with the effects of radiation on living things.

RADIOCHEMICAL ANALYSIS. The analysis of radioactive material for both the radioactive and non-radioactive constituents.

RADIOCHEMISTRY. That part of chemistry which deals with radioactive materials. It includes the production of radionuclides and their compounds, the application of chemical techniques to nuclear studies (*nuclear chemistry*), and the application of nuclear techniques to the investigation of chemical problems. It is to be distinguished from radiation chemistry, which is defined separately.

RADIOCOMMUNICATION: RADIO. Telecommunication using radiowaves not guided between transmitter and receiver by artificial boundaries such as wires or wave guides.

RADIOELEMENT. A radioactive element (not nuclide). The term is sometimes restricted to naturally occurring radioactive elements.

RADIOGALAXY. A galaxy emitting strong radio signals. The absolute power of such a galaxy may be as high as 10^{37} watts and a typical value for the minimum energy is about 10^{53} joules.

RADIOGENIC. Resulting from radioactive decay.

RADIOGRAPH. A shadow photograph produced by penetrating radiation after passing through an object. The radiation used is commonly X- or γ-rays, but neutrons are also employed and sometimes electrons and protons. See also: *Electron radiography. Gamma-ray radiography. Neutron radiography. Proton radiography. X-ray radiography.*

RADIOGRAPHIC DEFINITION. A qualitative term denoting the clarity of detail in a radiograph. A quantitative measure of the lack of definition is afforded by the *unsharpness*, which may be defined as the width of the band of changing density across the radiographic image of a sharp opaque edge.

RADIOGRAPHIC STEREOMETRY. The process of finding the position and dimensions of a feature within a body (often human), from measurements made on radiographs taken from two different directions. See also: *Tomography.*

RADIOGRAPHY, HIGH SPEED. See: *Flash radiography.*

RADIO INTERFEROMETER. See: *Radio telescope.*

RADIOISOTOPE. Of a given element: a radioactive isotope of that element.

RADIOISOTOPE SCANNING. A technique for the determination of the spatial distribution of radio isotopes, used mostly in medicine as a diagnostic aid.

RADIOLOCATION. See: *Radar.*

RADIOLOGY. The science and application of X-rays, γ-rays and other penetrating ionizing radiations, whether in medicine or elsewhere.

RADIOMETEOROGRAPH. See: *Radiosonde.*

RADIOMETER. An instrument for the detection and usually the measurement, of radiant electromagnetic energy. See also: *Actinometer. Bolometer. Thermopile.*

RADIOMETER ACTION. The force exerted by a gas on a fixed surface when the temperature distribution over the surface is uneven. It is only observed at gas pressures of 10^{-1} or 10^{-2} mm of mercury. (Note: 1 mm of mercury equals $133 \cdot 322$ N/m^2.)

RADIOMETER, VANE. (1) The *Crookes* radiometer, in which four vanes are free to rotate in a vacuum about a vertical axis, their faces being blackened on alternate sides: rotation occurs since molecules striking the dark faces, which become hotter than the other faces, rebound with a higher speed. The speed of rotation is a measure of the intensity of the radiation. (2) The *Nichols* radiometer, in which two vanes only are employed, which are mounted on a torsion fibre which carries a mirror, the deflection of which is a measure of the intensity of the radiation.

RADIONUCLIDE. A radioactive nuclide.

RADIO PILL. A pill used to transmit information from the gastrointestinal tract on the pressure and temperature therein.

RADIO SCINTILLATIONS. Random fluctuations observed in the emission received from radio stars. The phenomenon is not related to the nature of the sources but arises from non-uniformity in the distribution of electrons in the ionosphere.

RADIOSENSITIVITY. The relative susceptibility of living cells, tissues, organs, or organisms to the effects of ionizing radiation.

RADIOSONDE. A small free balloon carrying instruments which transmit radio information to the ground on upper-air conditions. It is sometimes known as a *radiometeorograph.*

RADIO SOURCE. A small area of the sky from which the radio radiation is appreciably larger than the background radiation from the surrounding regions. In the widest sense of the term it includes such objects as the Sun and certain planets, but it is usually restricted to sources outside the solar system, i.e. the so-called radio stars.

RADIO STARS. Discrete sources of radio emission, i.e. *localized* radio sources, which are outside the solar system. The term radio star was coined when such sources were thought to be similar to visible stars which emitted radio waves rather than visible light, and is now being displaced by the term *radio source.* Novae are now known to be radio stars. See also: *Nova. Pulsar.*

RADIO TELESCOPE. A highly directive aerial system which is directed towards the sky so as to receive radio emission from extra-terrestrial sources. It may be wholly or partly steerable (commonly using a parabolic reflector) or may consist of a fixed array, and the resolution may be extended by using two aerials as a *radio interferometer.* Radio telescope design is normally a compromise between resolving power, sensitivity, ease of steering (for steerable systems) and cost.

RADIOTHERAPY. The use of ionizing radiations for medical treatment. The radiations used range from infra-red to X- and γ-rays, although the use of the term usually implies only the last two. X-rays are perhaps the most commonly employed (from conventional X-ray tubes, cyclotrons, betatrons or other accelerators), but γ-rays from radium or radioisotopes such as cobalt-60, are also often used. The use of α-rays from radon or polonium is also common; and in less common use are *electrons, neutrons,* and *protons.*

RADIOTHERAPY, CROSS-FIRE TECHNIQUE IN. The irradiation of a deep-seated region in the body from several directions so as to reduce damage to surrounding tissues for a given dose. A variation of this technique involves the movement of the radiation beam about the region to be irradiated (*moving-field therapy*).

RADIOTHERAPY, ENTRY AND EXIT PORTAL IN. (1) *Entry portal:* the area through which a beam of radiation enters the patient's body. (2) *Exit portal:* the area through which a beam of radiation leaves the body. See also: *Beam direction indicator (radiotherapy).*

RADIO WAVES. Electromagnetic radiation in the wavelength range from a few centimetres to some tens of metres.

RADIO WAVES, ABSORPTION OF. The absorption of the waves in their passage through a medium. It arises mainly from the interaction of the waves with free electrons in the medium and is therefore very marked in metals. The magnitude of the attenuation may be expressed in nepers or decibels, but the *absorption coefficient* (absorption per unit distance) is also used. See also: *Absorption coefficient for radiation.*

RADIO WAVES, ATTENUATION OF. The reduction in the intensity of the waves, from all causes, in their passage through a medium. The principal causes are absorption and scattering (sometimes called *deviative absorption*), and the attenuation coefficient may be expressed in the same way as the absorption coefficient. See also: *Attenuation coefficient for radiation.*

RADIO WAVES, DIFFRACTION OF. See: *Diffraction.*

RADIO WAVES, DISPERSION OF. The variation of the phase velocity with frequency in the passage of radio waves through a dispersive medium, i.e. one in which the phase velocity is different from the group velocity. When the latter is greater than the former the dispersion is said to be *anomalous.* See also: *Anomalous dispersion. Dispersion.*

RADIO WAVES, GUIDED. See: *Waveguide.*

RADIO WINDOW. The range of radio frequencies to which the Earth's atmosphere is transparent. It lies between a few mm and about 20 m.

RADIUM APPLIANCES. Appliances containing radium, for use in radiotherapy. The most common are the *radium needle*, for interstitial insertion, which as its name implies is a needle-shaped sealed tube; and the *radium tube*, for external application, which is a blunt-ended tube. Such tubes may also be incorporated in moulded plaster shapes (*applicators*) for surface application, such shapes being known as *radium moulds.*

RADIUM STANDARD. See: *Radioactivity standard.*

RADIUS OF GYRATION. See: *Moment of inertia.*

RADON. See: *Radioactive emanation: Emanation.*

RAIE BLANCHE. A narrow absorption maximum which occurs close to an X-ray absorption edge. It appears white on paper or film, hence its name.

RAIES ULTIMES. The strongest lines in the spectrum of an element, being the last lines to disappear as the amount of the element decreases and thus serving as a sensitive test for its presence. Also called *sensitive lines* and *ultimate lines*.

RAIN. The most common type of atmospheric precipitation. It occurs when the water droplets formed by condensation from moist air attain a size of about 100 μm. Up to about 250 μm the rain is classified as *drizzle*.

RAINBOW. A set (often two, occasionally three) of concentric spectrally coloured rays seen when one looks into a spray or mist which is illuminated from behind by white light, the best-known example being seen when the Sun shines on rain drops in the atmosphere. The colours of the rainbow are caused by dispersive refraction and internal reflection in the raindrops, the primary and secondary rainbows being associated with one and two internal reflections respectively. See also: *Spectrum*.

RAM. See: *Extrusion*.

RAMAN EFFECT. The appearance of additional weak lines in the spectrum of the light scattered by a substance illuminated by monochromatic light. The lines occur close to, and on each side of, the main spectral lines, and arise from the increase or decrease of frequency when the incident light-quanta lose energy to, or gain energy from, the molecular vibrations (or rotations) of the substance (liquid, solid, or gas) concerned. By analogy with the terminology used in fluorescence the lines corresponding to a loss of energy are called *Stokes lines*, and those corresponding to a gain of energy are called *anti-Stokes lines*. The latter are even weaker than the former.

RAMAN SPECTROSCOPY. The study of the Raman lines or spectra to gain information regarding the scattering substance. The lines are characteristic of the molecular species present.

RAMJET. See: *Jet, ram*.

RAMSAUER EFFECT: RAMSAUER–TOWNSEND EFFECT. The transparency of a gas to a restricted range of low energy electrons (the *Ramsauer well*). The effect occurs in all rare gases and in most hydrocarbon gases. It is responsible for negative resistance in irradiated thermionic diodes and inert gas-filled triodes, and is of importance in gaseous discharges.

RANDOM DATA. Data which cannot be expressed by an explicit mathematical relationship, but which vary in a random fashion with time. For mathematical analysis the data are divided into *non-stationary data*, where the statistical properties change with time; and *stationary data*, where these properties are a function of time differences and not of particular times. The analysis of random data finds applications in such fields as vibration, acoustics, seismology, and communication. See also: *Deterministic data*.

RANDOM WALK. This originally referred to finding the probability that a man, starting from a given origin, walking a given distance in a straight line, then walking the same distance in any direction whatever, and repeating the process a given number of times, shall find himself at a specified distance from the origin. The problem has since been extended to include random flights. A wide variety of physical problems (e.g. diffusion and transport phenomena) can be treated as random walk problems.

RANGE, EXTRAPOLATED. Of charged ionizing particles: the intercept on the range axis of a straight line drawn through the descending portion of a graph showing the number of particles (originally monoenergetic) against the range.

RANGE OF IONIZING PARTICLES. The displacement of charged ionizing particles at the point where they cease to ionize, specified for particles of a given type in a given material. For heavy particles the mean range may be calculated from the stopping power. See also: *Bethe–Bloch formula. Stopping power*.

RANGE OF NUCLEAR FORCES. The distance (about 10^{-13} cm, i.e. the nuclear radius) over which the forces are significant.

RANGE, RESIDUAL. At any point of the path of a charged ionizing particle: the further displacement that will occur before the particle ceases to ionize, specified for particles of a given type in a given material.

RANGE STRAGGLING. Of charged ionizing particles of a specified type in a given material: the fluctuation of range among the particles.

RANKINE BALANCE. A torsion balance for the comparative measurements of the susceptibilities of feebly paramagnetic and diamagnetic materials, in which a cylindrical permanent magnet is mounted vertically in a system which takes up a position which is practically independent of the Earth's magnetic field. Measurements are made of the deflection of this system when the specimens to be examined are introduced in turn.

RANKINE CYCLE. A modification of the Carnot cycle used as a standard of efficiency for steam power plant. It refers to a steam generator, engine or turbine, condenser and feed pump.

RANKINE–HUGONIOT EQUATIONS. A set of equations giving the density, pressure and velocity on the two sides of a shock wave.

RAOULT LAW. Relates the vapour pressure of a solution p to that of the pure solvent p_0 by the expression $p = p_0 x$, where x is the mole fraction of solvent.

RARE EARTHS. Two groups of elements having remarkably similar properties within each group. The first group comprises elements with atomic numbers 21 (scandium) and 39 (yttrium), and those with atomic numbers from 58 (cerium) to 71 (lutetium), the latter series possessing properties that seem to place them in the same space in the Periodic Table as that occupied by element 57 (lanthanum). The second group is composed of elements with atomic numbers from 90 (thorium) to 103 (lawrencium) whose properties seem to place them in the same space as element 89 (actinium). The lanthanum series is now usually referred to as the *lanthanides* and the actinium series as the *actinides*. Both groups correspond to the filling up of electron sub-shells, the $4f$ shell in the case of the lanthanides and the $5f$ or $6d$ in the case of the actinides. See also: *Electron shell. Appendix I.*

RAREFIED GAS LAWS. Laws which relate to conditions in which the mean free path in a gas is comparable with the dimensions of an immersed object or surrounding enclosure. In *slip flow* the mean free path is small but not negligible compared with these dimensions, and in *free molecule flow* it is large.

RARE GAS: INERT GAS: NOBLE GAS. Any element of Group 0 of the periodic system, namely helium, neon, argon, krypton, xenon, and radon. The electron shells are completely filled and hence these gases are practically devoid of chemical properties, but a number of crystalline compounds have been prepared with fluorine, oxygen and certain metals.

RATIONAL EQUATIONS. For electric or magnetic quantities: equations in which the presence of cylindrical or spherical symmetry is indicated by the presence of a factor 2π or 4π respectively. See also: *Units, MKSA. Units, rationalized.*

RATIONAL FUNCTION. A function p/q, where p and q are each polynomials.

RATIONAL INDICES, LAW OF. In crystallography: expresses the fact that for any individual crystal the intercepts made by a crystal face on the crystallographic axes are always in the proportion $a/h : b/k : c/l$, where a, b, c are the axial lengths of the crystal and h, k, l are the Miller indices. The law is a consequence of the regular internal structure of crystals.

RATIONAL NUMBER. A number (e.g. 3/4) which can be expressed as the ratio of two integers.

RAY ELLIPSOID: FRESNEL ELLIPSOID. A geometrical construction for the determination of propagation velocities and vibration directions in a doubly refracting crystal. It takes the form of an ellipsoid, the semi-axes of which are proportional to the reciprocals of the principal indices of refraction, and the directions of which are parallel to the corresponding vibration directions. See also: *Optical indicatrix: Fletcher indicatrix. Normal surface. Ray surface.*

RAYLEIGH COEFFICIENT. A non-dimensional coefficient characterizing the onset of convection currents when heat is transmitted across horizontal fluid layers.

RAYLEIGH CRITERION. (1) For an optical magnifying system: states that, for the resolution of two points to be just possible, the central maximum of the Airy pattern of one shall fall on the first dark ring of the other. (2) For an optical system concerned with the separation of spectral lines, etc.: states that, for the resolution of two wavelengths (of equal intensity) to be just possible, the central maximum of one shall fall on the first minimum of the other. See also: *Resolving power.*

RAYLEIGH DISK. A thin circular disk suspended vertically by means of a torsion fibre. When placed at an angle in a fluid stream either direct or alternating (as with sound waves) it experiences a torque which is proportional to the particle velocity for direct flow, or the r.m.s. particle velocity for alternating flow. The disk is deflected so as to lie at right angles to the direction of particle motion.

RAYLEIGH–JEANS LAW. Gives the distribution of energy in the spectrum of a black body as a function of temperature and wavelength. It is approximately true only for large wavelengths and has been replaced by Planck's radiation formula.

RAYLEIGH–RITZ METHOD. A method used to obtain approximate solutions of certain boundary value problems, namely those described by a differential equation, or a system of such equations, with prescribed boundary conditions.

RAYLEIGH SCATTERING. The scattering of radiation by objects small compared with the wavelength of the radiation: in effect, scattering by points. The scattered intensity is inversely proportional to the fourth power of the wavelength, which explains, among other things, the reddish colour of sunset and the blue colour of the sky.

RAYLEIGH–TAYLOR INSTABILITIES. Instabilities occurring when one fluid is accelerated by another less dense one.

279

RAYLEIGH WAVE. In an elastic medium: a surface wave in which the particles of the medium are displaced at right angles to the surface.

RAYS. Lines which represent the directions in which electromagnetic radiation (commonly light) is travelling. The lines are perpendicular to the wave fronts. They are straight in homogeneous media, curved in non-homogeneous media, and are bent (refracted) at interfaces between different media. See also: *Geometrical optics.*

RAY SURFACE. One of a number of geometrical constructions for the determination of propagation velocities and vibration directions in doubly refracting crystals. It shows the distribution of ray velocities. See also: *Optical indicatrix: Fletcher indicatrix. Ray ellipsoid: Fresnel ellipsoid. Normal surface.*

RAY TRACING. The tracing of a ray through an optical or electron-optical system which has only plane or spherical interfaces, where it is not feasible or desirable to set up accurate equations.

R-CENTRE. See: *Colour centres.*

REACTANCE. The imaginary part of impedance.

REACTANCE AMPLIFIER: PARAMETRIC AMPLIFIER. An amplifier for which the equations describing its operation contain one or more time-dependent reactance parameters. The operation of the amplifier depends on the fact that a time-varying reactance can exhibit negative resistance under some conditions and can act as a frequency converter under others.

REACTION PROPULSION. Propulsion which depends on the ejection of high-velocity matter, i.e. jet and rocket propulsion.

REACTION RATE. See: *Chemical reaction, specific reaction rate constant of.*

REACTIVE COMPONENT. In electrical engineering: the component of the vector representing an alternating quantity which is in *quadrature* (at 90°) with some reference vector, e.g. the reactive component of the current, or *reactive current* (in quadrature with the voltage), the *reactive voltage* (in quadrature with the current), the *reactive volt-amperes* (*var*) (the product of the reactive voltage and the current, or of the reactive current and the voltage). The reactive component is also known as the *quadrature component, wattless component,* and *idle component.*

REACTIVITY. In nuclear technology. See: *Nuclear reactor, reactivity of.*

REACTOR. (1) A nuclear reactor. (2) An electrical reactor, i.e. a device for introducing reactance into a circuit. (3) A chemical reactor, i.e. a vessel in which chemical reactions may take place, normally for industrial purposes.

REACTOR OSCILLATOR. See: *Nuclear reactor oscillator.*

REACTOR PHYSICS. The physics of nuclear reactors.

REAL NUMBER. Any rational or irrational number. See also: *Complex number.*

REAL TIME PROCESS. A process which produces results in near synchronism with the data or effects being studied, as in the computer-controlled tracking of satellites, or other data-handling processes.

RECALESCENCE. The evolution of heat which occurs when iron or steel cools through the critical range, i.e. the range in which the reversible change occurs from austenite to ferrite, pearlite, or cementite.

RECIPROCAL LATTICE. A geometrical concept which is widely used in crystal diffraction problems. It may be regarded as representing the periodic spatial distribution of the crystal's reflecting ability, a diffraction pattern being a distorted projection of this lattice, the amount of distortion depending on the technique employed. The reciprocal lattice point corresponding to a given set of planes is at a distance from an arbitrary origin which is inversely proportional to the spacing of the planes, in a direction normal to them. See also: *Reflection sphere.*

RECIPROCAL SPACE: MOMENTUM SPACE. The space in which the reciprocal lattice is plotted. It is the same as that used in the band theory of solids.

RECIPROCITY LAW. In photography: states that, all other conditions remaining constant, the exposure time required to produce a given photographic density is inversely proportional to the intensity of the radiation. The law does not hold for high and low light intensities, where *reciprocity-law failure* is said to occur.

RECIPROCITY PRINCIPLE. For an electric network composed of passive linear impedances: states that the ratio of the e.m.f. introduced into any branch to the current measured in any other branch (i.e. the *transfer impedance*) is equal in magnitude and phase to the ratio that would be observed if the positions of the e.m.f. and current were interchanged. A similar principle holds for the loads and deflections in *elastic structures* and for *electroacoustic transducers* coupled by a fluid medium.

RECOIL PARTICLE. A particle that has been set into motion by a collision with another particle, by the ejection of another particle, or by any process in which the particle acquires kinetic energy.

Common examples are *recoil atoms or nuclei* and *recoil electrons*. See also: *Aggregate recoil. Compton effect.*

RECOMBINATION. Of ions: the disappearance of positive and negative ions by mutual neutralization. The *recombination coefficient* is the ratio of the neutralization rate to the square of the ion density.

RECRYSTALLIZATION. The reforming of crystals; usually by solution and evaporation in the case of chemical substances, but in metals the term denotes the replacement of certain crystals by other crystals which grow at their expense under the influence of temperature.

RECTIFICATION. (1) In electricity: the process of obtaining a unidirectional voltage from an alternating supply. It may be achieved by suppressing those parts of the wave form whose polarity is opposite to that required (*half-wave rectification*) or by reversing the polarity `of these parts (*full-wave aectification*). (2) In chemistry: the purification of r liquid by distillation.

RECUPERATOR. A continuous heat exchanger for restoring waste heat to a working system (e.g. to a furnace). See also: *Heat exchanger. Regenerator.*

RECURRENCE PHENOMENA. For cosmic rays: variations of intensity which occur with the periodicity of the solar rotation. The phenomena are also described as the *twenty-seven-day recurrence effect.*

REDOX PROCESS. An abbreviation for reduction-oxidation process, i.e. a process which involves both reduction and oxidation steps. An example of such a process is the solvent extraction process for nuclear reactor fuels.

RED SHIFT. (1) The systematic displacement of spectral lines from distant galaxies towards the red end of the spectrum. (2) The relativistic frequency shift which may be detected by using the Mössbauer effect and by astronomical observations.

REDUCED EQUATION OF STATE. See: *Corresponding states, law of.*

REDUCED MASS. For a body of mass m revolving about a body of mass M: the value by which m must be replaced when the equations of motion are transferred from coordinates having their origin at M to centre-of-mass coordinates. It is given by $Mm/(M + m)$ and approaches m as M approaches infinity. The reduced mass is used in atomic and molecular spectroscopy and in problems involving the motion of celestial bodies.

REDUCED PRESSURE. See: *Corresponding states, law of.*

REDUCED TEMPERATURE. See: *Corresponding states, law of.*

REDUCTION. The removal of oxygen from a substance. More generally, the gain of electrons by an atom. See also: *Oxidation.*

REDUNDANCY TECHNIQUES. In computing systems: the reduction in the probability of computing errors arising from malfunctions of computer components by the introduction of additional elements which repeat selected basic operations.

REFERENCE SPHERE. Of a crystal projection: See: *Crystal projection.*

REFERENCE STARS. See: *Carte du ciel: Astrographic chart.*

REFLECTANCE. See: *Radiant total reflectance.*

REFLECTION. The change in direction, within a given medium, of radiation travelling in the same medium, when the radiation encounters the surface of another medium. The radiation may be acoustic or electromagnetic or, in general, may be any regular wave motion. After reflection at a *smooth surface* (i.e. one in which the irregularities are small compared with the wavelength of the radiation) the direction of propagation remains sharply defined, the incident and reflected wave trains travel in directions making equal angles (the angles of *incidence* and *reflection* respectively) with the normal to the reflecting surface, and are co-planar with it. See also: *Refraction.*

REFLECTION COEFFICIENT: REFLECTION FACTOR: REFLECTION POWER: REFLECTIVITY. The ratio of the reflected wave to the incident wave, expressed in terms of intensity, power, field strength, etc., according to the type of radiation involved. For neutrons the term is sometimes used to denote the albedo.

REFLECTION CURVE. For X-rays: See: *Integrated reflection.*

REFLECTION, DIFFUSE. Reflection in many directions as from a rough surface or a suspension, i.e. scattering. The term is also used in diffraction analysis to denote scattering arising from crystalline imperfections, disorder, and atomic vibrations. See also *Albedo. Scattering, diffuse.*

REFLECTION PLATE. A flat surface attached to a wind-tunnel model of the "half-model" type. It provides an aerodynamic image of the half-model and thus effectively doubles the size of the model and the Reynolds number of the test.

REFLECTION, SPECULAR. Reflection as at a mirror, i.e. by a smooth surface.

REFLECTION SPHERE. In a reciprocal lattice: a sphere whose surface includes the origin of the lattice and whose centre lies on the line of a beam which is capable of being diffracted by the corresponding crystal (typically a beam of X-rays, electrons or neutrons). Any crystal plane whose reciprocal lattice point lies on the surface of the sphere will "reflect" the beam in question. The radius of the sphere is inversely proportional to the wavelength of the radiation.

REFLECTOR. (1) A material or surface which reflects incident radiation. (2) Of a nuclear reactor: a part of the reactor placed adjacent to the core so as to return some of the escaping neutrons to the core by scattering collisions.

REFRACTION. The change in direction of radiation travelling in a given medium when it crosses the surface of another medium. The radiation may be acoustic or electromagnetic or, in general, may be any regular wave motion. Refraction is associated with a change of speed from one medium to the other. See also: *Reflection. Refraction, Snell laws of. Refractive index.*

REFRACTION, SNELL LAWS OF. State: (1) that, when radiation passes from one medium to another, the directions of incidence and refraction are co-planar with the normal to the surface between the media at the point of incidence; and (2) that the refractive indices of the two media n_1 and n_2 are related by the expression $n_1/n_2 = \sin i_2/\sin i_1$, where i_1 is the angle of incidence and i_2 that of refraction; and n_1 and i_1 refer to the first medium, n_2 and i_2 referring to the second. See also: *Total reflection.*

REFRACTIVE INDEX. Of a given medium: this is commonly defined as the ratio of the sine of the angle of *incidence* (*in vacuo*) to the sine of the angle of *refraction*, where these angles are the angles between the directions of the incident and refracted wave trains and the normal to the surface of the medium concerned, with which normal these directions are co-planar. The refractive index may also be defined as the ratio of the phase velocity of the radiation in free space to that in the medium concerned. Measurements of refractive index are, however, usually carried out in air rather than *in vacuo* since the difference is very small. See also: *Dielectric, refractive index of. Refraction, Snell laws of.*

REFRACTIVE INDEX, COMPLEX. A refractive index which is necessary for materials which absorb strongly, typically metals. It is given by $n(1 - ik)$, where n is the customary refractive index and k the *absorption index* (equal to $\lambda a/4\pi$, where a is the absorption coefficient and λ the wavelength). This complex index is of importance in the study of metallic reflection.

REFRACTIVITY, MOLECULAR. Of a material, for a given wavelength and state of aggregation: is given by

$$\frac{n^2 - 1}{n^2 + 2} \cdot \frac{M}{\varrho},$$

where n is the refractive index, M is the molecular weight, and ϱ the density.

REFRACTIVITY, SPECIFIC. The molecular refractivity divided by the molecular weight, i.e.

$$\frac{n^2 - 1}{n^2 + 2} \cdot \frac{1}{\varrho},$$

where n is the refractive index and ϱ the density.

REFRACTOMETRY. The measurement of refractive index.

REFRACTORY MATERIAL. A material with a high melting or softening point. The lower limit of this point is sometimes taken as $1500°C$. See also: *Ceramics. Metals, refractory.*

REFRACTORY PERIOD, ABSOLUTE. Of a nerve: the short period, following the passage of an action potential down a nerve, during which the nerve is quite inexcitable. See also: *Action potential. Refractory period, relative. Resting potential.*

REFRACTORY PERIOD, RELATIVE. Of a nerve: the period following the absolute refractory period, during which a nerve shows reduced excitability. See also: *Action potential. Refractory period, absolute. Resting potential.*

REFRIGERANT. A substance suitable for use as a working medium in a commercial refrigerator, e.g. ammonia, carbon dioxide, sulphur dioxide, freon.

REFRIGERATING CAPACITY. The rate of heat absorption of a cooling system, often expressed in "tons of refrigeration". One ton of refrigeration is equal to 200 Btu/min, or $211 \cdot 012$ kJ/min.

REFRIGERATING CAPACITY, SPECIFIC. The number of tons of refrigeration produced per horse power of input, i.e. per $745 \cdot 700$ W.

REFRIGERATION. The production of cold for such purposes as food preservation, air-conditioning, liquefaction of gases, and cryogenics in general. It is commonly achieved by the compression of a vapour, followed by condensation, and throttling to the original pressure, but other systems are also in use, including *air cycle refrigeration*, in which the refrigerant remains in the gaseous phase throughout the cycle; *absorption refrigeration*, in which the refrigerant vapour is absorbed by a liquid; and *thermoelectric refrigeration*, in which use is made of the Peltier effect. See also: *Adiabatic demagnetization, cooling by. Refrigerator, dilution. Refrigerator, magnetic.*

REFRIGERATION, TON OF. See: *Refrigerating capacity.*

REFRIGERATOR. A machine or plant by which mechanical, thermal, or electrical energy is used to produce and maintain a low temperature.

REFRIGERATOR, COEFFICIENT OF PERFORMANCE OF. The ratio of the heat extracted to the heat equivalent of the net energy supplied.

REFRIGERATOR, DILUTION. A refrigerator originally based on the cooling produced when a solution of ^{3}He in ^{4}He is diluted adiabatically, the principle being subsequently extended to make use of the separation of a mixture of 6·3% ^{3}He in ^{4}He into a concentrated ^{3}He phase and a dilute phase at temperatures below 0·8 K, the cooling being effected by dissolving the former phase in the latter. Temperatures of less than 10 mK have been reached in this way, the lower limit appearing to be about 4·5 mK.

REFRIGERATOR, MAGNETIC. A device for the maintenance of temperatures below 1 K by cyclic magnetization and demagnetization. An essential feature is the use of *thermal valves* which depend on the fact that pure superconducting metals have markedly different conductivities in their normal and superconducting states at temperatures well below the transition temperature. See also: *Magnetocaloric effect.*

REGELATION. The melting of ice by pressure, followed by re-freezing when the pressure is removed. Examples occur in ice skating and in the flow of glaciers.

REGENERATIVE FLYWHEEL. A device used for the storage of energy by a rotating flywheel. It has been used for some time in power torpedoes and punch presses and has recently been used in railway and road transport vehicles.

REGENERATOR. A heat exchanger for restoring waste heat to a working system (e.g. to a furnace) or for cooling, in which the two fluids concerned flow alternately through the same channels. See also: *Heat exchanger. Recuperator.*

REGGE MODEL. A model which, starting as a formal mathematical procedure for obtaining asymptotic bounds on scattering amplitudes, and being developed to aid in understanding high energy experiments, now plays an important part in providing a comprehensive theory of elementary particles and their interactions.

REGIME THEORY. The theory of the transport of silt by rivers through stable channels (termed "regime" channels) in alluvial deposits. Such channels are considered to be in a state of balance, the amount of silting being equal to the amount of scouring.

RÉGLETTE. A short subdivided scale attached at each end of a surveying wire or tape.

REGULAR FUNCTION: HOLOMORPHIC FUNCTION: ANALYTIC FUNCTION. A function $f(z)$ of the complex variable $z = x + iy$ is regular, analytic, or holomorphic at a point on the z-plane if the function and its first derivative are finite and single-valued there. If this property applies to all points within a given region of the complex plane, $f(z)$ is a regular (analytic or holomorphic) function throughout the region. Points at which there is no derivative are known as *singular points* or *singularities*.

REGULAR SYSTEM. In crystallography: another name for the cubic system.

REHBINDER EFFECT. The change of mechanical properties occurring when a material is immersed in a liquid containing a surface-active substance such as oleic acid or cetyl alcohol. It arises from the adsorption of such substances on the surface of the material.

RELATIVE BIOLOGICAL EFFECTIVENESS (RBE). Of an ionizing radiation for a particular living organism or part of an organism: the ratio of the absorbed dose of a reference radiation producing a specified biological effect to the absorbed dose of the radiation of interest producing the same effect. The term is now restricted to use in radiobiology.

RELATIVE DENSITY: SPECIFIC GRAVITY. Of a given substance: the ratio of the density of the substance to that of a standard substance, usually water, at a specified temperature. See also: *Atmosphere, relative density of.*

RELATIVE PERMITTIVITY: DIELECTRIC CONSTANT: SPECIFIC INDUCTIVE CAPACITY. Of a medium: the ratio of the capacitance of a capacitor with the medium between the electrodes to that of a capacitor with a vacuum between the electrodes. Alternatively: the ratio of the electric flux density produced in the medium to that which would be produced in a vacuum by the same electric force. See also: *Permittivity: Absolute permittivity.*

RELATIVISTIC INVARIANCE. Describes the property, as exhibited by Maxwell's equations for example, of being applicable whether the observer is travelling at constant speed or is at rest.

RELATIVISTIC MASS. The mass of a particle

moving at high speed, according to the special theory of relativity. It is equal to $m_0(1 - v^2/c^2)^{-\frac{1}{2}}$, where m_0 is the mass of the particle at rest (the *rest mass*), v is the speed of the particle, and c the speed of light.

RELATIVITY THEORIES. (1) *Special theory:* a theory based on the postulates that all the laws of physics are the same in all inertial frames of reference, and that the velocity of light is the same in all such frames. These postulates lead to the conclusion that, relative to a stationary observer, a moving body appears to increase its mass, and to become foreshortened in the direction of motion, by an amount (the Lorenz contraction) that becomes appreciable as the speed approaches that of light. It also leads to the concept of the equivalence of mass and energy. See also: *Lorenz contraction. Mass–energy equivalence. Relativistic mass. Time dilation.* (2) *General theory:* a generalization of the special theory which relates the measurements made by observers who are accelerated relative to each other, i.e. who are no longer in an inertial system. The theory has special relevance for gravitational theory, leading to the field theory of gravitation. See also: *Einstein displacement. Gravitational field. Perihelion, advance of. Red shift.*

RELAXATION HEATING. A method of supplying a controlled amount of heat to a thermally isolated sample at a temperature below 1 K, involving the simultaneous application of steady and oscillating magnetic fields.

RELAXATION LENGTH. For a physical quantity which decreases exponentially with distance: the distance over which the quantity decreases by a factor e.

RELAXATION METHOD. The solution of certain complicated sets of simultaneous linear equations (originally for engineering problems but subsequently extended to eigenvalue problems) by successive approximation.

RELAXATION PHENOMENON. Any phenomenon in which a system requires an observable length of time in which to achieve equilibrium following a sudden change in the condition of the system, imposed for example by stress or other forces. The study of the time-dependence of the approach to equilibrium is essential where relaxation occurs. A few examples of relaxation phenomena are afforded by nuclear relaxation, paramagnetic relaxation, and thermal relaxation. See also: *Relaxation length. Relaxation time.*

RELAXATION TIME. For a physical quantity which decreases exponentially with time: the time in which the quantity decreases by a factor e.

RELAY. An electrical device which uses the vari-ation of current in one circuit to control the current in another, e.g. by switching it on and off. The current in the first circuit may itself depend on the variation of a physical property such as thermal expansion, as in some forms of thermostat.

RELIABILITY FACTOR. In crystal structure determination. See: *Agreement residual.*

RELUCTANCE. See: *Magnetic reluctance.*

RELUCTIVITY. The reciprocal of permeability.

REM. A unit of dose equivalent. It is numerically equal to the absorbed dose in rads multiplied by the quality factor and other factors used in assessing dose equivalent. See also: *Dose equivalent. Rad. Sievert.*

REMANENCE: RESIDUAL MAGNETISM. The residual magnetic flux density when the magnetizing field is reduced to zero from saturation.

REMANENT MAGNETISM. The magnetization left in a specimen after the removal of a magnetic field.

RENEWAL THEORY. A mathematical model for the study of physical situations in which a sequence of events occurs at random time intervals. It is applied, for example, to the theory of counters, to storage problems, and to the reliability of complex equipment.

REPLACEMENT COLLISION. The replacement of one atom by another after a collision, e.g. during bombardment by ionizing radiation.

REPLICA. For electron microscopy: a thin replica, usually of carbon, metal, or plastic, which reproduces the surface features of a specimen for examination by the transmission electron microscope. See also: *Electron microscope* and entries following. *Shadow casting.*

REPROPORTIONATION. See: *Disproportionation.*

RESIDUAL MAGNETISM. See: *Remanence: Residual magnetism.*

RESIDUAL RAYS: RESTSTRAHLEN. Almost monochromatic infrared rays selectively reflected by certain materials (notably quartz and the alkali halides), radiation of neighbouring wavelengths being strongly absorbed. The wavelength of residual rays varies from $8\cdot8\ \mu$m for quartz to 96 μm for potassium iodide.

RESISTANCE, DIFFERENTIAL. See: *Resistance, incremental.*

RESISTANCE, ELECTRICAL. The property of a material by which it obstructs the flow of an electric current through it, by dissipating the energy of the current in some other form (e.g. heat). In a conductor it arises mainly from the scattering of conduction electrons by atoms and by crystalline defects of various kinds. The unit of resistance is the ohm, which is separately defined. See also: *Dielectric. Impedance. Insulator. Matthiesen rule.*

RESISTANCE, INCREMENTAL. The ratio of increments of potential difference and current, i.e. $\delta V/\delta I$, where V is the voltage and I the current. It is of importance in the case of a material with a non-linear voltage-current relationship (e.g. where a.c. and d.c. flow in the same circuit). It is also known as the *differential resistance.*

RESISTANCE, NEGATIVE. A property of some circuit components and networks which enables them to supply energy to a system connected to their terminals.

RESISTANCE, RESIDUAL. The temperature-independent part of electrical resistance, which persists at low temperatures and arises from the scattering of conduction electrons by defects of various kinds. See also: *Matthiessen rule.*

RESISTANCE, SKIN. See: *Skin effect.*

RESISTANCE, SPECIFIC: RESISTIVITY: VOLUME RESISTIVITY. Of a given material: the resistance of a unit length of unit cross-sectional area, expressed in units of ohm-cm or ohm-metre. It is the reciprocal of conductivity.

RESISTANCE, STANDARDS OF. Specially designed and constructed resistors whose resistance is stable and accurately known. Such resistors, set up in the National Physical Laboratory and the National Bureau of Standards, have shown drifts as little as one part per million in 10 years. See also: Entries under *Ohm.*

RESISTANCE, TEMPERATURE COEFFICIENT OF. The constant β in the expression $\sigma_t = \sigma_0(1 + \beta t)$, where σ_t is the specific resistance at temperature t and σ_0 is that at $0°C$. It is approximately constant over limited ranges of temperature. See also: *Superconductivity.*

RESISTANCE TO FLUID FLOW. The resistance to the motion of a body through a fluid or of a fluid past a body. See also: *Drag.*

RESISTIVE WALL AMPLIFIER. An electron-beam amplifier in which an impressed radio frequency signal is amplified by the interaction between the space-charge field induced by the beam and a resistive surface arranged in close proximity to the beam.

RESISTIVITY. Another name for specific resistance.

RESISTOR. An element of electrical apparatus whose primary function is to introduce resistance into an electric circuit. It may have either a fixed or variable resistance according to the uses to which it is put.

RESOLUTION. (1) Of an optical system: the minimum separation of two adjacent points or spectral lines that is detectable by the system. It is expressed quantitatively as the resolving power. (2) Of an instrument separating atomic particles: the minimum difference in mass, energy or momentum that is detectable by the instrument. It is expressed quantitatively as the resolving power. (3) Of photographic emulsion: the ability to record fine detail. It is expressed as the number of lines per mm that can be distinguished. (4) The separation of a racemic mixture into its optically active components. (5) The separation of a vector into its components. See also: *Resolving power.*

RESOLUTION LIMIT. See: *Resolving power.*

RESOLUTION TIME: RESOLVING TIME. Of a counter or counting system: the minimum time interval between two distinct events which will permit both to be counted.

RESOLVING POWER. The quantitative expression of the resolution. (1) For an optical magnifying system the resolving power is expressed as the minimum detectable angular separation of point sources, i.e. as the *resolution limit.* According to the widely used *Rayleigh criterion* this is given by $0·61 \lambda/NA$, where λ is the wavelength and NA the numerical aperture. (2) For an optical system concerned with the detection of spectral lines, etc., the resolving power (known as the *chromatic resolving power*) is expressed as $\lambda/\delta\lambda$, where λ is the wavelength and $\delta\lambda$ the minimum difference in wavelength that can be detected. This is also related to the *Rayleigh criterion.* (3) For instruments concerned with the separation of atomic particles (e.g. α- and β-spectrometers) the resolving power is sometimes expressed as $E/\delta E$, $P/\delta P$ or $M/\delta M$, and sometimes as their reciprocals, where E is energy, P is momentum, and M is mass, δE, δP, and δM being the minimum detectable differences as above. See also: *Rayleigh criterion. Resolution.*

RESONANCE. Of an oscillatory or vibratory system: the marked increase in the amplitude of oscillation or vibration when the system is subjected to an impressed frequency that is the same as (or very close to) the natural frequency of the system. Resonance occurs in acoustical, mechanical, atomic, electrical, magnetic, optical, and radio systems, for example; and the term has also been extended to cover, by analogy, certain chemical and nuclear phenomena, which are discussed separately. See also: following entries.

RESONANCE BROADENING. Of *spectral lines*: broadening arising from the exchange of energy by resonance between like atoms. See also: *Spectral lines, broadening of.*

RESONANCE ENERGY. See: *Resonance in nuclear processes. Resonance of chemical bonds. Resonance "particle".*

RESONANCE IN NUCLEAR PROCESSES. A term used to denote a critical variation with energy in the cross-section for scattering or for some other specified process. The cross-section shows a sharp peak at a certain energy of an incident particle, known as the *resonance energy*, and the corresponding energy level of the compound nucleus formed is called the *resonance level*. See also: *Neutrons, resonance. Neutrons, resonance absorption of. Neutrons, resonance escape probability for.*

RESONANCE INTEGRAL. An integral giving a measure of the probability that neutrons will be absorbed in the process of slowing down in an infinite medium, i.e. a measure of the inverse of the resonance escape probability. See also: *Neutrons, resonance escape probability for.*

RESONANCE OF CHEMICAL BONDS. A term used to denote the concept that, for certain molecules and certain bindings of those molecules, the state of binding oscillates or "resonates" between two or more bond arrangements and partakes of the properties suggested by each. The difference in electronic energy between that of the resonance hybrid and that of the most stable localized bond is known as the *resonance energy*.

RESONANCE "PARTICLE". A "particle" with a lifetime of only about 10^{-23} s. It decays via a strong interaction into other particles which may also be unstable. These resonance particles are detected as peaks in graphical experimental plots, e.g. of cross-section against mass (energy). They are often described as *"resonances"*.

RESONANCE RADIATION. Fluorescence radiation which has the same frequency as that of the incident radiation. Spectral lines arising from resonance (as when the radiation from one atom induces resonance radiation in another) constitute a *resonance spectrum*. See also: *Fluorescence.*

RESONANCE RADIATION, IMPRISONMENT OF. The delay in the emission of resonance radiation caused by a chain of emission and absorption within a material, the resonance radiation emitted by one atom being absorbed by another, for a succession of atoms.

RESONANT FREQUENCY. The natural frequency of a resonant circuit, given by $1/(2\pi \sqrt{LC})$, where L is the inductance and C the capacitance of the circuit.

RESONANT LINE. A parallel wire or concentric transmission line open or short-circuited at the ends and an integral number of quarter-wavelengths in length. It is used for stabilizing the frequency of short-wave oscillators, and in aerial systems.

REST ENERGY. The energy corresponding to the rest mass of a free particle, i.e. m_0c^2, where m_0 is the rest mass and c the speed of light. See also: *Electron rest mass. Relativistic mass.*

RESTING POTENTIAL. A term originally referring to the steady electrical potential associated with electrically active tissues (nerves, muscles, etc.) in their quiescent state. However, all living cells, whether of electrically active tissue or not, are found to maintain a resting or steady potential. This may be regarded as indicating a polarization of the cell membrane (the inside is negative with respect to the outside). See also: *Bioelectricity.*

RESTITUTION, COEFFICIENT OF. The ratio of the relative velocities of two colliding bodies after impact to that before, i.e. $(v_2 - v_1)/(u_1 - u_2)$, where u_1 and u_2 ($u_1 > u_2$) are the velocities of bodies 1 and 2 before impact and v_1 and v_2 ($v_1 < v_2$) their velocities after impact, measured along the line of centres at the instant of collision. This coefficient defines the extent to which the collision is elastic, being equal to unity for a purely *elastic collision* and zero for a purely *plastic* one. For two colliding bodies of the same material the coefficient is sensibly constant and is characteristic of that material.

REST MASS. See: *Electron rest mass. Relativistic mass.*

RETARDATION PLATE: WAVE PLATE. A plate of birefringent material cut so that a desired path difference is introduced between the ordinary and extraordinary vibrations of a ray of light travelling perpendicularly through it. See also: *Half-wave plate. Quarter-wave plate. Quartz wedge.*

RETENTIVITY. The value of the intensity of magnetization corresponding to remanence.

RETINA. The light-sensitive portion of the eye, containing two types of light receptor, termed *rods* and *cones* respectively.

RETROGRADE MOTION. Of a celestial body: motion such that the longitude (real or apparent) decreases with time. See also: *Direct motion. Planet, stationary points of.*

RETRO-REFLECTOR. A device which will reflect light or radiowaves back along the direction of incidence, e.g. a corner-cube.

RETURNED PULSE. In a pulsed radar system: See: *Direct pulse.*

REVERBERATION. The persistence of sound at a given point after the sound has been cut of at the source. It may arise from resonance, echoes, multiple reflections, etc.

REVERBERATION TIME. For a sound of a given frequency: the time required for the average (steady) sound energy density to decrease to 10^{-6} of its initial value, after the source has stopped, i.e. to decrease by 60 decibels.

REVERSIBILITY, MICROSCOPIC. Refers to the principle that states that, at equilibrium, each microscopic transition must be accompanied by an inverse process. Thus the total number of molecules leaving one quantum state for another in a given time must be equal to the number travelling in the reverse direction in the same time. The principle refers to transitions between two states and not to total rates of transition. See also: *Detailed balancing principle.*

REVERSIBILITY, OPTICAL. Refers to a principle that states that, if a ray travels from one point to another through an optical system, along a given path, then a ray can also proceed in the reverse direction along the same path.

REVERSIBLE PROCESS. A process whose effects can be reversed, so as to return the system involved to its original thermodynamic state.

REVERSING LAYER. Of the solar chromosphere: the lower region of the chromosphere where the Fraunhofer lines are formed by reversal of the bright emission lines of the photosphere. See also: *Solar atmosphere. Spectral lines, reversal of.*

REYNOLDS NUMBER. A dimensionless number of importance in fluid flow. It is concerned with the effect of viscosity and is given by $\varrho u l/\mu$, where u is the fluid speed, ϱ the fluid density, l a characteristic length, and μ the viscosity. It expresses the ratio of the inertial forces to the viscous forces and its value serves as a criterion for the stability of laminar flow, low values being associated with high stability. See also: *Dynamical similarity principle.*

RHEOLOGY. The study of deformation and flow in relation to the nature of the matter involved. It includes the study of elasticity, viscosity, and plasticity.

RHEOPEXY. A form of thixotropy in which the re-setting of a sol containing anisotropic particles (typically rods or plates) can be increased by a gentle rotation of the vessel containing the sol, which helps to reorientate these particles.

RHEOSTAT. A resistor whose resistance can be varied.

RHOMBOHEDRAL SYSTEM. In crystallography: another name for the trigonal system.

RHUMBATRON. A cavity resonator employed in the klystron oscillator for very high frequencies.

RHUMB LINE. A line indicating a constant compass course on a map. Such a line (which on the Mercator projection is straight) does not correspond to the shortest distance between two points, which is given by the great circle passing through the points. Rhumb lines are, however, very convenient for navigation over relatively small distances.

RICHARDSON EFFECT. A name sometimes given to the Einstein–De Haas effect.

RICHARDSON EQUATION: RICHARDSON–DUSHMAN EQUATION. An equation giving the relationship between the current density of electrons emitted from a hot metal, and the temperature of the metal. It may be written:

$$j = AT^2 \exp\left(-\phi/kT\right),$$

where j is the current density required, T is the absolute temperature, ϕ is the thermionic work function, k is Boltzmann's constant, and A is a constant. The equation is sometimes known as *Edison's equation.*

RICHTER SCALE. A scale of earthquake magnitude. See also: *Earthquake, magnitude of.*

RIDER. See: *Balance, rider for.*

RIEKE DIAGRAM. A chart used in the evaluation of microwave oscillator performance. Contours of constant power output and constant frequency are drawn on a polar diagram whose coordinates represent the components of the complex reflection coefficient at the oscillator load. It is a Smith chart without impedance coordinates.

RIEMANNIAN SPACE. A space of n dimensions comprising a set of points such that the "distance" between any neighbouring pair of points ($\mathbf{x}$) and ($\mathbf{x} + d\mathbf{x}$) is a scalar, ds, given by the formula: $ds^2 = g_{\mu\nu}\, dx^\mu\, dx^\nu$, where $\mu, \nu = 1, 2, 3, ..., n$, and the $g_{\mu\nu}$ are (i) the components of a covariant tensor of rank two called the *metrical tensor* and (ii) are symmetrical in their indices, i.e. $g_{\mu\nu} = g_{\nu\mu}$.

RIGHI-LEDUC EFFECT. The development of a temperature difference across a metallic strip

287

through which heat is flowing, when it is placed in a magnetic field perpendicular to its plane. The temperature gradient is given by $\nabla_t T = SH \times \nabla T$, where T is the temperature, S the *Righi–Leduc coefficient* and H is the magnetic field strength.

RIGHT ASCENSION (RA). Of a celestial body: the longitude of the body on the celestial sphere, measured eastwards from the *first point of Aries* (i.e. the point at which the ecliptic crosses the celestial equator at the vernal equinox) along the celestial equator. See also: *Hour angle. Longitude, celestial: Longitude ecliptic. Oblique ascension.*

RIGID BODY. Denotes a body the size and shape of which remain the same under the action of external forces, i.e. the distance between any pair of its constituent particles remains unaltered.

RIGID ROTATOR. A rotating body whose moment of inertia is constant and unaffected by the rotation.

RIME. Ice deposits formed under certain meteorological conditions by the rapid freezing of small supercooled drops of water.

RING NEBULA. Another name for planetary nebula. See also: *Nebula, planetary.*

RIP CURRENTS. Localized outward-flowing currents close to the shores of large lakes or oceans, resulting from the combined action of waves of translation (produced when deep water or oscillation waves enter shallow water) and of wind drag across the surface.

RIPPLES. Surface waves whose motion is controlled by surface tension. See also: *Waves, surface.*

RITZ COMBINATION PRINCIPLE. States that the multitude of frequencies shown by the line spectra of atoms or molecules may be expressed in terms of their wave numbers as the differences between relatively few terms taken two at a time in various combinations. Thus, for hydrogen, the wave numbers are given by $R/x^2 - R/y^2$, where R is Rydberg's constant and x and y are integers. Each term corresponds to a state of definite energy.

RITZ SERIES FORMULA. A formula giving the wave number N of the spectral lines of a given series. It may be written as $N = R(1/A^2 - 1/B^2)$, where R is the Rydberg constant, A is a constant for the series in question, and B has various integral values for the different lines in the spectrum.

ROCHE LIMIT. The minimum distance from a planet at which a satellite can revolve in a state of equilibrium. It is about 2·44 times the radius of the planet.

ROCHON PRISM. See: *Polarizing prism.*

ROCKET. An object or vehicle which is propelled by the thrust obtained from the ejection of high-velocity matter (the *propellant*), the whole of which is previously stored within the rocket itself. The most common propellant is high-temperature gas generated by the chemical combination of fuels and oxidants at high pressure in a combustion chamber, from which the gas is expanded in a nozzle and ejected; but other energy sources (e.g. nuclear) have been proposed. As a rocket carries its own energy supply it is, unlike a jet engine, not affected by the nature of the atmosphere surrounding it. See also: *Jet engine.*

RODS AND CONES. Of the eye: See: *Retina.*

ROLLIN FILM. See: *Helium film.*

ROLLING. Of a ship at sea or of an aircraft in the air: the movement of the ship or aircraft about the fore-and-aft axis. See also: *Pitching. Yawing.*

RÖNTGEN. A unit of exposure to X- or γ-rays in air. It expresses the total amount of electric charge produced per unit mass of air by all the ions of one sign when all the electrons liberated by the rays are completely stopped in air, and is defined as $2·58 \times 10^{-4}$ C kg^{-1} exactly. The röntgen is also used as a unit of ion dose. The SI unit is C kg^{-1}. See: *Appendix II.*

ROOT-MEAN-SQUARE DEVIATION. An average measure of the difference between two sets of numbers. For two sets of corresponding numbers $x_1, y_1; x_2, y_2; x_3, y_3; \ldots; x_n, y_n$, it is given by

$$[1/n \sum_{i=1}^{n} (x_i - y_i)^2]^{\frac{1}{2}}.$$

Minimization of the root-mean-square deviation is the basis of the method of least squares. See also: *Least squares, method of. Standard deviation.*

ROOT-MEAN-SQUARE (R.M.S.) VALUE. Of a periodically alternating quantity: the square root of the mean values of the squares of the instantaneous values of that quantity during one complete cycle. Probably its best known use is with respect to an alternating current or voltage.

ROTATING CRYSTAL METHOD. In crystal structure analysis: the examination of the diffracted beams from a single crystal which is illuminated by a monochromatic beam of X-rays or neutrons, the crystal being rotated about an axis perpendicular to the incident beam, so as to satisfy the conditions for Bragg reflection from the various sets of crystal planes in turn. See also: *Oscillating crystal method.*

ROTATIONAL ENERGY. Of a molecule: that portion of the total energy which is attributable to rotation. According to the equipartition principle it is $\frac{1}{2}kT$ (where k is Boltzmann's constant and T the absolute temperature) per molecule for each type of rotation. See also: *Equipartition of energy. Translational energy. Vibrational energy.*

ROTATIONAL WAVE. See: *Shear wave.*

ROTATION AXIS. An axis, characterized by rotation through a fraction of a revolution, which produces equivalent and identical positions in crystal form or structure. The fraction may be a half (*diad axis*), one-third (*triad axis*), one-quarter (*tetrad axis*), or one-sixth (*hexad axis*).

ROTATION BY MAGNETIZATION: EINSTEIN–DE HAAS EFFECT. The axial rotation of a rod of magnetic material experienced when a magnetic field is applied parallel to the axis. It is of importance in the determination of the gyromagnetic ratio. Sometimes known as the *Richardson effect.* See also: *Barnett effect: Magnetization by rotation.*

ROTATION CONSTANT. See: *Drude equation: Drude law.*

ROTATION (ROT) OF A VECTOR. See: *Curl. Differential vector operator.*

ROTON. The quantum of rotational motion in the Landau treatment of the elementary excitations of liquid helium, the quantum of longitudinal motion being the phonon.

ROUGHNESS. Of a surface: See: *Surface finish.*

ROWLAND CIRCLE. A circle on which lie the slit, diffraction grating and focused spectra of a concave grating. The diameter of the circle is equal to the radius of curvature of the grating. See also: *Paschen circle.*

ROWLAND GHOSTS. See: *Spectral ghosts.*

ROWLAND GRATING. A diffraction grating ruled on a concave mirror which focuses the spectra from the grating and eliminates chromatic aberration. A spectrograph employing such a grating need have only three parts—the slit, the grating, and the photographic plate.

ROWLAND MOUNTING. A mechanical device designed to ensure that the slit, grating, and plate holder of a concave grating shall all lie on a Rowland circle.

RR LYRAE VARIABLES. See: *Cepheid variables.*

R STAR. See: *Stars, spectral classes of.*

RUDORFF MIRRORS. Two mirrors slightly inclined to the plane of the screen in a Bunsen photometer, to permit the simultaneous viewing of both sides. See also: *Bunsen photometer.*

RULING ENGINE. A dividing engine specially adapted for ruling parallel lines at specified distances. Diffraction gratings, either plane or concave, may be ruled by such an engine.

RU POWDER. An abbreviation for raies ultimes powder. The powder contains a mixture of chemical elements in such proportions that at least seven lines of each of the most important elements will appear in the spectrum emitted when the material is burned in an arc.

RUSSELL–SAUNDERS COUPLING. See: *Coupling.*

RUTGER EQUATION. An equation relating the specific heats (in zero magnetic field) at the superconducting transition to the critical temperature; and to the slope (at the critical temperature) of the curve of critical field (the critical magnetic field needed to destroy superconductivity) versus temperature. It has the form

$$C_s - C_n = (VT_c/4\pi)\,(dH_c/dT)^2_{T=T_c},$$

where C_s is the specific heat in the superconducting state, C_n is that in the normal state, V is the specific volume, T_c is the critical temperature, H_c is the critical field, and T is the temperature.

RUTHERFORD SCATTERING: COULOMB SCATTERING. The scattering of a charged particle by the Coulomb field (electrostatic field) of a nucleus.

RV TAURI STARS. See: *Stars, variable, nomenclature of.*

RYDBERG CONSTANT. A fundamental constant R which enters into the wave-number formulae for all atomic spectra. For a hypothetical atom of infinite mass it is given by $R_\infty = 2\pi^2 me^4/ch^3$, where m is the rest mass of the electron, e is the electronic charge, c is the velocity of light, and h is Planck's constant. For a real atom the denominator becomes $ch^3(1 + m/M)$, where M is the mass of the atom, giving a value for R which is not seriously different from R_∞. Thus $R_\infty = 109\,737$ cm^{-1}, while for the hydrogen atom $R = 109\,678$ cm^{-1}.

RYDBERG SERIES. A general formula for the series of line spectra of the hydrogen atom in which allowance is made for quantum defects. It may be written

$$N = R[1/(n + a)^2 - 1/(m + b)^2],$$

where N is the wave number, R is the Rydberg constant, n and m are integers with m greater than n, and a and b are the quantum defects for each particular series. See also: *Hydrogen atom, line spectrum of.*

S

SABIN. A unit of equivalent sound absorption. See also: *Sound absorption, equivalent.*

SACCHARIDE. A sugar molecule with the general formula $(C_6H_{10}O_5)_n$. See also: *Polysaccharide.*

SACCHARIMETER. A polarimeter specially adapted for determining the strength of sugar solutions.

SACCHARIMETRY. The polarimetric measurement of the optical rotation of sugars.

SAFETY FACTOR. In structural engineering: the ratio of the breaking stress of a structure to the greatest stress the structure is predicted to encounter. It has sometimes been described as a factor of ignorance.

SAGITTAL FOCUS. See: *Astigmatic foci.*

SAHA EQUATION. An equation in which thermodynamic considerations are applied to electric discharges in thermal equilibrium. It may be written in various ways, for example:

$$\log (N_e^2/N_a) = -AV_i/T + 3/2 \log T + B,$$

where N_e is the concentration of electrons, N_a is that of gaseous atoms, V_i is the ionization potential, and T is the absolute temperature. A and B are constants, being equal to 5040 and 15·38 respectively.

SAINT ELMO FIRE. An electrical brush discharge, occurring during thunderstorms, between the atmosphere and the masts of ships, the trunks of trees, and buildings with appreciable projections. It is seen at night as a glow. It is also known as a corposant.

SALTATION. The discontinuous forward movement by solids driven along a gravity bed by a flow of fluid. It is associated with the transport of solids which are not carried in suspension by the fluid itself.

SALT BRIDGE. An inverted U-tube used mainly to connect the two halves of an electrolytic cell each of which contains one electrode. The U-tube is filled with a concentrated salt solution, e.g. potassium chloride or potassium nitrate.

SALTING OUT. The removal of a substance from solution by the addition of an appropriate ion to the solution, provided that complexes are not formed. An example is the precipitation of NaCl from a saturated solution, by the addition of HCl gas.

SAMPLING. The selection of a small sample from a large population, from a study of which knowledge may be obtained regarding a specified characteristic (qualitative or quantitative) of the population as a whole.

SARGENT DIAGRAM. A plot of log λ versus log E for β-emitters, where λ is the disintegration constant and E is the upper energy limit in the β-spectrum. Most of the points corresponding to the natural radioelements lie on one or other of two straight lines, an observation explained by the Fermi theory of β-decay. See also: *Fermi theory of beta-decay.*

SAROS. The period after which the relative positions of the Sun, Earth, and Moon recur. It is about 18 years and 11 days, and was used by the ancient Babylonians to predict eclipses.

SATELLITE. (1) *Astronomical:* a relatively small celestial body revolving round a planet. The sizes of the satellites in the solar system (of which there are at least 31, 12 belonging to Jupiter) range from 3000 miles (about 5000 km) in diameter (Ganymede of Jupiter) to less than 40 miles (about 65 km) (Jupiter VII to XII). (2) *Artificial:* an object placed by man in orbit round the Earth. If the object is to revolve at a height of a few hundred kilometres its speed must be about 30 000 km/h. The orbits of satellites are approximately elliptical and change in size, shape, and orientation owing to the asymmetry of the Earth's gravitational field, the influence of external forces (primarily due to the Sun and Moon), and the drag of the Earth's atmosphere. The latter effect causes the orbit to contract gradually, with the result that the satellite is eventually burned up in the atmosphere, but the effect is very small for orbits greater than about 1000 km above the Earth. Of the many new scientific results obtained from satellite observations one of the least expected was the discovery of the existence of radiation belts (the Van Allen belts) around the Earth. See also: *Communication satellite. Weather satellite.*

SATELLITES, SPECTRAL. (1) *In optical spectra:* weak lines which are displaced relative to the main spectral lines, commonly arising from the presence of an isotope of low natural abundance. (2) *In X-ray spectra:* weak lines observed close to strong lines on the high frequency side. They arise from double electron jumps in doubly ionized atoms.

SATELLITE TRIANGULATION. The use of artificial satellites as triangulation targets to perform a triangulation in space and determine the positions of observing stations.

SATURABLE REACTOR. A choke containing a ferromagnetic core, relying on the non-linear properties of the material for its operation. It is employed in magnetic amplifiers.

SATURATED AIR. Air containing the amount of water corresponding to the saturated vapour pressure of water at the same temperature.

SATURATED VAPOUR PRESSURE. See: *Vapour pressure.*

SATURATION. The state of containing or exhibiting the maximum amount of substance, energy, field, etc., or the action of bringing about that state. Some examples are given in the following entries. See also: *Colour, saturation of.*

SATURATION, MAGNETIC. The condition, approached only as the temperature approaches absolute zero, in which all magnetic spins are aligned in the same direction. In practice magnetic saturation usually signifies the condition in which the intensity of magnetization no longer continues to increase with increasing field strength. This is sometimes referred to as *technical saturation.*

SATURATION OF AN IONIZATION CHAMBER. The condition reached when all the ions formed by the incident ionizing radiation are collected and no ions are produced by collision. The corresponding current and voltage are known as the *saturation current* and *saturation voltage* respectively.

SATURATION OF AN IRRADIATED ELEMENT. The state reached when the disintegration rate of the nuclide formed is equal to its production rate. The corresponding activity is called the *saturation activity.*

SATURATION OF A SOLUTION. The condition in which the solvent can dissolve no more solute under the existing conditions of temperature and pressure.

SATURATION RATIO. Of moist air: See: *Humidity, relative.*

SATURN. The sixth planet in the solar system in order of distance from the Sun. Its diameter is about $9\frac{1}{2}$ times that of the Earth, its mass about 95 times, and its period of rotation is about $10\frac{1}{4}$ h. It has a unique ring system and possesses ten satellites, namely Janus, Mimas, Enceladus, Tethys, Dione, Rhea, Titan, Hyperion, Iapetus, Phoebe, the last being in retrograde motion.

SCABBING. In the explosive working of metals: See: *Spalling.*

SCALAR. A quantity which has magnitude only, as distinguished from a vector, which also has direction.

SCALAR FIELD. A region of space each point of which is described by a scalar, e.g. the temperatures in the atmosphere. See also: *Vector field.*

SCALAR POTENTIAL. A scalar function of position ϕ such that the field vector **F** is given by $\mathbf{F} = \nabla\phi$, where **F** is an *irrotational vector*, i.e. $\nabla \times \mathbf{F}$ is zero. **F** can represent the gravitational, electrostatic or magnetostatic field vectors, or the velocity vector of a fluid, etc. In these cases ϕ would represent the *gravitational potential, electrostatic potential, magnetostatic potential* or *velocity potential* respectively. See also: *Vector potential.*

SCALAR PRODUCT. Of two vectors **A** and **B**: is given by $\mathbf{A} \cdot \mathbf{B} = AB \cos \theta$, where A and B are the magnitudes of the vectors and θ is the angle between them. See also: *Vector product.*

SCALER. An instrument incorporating one or more scaling circuits and used for counting electrical pulses.

SCALING CIRCUIT. An electronic circuit that produces an output pulse whenever a prescribed number of input pulses has been received. A *binary scaling circuit* has 2 as the prescribed number, a *decade scaling circuit* having 10. See also: *Dekatron.*

SCANNING. The measurement or observation of some physical quantity at a succession of positions on a surface, or within a volume, either continuously or at a series of closely spaced intervals.

SCANNING COILS: DEFLECTION COILS. An assembly of one or more coils around a cathode ray tube, used for deflecting the electron beam by means of the magnetic field set up when current is passing through the coils. Also known as *Scanning yoke: Deflection yoke.*

SCATTERING. Of particles or radiation: a process in which a change in direction or energy of an incident particle or of radiation is caused by collision with a particle, a system of particles, or a discontinuity.

SCATTERING AMPLITUDE. See: *Scattering length.*

SCATTERING, COHERENT. Scattering in which a definite relation exists between the phases of the scattered and incident waves.

SCATTERING, COMPTON. See: *Compton effect.*

SCATTERING COULOMB: SCATTERING RUTHERFORD. The scattering of a charged particle by the Coulomb field (electrostatic field) of a nucleus.

SCATTERING, CRITICAL. The intense scattering of radiation by substances showing a second-order transition, at temperatures near the critical transition temperature. Examples are the *critical opalescence* of a gas near the critical point, which occurs even when the gas and its liquid are normally transparent; and the *critical magnetic scattering* of slow neutrons by a ferromagnetic crystal which occurs at temperatures near the Curie point, and which results from the large spontaneous fluctuations in magnetization which occur at these temperatures.

SCATTERING, DIFFERENTIAL. Refers to scattering in a particular direction. See also: *Cross-section, differential.*

SCATTERING, DIFFUSE. Of X-rays and neutrons: Scattering other than elastic and coherent scattering. It may arise from crystalline imperfections, disorder (of atomic or isotopic positions, or of spins), and atomic vibrations. See also: *Reflection, diffuse.*

SCATTERING, ELASTIC. Scattering in which the total kinetic energy is unchanged.

SCATTERING, FLUCTUATION. Scattering arising from fluctuations in the refractive index of the scattering medium. It may arise from phase transitions (critical scattering) or variations in density (small-angle scattering).

SCATTERING, INCOHERENT. Scattering in which no definite relation exists between the phases of the scattered and incident waves.

SCATTERING, INELASTIC. Scattering in which the total kinetic energy changes.

SCATTERING LENGTH. A measure of the scattering power of an atom used, for example, in the study of neutron diffraction by crystals. It is given by $|a|^2 = \sigma/4\pi$, where a is the scattering length, or *scattering amplitude*, and σ is the scattering cross-section. See also: *Atomic scattering factor.*

SCATTERING, MAGNETIC. Of neutrons: the scattering of neutrons by the unpaired electrons of atoms with a magnetic moment. It is additional to the nuclear scattering, which is characteristic of all atoms.

SCATTERING MATRIX. The array of observable quantities associated with the scattering of one system by another. It is used in scattering theory to describe the asymptotic behaviour of the wave-functions in the various available channels. See also: *Matrix.*

SCATTERING, MULTIPLE. Scattering in which an incident particle or photon undergoes a number of consecutive collisions.

SCATTERING, NUCLEAR. See: *Nuclear models. Optical model of particle scattering. Scattering, potential. Scattering, resonance.*

SCATTERING, POTENTIAL. Of a particle by an atomic nucleus: scattering in which the incident particle is considered to be reflected at the surface of the nucleus as though the latter were a hard sphere.

SCATTERING, RADIATIVE INELASTIC. Inelastic scattering in which some of the kinetic energy of an incident particle goes into the excitation of a target nucleus, followed by subsequent de-excitation through the emission of one or more photons.

SCATTERING, RAYLEIGH. The scattering of radiation by objects small compared with the wavelength of the radiation: in effect, scattering by points. The scattered intensity is inversely proportional to the fourth power of the wavelength, which explains, among other things, the reddish colour of sunset and the blue colour of the sky. See also: *Tyndall effect.*

SCATTERING, RESONANCE. Elastic scattering by a nucleus in which the excitation of a resonance level in a compound nucleus is involved. See also: *Resonance in nuclear processes.*

SCATTERING, SINGLE. Scattering in which an incident particle or photon undergoes a single collision.

SCATTERING, SMALL-ANGLE. The scattering of a beam of radiation at small angles by particles or cavities in a medium, whose dimensions are large compared with the wavelength of the radiation. The phenomenon is observed in the scattering of visible light, sound, nuclear particles, and indeed of any radiation which can be characterized by a wavelength, but the most important type is probably the small-angle scattering of X-rays by discontinuities whose size may range from 10 to 1000 Å. Such scattering has found particular application in the study of molecules of high polymers, of colloidal solutions, and of solids, for which qualitative or quantitative information may be obtained regarding the size and distribution of heterogeneities on a submicroscopic scale.

SCATTERING, THERMAL INELASTIC. Inelastic scattering in which a slow neutron or other particle exchanges energy with a molecule or crystal lattice.

SCATTERING, THOMSON. The scattering of photons by a free charged particle, in particular by an electron, according to the Thomson equation

$$\sigma = \frac{8}{3}\pi\left(\frac{e^2}{mc^2}\right)^2 \text{ or } \frac{8}{3}\pi(r_e)^2,$$

where σ is the scattering cross-section per electron, e is the electronic charge, m the electronic rest mass, c the speed of light, and r_e is the classical radius of the electron. The equation is valid only if the photon energy is much less than the rest energy (mc^2) of the electron.

SCHILLER LAYERS. Surface layers of certain crystals which give rise to a play of colours when suitably illuminated. The colours arise from the diffraction, reflection, and scattering of light by regularly oriented inclusions or cavities of microscopic size.

SCHLIEREN METHOD. A method for revealing inhomogeneities in a transparent medium, which depends on changes of refractive index and the resulting displacement or distortion of an optical image formed by light passing through the medium. The changes of refractive index may arise from the presence of shock waves or turbulence, or from thermal causes. Examples of the former are seen in flow visualization (as in wind tunnels) and, of the latter, in flame photography.

SCHMIDT CAMERA. A camera used when image sharpness combined with high light-gathering power is required, as in astronomical research or high altitude photographic reconnaissance. The image is formed by a concave spherical mirror corrected for spherical aberration by a specially shaped corrector plate or lens, the *Schmidt lens,* which is inserted in front of it. This lens is weakly converging near its centre and weakly diverging along its periphery.

SCHMIDT GRAPHICAL METHOD. A method for the approximate analysis of one-dimensional non-steady heat flow.

SCHMIDT LINES. Two parallel lines on the plot of nuclear magnetic moment against nuclear spin that show the relationship to be expected in the *one-particle model* of the nucleus, in which the nuclear properties are attributed to the effect of one nucleon. Most of the points observed experimentally lie between these lines which are therefore also called *Schmidt limits.*

SCHMIDT NUMBER. A dimensionless number given by the ratio of the kinematic viscosity to the mass diffusivity. It expresses the ratio of the diffusivity of momentum to that of matter through a fluid.

SCHOTTKY DEFECT: SCHOTTKY DISORDER. See: *Point defect.*

SCHRÖDINGER EQUATION. The basic equation of wave mechanics, which expresses the wave corresponding to the motion of a particle in a field of force. This motion is described by a *wave function* ψ, which is a complex function of position and time, given by the Schrödinger equation:

$$-\frac{\hbar^2}{2m}\nabla^2\psi + V\psi = -\frac{\hbar}{i}\frac{\delta\psi}{\delta t},$$

where $\hbar$ is Planck's constant divided by 2π, ∇^2 is the Laplacian operator, m is the mass of the particle, V its potential energy (usually a function of position), and t is time. This is the *time-dependent* Schrödinger equation. Where interest is centred on stationary states (e.g. in an atom) ψ may be replaced by a function ϕ, which is a function of position only, in the *time-independent* Schrödinger equation: $-(\hbar^2/2m)\nabla^2\phi + V\phi = E\phi$, where E is the total energy of the particle. Solutions of this equation exist only for specific values of E, known as *eigenvalues*, and to each such value there corresponds an *eigenfunction*, analogous to the energy of a Bohr orbit. The Schrödinger equations can also be represented in terms of the Hamiltonian operator H, as $H\psi = i\hbar(\delta\psi/\delta t)$ (time-dependent) and $H\phi = E\phi$ (time-independent). See also: *Electron cloud. Klein–Gordon equation. Matrix mechanics. Variation method.*

SCHRÖDINGER EQUATION, MANY BODY. An equation employed when a number of particles, forming a mutually interactive system, is to be considered. It is a complex form of the usual Schrödinger equation (which applies essentially only to a one-particle system) in which other particles can be regarded at most as contributing to the potential of the particle.

SHOTTKY EFFECT. See: *Shot noise: Shot effect.*

SCHULER PENDULUM. A pendulum which remains aligned to the local vertical when subjected to external forces. It has been used as the basis of systems (known as "*Schuler-tuned*" systems) for maintaining reference directions in inertial guidance equipment.

SCHUMANN PLATES. Photographic plates having practically no gelatine, used for photographing ultraviolet spectra below wavelengths of about 2000 Å ($0.2\ \mu$m) where gelatine is strongly absorbent.

This region of the spectrum is known as the *Schumann region*.

SCHWARTZCHILD RADIUS. The radius of a massive body below which, according to the general theory of relativity, no signal can escape from the body, i.e. at which the body becomes a black hole. It is equal to $2MG/c^2$, where M is the mass of the body, G the gravitational constant, and c the speed of light. The Schwartzchild criterion for the production of a black hole is, by coincidence, that derived by Laplace using classical mechanics. See also: *Stellar evolution*.

SCINTILLATION. A burst of luminescence of short duration, caused by the absorption of a single ionizing particle or photon.

SCINTILLATOR. A material which converts the energy of ionizing radiation into visible light. It may be an organic, inorganic, liquid, or plastic phosphor, or may be a noble gas.

SCLEROSCOPE. See: *Hardness*.

SCOTOPHOR. A device for measuring the dark adaptation of the human eye by means of test flashes of light after various periods of rest in the dark.

SCOUR. The erosion of the channel walls of a natural stream by the flow of the stream. See also: *Coriolis effect. Meander*.

SCREEN GRID. An electrode interposed between grid and plate in a thermionic valve.

SCREENING. Of an atomic nucleus by the surrounding electrons: the reduction of the effective charge of the nucleus. The amount by which the effective atomic number falls below the actual atomic number is called the *screening constant* or *screening number*.

SCREW. An assembly of core and screw thread.

SCREW AXIS. An axis of crystal symmetry characterised by rotation through a fraction of a revolution ($\frac{1}{2}$, $\frac{1}{3}$, $\frac{1}{4}$, or $\frac{1}{6}$) followed by translation along the axis by a fraction of a unit-cell edge.

SCREW DISLOCATION. See: *Dislocation*.

SCREW DISPLACEMENT. Of a rigid body: a rotation of the body about an axis· accompanied by a translation of the body in the direction of the same axis.

SCREW, PITCH OF. See: *Pitch*.

SCREW THREAD. A helical ridge, with a uniform section that is approximately V-shaped, square, or rounded, formed on a cylindrical core, the pitch and core diameter being standardized according to one of a variety of specifications, which are mostly being replaced by an internationally agreed unified screw thread.

SEA-LEVEL, MEAN. At a given point on the coast: the level of the water, against a fixed mark on the land, averaged out over a period of sufficient length to eliminate the effects of oscillatory variations due to waves, swells, and tides. To eliminate certain long-period tides, a period of 19 years is necessary. See also: *Geoid*.

SEARCH COIL. A small coil used to measure changes in magnetic induction from the change of flux through the coil.

SEARCHLIGHT. A powerful parallel beam of light emitted by a carbon arc situated at the focus of a paraboloidal mirror. The housing is fitted with louvres to prevent the direct light from the arc forming a divergent beam. Searchlights are used for picking out objects at night and for making observations on the upper atmosphere with a modulated beam.

SECOND. The CGS and SI unit of time. It was formerly defined as 1/31 556 925·975 of the tropical year for 1900 and 1/86400 of a mean solar day; but the SI definition is the duration of 9 192 631 770 periods of the radiation corresponding to the transition between the two hyperfine levels of the ground state of the caesium-133 atom. See also: *Atomic beam frequency standards. SI units*.

SECONDARY CELL. See: *Accumulator*.

SECONDARY ELECTRONS. Electrons emitted from a free atom or molecule, or from the surface of a body, as a result of bombardment by photons or charged particles (including other electrons).

SECONDARY FOCUS. See: *Astigmatic foci*.

SECONDARY IONIZATION. Ionization produced by ionizing particles which arise from the interaction of primary radiation with matter, i.e. ionization produced by secondary radiation. See also: *Primary ionization*.

SECONDARY RADIATION. Radiation produced by the interaction of primary radiation with matter. See also: *Primary radiation*.

SECOND QUANTIZATION. The process by which a classical field may be considered as an assembly of particles. See also: *Quantum field theory*.

SECOND SOUND. In liquid helium II: a temperature wave which results when the superfluid and normal components of liquid helium II oscillate out of phase with each other, so that the cold superfluid component collects at a point of low temperature, while the normal component collects at a point of "high" temperature half a wavelength away. See also: *First sound. Helium II, two-fluid model for*.

SECTOR DISK. A disk used when it is desired to cut down the effective intensity of a light source by a known fraction. The simplest form is a circular, opaque disk from which a sector or sectors of known area have been cut out. By interposing this disk into the path of the light, and rotating it at a speed higher than the critical flicker frequency, the intensity of the light is cut down to a fraction equal to the ratio of the open sector(s) to that of the whole disk.

SECULAR EQUATION. An equation of the form

$$y(\lambda) = |a_{ij} - b_{ij}\lambda| = 0,$$

which becomes a polynomial in λ when the determinant is expanded. The name comes from the early application of such an equation to the secular perturbations in planetary motion, but is now frequently applied in classical or quantum mechanical problems.

SECULAR EQUILIBRIUM. See: *Radioactive equilibrium.*

SECULAR INEQUALITIES. Of planetary orbits: progressive or very long period disturbances of planetary orbits affecting eccentricity and inclination about their mean values. The disturbances arise from changes in the relative positions of the planets in their orbits and are very small. Thus the eccentricity of the Earth's orbit should diminish from its present value of 0·017 to a value of 0·003 in about 24000 years.

SECULAR VARIATION. A variation of a geophysical or astrophysical quantity which arises from long term effects and is correspondingly slow. See also: *Annual variation. Magnetic variation, secular. Parallax, secular. Star position, apparent, annual variations of.*

SEDIMENTARY DEPOSITION. The formation of rocks (known as *sedimentary rocks*) by the accumulation of rock and mineral material from suspension in water, wind or ice, or from solution in water.

SEDIMENTATION. The settling of solids from suspension in liquids under the force of gravity. In general, the term is also applied to any settling process involving the separation of immiscible phases.

SEDIMENTATION ANALYSIS. The grading of solid particles according to their rate of settling when suspended in a liquid.

SEDIMENTATION CONSTANT. In centrifuge technique: the rate of sedimentation of a particle in a given medium under unit acceleration. See also: *Svedberg.*

SEDIMENTATION POTENTIAL. The potential difference set up by the Dorn effect.

SEEBECK EFFECT. The setting up of an e.m.f., and consequently of a current, when two different metals are joined at two points to make a closed electrical circuit and when these two points differ in temperature. It is the opposite of the Peltier effect. See also: *Thermoelectric effects.*

SEEING, ASTRONOMICAL. The quality of astronomical observation as affected by atmospheric inhomogeneities and unsteadiness. The term does not refer to the general atmospheric absorption of light. See also: *Twinkling.*

SEGER CONES. See: *Pyrometric cones.*

SEGREGATION. In alloys: the non-uniform distribution of impurities, inclusions, and alloying constituents. In castings it is a result of an unwanted decrease in temperature and may lead to serious defects.

SEIDEL ABERRATIONS. Of an optical system: five independent aberrations shown by von Seidel to exist in an optical system. These are: spherical aberration, coma, astigmatism, curvature of the field, and distortion. They refer to monochromatic light so that, for a complete list of aberrations, chromatic aberration must be added. See also: under the individual aberrations mentioned above.

SEISMICITY. Earth movements.

SEISMIC PROSPECTING. Geophysical prospecting by the use of artificially generated seismic waves. Variations in the speed of these waves, particularly of the longitudinal waves, may indicate changes in rock type or geological structure. See also: *Seismic waves, primary (P) and secondary (S).*

SEISMIC SURGE. A "tidal" wave arising from an earthquake. See also: *Storm tide: Storm surge.*

SEISMIC WAVES. Shock waves in the Earth which issue from the focus of an earthquake, or are deliberately generated by an explosion for experimental purposes. They are of two types: *bodily seismic waves,* which travel inside the Earth, and *surface seismic waves,* which travel over the surface of the Earth. See also: *Elastic waves. Tsunami.*

SEISMIC WAVES, PRIMARY (P) AND SECONDARY (S). Bodily seismic waves. The primary or P waves are longitudinal, and the secondary or S waves are transverse. The latter may be polarized. Where the particles move horizontally, the waves are referred to as SH waves. Where they move vertically they are known as SV waves.

SEISMOGRAPH. An instrument for recording the earth tremors caused by earthquakes, explosions, or storms at sea.

SEISMOLOGY. The study of earthquakes and their causes, together with the velocities and energies of the elastic waves resulting from them.

SELECTION RULE. Any rule that forbids the occurrence of certain transitions between states of atomic or sub-atomic systems. Such rules are based on the conservation of some physical quantity, e.g. angular momentum, parity, isobaric spin, strangeness, and are not rigorous in that "forbidden" transitions are sometimes found to occur, albeit with reduced probability. See also: *Transition, allowed. Transition, forbidden.*

SELF-ABSORPTION. The absorption of radiation by the emitter of the radiation, e.g. the selective absorption of radiation emitted by the hot central core of a gas in the cooler outer vapour of the gas, which is also known as *self-reversal*, or the absorption of ionizing radiation by the body of the emitter.

SELF-CONSISTENT FIELD METHOD: HARTREE METHOD. An iteration method used in the determination of solutions of the Schrödinger equation for atomic problems. See also: *Slater method.*

SELF-DIFFUSION. A term used to describe the migration of atoms in a crystalline solid, leading to a redistribution without change in chemical composition, or of the atoms or molecules of a pure liquid or gas. The *self-diffusion coefficient* may be measured by the use of radioisotopic tracers.

SELF-ENERGY. (1) Another term for rest energy. (2) The Helmholtz free energy associated with the polarization of a dielectric body.

SELF-REVERSAL. See: *Self-absorption.*

SELF-SHIELDING. Of an irradiated body: the shielding of the inner portions of the body by the absorption of radiation in its outer portions.

SELF-SUSTAINED. In nuclear technology: refers to the condition of a neutron-chain-reacting medium for which the effective multiplication factor is unity, i.e. to the critical condition. See also: *Chain reaction, neutron. Multiplication factor: Multiplication constant.*

SELLMEIER DISPERSION FORMULA. Refers to anomalous dispersion. It is assumed that a medium contains elastically bound particles with a natural frequency of vibration such that if light of this frequency is passed through the medium, the particles resonate and light is absorbed. The formula is

$$n^2 = 1 + \Sigma \frac{A_j \lambda^2}{\lambda^2 - \lambda_j^2},$$

where n is the refractive index, λ is the wavelength of the light, λ_j the natural frequency of the jth particle, and A_j is proportional to the number of particles with that frequency. When only two characteristic frequencies are involved, one in the infrared (λ_r) and one in the ultraviolet (λ_v), the Sellmeier formula reduces to the *Ketteler formula*

$$n^2 = 1 + n \frac{A_r}{\lambda^2 - \lambda_r^2} + \frac{A_v}{\lambda^2 - \lambda_v^2}.$$

See also: *Dispersion.*

SEMICONDUCTOR. A crystalline material whose electrical resistivity is intermediate between that of metallic conductors and that of insulators (lying in the range 10^{-2} to 10^{-9} ohm-cm at room temperature) and whose resistance falls with increasing temperature over a large range of temperatures. The conductivity may arise not only from the motion of negative electrons but of positive holes (created by thermal excitation). The use of semiconductors is widespread in such devices as rectifiers, amplifiers, detectors, photocells and transistors. See also: *Band theory of solids. Electron hole: Positive hole.* The following entries.

SEMICONDUCTOR, DEGENERATE. One which is so heavily doped as to resemble a metal.

SEMICONDUCTOR, DOPING OF. The addition of impurities to a semiconductor, or the production of a non-stoichiometric composition in it, to achieve a desired conductivity.

SEMICONDUCTOR, EXTRINSIC. A semiconductor whose electrical properties depend on the presence of impurity centres or of imperfections.

SEMICONDUCTOR, INTRINSIC. A material which is naturally semiconducting even in the pure state. Conduction is due to electrons promoted to the conduction band by heat and to the holes created in the valency band by such promotion.

SEMICONDUCTOR JUNCTIONS. (a) *An n-type semiconductor and a metal:* in such a junction, if the semiconductor is negative relative to the metal, conduction (in this case electron conduction) from the semiconductor to the metal is much easier than vice versa. (b) *A p-type semiconductor and a metal:* in such a junction, if the semiconductor is positive relative to the metal, the easy conduction is in the opposite direction, arising from the flow of positive holes from the semiconductor to the metal. (c) *A p-type semiconductor and an n-type semiconductor:* in such a junction, if the p-type semiconductor is made positive relative to the n-type, majority carriers can cross the junction in both directions. With the reverse polarity only a small current is possible.

SEMICONDUCTOR, N-TYPE. An extrinsic semiconductor in which the impurities or imperfections are such that the conduction electron density is in excess of the mobile hole density. See also: *Donor.*

SEMICONDUCTOR, P-TYPE. An extrinsic semiconductor in which the impurities or imperfections are such that the mobile hole density exceeds the conduction electron density. See also: *Acceptor.*

SEMICONDUCTOR, ORGANIC. An organic solid whose electrical conductivity is not ionic but electronic and in which the conductivity increases with increasing temperature.

SEMIMETALS. The As and Te groups of elements in the periodic classification, which are poor conductors of electricity. The elements comprise As, Sb, Bi, Te, and Se. See also: *Appendix I.*

SEMIPERMEABLE MEMBRANE. A membrane which allows the passage of a solvent, but not of the solute contained in that solvent.

SENSATION LEVEL. See: *Hearing loss.*

SENSITIVE FLAME. A gas flame which changes its height or shape when sound waves fall upon it.

SENSITIVE TIME. Of a counter, expansion chamber, etc.: the time during which the passage of an ionizing particle can be detected or rendered visible.

SENSITIVE TINT. A tint, exhibited by a Bravais biplate, which is sensitive to a slight change in polarization. See also: *Bravais biplate.*

SENSITIVE VOLUME. (1) Of a radiation detector: that part of the detector in which lies the origin of the output signal. (2) In radiobiology: that part of a cell or organ which is believed to be sensitive to ionizing radiation. See also: *Radiation, target theory of.*

SENSITIVITY. Of an instrument: the change of output per unit change of input, e.g. the ratio of deflection to current in a galvanometer; or, sometimes, the smallest input that can be detected or measured, e.g. the smallest mass that a balance can measure.

SENSITOMETRY. The measurement of the response, or sensitivity, to light of photographic materials. It involves a study of the treatment received by such materials (e.g. exposure and development) and the resultant blackening. See also: *Photographic density. Photographic exposure.*

SEPARATION ENERGY. See: *Binding energy.*

SEQUESTRATION. The formation of soluble complexes of metal ions by the addition of suitable reagents (*sequestering agents* or *sequestrants*) under conditions in which the complexes would otherwise be precipitated. The term is also used to mean the solution of precipitates which is effected by adding such reagents.

SERVOMECHANISM. An automatic control system, incorporating power amplification, by which the mechanical motion of an output member is constrained to follow closely the motion of an input member. Such a system is a *closed-loop system*, i.e. one in which the output (or some function thereof) is fed back and compared to some reference at the input, the difference being used to effect the desired control.

SET. A collection of entities, usually represented by braces {}. For example, $X = \{1, 2, 4\}$ is the set X which has the number 1, 2, and 4 as members and has no other members. Again $Y = \{y_i | i = 1, 2, ..., n\}$ is the set Y which contains n objects of a particular type y, which may or may not be distinct. (Note: in this second example it is not known what sort of objects the y_i might be. They could be numbers, or ships, or ideas, or even other sets.)

SETTLER. Any equipment which removes suspended particles from gases or liquids on an industrial scale. Such equipment commonly depends on gravitational or centrifugal force for its action. See also: *Electrostatic precipitation.*

SEXTANT. An instrument for measuring the altitude of a celestial object, whose main use is for navigation. The *marine sextant* consists essentially of a frame carrying a graduated 60° scale, a fixed half-clear mirror (the horizon mirror), and a movable fully silvered mirror (the index mirror) which is carried on an arm pivoted at the centre of the scale. The reflection of a celestial body in the index mirror is brought into "coincidence" with the horizon via the horizon mirror and the altitude of the body read off. The marine sextant may also be used for taking bearings on celestial and terrestrial objects. In the *air sextant* the absence of a horizon is compensated for by the incorporation of a spirit-level type bubble which defines the vertical, and the image of the celestial object is set in the centre of the bubble. A mechanical averaging mechanism is fitted, as a single sight could well be in error owing to the accelerations to which aircraft are subject.

SEYFERT GALAXIES. A small number of galaxies (about 1–2% of all galaxies) of very high luminosity, including intense infrared sources. They have compact bright starlike nuclei, with spectra quite unlike those of ordinary stars.

SHADOW CASTING. Of an electron microscope replica: the deposition of a metallic layer, by evaporation at a very acute angle, on the replica. This enhances contrast since each prominence on the surface casts a "shadow" on the side remote from the source of the metal atoms. See also: *Replica.*

SHEAR MODULUS: MODULUS OF RIGIDITY. Of an elastic body: the tangential force necessary to produce unit angular deformation. See also: *Elastic modulus*.

SHEAR STRENGTH. The shear stress required to cause plastic flow.

SHEAR STRESS. The stress required to produce *shear*, i.e. to cause a plane in a solid body to be displaced parallel to itself.

SHEAR WAVE. In an elastic medium: a form of elastic wave motion in which the particles of the medium are displaced in a direction at right angles to the direction of propagation. Also known as *distortional wave, rotational wave, transverse wave*.

SHERARDIZING. The application of a zinc coating to metals (usually steel) by heating them with zinc and zinc oxide dust in a closed rotating drum at 350–375°C.

SHIP RESISTANCE. The force required to keep a ship on course in steady motion. It may be divided into *viscous resistance* (arising from the viscous property of water) and *wave or wave-making resistance* (arising from the wave-making property of the air–water free surface), a division which forms the basis of the *Froude method* of extrapolating the resistance of a ship model to that of a full-scale vessel.

SHOCK TUBE. A uniform tube divided by a diaphragm into two compartments containing gases at different pressures. By bursting the diaphragm (either mechanically or owing to the pressure difference) a shock wave moves into the low pressure part of the tube and an expansion wave into the high pressure part.

SHOCK WAVE. (1) A sound wave in gas characterized by a very steep, almost discontinuous, rise in pressure which occurs when a region of high pressure overtakes a region of low pressure, with a consequent rapid compression of the gas. See also: *Expansion wave*. (2) A similar wave in a solid.

SHOOTING STAR. See: *Meteors*.

SHOT NOISE: SHOT EFFECT. In thermionic emission: the fluctuation in the emission arising from its statistical nature, i.e. from the random emission of the electrons. It is also known as the *Schottky effect*. See also: *Electrical noise*.

SHUNT. An electrical bypath connected so as to take part of the current which would otherwise flow through the equipment that is so shunted.

SHUTTER. (1) In photography: any device which allows a pre-determined amount of light to reach the sensitive material at the image plane. (2) In nuclear reactor technology: a movable plate of absorbing material used to cut off a beam of neutrons,

γ-rays or other radiation. See also: *Q-switching: Q-spoiling*.

SIDEBAND. The frequency band on either side of the carrier frequency within which fall the frequencies of the waves produced by modulation. The term is also applied to the wave components lying within such a band.

SIDEREAL DAY. The interval between two consecutive transits of the first point of Aries across any selected meridian. Neglecting the small effects of precession, etc., it is the interval between two consecutive transits of the same fixed star. It may be defined for most purposes as the period of rotation of the Earth on its axis, and expressed in mean solar time it is 23 hours, 56 minutes, and 4·0906 seconds. Like the solar day the sidereal day is divided into 24 hours (*sidereal hours*) each of which is in turn divided into 60 minutes, each minute being subdivided into 60 seconds. See also: *Day*.

SIDEREAL MONTH. The interval between successive passages of the Moon through the same point in its orbit relative to the fixed stars. Its length is just over 27·3 mean solar days. See also: *Month*.

SIDEREAL PERIOD. The interval between two successive identical positions of a celestial body with reference to the fixed stars.

SIDEREAL YEAR. The interval between two successive passages of the centre of the Sun past any one star situated in the ecliptic and devoid of proper motion. It is slightly longer than the tropical year and amounts to 365·2564 mean solar days, decreasing by 10^{-7} mean solar days per century. See also: *Year*.

SIDERITE. See: *Meteorite*.

SIDEROLITE. See: *Meteorite*.

SIDESLIP, VELOCITY OF. The component of the velocity of the centre of gravity of an aeroplane in a direction perpendicular to the vertical plane of symmetry of the aeroplane and relative to the undisturbed air in the vicinity of the aeroplane. Sideslip is positive when it is directed to starboard.

SIEMENS. The SI unit of conductance, equal to the reciprocal ohm or *mho.* See: *Appendix II*.

SIEVERT. The SI unit of dose equivalent of (ionizing radiation), which is to replace the rem. It is equal to 1 joule per kilogram, hence 1 rem is equal to 10^{-2} sievert. See: *Rem*.

SIGHT RULE. See: *Alidade*.

SIGMA PARTICLE. See: *Hyperon*.

SIGNAL GENERATOR. A wide-range radio-frequency oscillator, usually with provision for audio or video frequency modulation, used for test purposes.

SIGNIFICANCE. An estimate of the likelihood that a given measurement can be regarded as lying outside a group of measurements distributed according to some known law. For the normal distribution, for example, the probability P, that the deviation of a measurement from the mean is greater than the standard error σ is 0.317, that it is more than 1.96σ is 0.05, and that it is greater than 2.58σ is 0.01. If P is 0.05 the probability is unlikely, if it is 0.01 it is highly unlikely, i.e. the deviation is significant in each case.

SILICA GEL. An active form of amorphous silica which acts as a powerful desiccant.

SILICATES. Salts of the silicic acids. Their dominant structural feature is the grouping of four oxygen atoms at the vertices of a tetrahedron, about a silicon atom. Silicates constitute about 95% of the Earth's crust and number about one-third of all mineral species.

SILICON CELL. A photovoltaic barrier layer cell, utilizing a $p-n$ junction, used in space craft as a source of power.

SILICON CHIP. A small silicon slice to which appropriate impurities have been added, and which is used, for example, in integrated circuits.

SILICONES. Synthetic compounds of silicon and various siloxanes (compounds with the basic structural unit Si—O—Si), used commercially as liquids, greases, elastomers, resins, and as ingredients of proprietary compounds such as paints and polishes.

SILSBEE HYPOTHESIS. A hypothesis which states that the superconducting properties of a superconductor are destroyed if the current exceeds a certain value. See also: *Superconductivity, critical field in.*

SILVER VOLTAMETER. A type of voltameter used in the realization of the international ampere. It was basically an electrolytic cell with a silver anode and a platinum cathode, the electrolyte being silver nitrate. See also: *Ampere, international.*

SIMON LIQUEFIER. A device for liquefying helium by a single adiabatic expansion. Helium gas is admitted at high pressure to a pressure vessel and cooled by liquid or solid hydrogen. The vessel is then thermally isolated and the helium gas allowed to expand adiabatically, upon which it liquefies in the vessel.

SIMON MELTING EQUATION. An expression relating pressure, temperature and melting point, often expressed as $p/a = (T/T_0)^c - 1$, where p and T are melting curve values of pressure and temperature, T_0 is the normal melting temperature at the triple point, c is a constant (about 4 for metals and 2 for non-metals), and a is the negative pressure which would have to be applied to the solid at absolute zero to make it melt, a pressure which is identified as the *internal pressure*, and is a measure of the cohesive forces of the solid. The Simon melting equation can be shown to be consistent with the *Lindemann melting-point formula*. See also: *Lindemann melting-point formula. Melting.*

SIMPSON RULE. A rule for the estimation of the area of an irregular figure. The area is divided into an even number of strips of equal width, and the rule states that the area is equal to one-third the strip width multiplied by the sum of the first and last boundary ordinates, plus 4 times the sum of the even ordinates, plus twice the sum of the odd ordinates. The rule is used for such diverse purposes as the determination of the areas of ponds and lakes or of the cross-sectional areas of beams and for the evaluation of definite integrals.

SINE BAR. An accurately made steel bar mounted on two identical rollers at a precisely known distance apart. It is used in setting out angles to close limits, by inserting blocks of precisely appropriate height under one of the rollers.

SINE TABLE. A development of the sine bar in which the bar is replaced by a base plate to which a table is hinged, the hinge taking the place of one of the rollers, the other being attached to the underside of the table.

SINGING SANDS. Sands which emit audible sounds when they undergo shear. Such sounds occur (a) when certain dry beach sands are trodden on or poked with a blunt probe, the frequency of the sound being about 1000 Hz, and (b) when, in some desert regions, successive portions of a sand avalanche down the face of a sand-dune are brought to rest and undergo internal shearing on being telescoped, a deep booming sound being emitted at a frequency of about 250 Hz. See also: *Acoustic emission from materials.*

SINGLE CRYSTAL CIRCUIT. See: *Solid circuit.*

SINGLET STATE. Of a spectral term: a term with a multiplicity of one.

SINGLET-TRIPLET SEPARATION. The difference in energy between singlet and triplet states of the same atomic or molecular electronic configuration.

SINGULARITY: SINGULAR POINT. In mathematics: See: *Analytic function: Holomorphic function: Regular function.*

SINK. See: *Source.*

SINTERING. The heating of a (usually) cold-compacted or die-pressed mass of powder at a temperature approaching, but lower than, the melting point, so as to decrease its porosity and increase its mechanical strength. See also: *Compact* (*powder metallurgy*).

SIREN. A device for generating loud sounds where considerable harmonic content is acceptable. One of the most common types involves the passage of a stream of air at high pressure through a set of equally spaced radial louvres in a rotating disk, the rotation being achieved by a turbine effect. Most sirens operate in the audible region but ultrasonic sirens are also known.

SI UNITS (SYSTEME INTERNATIONAL D'UNITES). A coherent rationalized system of units formed by the addition of three basic units, the degree Kelvin, the mole and the candela, to the rationalized MKSA system, together with two supplementary units, the radian and the steradian. The basic units are thus the metre, kilogram, second, ampere, degree Kelvin (now referred to only as the Kelvin), mole and candela. See also: *Units, MKSA. Units, prefixes for. Appendix II.*

SIZE-FACTOR COMPOUNDS. See: *Alloy.*

SKEWNESS. A term used in statistics to describe the asymmetry of a distribution of probability, and sometimes of frequency. A typical distribution with no skewness is the Gaussian or normal distribution.

SKIN EFFECT. The decrease in the depth of penetration of an electric current in a conductor as the frequency of the current increases, so that at high frequencies the current is restricted to a thin outer layer. The effect arises from the increase of internal self-inductance of the conductor with the depth below the surface. The *skin depth* or *penetration depth* is the depth below the surface at which the current density has decreased to $1/e$ of its value at the surface.

SKIN EFFECT, ANOMALOUS. An effect observed in pure metallic elements at liquid helium temperatures, whereby the surface resistivity is independent of the d.c. resistivity and is directly related to the shape of the Fermi surface.

SKIN FRICTION. See: *Surface friction: Skin friction.*

SKIP DISTANCE. The distance between the furthest point from a radio transmitter at which the ground wave can be received and the nearest point to the transmitter at which the sky wave can be received.

SKULL MELTING. A technique for the moulding of materials, in which the material is melted in a water-cooled copper hearth and then poured into the required mould. The process derives its name from the "skull" of solid material which lines the hearth and protects the copper from the melt.

SKY COMPASS. A compass designed for use in polar regions when the Sun is just below the horizon and the sky is too bright to permit stellar observations, and where a magnetic compass is of little value. It depends on the fact that the light scattered from the zenith sky is highly polarized in a plane containing the Sun, the observer, and his zenith, and operates by locating this plane, and hence permitting the Sun's azimuth to be determined.

SKY WAVE: IONOSPHERIC WAVE. A radio wave that is reflected by the ionosphere before reception.

SLAG. The vitreous mass which separates from fused metals during the melting of ores.

SLATER METHOD. A method, involving anti-symmetrical functions, used in the treatment of the problem of the many-electron atom where the relative values of Coulomb and exchange energy are required. See also: *Self-consistent field method: Hartree method.*

SLENDER BODY THEORY. See: *Fluid flow, slender body theory of.*

SLIP. (1) In a metal: one of the three main ways in which plastic deformation can take place. It is characterized by the movement of slabs of material, each many atomic layers thick and occurring within each grain, which slip or glide over each other on relatively few, more or less parallel, *slip planes*. Sometimes there is more than one such set of planes. Slip is also known as glide. See also: *Critical shear stress. Easy glide.* (2) Of a propeller: the difference between the experimental mean pitch and the effective pitch. See also: *Pitch.*

SLIP BAND. Dark bands, more or less parallel, which are seen at the surface of a metal in which the slip processes have continued to the surface, and result from a step structure on the surface.

SLIP BAND, SECONDARY. See: *Deformation band.*

SLIP COEFFICIENT. For a propeller: the amount by which the forward velocity is exceeded by the product of the pitch and the rate of revolution of the propeller.

SLIP ELEMENTS. The crystallographic indices of the plane and direction of slip.

SLIP FLOW. See: *Rarefied gas laws.*

SLIPSTREAM. Of a propeller: the stream of air driven back by the propeller.

SLIT. The opening (usually long and narrow) by which radiation enters or leaves instruments such as spectrometers, diffractometers, collimators, etc.

SLOTTED AEROFOIL. An aerofoil having one or more air passages (*slots*) located ahead of the leading edge of the main aerofoil surface. The effect of the slots is to delay boundary layer separation, with a consequent increase in lift and reduction in liability to stall.

SLOWING DOWN AREA OF NEUTRONS. See: *Fermi age.*

SLOWING DOWN DENSITY OF NEUTRONS. The number of neutrons per unit volume and unit time which, in slowing down, pass a given energy value. See also: *Fermi age equation.*

SLOWING DOWN LENGTH OF NEUTRONS. See: *Fermi age.*

SLOW MOTION PICTURES. Cinematograph pictures made by exposing film at up to 240 000 frames/s (or higher for special purposes), which are then projected at the usual rate of 24 frames/s. The events portrayed thus appear to take place at a slower rate than the actual events. Slow motion pictures are used to study high-speed events such as the movement of birds, vibrations, missiles in flight, and so on.

SLOW WAVES. The name given to electromagnetic waves when the velocity of propagation is less than that of light. They are of particular importance in electronic devices which depend on their interaction with charged particles, such as linear accelerators, magnetron oscillators, and travelling-wave-tube amplifiers.

SMALL CIRCLE. The intersection of the surface of a sphere by a plane which does not pass through the centre of the sphere. See also: *Great circle.*

SMECTIC PHASES. See: *Liquid crystals.*

SMIGELSKAS EFFECT. See: *Kirkendall effect.*

SMITH CHART. A chart used in the solution of transmission line and wave guide problems. Contours of constant resistance, reactance, and standing-wave ratio are drawn on a polar diagram whose coordinates represent the components of the complex reflection coefficient. See also: *Rieke diagram.*

SMITH–PURCELL EFFECT. The emission of electromagnetic radiation (infrared, visible or ultraviolet) when a free relativistically accelerated electron beam is fired at grazing incidence over the surface of a metallic diffraction grating which is maintained at a constant potential. It arises from the oscillation of the induced positive charges.

SMOOTH SURFACE. See: *Reflection.*

SNAP (SYSTEMS FOR NUCLEAR AUXILIARY POWER). A United States programme for the development of nuclear auxiliary power units for the space programme, using reactors and radioisotopes as energy sources with thermoelectric and turboelectric conversion systems. The programme was subsequently extended to include the design and construction of radioisotope-fuelled generators for terrestrial use. See also: *Power sources, isotopic.*

SNELL LAWS. See: *Refraction, Snell laws of.*

SNOEK EFFECT. The stress-induced ordering of interstitial atoms in body-centred cubic metals. This gives rise to changes in the mechanical relaxation spectrum of the metal.

SNOW. (1) A finely divided, crystalline (hexagonal) form of water produced in the atmosphere by the precipitation of water vapour. (2) The random pattern of white dots seen on a cathode-ray tube screen with a weak or zero signal. It is a manifestation of thermal and shot noise.

SOAPS. Soluble salts of the higher fatty acids, commonly the sodium salts.

SOFT MAGNETIC MATERIAL. A material with a small coercivity, i.e. a few oersted or less. (Note: 1 oersted is equal to 79·6 A/m.) See also: *Hard magnetic material.*

SOFT RADIATION. See: *Hard and soft radiation.*

SOHNKE LAW. For cleavage fracture: See: *Fracture, cleavage.*

SOLAR ANTAPEX. The region on the celestial sphere from which the solar system as a whole appears to be receding. It is diametrically opposite to the *solar apex*. See also: *Parallax, secular.*

SOLAR APEX. The region on the celestial sphere towards which the solar system as a whole appears to be moving. It lies in the constellation of Hercules and has coordinates right ascension 271°, declination +31° approximately. The rate of motion is about 20 km/s.

SOLAR ATMOSPHERE. This consists of the visible part of the Sun's surface, the photosphere, on which sun-spots and other physical markings appear, which is the source of practically all the heat and light radiated from the Sun; an intermediate layer some 8000 or 10000 km thick, the chromosphere; and an outer layer, the corona, which extends to a distance of several solar diameters. Both the chromosphere and the corona emit a steady flux of radio waves in the frequency range 15000–30000 MHz. See also: *Chromosphere. Corona, solar. Photosphere. Reversing layer.*

SOLAR CONSTANT. The flux of solar energy received at the mean distance of the Earth from the Sun before modification by the Earth's atmosphere. It is about 140 mW cm^{-2}.

SOLAR DAY. (1) *Apparent:* the interval between two successive transits of the observed Sun (or *true Sun*) across the meridian. It is not constant. See also: *Time, equation of.* (2) *Mean:* the interval between two successive transits of a fictitious *mean Sun* (which moves at a constant rate equal to the average rate of the true Sun) across the meridian. One mean solar day is equal to 24 *mean solar hours*, each of which equals 60 *mean solar minutes*, each of which equals 60 *mean solar seconds*. See also: *Day. Time, equation of.*

SOLAR ENERGY. The energy emitted by the Sun. It consists of electromagnetic radiation of all wavelengths, with a rate of emission of about 4×10^{33} ergs/s, i.e. 4×10^{26} J/s, of which only a fraction of about 0.5×10^{-5} reaches the Earth's surface. It is believed to arise from two reactions, the carbon cycle and the proton–proton chain, which are defined separately. Solar energy has been utilized in direct thermal devices, as in solar stills and solar furnaces; in devices for the generation of electricity, using the thermoelectric effect and the photovoltaic effect (as in the silicon cell used in space craft); and in photo-chemical processes.

SOLAR FACULAE. Bright features observed on the surface of the Sun. They are seen at the solar poles, in the Sun's chromosphere (upper part of the solar atmosphere) or its photosphere (lower part). Chromospheric faculae are sometimes known as *flocculi.*

SOLAR FLARES. Catastrophic events, commonly associated with sun spots, occurring on the Sun and appearing as sudden faculae in the chromosphere. The associated bursts of luminous material projected into the corona are known as *solar surges.*

SOLARIMETER. A pyranometer or pyrheliometer designed to measure radiation received from the Sun. See also: *Actinometer.*

SOLARIZATION. (1) The reduction in transparency of glass, and sometimes its permanent coloration, by sunlight or ultraviolet radiation. (2) The reversal of a photographic image as a result of over-exposure.

SOLAR PROMINENCES. Hot tenuous clouds of gas which extend upwards through the chromosphere and high into the corona.

SOLAR SURGES. See: *Solar flares.*

SOLAR SYSTEM. That part of space surrounding the Sun, including the Sun itself and those bodies which are under the permanent or transitory influence of its gravitational field. These consist principally of the nine planets (Mercury, Venus, Earth, Mars, Jupiter, Saturn, Uranus, Neptune, and Pluto), the thirty-one (at least) natural satellites of these planets, and many thousands of minor planets or asteroids, comets and meteors.

SOLAR WIND. A highly ionized plasma of protons and electrons, in overall neutrality, which streams away from the Sun at 300–800 km/s.

SOLENOID. A cylindrical former around which wire is wound in one or more layers. Since a magnetic field is set up along the axis when a current is passed through the wire the solenoid is the basis of all forms of electromagnet and therefore of many electrically operated devices.

SOLID. That state or phase in which matter exhibits both definite shape and definite volume. Solids are characterized by a more or less regular arrangement of their constituent particles (atoms, molecules, ions, etc.) which is most marked in crystals but is not entirely absent even in so-called amorphous bodies. Solids resist any force that tends to alter their volume and form.

SOLID CIRCUIT. A complete electronic circuit, containing both passive and active components, fabricated within a small single crystal wafer of a semiconductor. It has been suggested that *single-crystal circuit* might be a better name.

SOLID, HOOKEAN. A solid in which stress is proportional to strain, in which elastic reaction and recovery are both "immediate" (i.e. attained with acoustic speed), and in which recovery is complete. See also: *Complex body. Fluid, Newtonian.*

SOLIDITY. Of a system of rotating aerofoils: the ratio of the total blade area to the disk area (i.e. the area swept out by the tips of the blades).

SOLID SOLUTION. A solid in which atoms of one type are partially replaced by atoms of another type, or in which such atoms are added without replacing existing atoms, and in which the type of crystal structure remains unchanged. Where the incorporation of new atoms occurs by replacement the solid solution is said to be *substitutional*, and where the new atoms occupy positions between the original atoms it is said to be *interstitial*. The new atoms may either be located at random or may occur in preferred locations, giving rise to an ordered phase. See also: *Intermetallic compound. Order–disorder transformation. Superlattice.*

SOLID-STATE DETECTOR. For ionizing radiation: a detector which depends on the motion of free electrons and holes produced by ionizing radiation in a solid (commonly a semiconductor). See also: *Counter, crystal. Counter, scintillation. Counter, semiconductor.*

SOLID-STATE PHYSICS. That branch of physics dealing with the structure, properties and behaviour of solids, both ideal and actual.

SOLIDUS. A line in an equilibrium diagram indicating the temperatures at which solidification is completed or melting begins in alloys of different composition. Thus, all phases below the solidus curve are solid. See also: *Liquidus.*

SOLLER SLITS. A stack of fine parallel slits, consisting essentially of a series of parallel flat plates, used as a collimator in X-ray or neutron diffraction to define a broad incident or diffracted beam with small angular spread. Convergent slits may also be used when a focusing effect is desired.

SOLSTICE, SUMMER. The instant at which the Sun in its apparent motion is at its maximum distance north of the celestial equator, when it touches the tropic of Cancer at the *First point of Cancer*. It occurs on or about 21 June.

SOLSTICE, WINTER. The instant at which the Sun in its apparent motion is at its maximum distance south of the celestial equator, when it touches the tropic of Capricorn at the *First point of Capricorn*. It occurs on or about 22 December.

SOLSTITIAL COLURE. The great circle passing through the celestial poles and those of the ecliptic, and therefore through both solstitial points. See also: *Equinoctial colure.*

SOLUBILITY. Of a solute in a solvent at a specified temperature and pressure: the composition of the saturated solution at the temperature and pressure in question. It may be expressed in a number of ways, for example, as the saturation concentration of the solute in the solvent. See also: *Solubility product.*

SOLUBILITY PRODUCT. For an electrolyte solution at a given temperature: the product of the ionic concentrations at saturation. It defines the degree of solubility of the electrolyte.

SOLUTE. The substance or substances that are dissolved in a solvent. See also: *Solvent.*

SOLUTION. A homogeneous phase of matter having more than one component and comprising at least two pure substances which may be separated out by methods involving phase changes (e.g. distillation, freezing, adsorption) but not by grosser methods such as filtration.

SOLUTION, BOILING-POINT ELEVATION OF. The elevation of the boiling point of a solvent on the addition of a solute. For dilute solutions it is inversely proportional to the sum of the molar concentrations of the solvent and solute.

SOLUTION, FREEZING-POINT DEPRESSION OF. See: *Cryoscopic constant.*

SOLUTION, IDEAL. A solution for which Raoult's law is valid.

SOLUTION, SATURATED. A solution which, at a specified temperature and pressure, can exist in stable equilibrium with an excess of solute, i.e. one which contains the maximum proportion of solute at that temperature and pressure.

SOLUTION, SUPERSATURATED. A metastable solution which contains more solute than does a saturated solution at the same temperature and pressure. It is liable, if disturbed, or if seeded with more solute, to change spontaneously to a saturated solution and excess solute.

SOLUTION, UNSATURATED. A solution which contains less solute than does a saturated solution at the same temperature and pressure.

SOLVATION. The attachment of molecules of a solvent to molecules or ions of a solute, e.g. hydrates (such as $CuSO_4 \cdot 5 H_2O$) which are present in solution.

SOLVENT. In general, that component of a solution which is present in excess. In solutions of solids or gases in liquids, however, it is convenient to regard the pure liquid as the solvent and the solid or gas as the solute.

SOLVENT EXTRACTION: LIQUID–LIQUID EXTRACTION. A process for the separation of the components of a liquid mixture by bringing it into intimate contact with an immiscible or partially miscible liquid (the *solvent*) in which one or more components of the mixture are soluble.

SOLVOLYSIS: LYOLYSIS. A general term descriptive of an ionic reaction between a solvent and a solute. It includes such reactions as hydrolysis, but not solvation.

SOMMERFELD FINE-STRUCTURE CONSTANT. A non-dimensional constant appearing in the theory of the fine structure of the atomic states of a single electron. It is equal to $2\pi e^2/hc$, where e is the electronic charge, h is Planck's constant and c is the speed of light. Its value is about $1/137$, or $7 \cdot 3 \times 10^{-3}$. See also: *Fundamental theory of Eddington*.

SOMMERFELD ORBITS. Additional electron orbits in the hydrogen atom which arose from the extension of Bohr's theory to include elliptical orbits.

SOMMERFELD WAVES. Electromagnetic surface waves guided by the plane surface between a dielectric and a good conductor.

SONAR. An acronym for *So*und *Na*vigation *A*nd *R*anging. It refers to a system of undersea ultrasonic signalling primarily used for submarine detection and for depth measurement. It is similar to radar in its operation and is also known as *Asdic*. See also: *Echo sounding*.

SONIC BOOM. The sound arising from shock waves emitted when an aircraft travels at a speed greater than that of sound in the same region. It is the acoustic analogue of Čerenkov radiation. The onset of the boom is quite sudden and often startling and the boom itself is audible at extremely large distances (up to 40 miles or more in some instances) from the generating aircraft.

SONOMETER. A sounding board supporting a wire stretched between two knife-edges and used for verifying the relation between the frequency of vibration of the wire and its length, diameter, density and tension.

SORBENT. See: *Chromatography*.

SORET EFFECT. A name sometimes used to describe thermal diffusion in solids or liquids.

SOUND. (1) A disturbance propagated as a wave motion in an elastic medium, of such a character as to be capable of exciting the sensation of hearing. By extension the term is sometimes applied to any disturbance propagated as a wave motion in an elastic medium. If the frequency is too high to be audible the disturbance is classed as *ultrasonic*. If it is too low the disturbance is termed *infrasonic*. (2) The sensation of hearing evoked by a disturbance propagated as in (1).

SOUND ABSORPTION COEFFICIENT. Of a surface or material for a given frequency, under specified conditions: the reflection coefficient under the condition specified subtracted from unity.

SOUND ABSORPTION COEFFICIENT, REVERBERANT. Of a surface or material for a given frequency: the sound absorption coefficient when the distribution of the incident sound is completely random.

SOUND ABSORPTION, EQUIVALENT. Of a room or an object in a room for a given frequency: that area of a surface having a reverberant absorption coefficient of unity which would absorb sound energy at the same rate as the room or object. The unit of equivalent absorption is termed an *absorption unit* and is given the name *SABIN* when taken as one square foot.

SOUND ANALYSIS. The determination of all the parameters of the sound field at a given point. It involves the determination of the relative amplitudes and phases of the Fourier components of the sound, together with the types and directions of propagation of the various waves making up the field. In practice, however, sound analysis has come to mean only the determination of the Fourier components of the wave without reference to the particular type of wave.

SOUND, ATTENUATION COEFFICIENT OR CONSTANT. The real part of the propagation coefficient. It determines the diminution in amplitude per unit distance or, for a recurrent structure, per section.

SOUND, BAND PRESSURE LEVEL OF. The pressure level of the sound energy within a specified frequency band. See also: *Sound pressure level*.

SOUND ENERGY. At a specified part of a medium: the total energy in that part of the medium less the energy (if any) which is not due to sound.

SOUND ENERGY DENSITY. At a point in a sound field: the sound energy per unit volume at the point.

SOUND ENERGY FLUX. Through an area: the mean flow of sound energy per unit time through the area.

SOUND FILM. A combined cinematograph film and sound track. The frequency response of such a film may extend from about 40–15 000 Hz and is thus comparable with that of the normal ear for which the corresponding range is about 25–18 000 Hz. See also: *Sound recording*.

SOUND INTENSITY. At a point in a progressive wave: the sound energy flux per unit area normal to the direction of propagation, i.e. the *sound energy flux density*. The term is also used in a more general sense to signify some quantity proportional to the sound energy.

SOUND LENS. See: *Sound, reflection and refraction of*.

SOUND PARTICLE VELOCITY. At a point in a sound field: the alternating component of the total velocity of movement of the medium at the point minus the velocity (if any) which is not due to sound.

SOUND, PITCH OF. See: *Pitch*.

SOUND PRESSURE. At a point in a sound field: the alternating component of the pressure at the point. It may be expressed, either as the instantaneous or r.m.s. value, in units of force per unit area. See also: *Pressure*.

SOUND PRESSURE LEVEL. The ratio, expressed in decibels, of the sound pressure to a specified reference pressure. See also: *Phon. Sound, band pressure level of*.

SOUND PROPAGATION COEFFICIENT OR CONSTANT. Of a plane progressive wave in a continuous isotropic medium or in a uniform transmission system: the natural logarithm of the complex ratio of the steady state sound pressure or particle velocity at one point to that at a second point at unit distance from the first in the direction of propagation.

SOUND, QUALITY OF. See: *Sound, timbre of*.

SOUND, RADIATION PRESSURE OF. The pressure exerted by a sound wave at a point on an obstacle, excluding forces due to streaming. See also: *Radiation pressure*.

SOUND RANGING. The location of the position of a gun from the time of reception of the *gun wave* (i.e. the sound wave produced at a gun when the gun is fired) at three or more points.

SOUND RECORDING. The production of a more or less permanent sound record by changing the state or configuration of a suitable material in sympathy with a sound signal, in such a way that the signal can subsequently be reproduced. The recording material may be a wax or plastic disk, a magnetic wire or tape, or a photographic film.

SOUND, REFLECTION AND REFRACTION OF. For the reflection and refraction of a plane progressive wave at a plane boundary between two media the same laws hold as in optics. *Sound lenses*, analogous to light lenses, may also be constructed.

SOUND REFLECTION COEFFICIENT. Of a surface or material for a given frequency, under specified conditions: the fraction of the incident sound energy which is reflected from the surface or material under the conditions specified.

SOUND RESONANCE. See: *Resonance*.

SOUND, TIMBRE OF. That subjective quality of a sound which enables a listener to judge that two sounds, having the same loudness and pitch, are dissimilar. It is sometimes known as *quality*.

SOUND TRACK. A narrow band, usually along the margin of a sound film, which carries the sound record. In a *variable-area* track the width varies with the wave form of the signal, and in a *variable-density* track the width is constant and the wave form controls the light transmission along the track.

SOUND, VELOCITY OF. This is given for a homogeneous medium by Newton's formula: $v = (E/\varrho)$, where v is the velocity, E the appropriate elastic modulus of the medium, and ϱ its density. For liquids the modulus is the bulk modulus, for solids it is Young's modulus, and for gases it is γp, where γ is the ratio of specific heats and p the pressure. Only for intense sound does Newton's formula require appreciable modification. The velocity of sound in air at 0°C is 331 m/s and in seawater is about 1500 m/s.

SOUND VIBRATIONS, FREE. Sound emitted from a source which is vibrating at its own natural frequency. See also: *Forced vibrations*.

SOUND VIBRATIONS, LONGITUDINAL. Sound vibrations corresponding to a longitudinal or compressional wave. See also: *Compressional wave*.

SOUND VIBRATIONS, TRANSVERSE. Sound vibrations corresponding to a transverse or shear wave. See also: *Shear wave*.

SOURCE. A point at which a fluid is continually emitted and from which the flow is radial and uniform. A negative source is known as a sink. The concept of sources and sinks is extended to cover the flow of energy in all its forms.

SPACE. (1) A set dealing with particular types of property or concerned with a particular kind of geometrical terminology. Thus the properties of continuity and convergence are associated with *topological space*, distance and angle with *Euclidean space*, to mention two examples. (2) Any region outside the Earth (i.e. outside its atmosphere). The term has sometimes been restricted to regions outside the Galaxy (i.e. to *outer space*). (3) That in which material bodies exist, thus *empty space* is a vacuum.

305

SPACE CHARGE. The charge existing in a given volume of space, which arises from a uniform distribution of charges of electricity of either sign. The charges may be located on atoms or molecules, or may arise from swarms of electrons in motion. A space charge produces its own field and therefore, on its formation, affects the previously existing field. Space charge is of importance in thermionic tubes, photoelectric cells, particle accelerators etc. owing to its effect on the flow of current, and also plays a part in electrical conduction in solids.

SPACE-CHARGE WAVES. Perturbations of space-charge density propagated along an electron beam as a wave motion, and carrying corresponding perturbations of the electron current, electron speed and electromagnetic field.

SPACE GROUP. One of a set of mutually consistent groups of symmetry elements which determine the relative positions in which atoms or groupings of atoms can occur in a unit cell. They are developed from the fourteen Bravais lattices by the application of mirror planes, rotation axes and inversion axes (the *macroscopic symmetry elements*) to give thirty-two crystal classes, and by the application to these classes of glide planes and screw axes (*microscopic symmetry elements*) to give 230 space-groups.

SPACE LATTICE: BRAVAIS LATTICE. An infinite three-dimensional array of points in space, such that each point has the same environment. Only fourteen distinct arrays or lattices of this kind are possible and these may be referred to one or other of the seven crystal systems. A *primitive lattice* (symbol P) has a lattice point at each corner of the appropriate unit cell. A *body-centred lattice* (symbol I) has, in addition, a point at the "centre" of the unit cell. The *face-centred lattice* (symbol F) has a point at the centre of each face as well as one at each corner, and the remaining type of lattice (symbol C) has, instead, points at the centres of only one pair of faces. Although a lattice is a geometrical concept and not a crystal, and therefore cannot scatter radiation, the structural motif (some particular atomic or molecular group) the repetition of which in three dimensions constitutes a crystal, may always be associated with a lattice point at each repetition i.e. the motif may be replaced by a representative point situated at a point of the lattice.

SPACE-TIME. A space of four dimensions which specify the space and time coordinates of an event. Einstein's two relativity postulates lead to a *space-time continuum* in which elements of length and time are no longer invariant under a transformation from one set of axes to another set which is in uniform relative motion with respect to the first. See also: *Relativistic invariance. Relativity theories.*

SPACE-TIME, SINGULARITY IN. A region in which the curvature of space-time is such that the known laws of physics break down, e.g. a black hole.

SPACE WAVE. A radio wave which is received after being reflected one or more times from the Earth's surface.

SPACING. See: *Interplanar spacing.*

SPALLATION. A nuclear reaction in which several particles or nuclei are ejected from a target nucleus, this nucleus being appreciably reduced in consequence both in mass number and atomic number.

SPALLING. In the explosive working of metals: fracture arising from the interaction between the tensile front of a reflected stress wave and its incident compression tail. This type of fracture is also known as *scabbing.*

SPAN. (1) Of an aerofoil: the length along a specified line measured normal to the mean air flow. (2) Of an aeroplane: the distance between the wing tips.

SPARK. An electric discharge taking place in air or other insulating material. See also: *Discharge, spark.*

SPARK CHAMBER. An instrument for rendering visible the tracks of ionizing particles. It originated in the observation that in a parallel-plate spark counter a discharge occurs at a point where the charged particle traverses the gap between the plates. See also: *Bubble chamber. Cloud chamber. Counter, spark. Streamer chamber.*

SPARK CHANNEL. That part of a spark discharge which follows the processes (avalanche formation, etc.) leading to the establishment of a highly conducting plasma.

SPARK GAP. An arrangement of two electrodes between which a spark passes when the applied voltage is sufficiently high.

SPARKING POTENTIAL. The minimum breakdown voltage of a spark gap. It is generally measured under conditions where an increasing d.c. voltage is applied to the gap until a spark occurs.

SPECIFIC. A qualifying word denoting the value of a physical quantity per unit mass, area, volume, length etc., according to circumstances, e.g. specific activity, specific heat, specific ionization, specific resistance, etc. It has been recommended by the IUPAC that, *for extensive quantities,* the word should be restricted to the meaning "divided by mass".

SPECIFIC EMISSION. Of radiation from a surface: the energy emitted at a given wavelength per unit area of the surface in unit time.

SPECIFIC GRAVITY BOTTLE: DENSITY BOTTLE. An accurately calibrated glass bottle used, by comparing the weights of the same volumes of a liquid and a reference liquid, to measure the specific gravity of the first liquid. See also: *Pyknometer*.

SPECIFIC GRAVITY: RELATIVE DENSITY. Of a given substance: the ratio of the density of the substance to that of a standard substance, usually water, at a specified temperature. See also: *Atmosphere, relative density of*.

SPECIFIC HEAT ANOMALY. A discontinuity in the curve of specific heat versus temperature (which normally shows a steady rise of specific heat with temperature), which is associated with order–disorder phenomena (e.g. the Curie point for a ferromagnetic metal), or with transition phenomena in atomic energy levels.

SPECIFIC HEAT: SPECIFIC HEAT CAPACITY. Of a given substance at a specified temperature T: the limit, as ΔT approaches zero, of the ratio $\Delta Q/\Delta T$, where ΔQ is the amount of heat which must be added to unit mass to increase the temperature of the substance from T to $T + \Delta T$. It is the heat capacity per unit mass. See also: *Heat capacity*.

SPECIFIC HEATS, RATIO OF. The ratio of the specific heat of a gas at constant pressure to that at constant volume. For adiabatic changes of a perfect gas $pv^\gamma = $ constant, where p is the pressure, v the volume, and γ the ratio of the specific heats.

SPECIFIC HEAT THEORIES. A term referring to a number of theories of the specific heat of solids, starting from the observation by Dulong and Petit that the atomic heat is constant over a large temperature range and leading through the Einstein theory and its modification by Nernst and Lindemann, its improvement by Debye (who allowed for the first time for the numerous possible modes of the thermal vibration of atoms), to modern work, following Born and von Kármán's detailed study of longitudinal and transverse wave velocities, on the precise frequency distribution. See also: *Atomic heat. Born–von Kármán boundary conditions. Debye theory of specific heats. Einstein theory of specific heats. Nernst–Lindemann theory of specific heats*.

SPECIFIC INDUCTIVE CAPACITY: DIELECTRIC CONSTANT: RELATIVE PERMITTIVITY. Of a medium: the ratio of the capacitance of a capacitor with the medium between the electrodes to that of a capacitor with a vacuum between the electrodes. Alternatively: the ratio of the electric flux density produced in the medium to that which would be produced in a vacuum by the same electric force. See also: *Permittivity: Absolute permittivity*.

SPECIFIC RESISTANCE. Of a given material: the resistance of a unit length of unit cross-sectional area, expressed in units of ohm-centimetre or ohm-metre. It is the reciprocal of conductivity and is also known as *Resistivity* or *Volume resistivity*.

SPECIFIC ROTATION. For an optically active material: the angular rotation of the plane of polarization per unit path of the light passing through the material, divided by the density of the material. Its variation with wavelength is given by the Drude equation.

SPECIFIC SURFACE. Of a particle or droplet: the ratio of the surface area to the volume.

SPECIFIC VOLUME. The volume per unit mass. It is the reciprocal of the density.

SPECTRA. See: *Spectrum*.

SPECTRAL BANDS, DEGRADATION OF. See: *Band head*.

SPECTRAL BANDS, DIFFUSE. Bands which occur in certain spectra (e.g. those of some liquids), which have no definite edge and do not appear to be capable of resolution into individual lines.

SPECTRAL DOUBLET. A spectral line with a multiplicity of two. The frequency separation of the two levels, expressed as a wave number, is called the *doublet interval*.

SPECTRAL GHOSTS. False images of a spectral line produced by irregularities in the ruling of diffraction gratings. They may be categorized as *Rowland ghosts*, which are grouped symmetrically on both sides of the true line, and *Lyman ghosts*, which are false spectra having non-integral orders.

SPECTRAL HARDENING. Of radiation or particles passing through a medium: the increase in average energy owing to the preferential loss at low energies by absorption, leakage or scattering.

SPECTRAL LINES, BROADENING OF. The broadening of spectral lines for instrumental reasons or for reasons such as *self-absorption, radiation broadening, pressure broadening, Doppler broadening, resonance broadening*, and *Stark broadening*, all of which are separately defined.

SPECTRAL LINES, FORBIDDEN. See: *Selection rule. Transition, allowed. Transition, forbidden*.

SPECTRAL LINES, REVERSAL OF. The reversal of the bright lines of an emission spectrum when they are turned into absorption lines, as in the case of Fraunhofer lines. See also: *Reversing layer*.

SPECTRAL NOTATION, ATOMIC. A symbolism for the designation of atomic energy levels, the differences between which give the frequencies of the spectral lines. The general designation of such a level (or *spectral term*) is nl, ^mL_J, where n is the total quantum number, l the azimuthal quantum number, m the multiplicity ($m = 1, 2, 3$, etc., for singlet, doublet, triplet, etc.), L the total orbital momentum quantum number ($S = 0$, $P = 1$, $D = 2$, $F = 3$, etc.), and J the total angular momentum quantum number (integral or half-integral). Thus the ground state for the hydrogen atom is designated by $1s$, $^2S_{\frac{1}{2}}$. Where details of fine structure are omitted a shortened notation is employed. The ground state for the hydrogen atom, for example, then becomes 1^2S, and the terms corresponding to the Lyman series become 2^2P, 3^2P, 4^2P, etc., the corresponding spectral lines being $1^2S - 2^2P$, $1^2S - 3^2P$, $1^2S - 4^2P$, etc. See also: *Electron shell. Quantum number. Spectrum, atomic.*

SPECTRAL NOTATION, MOLECULAR. A symbolism for the designation of molecular energy levels. Individual term values are more complicated than for atomic spectra and vary in type according to the spectral range involved. The far infrared spectrum is interpreted as a pure rotation spectrum, the energy levels being given by

$$[h^2/(8\pi^2 I)] [J(J + 1)],$$

where h is Planck's constant, I is the moment of inertia of the rotator, and J is the rotational quantum number, equal to $0, 1, 2, \ldots$ The near infrared spectrum is interpreted as a vibration spectrum, the energy levels being given by $h\nu_0(v + \frac{1}{2})$, where ν_0 is the classical frequency of the oscillator and v is the vibrational quantum number, equal to $0, 1, 2, \ldots$ The visible and ultraviolet spectrum is interpreted as an electronic spectrum, but the pure electronic spectrum is not normally observed alone since electronic transitions are usually accompanied by vibrational or rotational transitions which, with their interaction, must be allowed for. See also: *Spectrum, molecular.*

SPECTRAL PURITY. A term sometimes used for colorimetric purity. See also: *Colorimetric purity.*

SPECTRAL SERIES. A series of lines in a spectrum in which there is a regular relationship between the frequencies or wave numbers of the individual lines. The best known are probably the various series in the line spectrum of the hydrogen atom. See also: *Hydrogen atom, line spectrum of. Rydberg series.*

SPECTRAL SERIES, LIMIT OF. The mathematical limit obtained by setting the total quantum number (usually denoted by n) equal to infinity. See also: *Hydrogen atom, line spectrum of.*

SPECTRAL TERM. See: *Spectral notation, atomic.*

SPECTRA, MOLECULAR, BRANCHES IN. A branch, of which there are three (designated as P, Q, and R respectively), refers to one of a set of lines in a spectral band as characterized by the position of the associated parabola on the Fortrat diagram. If the wave number ν is expressed in the form $\nu = c + dm + em^2$, where c, d, and e are constants and m is an integer which numbers successive lines, the P branch corresponds to a negative value of m, the Q branch to a zero value, and the R branch to a positive value. The Q branch is also known as the *null branch* or *zero branch*.

SPECTRA, PERTURBATIONS IN. The occurrence of spectral lines in atomic or molecular spectra in places where they are not expected theoretically, or with anomalous intensities. The theoretical predictions are in error owing to the neglect of interaction terms which turn out to be significantly larger than had been assumed.

SPECTRA, RADIOFREQUENCY. The spectra observed for a number of types of low energy transition of both atoms and molecules. The main examples are found in *electron paramagnetic resonance, ferromagnetic resonance, microwave spectroscopy, nuclear magnetic resonance,* and *nuclear quadrupole resonance*, all of which are defined separately.

SPECTROCHEMICAL ANALYSIS. The determination of the presence or amount of specified chemical elements in a specimen of material by methods using ultraviolet, visible, and infrared regions of the spectrum.

SPECTROGRAM. A record produced by a spectrograph.

SPECTROGRAPH. An instrument for producing a record of a spectrum, the essential features of which are a source of radiation, a collimator, a means of sorting out the components of the radiation in order of wavelength, energy, momentum, mass, or other related quantity, and a means of recording. In an optical spectrograph the light is dispersed by a prism or a diffraction grating (plane or concave), and a record is made photographically, by photoelectric or photoconducting detectors, by thermopiles or bolometers, or by some form of image converter, according to the spectral range concerned and the type of problem involved. See also: *Eagle mounting. Littrow mirror. Mass spectrography: Mass spectrometry. Paschen circle. Rowland grating. Spectrometer. Spectroscope. Wadsworth spectrograph.*

SPECTROGRAPH, FLUORITE. A spectrograph employing a fluorite prism for use in the Schumann region of the spectrum between about 1200 Å to 2000 Å, i.e. 0·12–0·20 μm.

SPECTROGRAPH, QUARTZ. A spectrograph employing a quartz prism for use in the ultraviolet region of the spectrum, between about 1800 Å and 5200 Å, i.e. 0·18–0·52 μm.

SPECTROHELIOGRAPH. An instrument for obtaining one or more images of the Sun's surface at one or more selected wavelengths. It is used in the study of the effects of magnetic storms and sunspots.

SPECTROMETER. An instrument used for measuring the angular deviation of the components of a spectrum. The basic requirements are the same as for a spectrograph, together with provision for angular measurement and the incorporation of an appropriate form of detector. See also: *Alpha-ray spectrometry. Beta-ray spectrometer. Gamma-ray spectrometer. Neutron spectrometer. Mass spectrography: Mass spectrometry. Spectrograph. Spectroscope. X-ray spectrometry.*

SPECTROMETER, DIRECT READING. A direct vision spectroscope provided with a scale on which the wavelength corresponding to any part of the spectrum can be read directly.

SPECTROMETER, VACUUM. A spectrometer which is evacuated (or filled with a gas more transparent than air) for use in the ultraviolet region below about 2000 Å, i.e. 0·20 μm.

SPECTROPHOTOMETER. An instrument for measuring the energy distribution in some part of the visual spectrum. It consists essentially of a spectrometer combined with some type of photometer. Devices are also commonly introduced to bring together different parts of the spectrum into the same field of view. The name is often given to what is strictly a spectroradiometer, i.e. to an instrument working into the infrared and ultraviolet.

SPECTROPHOTOMETER, ABSORPTION. A spectrophotometer in which interest is centred on the amount of light absorbed at various wavelengths.

SPECTROPHOTOMETER, PHOTOGRAPHIC. A spectrophotometer in which the measurements are made via a photographic plate rather than a photometer. It is of importance where the source is unsteady or transient or when the light flux is weak, and finds its main applications in astronomy and in industry.

SPECTRORADIOMETER. An instrument for measuring the distribution of radiant energy throughout the spectral range. See also: *Spectrophotometer.*

SPECTROSCOPE. An instrument for the visual examination of a spectrum. The basic requirements are the same as for a spectrograph or spectrometer but the spectra are observed by means of a telescope.

See also: *Spectrograph. Spectrometer.*

SPECTROSCOPE, CONSTANT DEVIATION. A prism spectroscope in which the prism is cut so that the light always enters and leaves at the same angle, so that the collimator and telescope remain fixed, while the prism is rotated.

SPECTROSCOPE, DIRECT VISION. A small hand-held instrument suitable for the qualitative examination of simple emission and absorption spectra. The essential feature is an *Amici prism*, which provides dispersion without deviation, being composed of three (or more) compensating prisms made of crown and flint glass and cemented together.

SPECTROSCOPE, ÉCHELON. A spectroscope employing an échelon grating as a dispersing element.

SPECTROSCOPIC SPLITTING FACTOR. See: *Landé g factor: Landé splitting factor.*

SPECTROSCOPY. The production, observation or measurement, theory and interpretation of spectra of any kind. See also: *Spectrum.*

SPECTROSCOPY, ÉCHELLE. See: *Échelon, Échelle and Échelette gratings.*

SPECTROSCOPY, FAR INFRARED. The measurement of spectral lines in the infrared wavelength range up to about 750 μm. The sources most commonly employed are the mercury arc, globar rod, and molecular laser. Dispersion is by a grating or Fourier transform spectroscopy, and a Golay cell is usually employed as a detector.

SPECTROSCOPY, MICROWAVE. The measurement of spectral lines in the wavelength range of about 10 cm to 0·5 mm (i.e. in the frequency range of about 3000–600000 MHz). In its simplest form a microwave spectrometer consists of a klystron (to produce highly monochromatic radiation), an absorption cell and a suitable detector (e.g. a silicon–tungsten crystal detector), to measure the change of absorption with frequency. There are two basic kinds of spectroscopy in the microwave region, *electron paramagnetic resonance* and *gaseous microwave spectroscopy*, which are defined separately. See also: *Spectra, radiofrequency.*

SPECTROSCOPY, MODULATION. The measurement and interpretation of changes in optical spectra resulting from various perturbations applied to the material under study. One of the main applications is to the enhancement of weak structures in such spectra.

SPECTROSCOPY, PHOTOELECTRON. The analysis of the energies of electrons ejected from atomic systems to determine, for example, the energy with which a selected electron is bound in a system. The two major fields that have developed involve X-ray and ultra-violet irradiation, giving information regarding core electrons and outer electrons respectively.

SPECTROSCOPY, REFLECTION. The measurement of the variation with frequency of the specular reflectivity of a material, to determine either its optical constants or the positions and strengths of features in its absorption spectrum.

SPECTRUM. A visual display, photographic record, or plot of the distribution of any one particular type of radiation as a function of its wavelength, energy, momentum, mass, or other related quantity. The term also often designates the distribution itself. One may speak of an optical spectrum, an X-ray spectrum, a radio-frequency spectrum, and so on. The most common example of an optical spectrum is that seen in the rainbow, where the spectral colours of white light are displayed for all to see. See also: entries under *Spectra* and *Spectral*.

SPECTRUM, ABSORPTION. The spectrum observed when continuous radiation is passed through a medium which absorbs certain wavelengths selectively, thereby giving rise to dark bands or lines.

SPECTRUM ANALYSER. A device for determining the frequency-energy distribution of a radio signal or group of signals. It is of wide use in radar to monitor the spectrum of pulse transmitters.

SPECTRUM, ARC. A spectrum produced by an electric arc. The specimen to be investigated is placed between the electrodes of an arc or applied to them as a coating and the light emitted is examined spectroscopically. Owing to the high temperature nearly all compounds are dissociated and the spectra observed are those due to the component elements.

SPECTRUM, ATOMIC. The spectrum emitted by an excited atom. It is in general characterized by more or less sharply defined lines, in contrast to the bands associated with molecular spectra. See also: *Raies ultimes. Spectral notation, atomic. Spectrum, X-ray.*

SPECTRUM, ATOMIC, INTERCOMBINATION LINES IN. Lines found in the spectra of heavy atoms due to intercombination of singlet and higher order terms.

SPECTRUM, BAND. A name often used to describe a molecular spectrum, since the spectral lines of such a spectrum are grouped in such a way as to give an appearance of bands. See also: *Band head. Spectral bands, diffuse. Spectrum, atomic. Spectrum, molecular.*

SPECTRUM, CHANNELLED. An interference spectrum obtained when white light is reflected from a thin parallel-sided transparent plate, when interference occurs between the externally and internally reflected rays. It constitutes an example of amplitude splitting. In a spectroscope the spectrum is seen to be crossed by a series of dark fringes, known as *Edser–Butler bands*, at those wavelengths at which destructive interference occurs. See also: *Interferometer, classification of.*

SPECTRUM, CHARGE-TRANSFER. A spectrum associated with the transfer of electric charge from one atom or molecule to another, typically as a result of the absorption of radiation.

SPECTRUM, CONTINUOUS. A spectrum in which radiation is distributed over an uninterrupted range of wavelengths in contrast to a line or band spectrum. It arises either from radiating matter in a condensed state (which may approximate to a black body), from the ionization or dissociation of free atoms and molecules or their subsequent recombination, or from the retardation or acceleration of charged particles.

SPECTRUM, EMISSION. The spectrum produced by an emitting source as distinct from an absorption spectrum.

SPECTRUM, EQUAL ENERGY. An artificial spectrum in which the radiant energy is distributed equally over all wavelengths.

SPECTRUM, EXPLOSION. A spectrum of an exploding solid or liquid. It shows lines of high excitation states.

SPECTRUM, FINE STRUCTURE IN. The splitting of spectral lines in atomic emission spectra owing to term multiplicity, i.e. owing to the splitting of atomic energy levels. See also: *Multiplicity. Spectral notation, atomic. Spectrum, hyperfine structure in.*

SPECTRUM, FLAME. (1) The spectrum exhibited by a flame itself, e.g. by a hydrogen or an oxygen flame. (2) The spectrum obtained from a substance which is burned in a flame.

SPECTRUM, FLASH. The chromospheric emission spectrum of the Sun, visible for a few seconds just before and just after totality in a solar eclipse. This spectrum furnishes evidence of the great abundance of helium in the Sun.

SPECTRUM, HYPERFINE STRUCTURE IN. A splitting of the spectral lines in atomic emission spectra, which themselves arise from multiplicity, associated with spin-dependent energy level shifts, isotopic shifts, and other effects, e.g. that due to nuclear octupole interaction. See also: *Spectrum, fine structure in.*

SPECTRUM, INTERFERENCE. An optical spectrum produced by interference, as in a diffraction grating or interferometer. See also: *Diffraction grating. Interferometer. Spectrum, channelled.*

SPECTRUM, LINE. (1) A name often used to describe an atomic spectrum, which is characterized by obvious lines in contrast with a molecular spectrum. See also: *Spectrum, band.* (2) Any spectrum associated with discrete values of a property of interest (e.g. energy, mass, velocity, etc.) as distinct from a continuous spectrum.

SPECTRUM, MOLECULAR. The spectrum emitted by an excited molecule. The lines of such a spectrum are grouped together in such a way as to give an appearance of bands, in contrast to the sharp lines of an atomic spectrum. This is a consequence of the large number of transitions resulting from the existence of energy levels, characteristic of vibrational and rotational energy, which occur in addition to the electronic energy levels. See also: *Band head. Combination principle of Ritz. Spectral notation, molecular. Spectrum, band.*

SPECTRUM, NORMAL. (1) A spectrum in which the dispersion is linear with wavelength, typically a diffraction spectrum at small angles. (2) A spectrum obtained under conditions of normal dispersion as opposed to anomalous dispersion.

SPECTRUM, ORDER OF. Denotes the integral number of wavelengths by which the paths of two wave trains differ for constructive interference, when a diffraction grating is used in a spectrometer. See also: *Interference, order of.*

SPECTRUM, RECOMBINATION. A type of continuous spectrum produced by the overlapping of broad energy bands. It is the result of complicated recombinations between individual molecular systems.

SPECTRUM, RESONANCE. Spectral lines arising from resonance. See also: *Resonance radiation.*

SPECTRUM, SPARK. A spectrum produced by a spark discharge. In contrast to the arc spectrum the spark spectrum is due to ionized atoms, the *first spark spectrum* arising from singly ionized atoms, the *second* from doubly ionized atoms, and so on. See also: *Displacement law.*

SPECTRUM, X-RAY. Describes the combination of continuous and characteristic X-rays, consisting of the superposition of sharp lines on a continuous background. The sharp lines occur in groups characteristic of the transfer of outer electrons to each of the various inner shells ($K, L, M, \dots$), each line being denoted by the letter appropriate to the inner shell concerned together with symbols ($\alpha, \beta, \gamma, \dots$, and 1, 2, 3, ...) denoting the outer shells involved and the energy levels within these shells. See also: *Electromag-*

netic spectrum. Moseley law. Term diagram. X-rays.

SPECULAR DIFFUSE DENSITY. See: *Density, diffuse. Density, specular. Optical density. Photographic density.*

SPEECH SPECTROMETRY. The study of speech by means of a speech spectrogram, i.e. a visual record indicating the power density of speech (usually shown as the density of shading) as a function of time and frequency. Such spectrograms are used, among other things, in the development of speech communication systems and as an aid in teaching deaf people to talk, and may also be used in *speech synthesis,* i.e. to synthesise realistic speech-like sounds.

SPEECH SYNTHESIS. See: *Speech spectrometry.*

SPHERE GAP. A spark gap with spherical electrodes, used as a protective device against excess voltage or in the measurement of voltage.

SPHERICAL ABERRATION. An aberration in an optical system arising from the spherical shape of the optical surfaces, whereby rays from different zones of the objective strike the axis at different points instead of coming to a focus.

SPHERICAL HARMONIC. A polynomial which satisfies the Laplace equation in spherical coordinates.

SPHERICAL WAVE. A wave whose equiphase surfaces form a family of concentric spheres.

SPHEROID. In geodesy: a reference ellipsoid, approximating to the geoid, which is used for the computation of triangulation owing to the irregular nature of the geoid itself. See also: *Earth, figure of.*

SPHEROIDAL STATE. See: *Leidenfrost phenomenon.*

SPHEROIDAL TRIANGLE: GEODETIC TRIANGLE. In geodesy: three points on a spheroid together with the three geodesic lines which join them.

SPHEROMETER. An instrument for measuring the radius of curvature of a spherical surface (commonly a lens or mirror). One of the simplest types consists of a table having three fixed legs (with pointed feet) and a central micrometer screw. The desired radius may easily be obtained by measuring the distance between the point of this screw and the plane of the three feet when all four are touching the surface.

SPHYGMOMANOMETER. An apparatus for measuring the arterial blood pressure. It consists of an inflatable bag, wrapped round the upper arm and connected to a manometer.

SPIKE. (1) A disturbed region along the track of a high-energy particle. See also: *Displacement spike. Fission spike. Thermal spike.* (2) A fuel spike, i.e. nuclear fuel containing more highly enriched material than the majority of the other fuel in a reactor. It is also known as a *seed*. See also: *Nuclear reactor fuel, enriched*. (3) The passage of an action-potential down a nerve.

SPIN. (1) Of an elementary particle: its angular momentum in the absence of orbital motion. (2) Of an atomic nucleus: its spin angular momentum plus the contributions from the orbital motions of the nucleons in it. See also: *Angular momentum in atoms and nuclei. Helicity. Isobaric spin. Quantum number, spin*.

SPIN FLIP. The reversal of the spin direction, in certain circumstances, of nucleons or electrons.

SPIN GLASSES. Alloys (not true glasses) consisting of a small amount (up to 20%) of a magnetic element dissolved as a random solid solution in a non-magnetic host, e.g. iron in gold or manganese in copper. Such alloys exhibit, at sufficiently low temperatures, magnetic transitions in which the random spins of the magnetic atoms are frozen, in contrast with the ordering of spins that takes place in ferromagnetism or anti-ferromagnetism below the Curie or Néel temperature respectively.

SPIN-LATTICE RELAXATION. The process by which a system of spins (e.g. paramagnetic, ferromagnetic, nuclear) which has been excited, thereby gaining energy, comes into thermal equilibrium with its surroundings.

SPIN ORBITAL: SPIN WAVE FUNCTION. The complete wave function of an electron. It is the product of the orbital wavefunction and a wave function representing the orientation of the electron spin axis.

SPIN–ORBIT COUPLING. The interaction between the intrinsic and orbital angular momentum of a particle.

SPIN RESONANCE. Another name for electron spin resonance, ferromagnetic resonance, and paramagnetic resonance, as opposed to nuclear magnetic resonance. See also: *Electron paramagnetic resonance: Electron spin resonance. Ferromagnetic resonance*.

SPINS, PARALLEL AND ANTI-PARALLEL. Terms descriptive of the sense of the spins, two spins with the same sense being *parallel* and two with opposite senses being anti-parallel. See also: *Helicity*.

SPIN TEMPERATURE The temperature associated with a spin system which has been excited by absorption of energy but has not yet come into thermal equilibrium with its surroundings, i.e. when spin-lattice relaxation is not complete. It is meaningful only when the spin system is in internal thermodynamic equilibrium and when this system has a relatively weak interaction with the lattice.

SPINTHARISCOPE. An instrument, incorporating a zinc sulphide screen, for observing individual scintillations produced by ionizing radiation.

SPIN WAVE FUNCTION. See: *Spin orbital: Spin wave function*.

SPIN WAVES. In crystalline solids showing long-range order in the atomic magnetic moments: waves of deviation of the orientations of the magnetic moments from those occurring in the ordered state. They are propagated through exchange or superexchange coupling between electron spins on neighbouring atoms.

SPIRAL GALAXY. A galaxy having spiral arms consisting of stars, gas and dust, emerging from a central nucleus. The mass of such a galaxy lies in the range 10^{10} to 10^{12} times that of the Sun.

SPIROMETER. An instrument for measuring the air capacity of the lungs.

SPLASH ALBEDO. See: *Albedo*.

SPORADIC REFLECTIONS. From the ionosphere: See: *Ionization, sporadic-E*.

SPREADING AGENT. A substance which, when dissolved in a liquid, causes it to spread as a film over the surface of another liquid.

SPREADING COEFFICIENT. Of a liquid B on a solid or liquid substrate A: the value of the expression $\gamma_A - \gamma_B - \gamma_{AB}$, where γ_A and γ_B are the free surface energies of phases A and B and γ_{AB} is the interfacial free energy at their boundary. For liquid–air and liquid–liquid surfaces these free energies are equal to the surface or interfacial tensions. Spontaneous spreading will only occur if the expression is positive.

SPURS. On a photographic or cloud chamber track of an ionizing particle: side-tracks due to low energy secondary electrons or delta rays.

SPUTTERING. The ejection of atoms from the surfaces of solids under ion bombardment. It arises from atomic collisions and not from thermal evaporation, and is characterized by ejection in preferential directions, which sometimes, but not always, arises from the occurrence of focused collision sequences. The phenomenon is often used for the production of thin metallic layers.

SQUALL. A strong wind which rises and dies away rapidly, lasting for only a few minutes and frequently associated with a temporary change in wind direction.

SQUID (SUPERCONDUCTING QUANTUM INTERFERENCE DEVICE). A device for measuring very small changes in magnetic flux, involving the use of a Josephson junction. It can equally well be used to detect a very small current or voltage.

SS CYGNI STARS. See: *Stars, variable, nomenclature of*.

S STAR. See: *Stars, spectral classes of.*

STABILIZATION OF SHIPS. The reduction in the movement of ships at sea. The term usually refers to the reduction of rolling rather than of pitching or yawing, since rolling is the most troublesome movement and its reduction will also lessen yawing. The most common method involves the use of lateral fins whose inclination may be varied so as to oppose the roll, but other methods have also been employed, e.g. gyroscopic methods and methods involving the transfer of water or oil from one side of the ship to the other.

STACKING FAULT. See: *Surface defect.*

STAGNATION POINT. In fluid flow: a point where the velocity of the fluid is zero (usually relative to a solid body) and where a streamline divides into two (or more) streamlines continuing downstream.

STAGNATION TEMPERATURE. The temperature of a flowing gas at the point where the gas is brought to rest at a solid body.

STALLING. Of an aircraft: the condition in which a *stall* (i.e. the progressive breakdown of the flow over an aerofoil) seriously affects the handling of the aircraft.

STANDARD CELL. An electrochemical cell giving a constant, permanent and reproducible e.m.f. The standard cell now in almost universal use is the *Weston cadmium cell*, which is a primary cell having a positive electrode of mercury and a negative electrode of mercury and 10% cadmium, with saturated cadmium sulphate solution as electrolyte and mercury sulphate as depolarizer. It gives an e.m.f. of 1·01864 absolute volts at 20°C. (Note: 1 volt is equal to 10^8 absolute volts.)

STANDARD DEVIATION. A measure of the scatter of a series of observations or statistical data. It is the square root of the mean of the squares of the deviations of the observations or numerical data from their arithmetic mean. The square of the standard deviation is known as the *variance.* See also: *Error, standard.*

STANDARD ELECTRODE. A reference electrode giving a constant electric potential under specified conditions. The most common is the calomel electrode for which the voltage, relative to that of the standard (nominally null) hydrogen electrode, and as a function of composition and temperature, is known with high precision. The calomel electrode in normal use has a relative potential of 0·2676 V at 25°C. See also: *Electrode, calomel.*

STANDING WAVE METER. An instrument for measuring the standing wave ratio in a waveguide or transmission line.

STANDING WAVE RATIO. In a waveguide or transmission line: a measure of the amplitude of standing waves. It is defined as the ratio of the maximum to minimum voltage amplitude measured along the path of the waves. In some countries the reciprocal of this ratio is used.

STANDING WAVES. The waves in a "frozen" wave pattern which is formed by the interference of two progressive wave systems. Standing waves may occur in waveguides and transmission lines, may be optical or acoustic, may occur in streams of water, or generally wherever periodic waves are produced or scattered under suitable conditions. Standing waves are also known as *stationary waves.*

STANTON NUMBER. A dimensionless number characterizing the convection of heat in a fluid, liquid or gas. It is the inverse of the Prandtl number.

STAR. (1) In astronomy: a self-luminous celestial body. It derives its energy from thermonuclear transformations. See also under *Star. Stars. Stellar.* See: *Carbon cycle. Proton–proton chain.* (2) In a cloud or bubble chamber or in a nuclear emulsion: See: *Track of ionizing particle.*

STAR, ANOMALY OF. The position of one member of a binary system with respect to the other, expressed as an angular value.

STAR ASSOCIATION. A loose grouping of stars similar in some respects to an open cluster but occupying a considerably greater volume of space and more diluted by stars of the general field. Two types of association are common: the *O-associations* in which the most conspicuous objects are certain types of supergiant star, and the *T-associations* which contain T-Tauri type variable stars. See also: *Star cluster.*

STAR, BINARY, ASTROMETRIC. A binary star whose existence is deduced from variations in proper motion.

STAR, BINARY, COMPANION OF. See: *Star, multiple.*

STAR, BINARY, ECLIPSING. A binary star which, because of the small angular separation, cannot be resolved into its two component stars by direct observation but reveals its binary nature by the variation of the total light received from it, the variation arising from the periodic eclipsing of one component by the other.

STAR, BINARY, PRIMARY OF. That member of a binary system which is at a later stage of evolution than the other. It is usually associated with an originally larger mass.

STAR, BINARY, SECONDARY OF. That member of a binary system which is at an earlier stage of evolution than the other. It is usually associated with an originally smaller mass.

STAR, BINARY, SPECTROSCOPIC. A double star for which the separation is too small to permit visual observation, and whose character is revealed by Doppler shifts observed on the spectral lines.

STAR, BINARY: STAR, DOUBLE. A term referring to two close stars which describe an orbit around their common centre of gravity. Pairs of stars which appear to be close but are actually far removed from each other, and therefore show no orbital motion, are known as *optical binaries*.

STAR, BINARY, VISUAL. A binary star that can be recognized visually.

STAR CATALOGUE. A list giving the mean positions (right ascension and declination) of the stars, excluding nutation and aberration, at a given time. The positions at other times are obtainable from a knowledge of the annual variations, secular variations and proper motions.

STAR, CIRCUMPOLAR. A star which completes its diurnal motion about the celestial pole entirely above the observer's horizon.

STAR CLOUD. An aggregation of stars which appears as a cloud but is composed of millions of stars spaced over vast distances. Star clouds form part of the Milky Way. The so-called Magellanic clouds are mis-named as they constitute separate galaxies. See also: *Magellanic clouds*.

STAR CLUSTER. Any obvious condensation in the non-random distribution of stars. The two main types are the *galactic* or *open* clusters, consisting of relatively few stars (up to a few hundreds); and *globular* clusters which consist of many stars (up to some 10^5) which are spherically distributed about a central nucleus, with a concentration which is a maximum at the centre and falls off gradually towards a poorly defined boundary. The galactic clusters are in general located near the galactic plane, while the globular clusters are mostly grouped about the centre of the Galaxy in the direction of the constellation Sagittarius. See also: *Star association*.

STAR DRIFTS. The apparent motions of certain stars with respect to the fixed stars.

STAR, FIXED. One of a number of stars which maintain the same relative positions in the sky and form constant patterns or constellations which have not changed materially in historic time. See also: *Proper motion*.

STAR, GIANT. A star which is brighter than the main sequence stars of the same spectral class. See also: *Hertzsprung–Russell (H–R) diagram. Star, supergiant. Stars, spectral classes of*.

STAR, HIGH-VELOCITY. A member of a category of stars which appear from the Doppler shift of their spectral lines to have a high velocity (more than 100 km/s) relative to the solar system. The apparent velocity arises from the fact that the Sun and its neighbours are moving with high velocity around the galactic centre, a motion which is not shared by the high-velocity stars.

STARK BROADENING. The broadening of the spectral lines from a gas discharge on account of intermolecular fields arising from the presence of ions, dipoles, quadrupoles, etc.

STARK EFFECT. The displacement and splitting of the lines in atomic spectra, and the appearance of new lines, owing to the influence of a transverse electric field. In many respects it resembles the more complicated types of Zeeman effect. See also: *Lo Surdo tube*.

STAR, MULTIPLE. A system consisting of three or more stars which appear to be closely connected. Thus a triple system consists of a combination of a true binary star with a more distant star, known as the *companion* of the double star.

STAR POSITION, APPARENT, ANNUAL VARIATION OF. The annual rate of change, at a given epoch, of the apparent position of a star (in equatorial coordinates) due to luni-solar precession and proper motion. The rate of change of the annual variation itself is known as the *secular variation*. Both annual and secular variations are quoted separately for right ascension and declination.

STARS, DWARF. Small stars of extremely high density and low luminosity. The *white dwarfs* have luminosities much lower than those of the main sequence stars and the *sub-dwarfs* have luminosities which are only slightly lower. The sub-dwarfs lie in a short sequence on the Hertzsprung–Russell diagram, which is parallel to and just below the main sequence. See also: *Stellar evolution. Stellar luminosity. Stellar magnitude*.

STARS, FIELD. Stars that are distributed at random in space and do not belong to any particular star cluster. See also: *Galaxy, field*.

STARS, FUNDAMENTAL. Those stars whose right ascensions have been measured absolutely, i.e. by comparison with the Sun. They are used as reference points on the celestial sphere, from which the right ascensions of other stars may be found differentially.

STAR, SHELL. A star with an inert isothermal core, which consists almost entirely of hydrogen and accounts for about 10% of the star's total mass, round which there exists a thin shell wherein hydrogen is being consumed. The rest of the star consists of a vast envelope not hot enough to support a nuclear reaction. See also: *Stellar evolution*.

STARS, MAIN SEQUENCE OF. A sequence of stars in the Hertzsprung–Russell diagram. It represents those stars for which the luminosity is directly related to the spectral type. See also *Hertzsprung–Russell (H–R) diagram.*

STARS, PULSATING. Stars which are regularly expanding and contracting, and whose apparent brightness (and hence temperature) increases and decreases in sympathy, but with a slight delay of about a quarter of the period.

STARS, SCINTILLATION OF. The "twinkling" of the stars, caused by atmospheric turbulence. See also: *Seeing, astronomical. Twinkling.*

STARS, SPECTRAL CLASSES OF. The classification of stars according to the presence or absence of certain characteristic lines in their spectra. The classification in general use is the *Harvard–Draper sequence* in which the stars are denoted by letters according to their temperature. In descending order of temperature the sequence is as follows: W-O-B-A-F-G-K-M, plus variants of M designated as R, N and S. The temperatures range from up to 100 000 K for the W stars to some 3000 K for the M stars, the R, N and S stars being about 1000 K lower. These classes are often subdivided decimally (e.g. A_2, G_5), giant, supergiant, and dwarf stars of the same class being denoted by the use of the letters g, c, or d respectively preceding the class, e.g. gK_6, dG_5, the Sun being classified as dG_2. In addition the spectra of *gaseous nebulae* are classified as P-type and those of *novae* as Q-type. Refinements of the Harvard–Draper classification are also in use but are not commonly encountered, the best known being the *MKK* or *Yerkes system,* in which the effect of luminosity on stellar spectra is taken into account. See also: *Hertzsprung–Russell (H–R) diagram. Peculiar stars.*

STAR STREAMING. The phenomenon, arising from the rotation of the Galaxy, that the mean directions of motion of the stars lie in two preferential directions or "streams".

STAR, SUPERGIANT. A star which is considerably brighter even than a giant star. The supergiants are the most luminous stars of any spectral type. See also: *Hertzsprung–Russell (H–R) diagram. Star, giant. Stars, spectral classes of.*

STARS, VARIABLE. Stars whose apparent magnitudes vary in a way which is not explicable on the assumption that the stars are in eclipsing systems or in some sort of orbital motion. There are some twenty-six types of variable star which may conveniently be divided into three main classes: the *long-period variables*, with periods of 50–600 days (or more); the *short-period variables*, with periods from a few hours to 50 days; and the *irregular variables*, which seem to have no definite periods. Well over 1000 variable stars are now known. See

also: *Cepheid variables.*

STARS, VARIABLE, NOMENCLATURE OF. The first variable discovered in a constellation is denoted by the letter R, and the succeeding ones by S, T, ..., Z. After Z the two letters RR follow, then RS, RT, ..., RZ, then SS, etc., until ZZ is reached. Following this the letters AA, AB, etc., are assigned, and so on until QZ is reached, then the stars are named V with a number to indicate the order of discovery. Stars giving their names to classes of similar stars include the W Virginis, RV Tauri, SS Cygni, T Tauri, and RR Lyrae stars.

STARS, WOLF–RAYET. Class W stars, with spectra similar to those of novae, in which broad lines predominate. Most Wolf–Rayet stars are components of double stars and their temperatures may approach 100000 K.

STAR, SYMBIOTIC. A rare type of star that displays low-temperature absorption spectra and high-temperature emission spectra at the same time. Such a star is now believed to be a small, hot, blue star surrounded by an extensive variable envelope.

STAR, TANGENTIAL VELOCITY OF. The component of relative velocity at right angles to the line of sight from the Sun.

STAT-. A prefix used to denote CGS electrostatic units. See also: *Units, CGS electrostatic (e.s.u.).*

STATE. (1) Of a system: the condition of the system, identified by all its properties. (2) Of a molecule, atom or sub-atomic particle: the value of a specific property, e.g. an energy level. (3) Of matter in general: the state of aggregation, i.e. solid, liquid, gas, or plasma.

STATIC. See: *Atmospherics, radio.*

STATIC ELECTRICITY. Stationary electric charge produced on an insulator by rubbing, liquid flow, an incident stream of dust or powder and so on, or by the pick up of ions by minute particles. Static electricity may interfere with industrial operations by reason of unwanted attraction or repulsion. It may also give rise to sparks leading to fire in operating theatres or industrial and commercial premises.

STATIC ELECTRICITY, ELIMINATION OF. The dispersal of static charge by the production of ionization in the air between a charged surface and a nearby electrical conductor. The methods adopted usually involve the use of a corona or brush discharge or of a radioactive source.

STATIC MARGIN. Of an aircraft or missile: the longitudinal stability of the aircraft or missile. See also: *Aerodynamic centre.*

STATICS. The study of the behaviour of objects which are in equilibrium under the action of forces or torques, i.e. objects with (at least instantaneously) zero velocity. It treats of the composition and resolution of forces, and such properties as centre of gravity and moment of inertia. See also: *Dynamics. Kinematics. Mechanics.*

STATIC TEMPERATURE. At a point in a flowing medium: the temperature at that point, as measured by an infinitesimally small thermometer at rest relative to the medium and shielded from radiation.

STATIONARY DATA. See: *Random data.*

STATIONARY STATE. One of the discrete energy states in which a quantized particle or system may exist.

STATIONARY TIME, PRINCIPLE OF. See: *Fermat law: Fermat principle.*

STATIONARY WAVES. See: *Standing waves.*

STATISTICAL DISTRIBUTION. Another term for. *Frequency distribution.*

STATISTICAL MECHANICS. The statistical prediction of the macroscopic or bulk properties of matter from a knowledge of the properties of the constituent atoms or molecules and of the forces between them. The term is sometimes restricted to the treatment of systems in which equilibrium has been established. See also: *Boltzmann statistics. Bose–Einstein statistics. Fermi–Dirac statistics. Kinetic theory of gases. Partition function. Reversibility, microscopic. Uncertainty principle: Indeterminacy principle. Virial theorem.*

STATISTICAL WEIGHT. In the statistical investigation of a given quantity: a number assigned to each value or range of values of the quantity, equal to the number of times this value or range of values is found to be observed. In statistical mechanics it is the number of microstates corresponding to a given macrostate.

STATISTICS. The study of observed data in a systematic way to determine the general behaviour of a system, typically from a limited number of measurements or classifications. It involves the study of frequency distributions, the estimation of errors, tests for significance, assessment of probability, and related matters. See also: *Average: Mean* Entries under *Error. Frequency distribution. Probability. Quantum statistics. Significance. Standard deviation. Variance, analysis of.*

STEAM ENGINE. A heat engine in which the working fluid is steam.

STEAM POINT. The temperature at which liquid water and its vapour can exist together in equilibrium at a pressure of 760 mm of mercury. It is defined as $100 \cdot 00°C$ on the international temperature scale. See also: *Ice-point. Temperature scale, international practical.*

STEAM TABLES, INTERNATIONAL. A tabular representation of the properties of steam, in which are listed pressure, volume, entropy, enthalpy and temperature for saturated water (at the boiling point), saturated steam and steam vapour at various degrees of superheat.

STEEL. Steel was originally a malleable alloy of iron and up to about $1 \cdot 7\%$ carbon, but now the term refers to a whole range of alloy steels, some of which have very little carbon content. The alloys are added to improve mechanical properties, to decrease corrosion, to give low thermal expansion, and so on, and the additives include Al, B, Cr, Co, Cu, Pb, S, Se, Zr, Mn, Mo, Ni, Si, Nb, Ta, Ti, W, and V.

STEEPEST DESCENTS, METHOD OF. An approximate method for evaluating integrals of the form

$$I(k) = \int_C e^{kf(z)}g(z)\, dz$$

as k tends to infinity, where $f(z)$ and $g(z)$ are analytic functions of z, and where $f'(z) = 0$ at one point at least, called a *saddle point* or *col*. C is a contour which may be finite or infinite and may, for example, be along the real axis.

STEFAN-BOLTZMANN LAW. States that the radiant energy emitted by a black body is proportional to the fourth power of the absolute temperature. The constant of proportionality is known as the *Stefan constant, Stefan-Boltzmann constant* or *Boltzmann factor,* and is equal to about $5 \cdot 67 \times 10^{-8}$ $Wm^{-2}\ K^{-4}$ (or $5 \cdot 67 \times 10^{-5}\ erg\ cm^{-2}\ s^{-1}\ K^{-4}$). Its experimental value agrees well with that determined from Planck's radiation formula. See: *Appendix III.*

STEINMETZ LAW. An empirical law relating hysteresis loss to magnetic induction. It takes the form $W = \eta B^{1 \cdot 6}$, where W is the energy loss per unit volume and per cycle, B is the maximum magnetic induction attained during the cycle, and η is the *Steinmetz coefficient,* which is a constant for any one material. The power of $1 \cdot 6$ is, however, slightly variable from material to material and the law is not always reliable.

STELLARATOR. An early device concerned with the possible production of a controlled thermonuclear reaction, in which a hot plasma is contained within a strong toroidal magnetic field. The heat is generated solely by current flowing in an external conductor. See also: *Thermonuclear reaction, controlled.*

STELLAR BRIGHTNESS. See: *Stellar luminosity. Stellar magnitude.*

STELLAR DENSITY FUNCTION. Of a given region in space: the total number of stars per unit volume.

STELLAR EVOLUTION. A term denoting the life cycle of the stars from birth to death. The birth appears to be the coming together of a mass of gas and dust which shrinks in size under its own gravitation and radiates part of the gravitational energy, thereby becoming a luminous star. The centre becomes progressively hotter until thermonuclear reactions are possible, which then provide a stable (some 10^7–10^{10} years) source of energy for the star. The total luminosity of a star is a measure of the rate of consumption of nuclear fuel. When the fuel is completely exhausted the star shines entirely by virtue of gravitational contraction and the density becomes very high (10^5 to 10^8 g cm^{-3}). This process of contraction can end in the formation of a *white dwarf, neutron star* or *black hole,* depending on the original mass of the star. See also: *Black hole. Carbon cycle. Hertzsprung-Russell (H-R) diagram. Neutron star. Proton-proton chain.*

STELLAR LUMINOSITY. The total energy flux emitted by the surface of a star. It may be expressed not only as an energy flux but in units of the Sun's luminosity or as bolometric magnitude. See also: *Stellar magnitude.*

STELLAR MAGNITUDE. Stellar magnitude describes the brightness of a star on a logarithmic scale. The best known are the *apparent, absolute* and *bolometric* magnitudes. See also: The following entries.

STELLAR MAGNITUDE, ABSOLUTE. A measure of the luminous intensity of a star. It gives the energy flux density in a specified range of wavelengths at a standard distance of 10 parsec (3×10^{14} km) from the star in perfectly transparent space, and is given by $m_\lambda - 5 \log (r/10) - \alpha_\lambda(r)$, where m_λ is the apparent magnitude, r is the distance of the star in parsec, and $\alpha_\lambda(r)$ is the absorption along the line of sight in interstellar space.

STELLAR MAGNITUDE, APPARENT. A measure of the brightness of a star relative to a group of standard stars. It is defined as $-2 \cdot 5 \log B + k$, where B is the energy flux density due to the star, integrated over all appropriate wavelengths, at the top of the Earth's atmosphere, and k is a zero-point constant chosen by specifying the apparent magnitude for a number of standard stars. The apparent magnitude thus decreases numerically with increasing brightness. Where the brightness is measured by visual photometry the apparent magnitude is known as the *visual magnitude,* and is the magnitude usually meant when reference is made simply to the "magnitude of a star". Where it is measured photographically the corresponding apparent magnitude is termed the *photographic magnitude.* See also: *Colour index. UBV photometry.*

STELLAR MAGNITUDE, BOLOMETRIC. A measure of the total energy radiated by a star at all wavelengths. It is defined as $-2 \cdot 5 \log L +$ constant, where L is the total energy radiated per unit time and the constant is such that the bolometric magnitude equals the visual magnitude for certain stars of spectral class F. See also: *Bolometric scale.*

STELLAR MAGNITUDES, PHOTOGRAPHIC AND VISUAL. See: *Stellar magnitude, apparent.*

STELLAR POPULATIONS I AND II. Divisions into which stars can be classified according to their luminosity and spectral class. Population I comprises those stars which conform to the classical Herzsprung–Russell diagram and is typical of the region near the Sun and in the outer part of the Galaxy. Population II comprises those stars which do not conform to the classical Herzsprung–Russell diagram and is typical of the central regions of the Galaxy and of other spiral galaxies, and of globular clusters.

STELLAR SPECTRA. See: *Stars, spectral classes of.*

STELLAR TEMPERATURES. See: *Astrophysical temperatures.*

STEM. See: *Electron microscope, scanning transmission (STEM).*

STEPPED LEADER STROKE. In lightning: the initial discharge which determines the subsequent track of a lightning flash. It is characterized by its development in the form of a number of definite steps. See also: *Lightning.*

STEPPING MOTOR. An electric or electro-hydraulic motor which is driven by a train of pulses, each pulse of which rotates the motor through a given angle. Such motors, which find an important application in the numerical control of machine tools, may be employed to convert electric or electronic signals into mechanical motion, and may be made to obey instructions in numerical form similar to the instructions fed into digital computers.

STEP WEDGE. A block of material in the form of a series of steps used to judge the quality of a radiograph and in the radiographic calibration of an X-ray or other generator.

STERADIAN. One of two supplementary SI units of angular measurement, the other being the radian. It is defined as "the solid angle which, having its vertex at the centre of a sphere, cuts off an area of the surface of the sphere equal to that of a square with sides of length equal to the radius of the sphere". It is thus the solid angle subtended at the centre of a sphere of unit radius by a circle of unit area on the surface of the sphere. See: *Appendix II.*

STEREOCHEMISTRY. The study of the spatial arrangements of atoms in molecules and their relationship to chemical behaviour.

STEREOGRAPHIC NET. A chart which gives stereographically projected great and small circles of various radii, used in making measurements on a stereographic projection or in producing such a projection. It is also known as a *Wulff net.*

STEREOGRAPHIC PROJECTION. A plane projection used for displaying the positions of the poles of a crystal. These poles are projected on to the equatorial plane of the reference sphere, via lines joining them with the "south" pole for poles in the upper hemisphere, and with the "north" pole for those in the lower hemisphere. See also: *Gnomonic projection. Orthographic projection.*

STEREOISOMERISM. See: *Isomers.*

STEREOPHONY. (1) The localization of sounds by a person with normal binaural hearing. (2) The reproduction of sound through two or more channels to give an impression of the spatial distribution of the original sound sources.

STEREOSCOPY. The production of a sensation of depth by a suitable combination of two views of an object taken from slightly different positions. See also: *Anaglyph.*

STERIC HINDRANCE. The hindrance or prevention of certain chemical reactions or processes by the spatial arrangement and finite size of the atoms involved.

STERN–GERLACH EXPERIMENT. An experiment which gave the first demonstration of the quantization of angular momentum of the electron, by showing that an atomic beam when passed through an inhomogeneous magnetic field was not broadened into a band but was divided into separate components.

STICKING POTENTIAL. Of a cathode ray tube with a non-conducting screen: the limiting screen-to-cathode potential above which the number of secondary electrons emitted from the screen would be less than the number of primary electrons incident upon it.

STICKING PROBABILITY. In a nuclear reaction: the probability that a particle which has reached the surface of an atomic nucleus will be absorbed by it to form a compound nucleus.

STIFFNESS. In an elastic system: the steady force required to produce unit displacement.

STIGMATIC MOUNTING. Of a spectrograph: a mounting in which a point in the source gives rise to an equivalent point in the image.

STIGMATOR. A device for correcting the deflecting field of an electron microscope, which superposes on the field a second adjustable and non-symmetrical field.

STILB. A unit of luminance equal to 1 cd/cm². See also: *Luminance.*

STILES–CRAWFORD EFFECT. The reduction in the sensation of brightness produced by light entering the eye near the margin of the pupil, as compared with that produced by light entering through the centre of the pupil, other things being equal.

STIRLING CYCLE. A cycle of operations in which a gas is alternately compressed at ambient temperature and expanded at low temperature. It forms the basis of certain gas refrigerating machines and prime movers.

STIRLING ENGINE. An engine based on the Stirling cycle. Recent improvements suggest that such an engine might eventually be used instead of internal combustion engines. See also: *Fluidyne engine. Stirling cycle.*

STIRLING FORMULA. A formula for calculating $n!$ for large values of n. It states that, as n tends to infinity, $n!$ tends to $\sqrt{2\pi n}(n/e)^n$. It is often approximated by $\ln n! = n \ln n - n$. It is also stated in the form that $n!$ lies between $\sqrt{2\pi n}(n/e)^n$ and $\sqrt{2\pi n}(n/e)^n \left(1 + \dfrac{1}{12n - 1}\right)$.

STOCHASTIC PROCESS. A process or phenomenon for which the probability of an event occurring at one trial depends on the results of previous trials, as when an apparatus capable of small oscillations is disturbed at irregular intervals.

STOICHIOMETRIC SUBSTANCE. A substance having the exact "correct" composition, e.g. a compound containing the exact proportions of elements required by its formula.

STOKES. The CGS unit of kinematic viscosity. See also: *Kinematic viscosity. Appendix II.*

STOKES–EINSTEIN EQUATION. For the viscosity of a liquid containing a solid suspension: states that $\eta = (1 + KV)\eta_0$, where η is the viscosity of the liquid–solid mixture, η_0 is the viscosity of the liquid alone, V is the volume of suspended solids in unit volume of mixture, and K is a constant. For rigid spherical particles in a "dilute" suspension K is about 2·5.

STOKES LAW. (1) For the *viscous drag of a sphere:* states, in its simplest form, that the drag of a spherical body of radius r, passing at a uniform speed v through a viscous medium of viscosity η, is $6\pi v \eta r$. In this connection the drag may be regarded as the force required to maintain the uniform speed. (2) Of *fluorescence:* states that the wavelength of fluorescence radiation is greater than that of the radiation that excites the fluorescence. It does not always hold. See also: *Anti-Stokes emission. Raman effect.*

STOKES LINES. Excited spectral lines which correspond to a loss of energy, i.e. for which Stokes law holds. See also: *Anti-Stokes emission. Raman effect.*

STOKES THEOREM. States that the surface integral of the curl of a vector function is equal to the line integral of the function around a closed curve bounding the surface. It may be written

$$\int_s (\mathrm{curl}\, \mathbf{A} \cdot \mathbf{n})\, dS = \oint (\mathbf{A} \cdot d\mathbf{s}),$$

where $\mathbf{A}$ is the vector function concerned, $\mathbf{n}$ is a unit vector, dS an element of surface, and $d\mathbf{s}$ a vector element of length.

STOP BAND. The frequency range which is highly attenuated by an optical or electric filter, frequencies outside this range being freely transmitted. See also: *Pass band.*

STOP NUMBER: F NUMBER. Of a lens or lens system: the ratio of the focal length to the diameter of the entrance pupil. Since the amount of light passing through the system decreases as the square of this diameter the conventional *f*-numbers used in photography increase by steps of $\sqrt{2}$ i.e. 1, 1·4, 2 2·28, 4, 5·6, etc. so that each rise in the series requires a doubling of the exposure time, other things being equal.

STOPPING NUMBER. See: *Bethe–Bloch formula.*

STOPPING POWER. A measure of the effect of a substance upon the kinetic energy of a charged particle passing through it. It may be expressed as the energy loss per unit distance, i.e. the *linear stopping power*, given by $-\dfrac{dE}{dx}$, where E is the energy of the particle; per unit surface density, i.e. the *mass stopping power*, given by $-\dfrac{1}{\varrho}\dfrac{dE}{dx}$, where ϱ is the density; per atom per unit area normal to the motion of the particle, i.e. the *atomic stopping power*, given by $-\dfrac{1}{n_a}\dfrac{dE}{dx}$, where n_a is the number of atoms per unit volume; per molecule, i.e. the *molecular stopping power*, which is very nearly if not exactly equal to the sum of the atomic stopping powers of the constituent atoms; or per electron per unit area normal to the motion of the particle, i.e. the *electronic stopping power*, given by $-\dfrac{1}{n_e}\dfrac{dE}{dx}$, where n_e is the number of electrons per unit volume. The reduction in the stopping power of a dense material for relativistic particles, arising from the reduction in the effective electric field of the particles by the polarization of adjacent atoms, is known as the *density effect.* See also: *Bethe–Bloch formula.*

STOPPING POWER, RELATIVE. The stopping power of a material relative to that of a standard substance.

STORAGE CELL. See: *Accumulator.*

STORAGE RING. A synchrotron-like magnet ring structure, with a stationary magnetic field, wherein particles of suitable energy may be accumulated. Such rings are used for experiments with colliding beams in which one single ring is filled with two counter-rotating intersecting beams of opposite charge, or (for identically charged particles) two tangent or interlaced rings are employed. Storage rings permit particle collisions at energies that cannot be reached by other means.

STORAGE TUBE. An electronic tube in which a signal conveying information can be stored by conversion into an electrostatic charge on a storage electrode. Such tubes are essential in television and radar.

STORM. Any violent atmospheric commotion e.g. thunderstorm, rainstorm, sandstorm, squall. The term is also extended to cover other more or less violent disturbances, such as a magnetic storm.

STÖRMER CONE. See: *Cosmic-rays, allowed and forbidden regions for.*

STORM SURGE: STORM TIDE. A "tidal" wave which is not strictly tidal in that it is not caused by external tide-generating periodic forces, but by the effects of wind, and variations in atmospheric pressure. It is also known as a *flood wave.* See: *Tsunami. Waves, episodic.*

STRAGGLING. Of charged particles: the fluctuation of a property such as range or energy, caused by the random character of the energy losses suffered when the particles pass through matter.

STRAIGHTNESS. In metrology: one of the two fundamental qualities in precision engineering, the other being flatness. Straightness is commonly checked by the use of a *straight edge* which is a beam of steel or cast iron having a thin straight edge which has itself been checked, e.g. against a surface plate. See also: *Flatness. Surface plate.*

STRAIN. The fractional deformation produced in a body by the application of stress. For an elastic body it is related to the stress by the appropriate modulus of elasticity. See also: *Elastic modulus. Hooke law. Stress.*

STRAIN GAUGE. An instrument for measuring the strain at the surface of a solid. It transforms small changes of length into proportional changes in some physical characteristic such as electrical resistance, inductance, capacitance, vibrational frequency, photoelasticity, piezo-electric properties, magneto strictive properties, and semi-conducting properties. The simplest and perhaps the most widely used is the resistance gauge in which the resistance of a small length of wire, attached to the surface, measures changes in length at the surface. The most sensitive of these is the semiconductor strain gauge. See also: *Moiré fringes.*

STRAIN HARDENING. The increase in hardness of a metal on account of cold work. See also: *Work hardening.*

STRANGENESS. A quantum number introduced

to take account of the behaviour of some elementary particles with paradoxically long lifetimes—the strange particles. It is conserved in electromagnetic and strong interactions but not in weak interactions. See also: *Quantum number, internal.*

STRANGE PARTICLES. Elementary particles which possess a strangeness different from zero, and have paradoxically long lifetimes.

STRATOPAUSE. See: *Stratosphere.*

STRATOSPHERE. That layer of the atmosphere which lies above the troposphere. It lies between about 11–17 km above sea-level to about 55 km. It is approximately isothermal and is free of weather phenomena. Its upper boundary is known as the *stratopause.* See also: *Atmosphere.*

STREAMER CHAMBER. A development of the spark chamber in which the trajectories of charged particles are located by streamers formed by their passage through a chamber consisting essentially of two parallel conducting electrodes with a gas container in between. See also: *Bubble chamber. Cloud chamber. Counter, spark. Spark chamber.*

STREAM FUNCTION. A scalar function of position used to describe the steady two-dimensional flow of an incompressible fluid.

STREAMING. The production of unidirectional flow in a medium arising from the presence of sound waves.

STREAMING: CHANNELLING. The increased transmission of electromagnetic or particulate radiation through a medium resulting from the presence of extended voids or other regions of low attenuation.

STREAMING POTENTIAL. The production of an electrical potential when a liquid is forced through a porous diaphragm or other permeable solid. It is the converse of electro-osmosis.

STREAMLINE. In fluid flow: a line such that the tangent to it at any point is in the direction of flow at that point.

STREAMLINE FLOW. Another name for laminar flow, i.e. flow in which there is no macroscopic mixing between adjacent layers.

STRENGTH. Of a solid: the stress at which the solid will yield or fracture. The strength is usually qualified by a word denoting the type of stress involved, as, for example, shear strength, tensile strength, compressive strength, impact strength.

STRESS. The force applied to a body per unit area. For an elastic body it is related to the strain by the appropriate modulus of elasticity. See also: *Elastic modulus. Hooke law. Strain.*

STRESS ANALYSIS. (1) The mathematical analysis of the stress distributions in a body. (2) The experimental determination of stress distribution by the use of polarized light. See also: *Photoelasticity.*

STRESS CONCENTRATION. The increase in stress arising from the presence of a constriction. The effect is used in the notched-bar test of impact strength. See also: *Impact tests.*

STRESS CORROSION. Corrosion which is increased by the presence of stress.

STRESS FUNCTIONS. Functions which define uniquely the field of force as a function of position. The stresses are obtained by the differentiation of these functions.

STRESS, INTERNAL. In general the stress across any given plane within a solid. The term is, however, often restricted to the residual stresses present in a body on which no external forces act.

STRESS–STRAIN CURVE. A curve which represents the behaviour of a body under the action of an increasing stress, usually tensile.

STRESS TENSOR. A tensor whose components are the stresses exerted across surfaces normal to the directions of variation of a single coordinate.

STRETCHER STRAINS: LÜDER BANDS. Surface markings observed on mild steel and certain non-ferrous metals when strained beyond the elastic limit.

STRING, VIBRATION OF. Normally refers to transverse vibration, for which the frequencies of vibration, for a string of negligible stiffness stretched between two rigid supports, are given by $(j/2L)\sqrt{(T/m)}$, where $j = 1, 2, 3$, etc., for the fundamental and overtones, T is the tension in the string (i.e. the force applied to keep it stretched), m is the mass per unit length of the string, and L is the length of the string. This is known as the *Mersenne law.*

STRIPPING EMULSION. A photographic emulsion, for use in autoradiography, which can be removed from its base and placed in contact with a specimen containing radioactive material.

STRIPPING REACTION. A nuclear reaction, typified by the Oppenheimer–Phillips process, in which a compound particle striking or passing near an atomic nucleus loses one of its constituents to the nucleus without the formation of a compound nucleus.

STROBOSCOPE. An instrument which allows a periodically moving object to be observed only at predetermined intervals (e.g. by illuminating the object by short flashes of light at regular intervals or by permitting observation of the object only through a rotating slotted disk) and so enables the object to appear stationary or to be moving slowly, according to the relationship between the intervals of observation and of the periodic motion. Strobo-scopic methods may be used to render possible the examination or measurement of phenomena which might otherwise not be observable.

STRONG FOCUSING: ALTERNATING GRADI-ENT FOCUSING. A system for focusing charged particles in which the particles pass alternately through non-uniform electric or magnetic fields having gradients of opposite sign. The system was originally developed for proton synchrotrons but has also found other applications.

STRONG INTERACTIONS. The dominant inter-actions between elementary particles. They are re-sponsible for nuclear forces, the forces between mesons and nucleons, the production of mesons in particle collisions, and similar phenomena; and are believed to conserve parity and strangeness, and to be charge-independent. Those particles which have strong interaction properties are known as *hadrons*. See also: *Hadrons. Interactions between elementary particles.*

STRUCTURAL FORMULA. A schematic ex-pression of the structure of a substance in which the interatomic bonds are represented by straight lines. See also: *Molecular diagram. Tautomerism.*

STRUCTURE AMPLITUDE. Of a diffracting crystal for a given direction: the amplitude, mea-sured at unit distance, of the wave scattered in any chosen direction by a unit cell, for an incident plane wave of unit amplitude. The incident wave may be of X-rays, electrons, or neutrons, and the amplitude is expressed in terms of electron density, electric potential, or nuclear density respectively. The structure amplitude, being derived from the measured intensity, takes no account of the phase of the scattered wave. See also: *Structure factor.*

STRUCTURE FACTOR. The term commonly used instead of structure amplitude when the phase of the scattered wave is known and can be in-corporated in the expression for that amplitude.

STRUCTURE-SENSITIVE PROPERTIES. Of a material: those properties which are affected by imperfections and impurities, such as the strength and ductility of metals or the electrical properties of semiconductors. Such properties are to be contrasted with the *intrinsic properties* of a material, such as specific heat or heat of sublimation.

STURM–LIOUVILLE EQUATION. A differen-tial equation used in the solution of eigenvalue problems in mathematical physics. It is of the form

$$\frac{d}{dx}\left[p(x)\frac{dy}{dx}\right] + [g(x) + \lambda w(x)]\,y = 0,$$

where λ is the eigenvalue, chosen so that the solu-tions y, the eigenfunctions, satisfy given boundary conditions, and $w(x)$ is a weighting function.

SUB-CRITICAL. In nuclear technology: refers to the condition of a neutron-chain-reacting medium for which the effective multiplication factor is less than unity. See also: *Chain reaction, neutron.*

SUBHARMONIC. A sinusoidal quantity having a frequency which is an integral sub-multiple of the fundamental frequency of a periodic quantity by which it is excited. See also: *Harmonic.*

SUBLIMATION. The vaporization of a solid without the intermediate formation of a liquid.

SUBMICROSCOPIC PARTICLES, POLARIZA-TION OF. See: *Polarization.*

SUB-MILLIMETRE WAVES. Electromagnetic ra-diation in the band between the millimetre waves of radio and the waves of the far infrared. At the longer wavelengths guided-wave methods are usual, and at the shorter wavelengths free-wave "optical" methods are more suitable. The range of sub-millimetre waves extends from about 200 μm to about 1 mm. A widely used source of sub-millimetre radiation is the medium pressure mer-cury lamp encapsulated in quartz, and a commonly used detector is the Golay cell. See also: *Micro-waves.*

SUBSONIC FLOW. See: *Compressible flow.*

SUBSTITUTION WEIGHING. Another name for *counterpoise weighing.*

SUBTRACTIVE COLOUR PROCESS. See: *Photography, colour.*

SUCKSMITH RING BALANCE. A magnetic balance of moderate sensitivity suitable for the measurement of small forces over a wide temper-ature range, and used mostly for measurements of paramagnetic susceptibility. The specimen is rigidly suspended from a phosphor-bronze ring carrying two mirrors which convert a small deflection of the specimen into a large deflection of an incident beam of light.

SUMMATION TONES. See: *Combination tones.*

SUN. The central body of the solar system, con-taining 99·9% of its mass. It is an incandescent, gaseous sphere in which the commonest element is hydrogen, and rotates with a sidereal period of about 25·3 days. It is a star of spectral type G and absolute magnitude 5. Its diameter is about 1·4 $\times$ 10^6 km and its mass about 2 $\times$ 10^{30} kg (about 333 000 times that of the Earth). The effective *solar* *temperature,* i.e. the temperature of a spherical black body of the same diameter which would emit the same amount of radiant energy, is about 6000 K. See also: entries under *Solar.*

SUNSHINE RECORDER. An instrument for recording the duration of bright sunshine. The most common type is that of *Campbell-Stokes* which consists of a glass sphere that focuses the Sun's image on to a prepared card on which the hours are marked, the duration of the sunshine being determined from the length of the burnt track on the card.

SUNSPOT. A disturbance of the Sun's surface having a relatively dark circular or elliptical centre (the *umbra*) surrounded by a less dark area (the *penumbra*). Sunspots tend to occur in groups, are relatively short-lived, and occur in cycles of 11 or 12 years. They are restricted to regions of 10–40° north or south latitude, sinking from higher to lower latitudes during a sunspot cycle. Sunspots have intense magnetic fields and are associated with magnetic storms on the Earth.

SUNSPOT NUMBER. A number devised by Wolf which provides an index of sunspot activity. It is equal to $k(10g + f)$, where g is the number of groups of sunspots present, f is the total number of spots, and k is a constant depending on the instrument used for observation.

SUPERALLOY. A term, mainly used in the USA to denote one of a series of alloys based on Fe, Co, Ni, or a mixture of these elements, which are of outstanding strength in the temperature range 600–1000°C.

SUPERANTIFERROMAGNETISM. The exhibition of abnormally high magnetic susceptibility by a finely divided antiferromagnetic solid, an effect which is particularly noticeable at low temperatures.

SUPERCIRCULATION. The additional circulation round an aerofoil which occurs when air or gas is blown along the upper surface at a rate greater than that needed to suppress boundary-layer separation. Supercirculation provides an appreciable additional lift and is used to meet aircraft take-off and landing requirements.

SUPERCONDUCTIVITY. The disappearance or near disappearance of the electrical resistance of certain metals, alloys, and compounds at temperatures below a *transition temperature* (usually a few degrees above absolute zero) which is characteristic of the substance concerned and is known as the *critical temperature* for that substance. One of the main applications of superconductivity is to the production of electromagnets having a very high magnetic field and hence to superconducting motors and machines. See also: *Josephson effect.*

SUPERCONDUCTIVITY, CRITICAL FIELD IN. For a superconductor at a given temperature: that value of the magnetic field above which superconductivity disappears. At a temperature T it is proportional to $1 - (T/T_c)^2$, where T_c is the critical temperature.

SUPERCONDUCTIVITY, LONDON EQUATION FOR. States that $[(4\pi\lambda^2)/c]$ curl $\mathbf{J} + \mathbf{H} = 0$, where $\mathbf{J}$ is the current density and $\mathbf{H}$ the magnetic field at a given point in a superconductor, c being the speed of light. The quantity λ is the effective *penetration depth* of the field into the superconductor, below which $\mathbf{H}$ is zero or negligible. It is about 10^{-5} cm.

SUPERCONDUCTIVITY, TWO-FLUID MODEL FOR. A model according to which the conducting electrons belong to two interpenetrating assemblies, the first of which is characterized by normal electrical resistance, while the second has zero entropy and zero resistance, the proportion of normal electrons decreasing with temperature. The model is analogous to, but not identical with, the two-fluid model for helium II.

SUPERCONDUCTOR. A substance exhibiting superconductivity. Two types of superconductor are recognized: *type I superconductors*, for which flux penetration occurs over a very narrow range of magnetic field; and *type II superconductors*, for which it occurs over a comparatively wide range.

SUPERCONDUCTOR, INTERMEDIATE STATE OF. The state of the superconductor when it is in an external magnetic field which approaches the critical field. It is characterized by the existence of domains of alternately normal and superconducting regions.

SUPERCONDUCTOR, PENETRATION DEPTH IN. See: *Superconductivity, London equation for.*

SUPERCOOLING. Of a liquid: the cooling of the liquid below its freezing point without solidification.

SUPERCRITICAL. In nuclear technology: refers to the condition of a neutron-chain-reacting medium for which the effective multiplication factor is greater than unity. See also: *Chain reaction, neutron.*

SUPEREXCHANGE. An exchange interaction between magnetic ions in an antiferromagnetic material, which involves the direct participation of an intervening anion, typically oxygen.

SUPERFLUIDITY. A condition in which apparently frictionless flow is exhibited. The term superfluid is applied (a) to a material in the superconducting state, where it refers to the frictionless flow of electrons; and (b) to helium II, where it refers to fluid flow as usually understood. See also: *Helium II, two-fluid model for. Superconductivity, two-fluid model for.*

SUPERHEATER. A device in which a saturated vapour is heated from an external source. It is normally included as an integral part of the steam generator in a steam power plant.

SUPERHEATING. Of a liquid: the heating of the liquid above its boiling point without boiling taking place. The condition may arise if the liquid is kept perfectly quiet and especially if covered with a film of oil. It may eventually boil with explosive violence and the true boiling point is then operative. See also: *Superheater.*

SUPERLATTICE. A new space lattice which may be produced (as in alloy systems) in a structure when the original space-lattice points cease to be identical. See also: *Order–disorder transformation.*

SUPERLEAK. A leak of liquid helium II through small holes impassable for normal liquids. It is also known as *Lambda leak.*

SUPERMULTIPLET. A spectral multiplet consisting of more than three lines.

SUPERNOVA. An extremely bright star which is fundamentally different from an ordinary nova in that its appearance is the result of a cosmic explosion in which the star is destroyed. The total range of a supernova outburst may exceed nineteen magnitudes, the average magnitude of a supernova being −14 to −16. Three supernova have been observed in the Galaxy, the remnants of only one of which are still visible, constituting the *Crab Nebula.* This was the first of the three. It was observed in China in A.D. 1054 and lasted as a supernova for about 2 years. Otherwise supernovae are only observable as relatively faint objects in distant galaxies. See also: *Nova.*

SUPERPARAMAGNETISM: COLLECTIVE PARAMAGNETISM. The phenomenon according to which the magnetization of assemblies of ferromagnetic particles so small that each particle consists of only one magnetic domain, in its relation to field and temperature, is comparable with that of paramagnetic atoms or molecules, except that the magnetic moment per particle may be several orders of magnitude larger. See also: *Superantiferromagnetism.*

SUPERPLASTICITY. The property of certain cooled eutectoids whereby they exhibit an extremely large and uniform elongation under tensile stress. These alloys are characterized by a very small grain size, and it is thought that the extreme plasticity arises mainly from grain boundary sliding and not appreciably from dislocation movements, i.e. the mechanism is the opposite of that obtaining in creep.

SUPERPOSITION PRINCIPLE. States that, if a physical system is acted upon by a number of independent influences, the resultant effect is the sum (vector or algebraic, whichever is appropriate) of the individual influences. The principle applies only to linear systems, i.e. systems whose behaviour can be expressed by linear differential equations.

SUPERSONIC FLOW. See: *Compressible flow.*

SURFACE ACTIVITY. The alteration of the surface tension of a liquid by a solute. See also: *Rehbinder effect.*

SURFACE DEFECT. A crystal defect such as a large-angle boundary between crystals in a polycrystalline metal, a twin boundary, or a *stacking fault* (i.e. a mistake in the stacking sequence of atomic layers which may lead, for example, to regions of face-centred cubic packing in a hexagonal close-packed structure or vice versa), which may be detected at the crystal surface.

SURFACE DENSITY. The amount of some physical quantity distributed over unit area of a surface. Two well known examples are mass per unit area and electric charge per unit area.

SURFACE ENERGY. The energy per unit area of an exposed surface. The total surface energy is in general greater than the surface tension, which is the *free energy* per unit area. See also: *Free energy.*

SURFACE FINISH. The condition of the surface of a manufactured article. It may depart from the ideal smooth surface by the presence of (a) *Roughness:* a primary pattern of hills and valleys resulting from the cutting, abrasive or other action of the process involved; (b) *Waviness:* a pattern resulting from such things as the vibration between tool and work; and (c) *Error of form:* a deviation of the whole from its intended shape.

SURFACE FRICTION: SKIN FRICTION. The tangential force exerted on the surface of a body by a fluid with which it is in contact, when body and fluid are in relative motion. It is sometimes also known as *surface traction.* It is a component of drag in aircraft and is responsible for aerodynamic heating in high-speed flight.

SURFACE ORIENTATION. In a liquid: the assumption of a non-random arrangement by surface molecules that are not spherically symmetrical, e.g. the tendency in polar liquids for the molecules (which possess electric dipole moments) to be preferentially arranged so as not to lie in the surface.

SURFACE PLATE. A plate of cast iron the top surface of which has been machined or scraped to a high degree of flatness. It is used as a datum surface for accurate work. A more modern version is made of black granite.

SURFACE POTENTIAL, ELECTRICAL. The potential difference between opposite surfaces of a dielectric when it is introduced into an electric field. The potentials arise from the formation of electric dipoles in the dielectric, a negative charge being induced on one surface and a positive charge on the other.

SURFACE PRESSURE. The reduction in the surface tension of a liquid produced by the presence of a surface film.

SURFACE TENSION. The free energy of a surface per unit area. It is the work necessary at constant temperature to increase the surface area by unit area.

SURFACE TENSION, DYNAMIC. The surface tension of a liquid when the surface is continually renewed, e.g. when a jet of fluid vibrates or when capillary waves spread over the surface.

SURFACE TRACTION. Another name for surface friction.

SURFACE WAVE. (1) Elastic: any wave which propagates on the surface of an elastic medium. (2) Electromagnetic: a wave propagated parallel to the Earth's surface. (3) Distortional: a wave propagated on the free surface separating two fluid phases (usually a liquid and a gas or vapour of low density).

SURFACE WAVE ACOUSTICS. The processing of electronic signals as elastic waves in minute strips of solid, which are complementary to the solid-state circuits in wide use in electronics. One important use is for delay lines, since these elastic waves propagate at 10^{-5} the speed of light.

SURVEYING. (1) The determination of the position of objects and features on the Earth's surface and their representation on a suitable scale upon a map, plan or drawing. See also: *Triangulation*. (2) The determination of the magnitude and direction of magnetic elements (e.g. declination and dip) on the Earth's surface and their representation as in (1).

SURVEYING TAPES AND WIRES. Tapes and wires of accurately known length, for use in the measurement of distance with varying degrees of precision. See also: *Geodimeter. Mekometer. Tellurometer.*

SURVEY, ORIGIN OF. A point, of which there must be only one in every independent survey, at which the geodetic latitude and longitude, and the separation of the geoid and spheroid, are given arbitrary values. The last may be zero, and the others equal to values astronomically observed.

SUSCEPTANCE. The imaginary part of the admittance. If the resistance is negligible the susceptance is the reciprocal of the reactance.

SUSCEPTIBILITY. (1) *Electric:* the ratio of the electric dipole moment per unit volume to the electric field strength. (2) *Magnetic:* the ratio of the magnetic moment per unit volume to the magnetic field strength. It is positive for a paramagnetic material and negative for a diamagnetic one.

SUSCEPTIBILITY, ADIABATIC. The magnetic susceptibility at constant entropy.

SUSCEPTIBILITY, MASS. The magnetic susceptibility per unit mass.

SUSPENSION. A dispersion of relatively coarse particles in a liquid. The particles, unlike colloidal particles, undergo sedimentation under the action of gravity.

SVEDBERG. A unit of sedimentation rate, used in centrifuge technique. It corresponds to a rate of sedimentation of 10^{-13} cm s^{-1} under an accelerating force of 1 dyne, i.e. of 10^{-5} N.

SWAGING. A method of reducing the diameter of a metal rod, or sometimes a tube, by driving it between two grooved and tapered dies, by a series of blows, the rod or tube being rotated about its axis during the process.

SWAN BANDS. Of carbon: a triplet spectral band system of molecular carbon, C_2, of wavelengths between 6700 Å and 4370 Å (i.e. 0·67 and 0·437 μm). The bands form a marked feature of cometary spectra, usually being most pronounced in the head. See also: *Comet spectra.*

SWEEP CIRCUIT: TIME-BASE CIRCUIT. A circuit which produces either a voltage or current waveform which increases linearly with time to a maximum (the *sweep*), followed by a very rapid decrease to the initial value (the *flyback*). Such circuits may be "one-shot" or repetitive.

SYMMETRIC WAVE FUNCTION. See: *Bose Einstein statistics.*

SYMMETRY. The property, characteristic of the elements of certain mathematical, atomic, molecular or nuclear systems, of being mutually related by one or more of a specified set of operations. See also: terms beginning *Crystal symmetry.*

SYMMETRY GROUP. See: *Particle symmetry group*.

SYNCHROCYCLOTRON. A variant of the cyclotron in which the accelerating frequency is varied in order to compensate for the increase in relativistic mass of the accelerated particle.

SYNCHROTRON. A variant of the cyclotron in which the magnetic field is modulated. In the *proton–synchrotron* both the magnetic and electric fields are modulated. See also: *Storage ring*.

SYNCHROTRON, ALTERNATING GRADIENT. A development of the conventional synchrotron in which the synchrotron beam is kept narrow by the use of strong focusing.

SYNCHROTRON RADIATION. Polarized emission of radiation arising from the gyration of relativistic electrons or positrons in a magnetic field, as in the synchrotron. It may account, among other things, for the radio emission from the galactic halo. See also: *Galaxy, magnetic field of*.

SYNERESIS. The exudation of liquid from a gel, often in response to an applied force, including the compression of the gel by its own weight.

SYNODICAL MONTH. The interval between successive conjunctions (or oppositions) of the Moon and the Earth. Its length is about 29·5 days. It is the original (lunar) month used by the ancient astronomers. See also: *Month*.

SYNODIC PERIOD. Of the Moon or a planet: the interval between two successive similar positions of the Moon or planet relative to the line joining the Earth and the Sun. The position usually chosen is conjunction or opposition.

SYNOPTIC CHART: WEATHER MAP. A map on which are marked synchronous observations of atmospheric pressure, temperature, and wind velocity, together with observations of frontal regions, of the state of the weather, of the clouds, and of the visibility. Such charts or maps are used as a basis for forecasting the weather.

SYNTHESIS, CHEMICAL. The production of a substance by the chemical combination of two or more simpler substances.

SYNTHETIC MATERIAL. An artificial material, i.e. one not derived immediately from the natural product. Such a material does not necessarily differ in any essential way from the natural material, e.g. synthetic ruby or sapphire.

SYPHON. In principle, an inverted liquid-filled U-tube used for transferring liquid from an upper to a lower level via a level higher than either. The transfer is effected by virtue of the pressure difference between the two original levels.

SYSTEM. (1) A portion of matter which is isolated from other matter by a real or imagined surface. (2) An assembly of plant or equipment designed to perform a specific function and considered as a whole.

SYSTÈME INTERNATIONAL D'UNITÉS: See: *SI Units*.

SYSTEM, HETEROGENEOUS. A system having more than one phase.

SYSTEM, HOMOGENEOUS. A system having only one phase.

SZILARD–CHALMERS REACTION. A process for the separation of a radioisotope by chemical means, based on the rupture of the chemical bond between a radioactive atom formed in a nuclear reaction and the molecule of which the atom formed a part. An easy chemical separation is then possible between the radioactive atom and the parent material.

t

TACHYONS. Hypothetical particles with imaginary mass but real values of energy and momentum, which would always have speeds greater than that of light. The speed of a tachyon would increase as the energy decreased.

TACTOID. A rod-shaped droplet or flat particle, appearing in colloidal solutions, which exhibits double refraction.

TALBOT BANDS. Dark bands which appear in the spectrum of white light when a thin glass plate (about 0·1 mm thick) is inserted in a spectroscope, between the prism and the telescope objective, from the violet side of the spectrum, until it covers half the objective lens. The bands are due to interference between the two sets of rays on account of the path difference produced by the glass plate. They do not appear if the plate is inserted from the red end of the spectrum.

TALBOT LAW. States that the apparent brightness of an object, when it is viewed through a sector disk which rotates sufficiently fast, is proportional to the ratio of the angular aperture of the open to the closed sectors. See also: *Sector disk*.

TAMMAN TEMPERATURE. The lowest temperature at which a solid will react with another at an appreciable rate.

TAMM–DANCOFF EQUATION. An approximation to the wave function of a system of interacting particles, especially nucleons and mesons, based on the superposition of a certain number of possible states, this number determining the degree of approximation involved.

TANDEM GENERATOR. A modification of the electrostatic generator in which a doubling of the energy of the particles is achieved, for the same accelerating potential, by accelerating negative ions from ground potential, stripping off their electrons and accelerating the resulting positive particles back to ground potential. See also: *Van de Graaff accelerator*.

TANGENTIAL FOCUS. See: *Astigmatic foci*.

TARGET. (1) A substance or object exposed to bombardment by particles or radiation, e.g. an X-ray tube target or a target nucleus in a nuclear reaction. A *thick* target is one whose thickness is such as to cause an appreciable loss of energy of the incident particle or radiation. A *thin* target is one which is not thick enough to cause such a loss. (2) Any object capable of reflecting, and therefore being detected by, a radar or sonar beam.

TARGET THEORY OF RADIATION PROCESSES. A theory explaining the biological effects of ionizing radiation on the basis of a target area (or small sensitive region within the cell) on which one, two, or more "hits" (ionizing events) may be necessary to bring about an effect.

TAUTOMERISM. The existence of a substance with one molecular formula but two structural formulae. Tautomeric compounds may thus exist as an equilibrium mixture of two forms which can give rise to two sets of derivatives. See also: *Isomers*.

TAYLOR THEOREM. States that, if the $(n-1)$th derivative of a function $f(x)$, written as $f^{(n-1)}(x)$, is continuous for $a \leq x \leq b$, and $f^n(x)$ exists for $a < x < b$, and if b is set equal to $(a + h)$, then

$$(a + h) = f(a) + hf^{(1)}a + \frac{h^2}{2!} f^{(2)}(a) + \cdots$$
$$+ \frac{h^{n-1}}{(n-1)!} f^{(n-1)}(a) + R_n,$$

where R_n is the remainder after n terms. Where R_n converges as n increases, the resulting infinite power series is known as the *Taylor series*. See also: *Maclaurin theorem*.

TECTONICS. The study of the deformation of the Earth's crust. See also: *Plate tectonics*.

TEKTITE. See: *Meteorite*.

TELECENTRIC SYSTEM. A telescopic system in which the aperture stop is placed at one of the foci of the objective lens.

TELECOMMUNICATION. Strictly: communication at a distance. The term is, however, usually restricted to the telegraphic or telephonic communication of signs, signals, images or sounds of any kind, by line, radio, or other system of electric signalling.

TELEGRAPHY, EQUATION OF. An equation describing the transmission of electrical impulses (telegraph, telephone or radio signals) along a pair of parallel wires or a coaxial cable.

TELEMETRY. Measurement at a distance. It is concerned with the transmission of measured physical quantities to a distant location. It began as a technique of communication by wire but now finds wide application in communications with satellites and space ships. Recent developments include the use of new types of transducer, and miniature radio transmitters powered by long-life miniature batteries.

TELEPHONY. The communication at a distance of speech or other sounds by electrical means over wire circuits. It involves essentially a transmitter, which converts sound variations to electrical variations; transmission circuits, which carry the electrical signals; and a receiver, which converts the electrical variations back into sound.

TELEPHOTOMETRY. The measurement of light sources at a distance, e.g. stellar photometry, the luminance distribution of street lighting and the calibration of searchlights.

TELERADIOGRAPHY. Radiography employing large source-to-object distances and small object-to-film distances. The technique improves definition, reduces distortion, and provides an image which has very nearly the same size as the object.

TELESCOPE. An optical instrument for the production and examination of an image of a distant object. In its simplest form it consists of an objective for producing the image (which may be a simple lens, a compound lens or a concave mirror) and an eyepiece (which may consist of a simple lens or a more complex system of several lenses) for the examination of the image. A *terrestrial telescope* gives an erect image (by the use of an additional lens—the erecting lens) and an *astronomical telescope* (which is rarely used as a visual instrument) an inverted one. See also: *Coelostat. Heliostat.* And the following entries.

TELESCOPE, ABERRATIONS OF. The main aberrations are the five Seidel aberrations (spherical aberration, coma, astigmatism, curvature of the field, and distortion) together with chromatic aberration. All these are defined separately. See also: *Seidel aberrations.*

TELESCOPE, ANNALATIC. A telescope in which the variation in focusing for different object distances is accomplished by moving an auxiliary lens between the (fixed) objective and the eyepiece. It is often used in surveying instruments.

TELESCOPE, CASSEGRAIN. A form of reflecting telescope with a concave paraboloidal primary mirror, and a convex hyperboloid secondary mirror placed before the primary focus so as to reflect the beam back through a hole in the centre of the main mirror.

TELESCOPE, COUDÉ. A form of reflecting telescope in which light from the primary mirror is brought to a focus at a point that remains stationary whatever the telescope's direction.

TELESCOPE, GALILEAN. The oldest form of telescope. It consists of a converging lens (the objective) and a diverging lens (the eyepiece), either of which can be simple or compound. This form of instrument is the basis of opera glasses, cheap field glasses and toy telescopes but is not suitable for high magnifications and has a very small field of view.

TELESCOPE, GRAZING-INCIDENCE. A telescope used in X-ray and gamma-ray astronomy which utilizes the fact that true reflection takes place at glancing angles. See also: *X-ray reflection.*

TELESCOPE, GREGORIAN. A form of reflecting telescope with a concave paraboloidal primary mirror and a concave ellipsoidal secondary mirror placed beyond the primary focus so as to reflect the beam back through a hole in the centre of the main mirror.

TELESCOPE, HERSCHELIAN. A form of reflecting telescope suitable for visual work. It has one mirror only, an off-axis paraboloid, which permits the eyepiece to be used just clear of the main beam.

TELESCOPE, INFRARED. (1) A telescope for the examination of infrared radiation from celestial bodies. Infrared photometers and bolometers are typically employed to study spectra, luminosities and fluxes. (2) A portable infrared image converter used, among other things, for seeing objects by their infrared emission, for example at night. See also: *Thermograph.*

TELESCOPE, NEWTONIAN. A form of reflecting telescope suitable for visual work. It has a parabolic main mirror the beam from which is reflected to the side of the telescope tube by a small plane mirror, so that it can be received in the eyepiece, which is normal to the axis.

TELESCOPE, PANCRATIC. A variable power terrestrial telescope usually incorporating an arrangement by which the erecting lens and eyepiece are both moved at the same time relative to the objective.

TELESCOPE, REFLECTING. A type of telescope in which the objective is a concave mirror, usually paraboloidal, the image being recorded on a photographic plate situated in the focal plane, or examined via one of a variety of optical arrangements. See also: *Telescope, Cassegrain. Telescope, Coudé. Telescope, Galilean. Telescope, Gregorian. Telescope, Herschelian. Telescope, Newtonian.*

TELESCOPE, REFRACTING. A telescope in which the final image is produced entirely by a system of lenses.

TELESCOPE, RESOLUTION OF. See: *Resolution. Resolving power.*

TELESCOPE, SCHMIDT. A reflecting telescope in which the primary mirror is corrected for spherical aberration by inserting a lens, or corrector plate, exactly as in the Schmidt camera. See also: *Schmidt camera.*

TELESCOPE, SIGHTING. A low-power terrestrial telescope either directly attached to a rifle or geared to a gun.

TELESCOPE, UNIVERSAL. A telescope with interchangeable optical parts which combines almost all the possible optical arrangements of a reflecting telescope in one instrument.

TELETHERAPY. The use of remotely controlled radioactive sources in radiotherapy. See also: *Radiotherapy.*

TELEVISION. The electrical transmission of visual scenes and images by wire or radio. These may or may not be accompanied by sound, but if so they are synchronized with the events depicted. A television system consists essentially of a camera tube, for the conversion of an optical image into an electrical image; a means of amplifying and transmitting the electrical signals produced by the camera tube with or without the addition of synchronized signals derived from any associated sound; and a detection and reproduction system, which normally receives the electrical video signals and converts them into the scanning movements of the electron beam in an appropriate cathode-ray oscilloscope. The audio signals, if any, are detected and reproduced more or less as in conventional radio, with provision for synchronization added.

TELEVISION, COLOUR. The electrical transmission of colour pictures by television. It is based on the fact that light of any colour can be matched by the combination of the three primary colours red, green and blue light in appropriate proportions. The object is imaged at the camera through red, blue and green filters on to photosensitive surfaces of camera tubes generating video signals corresponding to the red, green and blue component images. These are combined at the receiver to reconstitute an image of the original object in colour.

TELLURIC LINES. Those lines in the Fraunhofer spectrum which arise from absorption in the Earth's atmosphere. They lie mainly in the red and infrared region. See also: *Fraunhofer lines.*

TELLUROMETER. An instrument for measuring a large distance accurately from the phase shift of a 10 cm radar beam emitted from the instrument and re-radiated from a distant repeater. Distances of up to 30 miles (50 km) can be measured with an accuracy of 1 part in 10^5. See also: *Laser applications. Mekometer.*

TEMPERATURE. The degree of hotness or coldness. Its measurement involves the determination of some physical property (e.g. thermal expansion, electrical resistance, pressure, viscosity) of a suitable substance, which property is assumed to be related in a known fashion to the temperature as expressed on a scale which is based on the assignment of arbitrarily chosen temperatures to two fixed reference points (e.g. the ice point and the steam point). See also: *Temperature scale, empirical. Temperature scale, international practical. Temperature scale, Kelvin: Temperature scale, thermodynamic.*

TEMPERATURE, EFFECTIVE. (1) Of a stellar surface: the temperature of a black body radiating the same amount of energy per unit area as the star. (2) Of an extraterrestrial radio source: the temperature of a black body radiating the same amount of energy per unit area as the source, it being assumed that the radiation is thermal and that the radio spectrum has a corresponding intensity–wavelength distribution. (3) In medical physics: a subjective temperature which combines the effect of heat loss or gain by the body due to convection and radiation with that due to evaporation or condensation of moisture on the skin. It is an index of the degree of warmth which a person will experience for given combinations of dry bulb temperature and relative humidity.

TEMPERATURE FLUCTUATION NOISE. In electrical measuring instruments: fluctuations arising from random temperature fluctuation about the statistical mean. See also: *Electrical noise.*

TEMPERATURE, POTENTIAL. The temperature that a small mass of dry air would have if it were brought to a standard pressure (usually 1000 mbar) adiabatically. (*Note:* 1 mbar $= 10^2$ N/m².)

TEMPERATURE RADIATION. Synonym for black-body radiation. See: *Black body*.

TEMPERATURE SCALE, AVOGADRO. See: *Temperature scale, ideal gas*.

TEMPERATURE SCALE, EMPIRICAL. A temperature scale based on the assignment of arbitrary numerical values to the temperature of two fixed reference points, commonly the ice point and the steam point. The scale is defined by measuring some physical property (*thermometric property*) of a suitable substance (*thermometric substance*), which is assumed to be related to the temperature in a known fashion. The temperature interval between the two fixed points is divided into equal intervals. Three scales are of interest in this connection: the *Celsius* (*formerly centigrade*) *scale*, in which the ice point is at 0 deg and the steam point at 100 deg; the *Fahrenheit scale*, in which the ice point is at 32 deg and the steam point at 212 deg, and the *Réaumur scale*, in which the ice point is at 0 deg and the steam point at 80 deg. The measurement of temperature on any one of these scales is meaningful only with reference to a particular thermometric substance and property, the measured temperatures being different for different combinations of these. Calibration is therefore necessary for accurate measurement.

TEMPERATURE SCALE, IDEAL GAS. A temperature scale based on the behaviour of an ideal gas, for which the temperature is accurately proportional to the product of pressure and volume. Since it depends on the validity of Avogadro's law it is also known as the *Avogadro scale*.

TEMPERATURE SCALE, INTERNATIONAL PRACTICAL (IPTS.) A scale based on a series of primary and secondary fixed points, the temperatures of which are known accurately on the thermodynamic scale. The ice point is replaced by the triple point of water, set at $0.0100°$C or 273.16 K exactly, thus making the temperature of absolute zero $-273.15°$C. The list of primary fixed points comprises the temperatures of the following equilibrium states: the triple point of equilibrium hydrogen (13.81 K: $-259.34°$C), the equilibrium between the liquid and vapour phases of equilibrium hydrogen at a pressure of 33 330·6 Nm⁻² (17.042 K: $256.108°$C), the boiling point of equilibrium hydrogen (20.28 K: $-252.87°$C), the boiling point of neon (27.102 K: $-246.048°$C), the triple point of oxygen (54.361 K:

$-218.789°$C), the boiling point of oxygen (90.188 K: $-182.962°$C), the triple point of water (273.16 K: $0.01°$C), the boiling point of water (373.15 K: $100°$C), the freezing point of zinc (692.73 K: $419.58°$C), the freezing point of silver (1235.08 K: $961.93°$C) and the freezing point of gold (1337.58 K: $1064.43°$C). From 13.81 K to $630.74°$C (the melting point of antimony – one of the secondary fixed points) the standard instrument is the platinum resistance thermometer, and from $630.74°$C to $1064.43°$C it is the platinum -10% rhodium/platinum thermocouple. Above this the temperature is defined by the Planck radiation formula (which is defined separately), taking 1337.58 K as the reference temperature.

TEMPERATURE SCALE, KELVIN:
TEMPERATURE SCALE, THERMODYNAMIC. An ideal standard scale on which changes of temperature are strictly proportional to the quantities of heat converted into mechanical work during a series of Carnot cycles all working between the same two adiabatic limits on the pressure–volume diagram, and for which the lower temperature of one cycle is the upper temperature of the next, and so on. A constant volume gas thermometer, using an ideal gas, would perform in exact agreement with this standard and the scale may be realized in practice by a constant volume hydrogen thermometer if suitable corrections are applied. See also: *Kelvin*.

TEMPERATURE SCALE, RANKINE. An absolute temperature scale on which the zero is the absolute zero of temperature and one degree is the same as one degree Fahrenheit. On the Rankine scale the freezing point of water is then $491.7°$ and the boiling point $671.7°$, $0°$F corresponding to $459.7°$ Rankine.

TEMPERATURE, VIRTUAL. A meteorological parameter, used in barometric reduction calculations. It is the temperature which dry air must have at a given barometric pressure in order to have the same density as air with a specified moisture content at the same pressure, and at a given temperature.

TEMPER BRITTLENESS. Brittleness produced when certain steels (e.g. nickel–chrome steels) are held in, or slowly cooled through, a range of temperatures below the critical temperature, typically $600°$C–$300°$C.

TEMPER COLOURS. Colours exhibited by a thin film of oxide on the surface of plain carbon steel when it is heated in air. The variation of colour was formerly used as a rough measure of temperature.

TEMPERING. The reheating of a martensitic steel, formed by quenching, to some temperature below the transformation range, followed by cooling. It results in an increase of ductility and toughness, and a decrease of hardness. The term is sometimes also applied to other heat treatments. See also: *Martensitic transformation.*

TENSILE STRENGTH: ULTIMATE TENSILE STRESS. The tensile stress at which a body will fracture, or will continue to deform with decreasing load. It is sometimes termed the *breaking strength.*

TENSION. The force, usually in a wire, string or rod which is stretched between two points.

TENSOR ANALYSIS. A form of mathematical analysis concerned with relations and laws in n-dimensional space which are covariant, i.e. remain valid in passing from one system of co-ordinates to any other. A tensor of n^r components is said to be of *rank* (or *order*) r; and a tensor is termed *covariant*, *contravariant* or *mixed*, according to the law of transformation followed. One of the main applications is to the theory of general relativity, but tensor analysis is also applied to classical mechanics, quantum mechanics, fluid mechanics and elasticity among other things.

TENSOR, METRICAL. See: *Riemannian space.*

TERM DIAGRAM. An energy level diagram referring to atomic energy levels, i.e. to spectral terms from which possible transitions relating to optical (including ultraviolet and infrared) or X-ray spectra may be derived. See also: *Energy level diagram.*

TERMINAL VELOCITY. The constant velocity eventually achieved by a freely falling body in a given medium. For a sphere falling through a fluid of viscosity η, it is given by $2gr^2(\varrho - \varrho')/9\eta$, where g is the acceleration due to gravity, r the radius of the sphere, ϱ the density of the sphere, and ϱ' that of the liquid.

TERRESTRIAL PLANET. See: *Planet, minor.*

TERRESTRIAL RADIATION. The energy emitted into space by the Earth and its atmosphere.

TESLA. The MKSA and SI unit of magnetic flux density or magnetic induction. It is equal to 1 Wb/m². The CGS unit, the gauss (1 maxwell or line per cm²), is equal to 10^{-4} tesla. See: *Appendix II.*

TESLA COIL. A high-frequency induction coil in which the primary circuit includes a high-frequency spark gap instead of the usual interrupter.

TETRAD AXIS. See: *Crystal symmetry, rotation axis in.*

TETRAGONAL SYSTEM. A system of crystal symmetry containing those crystals which can be referred to a set of three mutually perpendicular axes, two of which are equal. The system is characterized by a single axis of four-fold symmetry, which may be of simple rotation or inversion.

TETRODE. A four-electrode electron valve or tube containing a cathode, anode, control electrode, and an additional electrode usually in the form of a grid or screen.

TEXTURE. In crystallography: preferred orientation in a polycrystalline solid, e.g. a metal. The type of texture may be indicated in a variety of ways, e.g. by specifying the indices of the plane of rolling in a rolled metal, together with the indices of the crystal direction lying parallel to the rolling direction. Thus the *cube texture*, found in rolled sheets of some face-centred cubic metals is (100)[001]. Some other textures are the *spiral texture* and the *fibre texture*. See also: *Fibre diagram. Layer lattice. Pole figure. Preferred orientation.*

TFTR (TOKAMAK FUSION TEST REACTOR). A development of the tokamak system being studied in the U.S.A. It will have conventional geometry and will probably employ neutral beam injection as a source of heat. The energetic beams involved could easily cross magnetic fields and enter the plasma, adding energy and particles to the plasma, and thus, by lowering the target for the plasma conditions needed, would make it easier to build a fusion reactor. See also: *Thermonuclear reaction, controlled. Tokamak.*

THEODOLITE. An instrument used in surveying to measure horizontal or vertical angles. It consists essentially of a telescope which is movable about a horizontal axis, and is attached to a base which is movable about a vertical axis, suitable scales being attached for the measurement of elevation and azimuth. A small theodolite which also carries means for observing the geomagnetic horizontal intensity is known as a *magnetic theodolite.*

THEORETICAL PHYSICS. The theoretical treatment of experimentally observed physical phenomena. The term does not mean academic as distinct from industrial or applied physics, and indeed much of the older branches of the subject are to be found in engineering textbooks.

THERM. The legal unit in Great Britain by which a gas company charges for gas. It is 100 000 British thermal units, i.e. 105 506 kJ. See also: *British thermal unit (B.t.u.).*

THERMAL ANALYSIS. The use of heating and cooling curves to detect endothermal or exothermal reactions associated with freezing points, melting points or phase transitions. The technique is widely

330

used in metallography in the construction of equilibrium diagrams.

THERMAL BOUNDARY LAYER. A boundary layer between a fluid and a body which impedes heat transfer between them.

THERMAL COMPARATOR. A device which enables comparative measurements of the thermal conductivities of solids to be made rapidly. It renders quantitative the well-known handling test by which conductors of heat can be distinguished from poor conductors by their colder feel. By the use of appropriate techniques the thermal comparator can also be used to determine the thermal conductivities of soft materials or even liquids, the thickness of surface deposits, platings, etc., and the assessment of surface finish or temperature.

THERMAL CONDUCTANCE. Of a material: the quantity of heat flowing per second across the material for unit temperature difference between its hot and cold faces. It is the reciprocal of thermal resistance.

THERMAL CONDUCTION. The transmission of heat through a medium. The heat energy is transmitted by atomic vibrations or, especially in metals, by the passage of free electrons. In semiconductors the movement of electron-hole pairs and radiative transfer may also play a part. See also: *Phonon. Wiedemann–Franz law.*

THERMAL CONDUCTIVITY. Of a material: the quantity of heat flowing per second across unit area of that material (the plane of the area being perpendicular to the temperature gradient), for unit temperature gradient. It is the reciprocal of the thermal resistivity. See also: *Wiedemann–Franz law.*

THERMAL DIFFUSION. In a mixture of two fluids across which a temperature gradient exists: the tendency for one fluid to concentrate in the warmer region and the other in the colder region. A similar effect occurs when there are more than two fluids, whereby a concentration gradient is established. Thermal diffusion (which is also known as the *Soret effect*) may be used as a method of isotope separation.

THERMAL DIFFUSIVITY. Of a material: a quantity which indicates the ability of the material to equalize temperature differences occurring within it. This quantity is given by $k/\varrho C$, where k is the thermal conductivity, ϱ the density, and C the specific heat. A high value of the thermal diffusivity indicates, therefore, that the material approaches equilibrium quickly.

THERMAL ELECTROMOTIVE FORCE. An electromotive force produced by the application of heat, as in the Seebeck and Thomson effects.

THERMAL EQUATOR. The line passing through the middle of the belt of high temperature which exists near the geographical equator. Its location varies with the season but, owing to the effects of land distribution and ocean currents, its mean location is to the north of the geographical equator.

THERMAL EQUILIBRIUM. That state of a system in which the net rate of heat exchange between its parts is zero.

THERMAL EXPANSION. The increase of a mass of substance (gas, liquid or solid) in length, area, or volume, as a result of an increase in temperature.

THERMAL EXPANSION COEFFICIENT. A coefficient expressing the rate of change of length (or volume) with change of temperature. It is the change of length (or volume) per unit length (or volume) for unit temperature difference, the pressure being kept constant. It may be stated for a gas, liquid or solid, as applicable; and for a solid the *linear expansion coefficient* (i.e. that relating to length) will be the same in all directions only if the solid is isotropic. Theoretical values for the expansion coefficients may be derived thermodynamically by making various assumptions regarding the potential energy. See also: entries under *Equation of state. Grüneisen law: Grüneisen second rule.*

THERMAL FATIGUE. (1) Failure occurring on the repeated heating and cooling of anisotropic materials, owing to their anisotropic thermal expansion. (2) Failure caused by stresses induced in materials when free thermal expansion is not permitted.

THERMAL IMAGING. The concentration of radiant energy, by means of an optical system, from a source of such energy. The solar furnace offers one example, and the arc image furnace another.

THERMAL NOISE. See: *Johnson noise.*

THERMAL REGULATION, BIOLOGICAL. The control of body temperature in the higher animals by control of heat exchange and metabolic heat production.

THERMAL RELAXATION. The establishment of thermal equilibrium in a body following a change in the thermal conditions. See also: *Relaxation phenomenon. Relaxation time.*

THERMAL RESISTANCE. Of a material: the ratio of the temperature difference between the hot and cold faces to the quantity of heat flowing per second between these faces. It is the reciprocal of the thermal conductance.

THERMAL RESISTIVITY. Of a material: the ratio of the temperature gradient in the material to the quantity of heat flowing per second across unit area of that material, the plane of the area being perpendicular to the temperature gradient. It is the reciprocal of the thermal conductivity.

THERMAL SHOCK. The shock experienced by a material when it is subjected to sudden changes in temperature. A brittle material may well fracture by one application of such a shock and the resistance to shock is sometimes specified in terms of the minimum shock that will cause fracture of a specimen of a given shape.

THERMAL SPIKE. The zone of local heating produced in a solid along the track of a high-energy particle. See also: *Displacement spike. Fission spike.*

THERMAL STRESS. Stress produced by temperature effects, e.g. from temperature gradients within a body.

THERMAL SWITCH. A heat switch.

THERMAL TRANSMISSION. Across a structure or through a solid: the quantity of heat flowing per unit time. It is also known as the *rate of heat flow*.

THERMAL TRANSMITTANCE. Of a body: the ratio of the intensity of thermal radiation transmitted by a body to the intensity incident upon the body.

THERMAL UTILIZATION FACTOR. For nuclear fission in an infinite medium or infinite nuclear reactor: the fraction of thermal neutrons that are absorbed in a fissile nuclide or in a nuclear fuel, as specified. See also: *Nuclear reactor, four-factor formula for.*

THERMAL VALVES. See: *Refrigerator, magnetic.*

THERMAL VIBRATIONS. The vibrations of the atoms in a solid about their mean positions, or the vibrations of polyatomic molecules. The amplitude of the vibrations increases with temperature; and, in a crystal, one manifestation is the diffuse scattering observed when X-rays or neutrons are diffracted by the crystal. See also: *Debye–Waller temperature factor. Scattering, diffuse.*

THERMIE. A unit of heat equal to the quantity of heat required to raise the temperature of 1 tonne of pure water from 14·5 to 15·5°C at normal atmospheric pressure. It was used in the French metre-tonne-second system (MTS), and is equal to 10^6 $cal_{15°}$, i.e. $4·186 \times 10^6$ J. See also: *Frigorie.*

THERMION. An ion or charged particle emitted by a heated body.

THERMIONIC CATHODE. In an electron tube: a cathode which emits electrons when heated. It may be coated with emissive material, usually in the form of oxides (typically of barium and strontium) having a low work function, or may be a *dispenser cathode* which is not so coated but is continuously supplied with emissive material from a separate element associated with it.

THERMIONIC CONSTANTS. A name sometimes applied to the work function and the constant A in the Richardson equation. See also: *Richardson equation: Richardson–Dushman equation.*

THERMIONIC EMISSION. The emission of ions or charged particles from hot bodies. See also: *Richardson equation: Richardson–Dushman equation.*

THERMIONIC GENERATION OF ELECTRICITY. The generation of electricity on a useful scale by thermionic emission. For a practical generator it is necessary to neutralize the space charge, for which the main method is the injection of an alkali metal, almost always caesium. See also: *Direct conversion of heat to electricity.*

THERMIONICS. The study of the emission of thermions.

THERMIONIC TUBE: THERMIONIC VALVE. An electron tube or valve in which one of the electrodes is heated to provide electronic or ionic emission.

THERMISTOR. A *therm*ally sensitive re*sistor* whose resistance falls rapidly with temperature. Thermistors are made from semiconducting material, and are used, for example, for measuring small temperature changes (as little as 0·005°C), controlling output voltage or current, and in compensating systems which normally have positive temperature coefficients, to prevent overheating.

THERMOCHROMISM. A reversible colour change induced in a substance by exposure to infrared radiation. A similar change induced by violet or ultraviolet light is known as *photochromism* or *phototropy.*

THERMOCOUPLE. A thermoelectric device, employing the Seebeck effect, used for the measurement of temperature. It consists typically of an electric circuit composed of two wires of dissimilar metals joined to each other at both ends, and containing provision for the measurement of e.m.f. One junction is kept at a constant known temperature and the other at the temperature to be measured, whose value is derived from the value of the e.m.f. generated at the junction, via a suitable calibration of the thermocouple. Thermocouples for high temperature measurement are commonly made from platinum and platinum–rhodium alloys (usually containing 10 or 13% rhodium), but for low temperatures other metals, e.g. copper–constantan, are usually employed. See also: *Thermopile.*

THERMODYNAMIC DIAGRAM. A diagram which serves to facilitate the application of the first and second laws of thermodynamics to power cycles, refrigeration heating, combustion heating and so on. It may be (a) a *pressure–volume diagram*, plotted for constant temperature or entropy; (b) an *enthalpy–entropy diagram*, plotted for constant pressure, temperature or volume (also known as a *Mollier diagram*); (c) a *temperature–entropy diagram*; (d) a *pressure–enthalpy diagram*; or (e) one of a series designed for use in combustion heating. In meteorology the temperature–entropy diagram is also used to provide solutions of the equations of thermodynamic transformation of dry or moist air, as is another diagram, the *temperature–log pressure diagram.*

THERMODYNAMIC EFFICIENCY. (1) Of a heat engine: the ratio of the work produced to the heat absorbed. (2) Of a combustion chamber: the ratio of the heat actually liberated by combustion to the heat liberated by the complete oxidation of all the fuel (when the remaining water in the fuel remains in the vapour form), i.e. to the net calorific value.

THERMODYNAMIC EQUILIBRIUM. That state of a system for which no finite rate of change can occur without a finite change, temporary or permanent, in the state of the environment. The state is in *stable equilibrium* if a permanent change in the state of the environment is necessary to achieve the finite rate of change referred to. See also: *Equilibrium.*

THERMODYNAMIC FUNCTIONS. Functions which depend only on the actual state of a system and not on its history. See also: *Thermodynamic potentials.*

THERMODYNAMIC POTENTIALS. A series of potentials which are analogous to potential energy in mechanics. They include the thermodynamic functions, namely internal (or total) energy, enthalpy, entropy, Helmholtz free energy, and Gibbs free energy (also known as free enthalpy).

THERMODYNAMICS. The study of interactions between systems and their effects on the states of these systems. Classical thermodynamics is particularly concerned with changes between equilibrium states. The subject originally dealt only with interactions involving heat, and the derivation of principles relating to these, but it has now been extended to include other applications of the same general principles.

THERMODYNAMIC SIMILARITY, PRINCIPLE OF. States that the distinguishing characteristics of any fluid are described by four independent quantities, namely the molecular weight, the critical pressure, the critical temperature, and the critical density. See also: *Corresponding states, law of.*

THERMODYNAMICS, IRREVERSIBLE. An extension of classical thermodynamics to permit the study of real dynamical processes (e.g. in biology).

THERMODYNAMICS, LAWS OF. These may be worded in a variety of ways. Those given below are among the more common. (a) The *zeroth law:* states that when any two systems are in equilibrium with a third they are in equilibrium with each other. (b) The *first law:* states that in a closed system, the total amount of energy of all kinds is constant. The term "energy" in this connection must now be taken to include mass. This law denies the possibility of perpetual motion of the first kind. (c) The *second law:* states that heat cannot pass from a colder to a hotter body without the intervention of some external force, medium or agency. This law is the basis of Carnot's theorem and denies the possibility of perpetual motion of the second kind. (d) The *third law*, or *Nernst heat theorem:* states that, at absolute zero, the entropy difference disappears between all those states of a system which are in internal thermodynamic equilibrium. It leads to a denial of the possibility of attaining absolute zero.

THERMOELASTIC EFFECT. The elastic relaxation arising from change of temperature.

THERMOELECTRIC COOLING: PELTIER COOLING. Cooling on a useful scale by use of the Peltier effect.

THERMOELECTRIC EFFECTS. Those effects in which a potential difference arises as a result of the flow of heat, and vice versa. Well-known examples are the Peltier effect, Seebeck effect, and Thomson effect, which are defined separately. See also: *Galvanomagnetic effects. Thermomagnetic effects.*

THERMOELECTRIC GENERATION OF ELECTRICITY. The generation of electricity on a useful scale by use of the Seebeck effect. See also: *Direct conversion of heat to electricity.*

THERMOELECTRIC POWER. The thermal e.m.f. per degree.

THERMOGRAPH. (1) A term used in meteorology to denote any type of recording thermometer.

(2) An image converter which converts an infrared image into a visible image. It is often used in medical diagnosis.

THERMOGRAVIMETRIC ANALYSIS. A technique used in analytical and preparative chemistry in which the change of weight of a sample is observed as a function of temperature. The weight–temperature curves (or *pyrolysis curves*) obtained are used, for example, to ensure that the injunction "heat to constant weight" may be obeyed.

THERMOJUNCTION. A junction between two dissimilar wires as used in the thermocouple and thermopile to utilize the Seebeck effect.

THERMOLUMINESCENT DATING. The dating of archaeological pottery by measuring the excess light emitted on heating. This excess light (i.e. light in excess of the ordinary thermal radiation) is produced by the release of trapped electrons, whose number is proportional to the radiation dose received since the original firing, i.e. the excess light is proportional to the age. See also: *Age determination by radioactivity*.

THERMOMAGNETIC EFFECTS. Thermoelectric effects which only occur in the presence of a magnetic field. Examples are the Ettingshausen effect, the Nernst effect, and the Righi–Leduc effect, which are defined separately. See also: *Galvano-magnetic effects. Thermoelectric effects*.

THERMOMECHANICAL EFFECT. The flow of helium II towards a region of higher temperature. See also: *Fountain effect*.

THERMOMETER. An instrument or device capable of measuring the relative degree of hotness of a body on a suitable scale in a reproducible fashion, usually by measuring some physical property of a suitable substance on the assumption that the property is related in a known fashion to the temperature. See also: *Temperature. Thermocouple. Thermopile.* Entries under *Pyrometer, Temperature scale*, and *Thermometer*.

THERMOMETER, BECKMANN. A mercury-in-glass thermometer having a very large bulb and hence being very sensitive over a range of a few degrees. It is used to measure small changes of temperature with precision, and incorporates a reservoir by which mercury may be added to, or removed from, the indicating thread to enable the thermometer to be used in different temperature ranges.

THERMOMETER, CLINICAL. A thermometer designed for measuring body temperature. It is a mercury-in-glass maximum thermometer in which the mercury is prevented from creeping back into the bulb as it cools by a constriction in the bore.

THERMOMETER, GAS. A thermometer in which a gas is the thermometric substance and pressure is the thermometric quantity, the volume of the gas being kept constant. See also: *Temperature scale, Kelvin: Temperature scale, thermodynamic*.

THERMOMETER, LIQUID-FILLED. The most common form of thermometer in everyday use. The thermometric property used is thermal expansion and the thermometric substance is typically mercury or alcohol, which is stored in a glass bulb and allowed to expand along a suitably calibrated glass bore. For very low temperatures, however, organic liquids such as toluene, pentane or propane are commonly used. Another form of thermometer employs a liquid (mercury, water or an organic liquid) sealed in a spiral of hollow metal tubing (a Bourdon tube) which uncoils as the liquid expands, and can be made to actuate a pointer moving over a dial, or to operate a pen recorder and so on.

THERMOMETER, MAXIMUM AND MINIMUM. A thermometer which registers the maximum and minimum temperatures achieved between one observation and the next. A type widely used is *Six's* thermometer, consisting of a bulb containing alcohol joined by a capillary tube to one limb of a U-tube containing mercury. The mercury in one limb rises as the alcohol expands and that in the other rises as the alcohol contracts, the positions of the highest points achieved being indicated by spring-loaded indexes resting on the surface of the mercury. Maximum thermometers (e.g. the clinical thermometer) alone are also known, as are minimum thermometers.

THERMOMETER, NOISE. A thermometer for determining temperature from the measurement of Johnson noise.

THERMOMETER, RESISTANCE. A thermometer in which the thermometric property is electrical resistance and the thermometric substance is a metal (commonly platinum) in the form of a coil of wire. Such a thermometer is widely used for the measurement and control of temperature at one or more fixed measuring points from some convenient control position.

THERMOMETER, SIX. See: *Thermometer, maximum and minimum*.

THERMOMETER, WEIGHT. An instrument for measuring the rise in temperature from a determination of the loss in weight when a liquid is expelled from a full container by heating, assuming a knowledge of the expansion coefficients of liquid and container. It can also be used *mutatis mutandis* for measuring the coefficient of expansion of the liquid.

THERMOMETER, WET-AND-DRY BULB. A hygrometer consisting of a similar pair of thermometers, the bulb of one of which is kept wet, and therefore cooled by evaporation. The difference in temperature between the two thermometers is a measure of the relative humidity of the air. The instrument is also known as a psychrometer. See also: *Hygrometer.*

THERMOMETRIC PROPERTY AND SUB-STANCE. See: *Temperature scale, empirical.*

THERMOMETRY, FIXED POINTS FOR. See: *Temperature scale, international practical.*

THERMOMETRY, LOW-TEMPERATURE. The measurement of temperatures below 100 K. The gas thermometer is the fundamental instrument but thermometers depending on the measurement of the velocity of sound in a gas, of vapour pressure, of nuclear resonance and of paramagnetic susceptibility have also been used.

THERMOMOLECULAR FORCE OR PRESSURE. The force exerted by the molecules of a rarefied gas in which a local source of heat is introduced. The force may be used for the measurement of low gas pressures as in the Knudsen gauge. See also: *Knudsen absolute manometer. Pressure gauge, Knudsen.*

THERMONUCLEAR REACTION. The fusion of light atomic nuclei to produce a nucleus which is heavier than either, together with excess energy. It is the main source of stellar energy. See also: *Solar energy. Stellar evolution.*

THERMONUCLEAR REACTION, CONTROL-LED. A reaction, not yet achieved, in which thermonuclear energy can be controlled and used, as is the energy of nuclear fission, as a source of power. Such a reaction would occur if a plasma could be contained, and heated to a sufficiently high temperature. The containment might be realized for example by the use of appropriately shaped magnetic fields, a number of such fields having been proposed, e.g. a variety of "pinches" and magnetic mirrors; and the required temperature might be achieved by joule, shock or compression heating, or by the injection into an already hot plasma of fast neutral particles. It is now generally accepted that the most promising approach is via a toroidal system in which the plasma is heated by a current flowing in it, following the tokamak device

pioneered by the Russians. Work on the development of tokamak-like systems is proceeding in the U.K. and the U.S.A., as well as in Russia itself. See also: *JET. Pinch effect. Stellarator. TFTR Tokamak.*

THERMOPILE. A thermoelectric device, employing the Seebeck effect, used typically for the measurement of thermal energy. It consists essentially of a number of thermojunctions connected in series, one set of junctions being subjected to thermal radiation, the other being kept at constant temperature, the measured e.m.f. being therefore proportional to the number of junctions. The metals usually chosen are antimony and bismuth. In the *linear thermopile*, which is constructed on the same principle, all the junctions of one kind lie along a central line which forms the axis of a cylindrical casing. The thermal radiation enters via a slit at one end of the casing to facilitate analysis of thermal spectra. See also: *Thermocouple.*

THERMOSTAT. A thermosensitive device designed to maintain a constant temperature, commonly in an enclosure. The term is also applied to a chemical bath kept at constant temperature. The temperature regulation is usually achieved by automatically cutting off the heating mechanism by a relay when the temperature exceeds the desired value and restoring it when the temperature falls below this value.

THERMOSYPHON. A heat exchanger in which the circulation of the working fluid is achieved by convection. Examples are the domestic radiator and the external radiator of an engine.

THERMOVAPORIMETRIC ANALYSIS. An analytical technique in which the gaseous products evolved on heating a substance are detected and measured, and continuously recorded as a function of temperature.

THÉVENIN–HELMHOLTZ THEOREM. States that an active electrical network, composed of linear bilateral circuit elements (i.e. circuit elements through which current may flow in either direction) and any number of voltage and current sources, can be replaced at any pair of terminals, on looking back into the network from that pair, by a single voltage source in series with a simple impedance. The magnitude of the voltage source is equal to the open-circuit voltage appearing across this pair of terminals, and that of the impedance is the impedance seen when looking back into the network from the two open terminals, when all voltage sources are replaced by their internal impedances and all current sources are open-circuited, their admittances being retained. The theorem is often known simply as the *Thévenin theorem.*

THICKNESS RATIO. In aerodynamics: the ratio of the maximum depth of an aerofoil, measured at right angles to the chord line, to the chord length.

THIN FILMS. Films whose thickness, broadly speaking, is less than 1 μm. They may be prepared by a variety of techniques including electrolytic deposition, electroless deposition, anodizing, and, more commonly, sputtering or evaporation. Thin films have mechanical, electrical, magnetic and optical properties which may differ from those of the bulk material and are used, commonly in the form of a deposit on a suitable substrate, for integrated circuits, resistors, capacitors, transistors, superconductors, and in interferometry, to name some examples. See also: *Epitaxy. Integrated circuit. Monolithic circuit. Transistor.*

THIXOTROPY. The property, shown by certain gels, of liquefying on being shaken and of reforming on standing.

THOMSON EFFECT. The occurrence of a potential difference between the ends of a single metal strip when a temperature gradient is set up along its length. Such a potential difference results in the production of heat when a current is passed along the strip, which is additional to the Joule heat, and may flow in the opposite direction to it. See also: *Thermoelectric effects.*

THOMSON SCATTERING. The scattering of photons by a free charged particle, in particular by an electron, according to the Thomson equation $\sigma = \dfrac{8}{3}\,\pi(e^2/mc^2)^2$ or $\dfrac{8\pi}{3}\,(r_e)^2$, where σ is the scattering cross-section per electron, e is the electronic charge, m the electronic rest mass, c the speed of light, and r_e is the classical radius of the electron. The equation is valid only if the photon energy is much less than the rest energy (mc^2) of the electron.

THORAEUS FILTER. A filter made of tin, copper and aluminium used in radiotherapy to permit the passage of certain wavelengths only.

THORIUM SERIES. The radioactive series of elements beginning with ^{232}Th and ending with ^{268}Pb. It is one of the three naturally occurring radioactive series, the others being the actinium series and the uranium series, which are separately defined. The series is also known as the *4n series* since the atomic number of each member is a multiple of 4.

THORON. See: *Radioactive emanation: Emanation.*

THREE-BODY PROBLEM. The problem of the motions of three particles which move under the influence of their mutual interactions. Even when restricted to the action of gravitational forces on constant masses it has so far been found to be insoluble in any general form, but solutions can be obtained in particular cases, e.g. where one body has no effect on the motions of the others. See also: *Many-body problem. Two-body problem.*

THYRATRON. A gas-filled thermionic valve or tube in which the initiation of current in an ionized gas or vapour is controlled by the voltage applied to a control electrode.

THYRISTOR. A semiconductor device for switching between high conductivity and high frequency modes.

TIDAL POWER. The power available as a result of the movement of large masses of water under the influence of tidal rise and fall. A common scheme for utilizing such power is to store water on the flood and operate a hydroelectric generator on the ebb, although it is also possible to operate on a rising tide as well as a falling one. Schemes for the utilization of wave motion as such have also been proposed. See also: *Wave power.*

TIDAL WAVE. (1) A tidal oscillation set up in the sea by external *tide-generating forces,* i.e. by the gravitational attraction of the Sun and Moon. It has a small amplitude (some metres at most) and a long wavelength (some kilometres). When a current is present whose speed is comparable to that of the wave (a comparatively rare occurrence), typically in an estuary, a tidal *bore* or *eagre* occurs, in which a wave with an approximately vertical surface, of height up to some 3 m, moves up the estuary with a speed greater than that of the tide as a whole. This phenomenon is most apparent at spring tides. (2) A term loosely used to denote a storm surge or a tsunami.

TIDE. The periodic rise and fall of the oceans and seas arising from the periodic variation of the combined gravitational attraction of the Sun and Moon. A *spring tide* occurs when the Sun's and Moon's attraction are in the same sense, and a *neap tide* occurs when they are opposed. In general high and low water occur twice each day and the spring and low tides occur twice each month, but various other short and long period variations also exist.

TIDE, DATUM OF. An arbitrary level from which the rise and fall of the tide may be measured. In harbour the tide may be measured in relation to the level of a dock sill, while for nautical charts it is taken either as the lowest level of the tide normally reached, or alternatively as the mean sea level.

TIDE-GENERATING FORCES. See: *Tidal wave.*

TILT BOUNDARY. See: *Grain boundary. Polygonization.*

TIME. (1) *Absolute or classical time:* time as postulated in Newtonian mechanics, which "flows uniformly on, without regard to anything external", and enables a date to be uniquely associated with any given event. (2) *Operational time:* a division or multiple of the interval between two characteristic events in an infinitely repeatable series. Here, time is an independent variable which can be infinitely divided and which extends to an infinite distance in either the positive or negative direction. (3)

Quantized time: time which is assumed to be no longer continuous and infinitely divisible. The concept arises from quantum mechanics, the quantum of time (termed a *chronon*), corresponding to the greatest observed radiation frequency, and being equal to 4.5×10^{-24} s. (4) *Civil time:* time which is related to the spinning of the Earth on its axis and its yearly rotation about the Sun, i.e. *astronomical time.* The mean solar time referred to the meridian of Greenwich is known as *Greenwich mean time, Greenwich mean astronomical time,* or *Greenwich civil time.* Before 1925 it was counted from noon, but it is now reckoned from midnight and is thus identical with *Universal time, Solar time* at any place is the hour angle of the true Sun at that place, while *mean solar time* refers to the mean Sun. *Sidereal time* is the hour angle of the first point of Aries. See also: *Day. Julian calendar. Month. Second. Year.*

TIME, ATOMIC. Time based on the atomic clock, and now adopted as Earth time. Since the atomic year is 1 s longer than the GMT year, leap seconds will be inserted about once a year to keep GMT within 0·7 s of Earth time. See: *Clock, atomic.*

TIME BASE. (1) A wave form from which time intervals can be determined in terms of the relative amplitudes at different parts. (2) The horizontal trace of the spot on a cathode-ray tube when the spot is moved so that its position is a simple function of time (usually linear). If a wave form under examination deflects the spot at right angles to this time base the spot will describe a path on the screen which gives the wave form as a function of time.

TIME-BASE CIRCUIT. See: *Sweep circuit: time-base circuit.*

TIME CONSTANT. Of an electrical circuit: the time taken for the current in an inductance or the charge on a capacitor to reach a fraction $\left(1 - \dfrac{1}{e}\right)$, i.e. about 63%, of its final value. These constants are L/R and RC respectively, where L is the inductance, R the resistance and C the capacity.

TIME DILATION. The apparent reduction in the rate of a moving clock with respect to an identical clock at rest, when seen by a stationary observer. It follows from the special theory of relativity that the ratio between the two rates is $\sqrt{(1 - v^2/c^2)}$, where v is the speed of the clock, and c is that of light. Experimental verification of the effect has been provided by observations on the decay times of high-velocity pi-mesons compared with those of pi-mesons at rest.

TIME, EQUATION OF. The difference between apparent and mean solar time. It has a maximum

(positive) value of nearly $14\frac{1}{2}$ min in February (expressed as mean time minus apparent time) and a minimum (negative) value of nearly $16\frac{1}{2}$ min in November, and vanishes four times a year. The equation of time is often represented, especially on globes, by a curved figure known as an *analemma.*

TIME LAG, FORMATIVE. Of an electrical discharge: See: *Discharge, time lag of.*

TIME LAG, STATISTICAL. Of an electrical discharge: See: *Discharge, time lag of.*

TIME, MEASUREMENT OF. See entries under *Clock.*

TIME-OF-FLIGHT METHOD. A method of analysing a pulsed beam of particles (e.g. ions, electrons, neutrons) according to their speed, in which measurements are made of the times taken for them to travel over a path of a given length.

TIME, UNITS OF. See: *Second.*

TIME ZONES. Zones (approximately 15° of longitude in width) introduced by international agreement to overcome the continuous variation of time with longitude which follows from the adoption of astronomical time as a basis for civil time. All civil affairs in a given zone are governed according to the local mean time on an agreed meridian in that zone. Variations such as Summer time or Daylight saving time are statutory variations on this theme, and result in the transfer of the "local" meridian into an adjacent time zone, usually situated to the east so as to delay the apparent time of sunset (and of dawn!).

TIN, CRY OF. See: *Acoustic emission from materials.*

TINT OF PASSAGE. A tint exhibited by a Bravais biplate, which is sensitive to a slight change in polarization. See also: *Bravais biplate.*

TINTOMETER. A colorimeter in which colours are compared with those of standard solutions or with coloured glass slides. See also: *Lovibond tintometer.*

TITIUS-BODE LAW. See: *Bode law.*

TITRATION. Usually signifies the addition of measured amounts of a standard solution (typically from a burette) to a known volume of a second solution until the chemical reaction between the two is just completed, so as to obtain the strength of the second solution; but the roles of the two solutions are sometimes reversed. The end point of the reaction may be determined by the change in colour of an indicator, by the use of a colorimeter or spectrophotometer, by measurement of electrical conductance or magnetic susceptibility, by following the course of the reaction from the response of a pair of electrodes immersed in the unknown solution, or by the response of a radio-frequency oscillator.

TOG. A unit of thermal resistance used in connection with the thermal insulation of clothing. It is the temperature difference in °C required to produce a heat flow of 10 W/m².

TOISE. An old French unit of length, probably dating from A.D. 790, approximately equal to 1·949 m. The Toise de Perou (1735) became the French standard of length in 1766 and was used in the production of the first brass metre bar and, later, of the Metre des Archives. See also: *Metre*.

TOKAMAK. A device concerned with the possible production of a controlled thermonuclear reaction, in which a hot plasma is contained within an axi-symmetric torus linked to an iron-core transformer. A varying current in the primary circuit of this transformer induces a current in the plasma, which serves not only to confine the plasma but to heat it. An external magnetic field is needed to keep the plasma macroscopically stable, and methods of auxiliary heating are also necessary. Tokamak research originated in Russia in the 1950's but at first attracted little interest. See also: *Thermonuclear reaction, controlled*.

TOLERANCE DOSE. Of ionizing radiation: a term formerly used to denote what is now termed *maximum permissible dose equivalent*. See: *Dose equivalent, maximum permissible*.

TOMOGRAPHY. The radiography of a selected section or plane of a body (usually human) by the synchronous motion of an X-ray tube and film during exposure, so that images not in the required plane are blurred. See also: *Radiographic stereometry*.

TONNE. A mass of 1000 kg, equal to about 2200 lb and therefore known as a metric ton.

TOPOLOGICAL SPACE. See: *Space*.

TOPOLOGY. The study of geometrical properties which are unchanged by continuous deformation. It is concerned with the concepts of neighbourhood and continuity, and spaces which deal with these properties and concepts are known as topological spaces. See also: *Space*.

TOPOLOGY, ALGEBRAIC. The association of algebraic structures, e.g. groups, with topological spaces in such a way that they are topologically invariant.

TORIC SURFACE. A surface generated by the rotation of an arc of a circle about a line in its own plane but not through the centre of the circle. This surface is also known as a *toroidal surface*. See also: *Torus*.

TORNADO. An intensely violent and destructive vortex-like distribution of winds with a central column or funnel up which air rushes at speeds never approached in any other phenomenon. The maximum horizontal wind speeds in some tornadoes are also higher than those observed in other phenomena. The horizontal dimensions of a tornado are small (up to 200 m across), and it is not to be confused with a tropical cyclone (i.e. typhoon or hurricane) whose breadths are thousands of times bigger. See also: *Cyclone. Waterspout*.

TORO. A small asteroid which orbits the Earth with a period of 1·6 years, whose orbit resonates with that of the Earth ard to a lesser extent with that of Venus. At its closest approach to the Earth it is about 0·13 astronomical units away.

TOROIDAL COIL. A coil in which the turns are uniformly distributed round a closed circular ring. See also: *Torus*.

TOROIDAL SURFACE. See: *Toric surface*.

TORQUE. The turning moment exerted by a tangential force acting at a distance from the axis of rotation or twist. It is equal to the product of the force and the distance in question.

TORR. See: *Pressure*.

TORRICELLIAN VACUUM. The vacuum occurring in the space above the mercury column in the glass tube of a mercury barometer.

TORSION. The strain set up in a cylindrical bar by a twisting force about the axis of symmetry.

TORSIONAL COMPLIANCE. The reciprocal of *torsional rigidity*.

TORSIONAL MODULUS. The torsional rigidity per unit length.

TORSIONAL RIGIDITY. For a circular elastic rod: the applied moment necessary to produce unit angle of twist. Its reciprocal is the *torsional compliance.*

TORUS. A surface generated by a circle rotating about an axis in its own plane, which does not pass through the centre. It is the shape of a doughnut with a hole in it. If r is the radius of the circle, and k the distance from the centre to the axis of revolution, the volume of the torus is $2\pi^2 kr^2$ and its surface area is $4\pi^2 kr$. It is also known as an *anchor ring*.

TOTAL DIFFERENTIAL. Another name for total derivative. See: *Derivative, total.*

TOTAL HEAD. The *head* (i.e. the height of a liquid column necessary to develop a given pressure at the base) measured by a pitot tube, i.e. the sum of the hydrostatic pressure and the kinetic head.

TOTAL RADIATION COEFFICIENT. The Stefan or Stefan–Boltzmann constant. See: *Stefan–Boltzmann law.*

TOTAL REFLECTION. Of electromagnetic radiation or sound: the complete reflection of a beam of radiation at the interface between two media of refractive index n_1 and n_2 respectively (where the radiation is travelling in the medium of refractive index n_1) when the angle of incidence exceeds the angle given by $\sin^{-1}\dfrac{n_2}{n_1}$ (the *critical angle*), and n_1 is greater than n_2. Total reflection may be described as *internal* or *external* according to circumstances. For example, for X-rays falling upon a solid, the total reflection is external, since the refractive index is smaller in the solid than in air. For light totally reflected at the inside surface of a prism, for example, it is internal. In the latter case there is some evidence that the light actually penetrates the interface, or is propagated along it, before "reflection".

TOUGHNESS: IMPACT STRENGTH. Of a material: the maximum energy of deformation which may be produced by impact loading. The *static toughness* is given by the area under the true stress–strain curve; and an index of the *dynamic toughness* is given by the energy absorbed when a specimen of specified dimensions is subjected to a blow sufficient to produce fracture. See also: *Impact tests.*

TOWNSEND IONIZATION COEFFICIENTS. See: *Ionization coefficient, primary: Townsend first ionization coefficient. Ionization coefficient, secondary.*

TRACER. See: *Isotopic tracer.*

TRACKING. See: *Dielectric breakdown.*

TRACK OF IONIZING PARTICLE. The path of the particle as revealed, for example, in a cloud or bubble chamber, or in a photographic emulsion. When a number of tracks (*prongs*) originate at a common point the resulting appearance is known as a *star* (or *emulsion star* if the effect occurs in a nuclear emulsion). Stars are also produced by successive disintegrations of an atom in a radioactive series or by spallation reactions. Tracks in a nuclear emulsion are classified as *shower* (or *thin*) *tracks*, *grey tracks* and *black tracks* in increasing order of grain density.

TRADE WINDS. A system of steady winds, confined roughly between 35°N and 35°S latitude, which are roughly north-eastern in the northern hemisphere and south-eastern in the southern. At varying heights they give way to the *antitrades* which blow in the opposite direction. See also: *Antitrades. Doldrums.*

TRAJECTORY. The path followed by a material particle or massive body in an electric or gravitational field.

TRANSACTINIDE ELEMENTS. See: *Transuranic elements.*

TRANSCEIVER. (1) A combined radio transmitter and receiver. (2) An equipment that is capable of facsimile transmission and reception.

TRANSCENDENTAL FUNCTION. A function of a variable which is not rational nor algebraical, but is usually expressible as an infinite series or product.

TRANSCENDENTAL NUMBER. A number which cannot be expressed as the root of an algebraic equation. Examples of such numbers are e and π. See also: *Algebraic number.*

TRANSCONDUCTANCE. See: *Conductance, mutual.*

TRANSDUCER. (1) An instrument which expresses the magnitude of one physical quantity in terms of another. It is commonly used for remote measurement when the related quantity is more suitable for transmission than the quantity of interest, e.g. in telemetry. (2) A device by means of which energy can flow from one or more transmission systems to one or more other transmission systems. (3) A device for converting energy from one form to another, e.g. the conversion of electrical energy to acoustic energy in a microphone.

TRANSENSOR. A passive telemetric transducer, i.e. one using no batteries or external wires, for measuring physical quantities such as pressure, temperature, displacement, etc., at short ranges from a powered detector. Such transensors may be very small, and have found wide application in biological measurements, the detector being situated outside the body.

TRANSFER IMPEDANCE. See: *Reciprocity principle.*

TRANSFORM. See: *Fourier transform. Laplace transform.*

TRANSFORMATION. (1) In a material: the change in the configuration of the atoms or molecules of the material by the alteration of the external constraints, also known as *phase transformation.* The new phase is characterized by its chemical composition, its state of aggregation and, particularly in metals and alloys, by its crystal structure. See also: *Metallic transformations.* (2) A nuclear disintegration.

TRANSFORMATION CURVE, ISOTHERMAL. See: *TTT diagram.*

TRANSFORMATION, DIFFUSIONLESS. A martensitic transformation. See: *Martensitic transformation. Metallic transformations.*

TRANSFORMATION, GALILEAN. The transformation of the coordinates from one system to another moving with constant relative velocity, according to non-relativistic kinematics.

TRANSFORMATION, MATHEMATICAL. A change of variables in an algebraic expression. *Linear transformations* are concerned with vector spaces, *equivalence transformations* with operators, *congruent transformations* with quadratic forms and *conformal transformations*, which are defined separately, with geometrical properties.

TRANSFORMER. A device for changing power from one voltage or current level to another. It consists essentially of two or more magnetically coupled coils or windings, the varying flux in one, the primary winding, inducing voltages in the others, the secondaries. The magnetic coupling is usually increased by a core of high-permeability material (normally laminated) which threads all the windings.

TRANSIENT. A transient oscillation, i.e. one which is set up by a sudden disturbance and quickly damped out. It may be mechanical, electrical or acoustic.

TRANSIENT EQUILIBRIUM. See: *Radioactive equilibrium.*

TRANSISTOR. A semiconductor device, with three or more electrodes, capable of producing amplification or rectification. Transistors are much smaller, longer-lasting and more robust than corresponding thermionic devices, and require no heating current. They are, however, temperature sensitive. See also: Entries under *Semiconductor.*

TRANSIT. Of a celestial body: the passage of the body across the meridian of an observer. Measurements of the times at which such transits occur are commonly used to determine longitude.

TRANSIT CIRCLE: MERIDIAN CIRCLE. An instrument for the precise determination of the right ascension and declination of a celestial object. It is the fundamental instrument of positional astronomy and consists ideally of a rigid refracting telescope mounted with its optical axis accurately constrained in the meridian plane. The telescope may be rotated about an east–west line in this plane and its setting accurately determined. The position of the object under observation is determined from a recording of the time and zenith distance at which it crosses the meridian. In the *mirror transit circle* the telescope is replaced by a rotatable mirror and the object under observation is viewed by one or other of two horizontally mounted telescopes.

TRANSITION. (1) A phase transformation or a change of physical property. (2) A change in a nuclear, atomic state or molecular state.

TRANSITION, ALLOWED. A transition between two nuclear or atomic states for which the changes in the appropriate quantum numbers have, according to the relevant selection rules, a high probability of occurring.

TRANSITION, BRITTLE–DUCTILE. The change from a tough, ductile condition to a fragile, brittle one, when certain metals are cooled through specific ranges of temperature. The change is reversible and the two conditions of the metal are characterized by the way in which it fractures. See also: *Fracture.*

TRANSITION ELEMENTS. Elements which contain inner *d*-levels in their electronic structure rather than outer *p*-levels. In building up the Periodic Table there is a "transition period" after the *s*-levels are filled, and before the *p*-levels are filled, during which electrons are added to the inner *d*-levels. The *first transition group* extends from Sc to Zn, the *second* from Y to Cd and the *third* from La to Hg, including a further interruption for the lanthanide elements and involving the 4*f* levels. Transition elements of the same group are characterized by having similar physical and chemical properties. See also: *Actinide elements. Lanthanide elements. Rare earths. Appendix I.*

TRANSITION, FORBIDDEN. A transition between two nuclear or atomic states for which the changes in the appropriate quantum numbers have, according to the relevant selection rules, a very small probability of occurring.

TRANSITION POINT. The temperature at which a substance changes phase.

TRANSITIONS OF THE FIRST, SECOND AND HIGHER ORDERS. Phase transformations in which there are discontinuities in the first, second and higher derivations of the Gibbs free energy respectively. Transitions of the *first order* include ordinary changes of state and are marked by the existence of latent heat. Transitions of the *second order* are characterized by a discontinuity in the specific heat. Examples are the transition of helium I to helium II at 2·19 K and the transition from ferromagnetism to paramagnetism. See also: *Lambda point. Scattering, critical.*

TRANSIT TIME. Of a charged particle in an electric field: the time taken for the particle to travel between two specified points, e.g. from one electrode to another.

TRANSLATIONAL ENERGY. Of a molecule: that portion of the total energy which is attributable to translation. According to the equipartition principle it is $\frac{1}{2}kT$ (where k is Boltzmann's constant and T the absolute temperature) per degree of freedom, i.e. a total of $\frac{3}{2}kT$ per molecule. Quantum theory gives essentially the same results. See also: *Equipartition of energy. Rotational energy. Vibrational energy.*

TRANSMISSION FACTOR, OPTICAL. The ratio of the transmitted to the incident luminous flux for a parallel beam of light. Also known as *Transparency* and *Transmittance*. See also: *Light, transmission coefficient for.*

TRANSMISSION LINE. A system of conductors forming a continuous path from one place to another and capable of transmitting electromagnetic energy along this path. The term usually excludes waveguides.

TRANSMISSION LINE, ATTENUATION COEFFICIENT OF. The real part of the propagation coefficient, also called the *attenuation constant*. It determines the diminution in amplitude per unit distance.

TRANSMISSION LINE, PROPAGATION COEFFICIENT OF. For a uniform line of infinite length or terminated by a network simulating an infinite length: the natural logarithm of the complex ratio of the steady state amplitudes of a wave at a specified frequency, at points in the direction of propagation which are separated by unit length. It is also called the *propagation constant.*

TRANSMITTANCE. See: *Transmission factor (optical).*

TRANSMUTATION. The change of one nuclide to another by a nuclear reaction.

TRANSONIC FLOW. See: *Compressible flow.*

TRANSPARENCY. See: *Transmission factor (optical).*

TRANSPORT EQUATION: BOLTZMANN EQUATION. Gives the distribution function, f, for a system of particles which are not in equilibrium. For monatomic particles it is of the form

$$\frac{\partial f}{\partial t} + \mathbf{V} \cdot \operatorname{grad} f + \mathbf{F} \cdot \operatorname{grad}_v f = \left(\frac{\partial f}{\partial t}\right)_c$$

where $\mathbf{V}$ is the velocity, $\mathbf{F}$ is the acceleration, $\operatorname{grad}_v f$ is the gradient of f in velocity space and $\left(\frac{\partial f}{\partial t}\right)_c$ is the rate of change of f due to collisions. See also: *Boltzmann distribution law.*

TRANSPORT NUMBER. The fraction of the total current carried by ions of one kind during electrolysis.

TRANSPORT THEORY. A theory for the treatment of the migration of nuclear particles, phonons, etc., based on the Boltzmann equation. It is used in conditions to which Fick's law does not apply.

TRANSURANIC ELEMENTS. Elements whose atomic numbers are greater than that of uranium. The term originally referred to the actinides beyond uranium, but following the discovery of elements beyond 103 (lawrencium) it now covers the elements beyond the actinides, i.e. the *transactinide elements*. The highest element so far established is hahnium (105) but tentative claims for the existence of elements beyond this have been made.

TRANSVERSE WAVE. (1) An electromagnetic wave in which the electric and magnetic field vectors are normal to each other and to the direction of propagation. (2) An elastic wave in which the particles of the medium are displaced in a direction at right angles to the direction of propagation. See also: *Shear wave.*

TRAVELLING WAVE TUBE. An electron beam tube containing a delay line, usually in the form of a helix, in which energy is imparted from the beam to an electromagnetic wave in the line.

TRAVERSE. A method of surveying used when triangulation is not feasible on account of the terrain concerned being flat or heavily forested, for example. It is based on a set of connected lines (*traverse lines*) whose directions and lengths are accurately measured.

T-R BOX: T-R SWITCH. A transmission system designed to permit a common aerial to be used for transmission and reception of radio and radar signals. Also known as a *Duplexer*.

TREVELYAN ROCKER. A device in which vibrations are maintained by periodic expansion and contraction arising from the interruption of a supply of heat by the vibrations themselves.

TRIAD. See: *Polyad.*

TRIAD AXIS. See: *Crystal symmetry, rotation axis in.*

TRIANGULATION. A method of surveying in which the region to be surveyed is divided into a number of connected triangles, the angles of which are accurately measured. The side of one of the triangles is adopted as a *base line*, the accurate measurement of whose length and astronomical azimuths enables the lengths and azimuths of the sides of all the triangles to be computed. Until recently most triangulations have been computed on local origins and different spheroids, but the ultimate aim of the geodesist is to compute all triangulation all over the world on one datum, viz. a single spheroid with one origin. See also: *Base lines (geodesy). Geodimeter. Mekometer. Tellurometer. Trilateration, radar.*

TRIBOLOGY. The study of friction, wear, and lubrication, hence such terms as *triboelectricity, tribophysics.*

TRICLINIC SYSTEM. The crystal system exhibiting the least symmetry. Crystals in this system possess no axis or plane of symmetry but may possess a centre of symmetry. They have no faces at right angles to each other and are referable only to three unequal axes, none of which is at right angles to another. Also called *anorthic system.*

TRIGGER CIRCUIT. A circuit which triggers a desired event when its electrical condition is changed. See also: *Pulse generator. Thyratron.*

TRIGONAL SYSTEM. The crystal system which is characterized by the possession of a single axis of three-fold symmetry, either of simple rotation or inversion. It is sometimes regarded as a sub-division of the hexagonal system. Crystals in this system may be referred either to hexagonal axes or to a set of three equal rhombohedral axes which are equally inclined to each other (but not at 90°), and is therefore sometimes known as the *rhombohedral system*. See also: *Hexagonal system.*

TRIGONOMETRIC FUNCTIONS. Functions of an angle such as the sine, cosine, tangent, etc., which can be described in terms of the ratios between the sides of a right-angled triangle.

TRILATERATION, RADAR. A form of triangulation in which the geographic coordinates of survey stations are determined from distances between pairs of stations as measured by radar. See also: *Base lines (geodesy). Geodimeter. Mekometer. Tellurometer.*

TRIMER. A triple molecule formed from three single molecules of the same kind. See also: *Monomer. Polymer.*

TRIODE. A three-electrode valve or tube containing an anode, cathode, and a control electrode in the form of a grid, interposed between the anode and cathode to control the flow of electrons.

TRIPACK INTEGRAL. Three layers of emulsion, sensitive to blue, green and red light respectively, coated one above the other, and used in subtractive processes of colour photography. See also: *Colour photography.*

TRIPLE POINT. The point, characterized by temperature and pressure, at which the solid, liquid and gaseous phases of a pure component are coexistent. According to the phase rule the system is invariant, i.e. the conditions are uniquely fixed.

TRIPLET, SPECTRAL. A spectral line with a multiplicity of three. See also: *Multiplicity.*

TRISTIMULUS SPECIFICATION. See: *Colorimetry, additive.*

TRISTIMULUS VALUES. See: *Colorimetry, additive.*

TRITON. The nucleus of tritium, $_1^3H$.

TROCHOID. The path described by a point on the radius of a circle or on the radius produced, as the circle rolls in its own plane on a fixed straight line. When the point is on the circle itself, the path described is a *cycloid*. When the point is on the radius produced, the path has a loop between neighbouring arches and is known as a *prolate cycloid*. When the point is on the radius the path never touches the fixed line and contains no loop. It is known as a *curtate cycloid.*

TROCHOTRON. A multi-electrode tube in which the electron beam follows a trochoidal path under the influence of crossed electric and magnetic fields. The electron beam can be guided to successive electrodes thus forming an electronic switch which can be used, for example, in a scaler.

TROJANS. Two clusters of asteroids or minor planets (numbering about 15 in all) which travel in the same orbit as Jupiter, and oscillate about Lagrangian points. One cluster is about 60° of longitude ahead of Jupiter and the other about 60° behind, the Lagrangian points concerned being equidistant from the Sun and Jupiter. The Trojans are named after the heroes of the Trojan wars.

TROPICAL MONTH. The interval between successive passages of the Moon across the same meridian of longitude. Its length is very slightly less than that of the sidereal month (27·3216 mean solar days as against 27·3217). See also: *Month.*

TROPICAL YEAR. The interval between two successive passages of the Sun through the first point of Aries, or the interval between two similar equinoxes or solstices. It is equal to 365·242196 mean solar days. See also: *Year.*

TROPOPAUSE. See: *Troposphere.*

TROPOSPHERE. That layer of the atmosphere which lies below the stratosphere and extends down to sea level. It is in this region that the normal weather changes occur, and within it the temperature decreases with increasing height. The boundary between the troposphere and the stratosphere is known as the tropopause. See also: *Atmosphere.*

TRUE SUN. See: *Solar day.*

TSUNAMI. A wave (commonly but incorrectly called a tidal wave) that is similar in form to a storm surge but arises from an upheaval in the sea bottom often due to Earthquake. It can cause great damage when it approaches a shore or harbour. See also: *Storm surge: Storm tide. Waves, episodic.*

T TAURI STARS. See: *Stars, variable, nomenclature of.*

TTT DIAGRAM. A diagram which is used as an adjunct to the equilibrium diagram in the investigation of alloy systems. The temperature is plotted against log time and the diagram, which consists of S-shaped curves, is obtained by a series of operations involving the cooling of specimens so rapidly to a fixed temperature that transformation during the temperature change is prevented, and then following the subsequent isothermal transformation by physical methods. TTT stands for time, temperature and transformation. The diagram is also known as an *isothermal transformation diagram.*

TUBE OR VALVE. See: *Electron tube: Electron valve.*

TUBULAR MOTOR. See: *Linear motor.*

TUNING FORK. A steel fork with two prongs and a central handle, designed to retain a constant frequency of oscillation when struck.

TUNNEL DIODE. A p–n junction in which both sides are so heavily doped with impurities that the semiconductor resembles a metal. For low values of forward bias, electrons from the n side are enabled to tunnel directly into the empty energy levels on the p side, and for small forward voltages the junction behaves as if it were ohmic.

TUNNELLING: TUNNEL EFFECT. The passage through a potential barrier of a particle whose energy is less than the barrier height. It is impossible according to classical mechanics but has a finite (but small) probability according to wave mechanics. The wave associated with the particle may be thought of as being nearly totally reflected by the barrier with a small fraction being transmitted through it.

TURBIDIMETRY. See: *Nephelometry.*

TURBIDITY. The cloudiness in a liquid caused by the presence of a finely divided suspension. The term sometimes denotes the suspended matter itself.

TURBINE. A rotary motor in which a wheel or drum with curved blades is driven by a moving fluid. The fluid may be air, gas, steam, water, etc., the earliest turbines being the windmill and the water wheel. The flow may be axial or radial, and the design of the turbine rotor and blades modified accordingly.

TURBINE, GAS. An internal combustion engine in which mechanical energy is obtained from the expansion of a high-temperature, high-pressure gas stream provided by the combustion of a suitable fuel with compressed air. The expanded gas drives the compressor via a turbine, and may either drive a second turbine to provide shaft power, or may be expanded to atmospheric pressure in the exhaust system to provide a jet of high velocity gas for propulsion. In the *turbo-jet* both methods are used, with the emphasis on the first. Gas turbines are used for aircraft propulsion; marine, locomotive and automobile propulsion; stationary power plants; and as ancillaries in industrial plant.

TURBINE LIQUEFIER. A gas-liquefying machine cooling by expansion and using a turbine to provide the expanding mechanism. See also: *Joule effect.*

TURBINE PUMP COMBINATION. A device for transmitting power from an engine by the circulation of a hydraulic fluid between a centrifugal pump on the driving shaft and a turbine on the driven shaft.

TURBINE, STEAM. A machine in which expanded steam is used to supply the motive power.

TURBO-JET. See: *Turbine, gas.*

TURBO-MACHINE. (1) A machine for converting fluid energy into rotor energy, i.e. a turbine. (2) A machine for converting rotor energy into fluid energy, by a turbine working in the "wrong" direction. Such machines include compressors, fans and pumps.

TURBO-PROP. A gas turbine which provides power direct to a propeller shaft.

TURBULENCE. An irregular eddying motion characteristic of fluid motion when laminar flow tends to disappear. Unlike laminar flow, it leads to macroscopic mixing. Turbulence occurs not only in gases and liquids but also in solar and stellar atmospheres. See also: *Eddy. Fluid flow, laminar. Fluid flow, turbulent.*

TURBULENCE, CRITICAL VELOCITY FOR. The velocity of flow in a fluid above which laminar flow becomes unstable and changes into turbulent flow.

TURBULENCE, SOUND FROM. Sound produced by pressure variations arising from the eddies accompanying turbulent motion. Familiar examples are the roar of a gale and the noise (*aerodynamic noise*) made by turbulence at the surface of an aircraft fuselage. The human blood stream is also turbulent and gives rise to sound which can be measured by suitable instruments.

TURING MACHINE. A theoretical or idealized computer, first put forward by Turing in 1937. He considered a "universal automaton" that would perform any calculation that any special single-purpose automaton could, so providing the theory underlying the electronic digital computer. The Turing machine concept provided the basis for the War-time Colossus series of computers, the first of which was built in 1943, details not being released however until 1975 and then not in full. See also: *Colossus.*

TWILIGHT. The period between sunset and darkness, or darkness and dawn. *Civil twilight* is defined as the period during which the altitude of the Sun changes from $-50'$ to $-6°$ or vice versa, these angles taking account of refraction and scattering by the atmosphere. The precise time of any instant during twilight is often expressed on the *crep scale*, and is defined on this scale as (time of day − time of sunset)/(duration of twilight), or (time of sunrise − time of day)/(duration of twilight). One of the uses of the crep scale is in the study of certain biological effects.

TWILIGHT PHENOMENA. Optical phenomena caused by the scattering of sunlight at various levels in the atmosphere. They include the *purple light* at sunset, arising from scattering by small dust particles at high levels; *crepuscular rays* produced by scattering at clouds or mountains when the Sun is below the horizon; the *twilight arch*, a segment of the Earth's shadow seen in the east at sunset; and the *twilight airglow*. See also: *Airglow. Crepuscular rays. Green flash. Zodiacal light.*

TWIN BOUNDARY. The (plane) boundary between a pair of twin crystals. See: *Twinning.*

TWINKLING. The scintillation of the stars, arising from turbulent strata high in the Earth's atmosphere. See also: *Seeing, astronomical*

TWINNING. Of crystals: the process by which two crystals of the same composition (*twins*) are in intimate contact with each other and are related either (a) by reflection in a plane of the crystal structure (the *twinning plane*) which is common to both crystals, or (b) by rotation, usually of 180°, about an axis (the *twinning axis*) common to both crystals. These types are known as *reflection twins* and *rotation twins* respectively; and, in crystals of high symmetry, twins may be of both types simultaneously.

TWO-BODY PROBLEM. The problem of the motion of two particles which move under the influence of their mutual interactions, and of nothing else. It is the only many-body problem capable of exact solution and is, indeed, the foundation of celestial mechanics. See also: *Many-body problem. Three-body problem.*

TWO-ELECTRON BOND. See: *Heitler–London theory.*

TYNDALL EFFECT. The scattering of light by a suspension of very small particles, for example, the scattering by a sol. The light is polarized and the sol appears cloudy. See also: *Scattering, Rayleigh.*

TYNDALLIMETRY. See: *Nephelometry.*

TYNDALL–RÖNTGEN EFFECT: OPTIC-ACOUSTIC EFFECT. The production of periodic pressure fluctuations and sound emission from a radiation-absorbing gas or vapour exposed to periodically interrupted thermal radiation. It arises from the successive heating and cooling of the gas or vapour.

TYPE-TOKEN RATIO. The ratio of the number of different words in a passage of written or spoken speech to the total number of words in the passage.

TYPHOON. See: *Cyclone*.

u

UBV PHOTOMETRY. The photometric measurement of the apparent magnitude of a star through three colour filters: ultraviolet (3600 Å), blue (4400 Å) and "visual" (5500 Å). Systems involving more colours are also used, i.e. red (R, 7000 Å), infrared (IR, 10 000 Å) and bolometric (bol, total radiation). The difference between the magnitude measured at one wavelength and that at a longer wavelength is known as the colour index, e.g. B-V and U-B. (Note 1 Å $= 10^{-10}$ m).

U-CENTRE. See: *Colour centres.*

ULTIMATE BALL HARDNESS. See: *Meyer hardness number.*

ULTIMATE LINES. See: *Raies ultimes.*

ULTIMATE TENSILE STRESS: TENSILE STRENGTH. The tensile stress at which a body will fracture, or will continue to deform with decreasing load. It is sometimes termed the *breaking strength.*

ULTRAFILTRATION. The filtration of colloids and certain viruses by the use of filters with very small pore sizes. These filters are composed of animal, vegetable or synthetic membranes, suitably supported; or may be sorption filters, of which the zeolites are the most important.

ULTRAMICROSCOPE. A microscope in which a convergent pencil of bright light reveals objects, such as colloidal particles, by the diffraction ring systems to which they give rise, although the particles themselves are not resolvable by ordinary methods.

ULTRASONIC AMPLIFICATION. In semiconductors: the amplification of ultrasonic waves by their interaction with free electrons in a semiconductor. The interaction sets up periodic distortions of the lattice which give rise to a periodic fluctuation of potential.

ULTRASONIC MACHINING. The use of ultrasonic vibrations (as in a vibrating tool tip, or a stationary tip with vibrating work material) for machining purposes. Applications to metal forming (e.g. press forming, wire drawing and extrusion), in which vibration of the relevant part of the tooling reduces frictional effects between the tool and the work material, also occur.

ULTRASONICS, MEDICAL APPLICATIONS OF. The use of ultrasonic waves for diagnosis, therapy and surgery, employing pulse-echo techniques, Doppler techniques and focusing techniques (which are capable of inducing trackless lesions).

ULTRASONIC WAVES. Sound waves having frequencies which are too high for audibility. See also: *Sound.*

ULTRAVIOLET CATASTROPHE. The radiation of an infinite amount of heat as the wavelength of radiation tends to zero, whatever the temperature, according to the Rayleigh–Jeans law for the energy distribution of a black body. The conclusion no longer follows when this law is replaced by the Planck radiation formula. See also: *Infrared catastrophe.*

ULTRAVIOLET RADIATION. That part of the electromagnetic spectrum which extends from wavelengths of about 4000 Å (0·4 μm) down to below some 400 Å (0·04 μm), on the border of the X-ray region. The range below about 2000 Å (0·2 μm), for the investigation of which a vacuum is necessary, is known as the *vacuum ultraviolet.*

UMBRA. The dark central portion of the shadow of an object cast by a source of finite size, e.g. the shadow of the Earth or Moon cast by the Sun, as in an eclipse. The outer, less dark, part of the shadow is known as the *penumbra.*

UMKLAPP PROCESS. An interaction between excitations in a crystal in which crystal momentum is not conserved. It may be a phonon–phonon process (the source of thermal resistance in a pure dielectric crystal), or an electron–phonon process (an important contribution to the electrical resistivity in metals). The term Umklapp means, in German, "flop-over".

UMLADUNG. The absorption of slow beams of positive ions in gases by the capture of electrons from gas molecules, which neutralizes the ions.

UNBALANCED LOAD. (1) In an interconnected supply system, such as polyphase a.c.: a load such that the currents supplied by the individual voltage sources are unequal in magnitude or phase. (2) An unequal set of impedances which takes such unbalanced currents.

UNCERTAINTY PRINCIPLE: INDETERMINACY PRINCIPLE. A principle enunciated by Heisenberg which states that no measurement can determine both the position of a particle and its momentum (or any other pair of conjugate variables, such as time and energy) so accurately that the product of their errors is less than Planck's constant. It is a necessary result of the quantum theory. See also: *Complementarity principle. Conjugate variables.*

UNCOUPLING PHENOMENA. (1) In atoms: the uncoupling of spin and orbital angular momentum vectors, as in the Paschen–Back effect, where the cause is a strong magnetic field. (2) In molecules: deviations from Hund's coupling cases which sometimes occur with increasing rotation. See also: *Coupling.*

UNDERCOOLING. Another name for *supercooling.*

UNIAXIAL CRYSTAL. A crystal for which there is only one direction (the optic axis) along which light travels with a single velocity, i.e. with no double refraction.

UNIFORMITY RATIO. Of illumination: See: *Illumination, uniformity ratio of.*

UNIMOLECULAR FILM. A monolayer.

UNIPOTENTIAL LENS. An electrostatic lens in which focusing is brought about by the application of a single potential to one or more of its elements.

UNIT CELL. In crystallography: a unit of structure of a crystal which, when repeated regularly in three dimensions, builds up the extended crystal structure. It is usual to choose the smallest unit which will do this.

UNITED ATOM METHOD. An alternative approach to that of the Heitler–London theory, in which a diatomic molecule is regarded as consisting of one nucleus, formed by the coalescence of the nuclei of the two constituent atoms, surrounded by the appropriate electronic configuration. This is the "united atom". See also: *Heitler–London theory.*

UNITS, ABSOLUTE ELECTRICAL. Electrical quantities measured in terms of basic units of mass, length and time, plus one additional electrical quantity. The CGS e.m.u. system was originally adopted for this purpose, together with the permeability of free space, to which was assigned the value of unity. The word "absolute" in this context means merely that the measurement is made by reference to certain fundamental units. See also: *Abampere. Abohm. Abvolt.*

UNITS, CGS. A coherent system of units of which the basic units are the centimetre, gramme, and second. For application to electrical and magnetic phenomena the electrostatic CGS system, the electromagnetic CGS system, or the mixed (Gaussian) system may be used. See: following entries.

UNITS, CGS ELECTROMAGNETIC (E.M.U.). A system of units formed by adding to CGS units a definition of electric current (the *biot* or *abampere*) based on the interaction law for the force between two currents, the permeability *in vacuo* being set equal to unity. Such units are prefixed by ab-, as in abampere, abcoulomb, abvolt, etc. See also: *Biot.*

UNITS, CGS ELECTROSTATIC (E.S.U.). A system of units formed by adding to CGS units a definition of electric charge (the *franklin* or *statcoulomb*), based on the Coulomb law for the force between two charges, the permittivity *in vacuo* being set at unity. Such units are prefixed by stat-, as in statampere, statcoulomb, statvolt, etc. See also: *Franklin.*

UNITS, CGS MIXED OR GAUSSIAN. A system of units formed by adding to CGS units the electric quantities from the CGS electrostatic system and the magnetic quantities from the CGS electromagnetic system. The velocity of light *in vacuo* appears in some of the equations relating magnetic and electric quantities.

UNITS, GRAVITATIONAL A system of units in which force is expressed as the weight of a standard body, e.g. pound-force (or -weight) and kilogramme-force (or -weight). Units in this system vary from place to place, but in practice the variation is often ignored; however a fixed unit may be obtained by ascribing a definite value (9.80665 m s^{-2}) to the acceleration due to gravity. The use of such a system is gradually decreasing. See: *Appendix II.*

UNITS, IMPERIAL. A system of units in which the basic units are the yard, pound and second, often known, however, as the *foot-pound-second*

347

system. No corresponding units of electric current or charge exist.

UNITS, INTERNATIONAL. Electrical units based on the "mercury" ohm and the "silver" ampere. See also: *Ampere, international. Ohm, international.*

UNITS, METRIC. Any system of units in which the basic units are based on the metre, gramme (or kilogramme) and second.

UNITS, MKS. A coherent system of units of which the basic units are the metre, kilogramme and second. For application to electrical and magnetic phenomena the ampere is added, the corresponding system being known as the *MKSA system.*

UNITS, MKSA. A coherent system of units formed by the addition of one basic unit, the ampere, to the MKS system. In its rationalized form (that usually employed) it is also known as the *Giorgi* system. The value of the permeability of free space is $4\pi \times 10^{-7}$ H/m in the rationalized form and the corresponding value for the permittivity of free space is $(4\pi \times 9 \times 10^9)^{-1}$, i.e. $8\cdot854 \times 10^{-12}$ F/m. There is a general one-to-one ratio between related electrical and mechanical quantities in this system.

UNITS, PREFIXES FOR. The prefixes internationally agreed for forming the names of decimal multiples and submultiples of SI units. These prefixes signify multiplication by the powers of 10 indicated: tera (T) 10^{12}; giga (G) 10^9; mega (M) 10^6; kilo (k) 10^3; hecto (h) 10^2; deca (da) 10^1; deci (d) 10^{-1}; centi (c) 10^{-2}; milli (m) 10^{-3}; micro (μ) 10^{-6}; nano (n) 10^{-9}; pico (p) 10^{-12}; femto (f) 10^{-15}; atto (a) 10^{-18}.

UNITS, RATIONALIZED. Units in which, by introducing the factor 4π into the values of the permeability and permittivity of free space (with appropriate units), the factors 4π and 2π appear only in electromagnetic equations where they could be expected from the symmetry, i.e. 4π appears only in electric or magnetic systems with spherical symmetry and 2π in those with cylindrical symmetry.

UNITS, SI. See: *SI units.*

UNIT TRIANGLE. In the stereographic projection of a crystal: any triangle which includes all crystal-

lographically distinct vectors existing in the symmetry class of the crystal concerned. Thus, in the cubic system, there are 48 equivalent unit triangles for full symmetry.

UNIT VECTOR. A vector of unit magnitude, used to express the direction of a vector quantity. Thus a vector quantity **a** may be written $a\mathbf{k}$, where **k** is a unit vector and a is the magnitude of the vector.

UNIVERSAL CONSTANTS. Another name for fundamental constants. See: *Fundamental physical constants.*

UNIVERSAL TIME. See: *Time.*

UNMODIFIED X- OR γ-RAY SCATTER. The coherent component of the scattered radiation, at the original wavelength, when X- or γ-rays interact with extranuclear electrons. See also: *Compton effect.*

UNSHARPNESS. A measure of the lack of definition in a radiographic image. See also: *Radiographic definition.*

URANIUM SERIES. The radioactive series of elements beginning with ^{238}U and ending with ^{206}Pb. It is one of the three naturally occurring radioactive series, the others being the actinium series and the thorium series, which are separately defined. The series is also known as the *4n + 2 series,* since the atomic number of each member can be expressed in this way.

URANUS. The seventh planet in the solar system in order of distance from the Sun. Its diameter is about 4 times that of the Earth, its mass about $14\frac{1}{2}$ times, its density about $0\cdot3$ times, and its period of rotation (retrograde) doubtful, the "standard" figure of $10\cdot8$ h being appreciably too small. It has five satellites and it is noteworthy that its equator is almost perpendicular to the ecliptic. It has recently been found to possess a ring system.

UTILIZATION FACTOR: UTILIZATION COEFFICIENT. Of illumination: See: *Illumination, utilization factor of.*

V

VACANCY–INTERSTITIAL PAIR. See: *Point defect.*

VACANT LATTICE SITE. See: *Point defect.*

VACUUM. (1) A portion of space devoid of matter. In practice, a region of space in which the pressure is very low. (2) A concept in the quantum theory of fields exemplified in Dirac's hole theory, according to which the "vacuum" consists of an infinite sea of electrons with negative energy. See also: *Dirac equation. Hole theory.*

VACUUM FLASK: DEWAR VESSEL. A double-walled vessel used for the storage of hot or cold liquids or solids, in which the space between the walls is evacuated and the walls themselves may be silvered, to reduce heat flow to (and from) the inside.

VACUUM GAUGE. See: entries under *Pressure gauge.*

VACUUM PUMP. A pump for the production of low gas pressures. See: entries under *Pump.*

VALENCY BAND. (1) The range of energy states in a solid crystal in which lie the energies of the valency electrons. (2) The band below the conduction band in an insulator or semiconductor. See also: *Band theory of solids.*

VALENCY: VALENCE. A measure of the ability of an atom of an element to combine with atoms of other elements. It is, in general, the number of unpaired electrons in the outer electron shell, but such a description is not always adequate. See also: *Electron, valency.* Various entries under *Bond.*

VALVE. A device which may permit flow to take place in one direction only. The term is applied to electron tubes and to devices for the control of liquids or gases, but not to electrical rectifiers other than electron tubes, i.e. not solid state devices. See also: *Electron tube: Electron valve.*

VAN ALLEN BELTS. Two extraterrestrial zones of high-energy particles situated at heights of about 2500–5000 km and 12000–20000 km respectively. The inner zone, composed mainly of protons of about 100 MeV energy but also containing electrons of lower energy, lies between geomagnetic latitudes 30°N and 30°S. The outer zone, believed to be composed mainly of electrons of energy 20–100 keV, extends roughly between geomagnetic latitudes 50°N and 50°S, but is extremely variable in position, shape and intensity.

VAN ARKEL PROCESS. A process for refining metals by the thermal dissociation of volatile halides on a heated surface.

VAN DE GRAAFF ACCELERATOR. An electrostatic generator in which use is made of the principle of the Wimshurst machine to produce very high voltages which may be used to accelerate charged particles (electrons or atomic particles) to high energies. Electric charge is sprayed on to an insulated moving belt at one end, and collected and stored at the other until the desired potential is achieved, when a continuous discharge through a suitable accelerating tube takes place, under conditions such that the rate at which charge is delivered is equal to that at which it leaves. The Van de Graaff generator is employed not only in particle accelerators for high-energy research but for high-voltage X-ray machines, typically working at 2 MeV, which are used mainly for medical and industrial radiography. See also: *Tandem generator.*

VAN DER WAALS ADSORPTION. Reversible adsorption between a solid and a gas, the forces involved being purely physical.

VAN DER WAALS EQUATION OF STATE. An equation of state for a real gas, which takes into account finite molecular size and intermolecular attractive forces. It may be written:

$$\left(p + \frac{a}{V^2}\right)(V - b) = RT,$$

where p is the pressure, V the volume per mole, R the gas constant per mole and T the absolute temperature, as in the equation of state for an ideal gas; and a and b are constants depending on the gas, a relating to intermolecular attractive forces and b to finite molecular size. See also: *Equation of state, Clausius.*

VAN DER WAALS FORCES. Weak interatomic or intermolecular attractive forces, arising mainly from interactions between dipoles, quadrupoles, etc., either permanent or induced, or resulting from fluctuations in the charge distribution. The latter type, known as *dispersion forces*, produce the largest contribution to the Van der Waals interaction between molecules.

VAN DER WAALS RADIUS. See: *Atomic radius.*

VAN'T HOFF REACTION ISOCHORE. For the variation with temperature of the equilibrium constant for a gaseous reaction in terms of the change in heat content: states that $(d \ln K_p)/dT = \triangle H/(RT^2)$, where K_p is the equilibrium constant at constant pressure, H is the heat of reaction at constant pressure, R is the gas constant, and T is the absolute temperature.

VAPOUR. A gas which is at a temperature below its critical temperature and can therefore be liquefied by the application of pressure. See also: *Gas.*

VAPOUR DENSITY. The mass per unit volume of a gas or vapour under specified conditions, usually expressed as the ratio of this density to that of oxygen under the same conditions. The concept is useful only when perfect gas behaviour may be assumed.

VAPOUR PRESSURE. The pressure exerted by the vapour of a substance which is in equilibrium with its condensed phase. This phase may be solid, liquid or supercooled liquid, with correspondingly different values of vapour pressure. The vapour pressure, also known as the *vapour tension* or *saturated vapour pressure*, is a characteristic monotonically increasing function of temperature, with a discontinuous derivative at the melting point (triple point).

VAPOUR PRESSURE, GENERAL EQUATION FOR. An equation for the vapour pressure of a solid, based on the assumption that the ideal gas laws hold, which may be written as

$$\log p = -\frac{L_0}{RT} + \frac{5}{2} \log T + \left(\int_0^T \frac{dT}{RT^2} \right) \times$$

$$\times \left(\int_0^T (C_1 - C_2)\, dT \right) + C_3,$$

where L_0 is the molecular latent heat of vaporization at absolute zero, R is the gas constant, T is the absolute temperature, C_1 is the internal molecular heat of the vapour due to rotation and vibration, C_2 is the molecular heat of the solid, and C_3 is a constant known as the *chemical constant* or *vapour pressure constant.* The equation for a liquid is slightly different.

VAPOUR TRAP. A device for removing vapour from a low-pressure pumping system to allow the production of high vacua. Common types involve the use of a Dewar vessel, a refrigerated baffle, or a baffle of activated charcoal.

VARIABLE. See: *Derivative. Variate.*

VARIABLE-AREA TRACK. See: *Sound track.*

VARIABLE-DENSITY TRACK. See: *Sound track.*

VARIABLE-MU (μ) VALVE OR TUBE. An electron valve or tube of which the effective mutual conductance can be altered smoothly over a wide range of grid bias.

VARIANCE. (1) Of a series of observations or statistical data: the square of the standard deviation. (2) Of divariant or thermodynamic systems: a term sometimes used instead of "degrees of freedom" to avoid confusion with the mechanical degrees of freedom used, for example, in molecular theory.

VARIANCE, ANALYSIS OF. The statistical use of variance measurements to test for significant non-random causes of experimentally observed results. It provides, essentially, a test of the homogeneity of a set of data, and aims, if the data are shown to be not homogeneous, to isolate the factor or factors causing the heterogeneity and to establish their relative importance. For this analysis to be possible it is necessary that the experiments be very carefully planned, so as to justify the assumptions on which the analysis is made. See also: *Frequency distribution. Significance. Standard deviation. Statistics.*

VARIATE. A statistical *variable*, i.e. the numerical value of a quantity which forms the subject of statistical analysis.

VARIATION METHOD. The most important of the general methods for finding approximate solutions of the Schrödinger equation. It is based on the *minimum energy theorem* which states that an approximate energy, found by integrating the Hamiltonian operator H over any approximate wave function ϕ, is never less than the true ground state energy, i.e.

$$\left(\int \phi^* H \phi \, dV \right) \Big/ \left(\int \phi^* \phi \, dV \right) \geqq E_0,$$

where ϕ^* is the conjugate complex of ϕ, V the potential energy and E_0 the ground state energy. The underlying idea in this method is to replace the problem of solving a differential equation by that of finding the maximum or minimum value of an integral. It has many other applications, e.g. the Rayleigh–Ritz method.

VARIOMETER. See: *Magnetic variometer.*

V-CENTRE. See: *Colour centres.*

VECTOR. A quantity possessing both magnitude and direction, as distinguished from a scalar, which has magnitude only. Vectors add according to the parallelogram law, i.e. if **A** and **B** are the two sides of a parallelogram, both drawn from the origin, then the vector **A** + **B** is represented in magnitude and direction by the diagonal of the parallelogram passing through the origin.

VECTOR FIELD. A region of space each point of which is described by a vector. Thus, in three dimensions, each point is described by three quantities, the *components* of the vector along the coordinate axes. Examples are wind velocities in the atmosphere, and electrostatic or electromagnetic fields. A vector field may be expressed in terms of two potentials, the scalar and vector potential. See also: *Scalar field. Vector potential.*

VECTOR, IRROTATIONAL. A vector whose curl is zero, which is therefore the gradient of a scalar function of position. See also: *Scalar potential.*

VECTOR POTENTIAL. Of a vector field: the vector **A**, of zero divergence, in the expression **F** = *grad* ϕ + curl **A** for a vector field **F** which is finite, uniform and continuous, where ϕ is the scalar potential. See also: *Scalar potential.*

VECTOR PRODUCT. Of two vectors **A** and **B**: a vector which is perpendicular to both **A** and **B**, whose magnitude is $AB \sin \theta$, where θ is the angle between them, and whose direction is that of a right-handed screw rotating from **A** to **B**. It is commonly denoted by **A** × **B** or **A** ∧ **B**, although [AB] and [A · B] are also used. It may be noted that **A** × **B** is not identical with **B** × **A**: they have the same magnitudes but opposite directions. See also: *Scalar product.*

VECTOR PRODUCT, TRIPLE. The vector product of a vector and a vector product of two other vectors. It is commonly represented by **A** × (**B** × **C**) or **A** ∧ (**B** ∧ **C**).

VEERING (METEOROLOGY). The clockwise rotation of the wind direction. The opposite of backing.

VEGARD LAW. States that the unit cell dimensions in a binary substitutional solid solution vary linearly with the percentage atomic composition. The law is only approximately true.

VELOCITY. A vector quantity specifying the rate of change of position of a body, together with its direction of motion.

VELOCITY ANALYSER: VELOCITY SELECTOR. A device for separating a stream of particles into groups according to velocity. It may be a time-of-flight device or diffracting crystal, as in a neutron monochromator, or it may work on the same principle as the mass spectrometer, to name some examples.

VELOCITY CONSTANT. Of a chemical reaction: See: *Chemical reaction, specific reaction rate constant of.*

VELOCITY HEAD: KINETIC HEAD. For a perfect fluid in steady flow one half the ratio of the square of the flow velocity to the gravitational acceleration. It is the height of a column of fluid giving a hydrostatic pressure of $\frac{1}{2} v^2$, where is the density and v the flow velocity. Pressure head + Velocity head + Elevation head = constant, according to the Bernouilli equation. See also: *Elevation head. Pressure head.*

VELOCITY MODULATION. See: *Modulation.*

VELOCITY POTENTIAL. For moving particles: a scalar function of position whose gradient gives the particle velocity at any point. It may be relevant to fluid flow, the propagation of sound, the movement of electric charge and so on. See also: *Scalar potential.*

VENA CONTRACTA. The point of minimum cross-sectional area in a jet of fluid discharged from an orifice.

VENTURI METER. A flow meter for use in closed pipes. It consists of a constriction inserted in the line of piping (a *venturi tube*), together with means for measuring the loss of pressure head over the convergent part of the constriction.

VENUS. The second planet from the Sun in the solar system, and the nearest to the Earth. Its diameter is 0·95 times and its mass about eight times that of the Earth. Its period of rotation is about 8 months (retrograde). It has no satellite. Its atmosphere consists largely of CO_2 and its surface composition resembles that of the Earth, the surface temperature being about 740 K.

VERDET CONSTANT. The constant of proportionality relating the rotation of the plane of polarization, in the Faraday effect, to the optical path length and magnetic field strength. See also: *Faraday effect.*

VERNIER ACUITY: CONTOUR ACUITY. The power of the eye to distinguish a displacement between two parts of a line, as in reading a vernier. It is expressed in terms of the least detectable angular separation between the two parts and varies, for unaided vision, from 3 to 12 s of arc.

VERTICAL ANGLE. The angle which a vertical circle on the celestial sphere makes with the prime vertical. See: *Prime vertical.*

VESTA. See: *Asteroid.*

VIBRATION. A term originally denoting the elastic oscillation of a body, but now extended to cover oscillations in general. See also: *Forced vibrations.* Entries under *Oscillation.*

VIBRATIONAL ENERGY. Of a molecule: that portion of the total energy which is attributable to vibration. According to the equipartition principle it is kT (where k is Boltzmann's constant and T the absolute temperature) per molecule for each mode of harmonic vibration, i.e. twice the kinetic energy per degree of freedom, account being taken of the potential energy. See also: *Equipartition of energy. Rotational energy. Translational energy.*

VICINAL FACES. Macroscopic crystal faces having high Miller indices instead of the low indices characteristic of normal faces. As they are, of necessity, distributed in accordance with the crystal symmetry, they are of value in indicating the true symmetry of a crystal.

VIDICON. A type of television camera tube in which a low-velocity electron beam scans a photo-conducting mosaic on which the optical image is projected. Different photoconductive materials are available which are sensitive to X-rays, ultraviolet rays and infrared rays, as well as to visible light, so that the "optical" image may arise from any of these radiations. Consequently a variety of medical and industrial applications of the Vidicon have been made, including X-ray radiography and infrared image conversion. See also: *Camera tube.*

VILLARD CIRCUIT. A half-wave voltage doubling circuit in which the voltages of two capacitors are added to that of a transformer secondary, producing a unidirectional voltage which, in alternate half cycles, approaches twice the peak voltage of the transformer itself.

VILLARI EFFECT. The changes in magnetic induction caused by mechanical stress. As the magnetic field strength is increased the sign of the effect may change. This is known as *Villari reversal.*

VIRIAL COEFFICIENTS. The empirical coefficients B, C and D in the equation of state for a real gas expressed in the form $pV = RT + Bp + Cp^2 + Dp^3$, where p is the pressure, V the volume per mole, R the gas constant per mole and T the absolute temperature. The temperature at which B changes sign (i.e. is zero) is that at or near which Boyle's law provides a good approximation to the true equation of state, and is known as the *Boyle temperature* of the gas concerned.

VIRIAL THEOREM. For a system of n particles acted upon by a force: states that the virial of the system,

$$-\tfrac{1}{2} \overline{\sum_{i=1}^{n} \mathbf{r}_i \cdot \mathbf{F}_i}$$

(where $\mathbf{r}_i$ is the position vector of the ith particle, $\mathbf{F}_i$ is the force acting on that particle, and the bar represents a suitable time average), is equal to the average kinetic energy of the system. Also, the virial of the ith particle, given by $-\tfrac{1}{2}\mathbf{r}_i \cdot \mathbf{F}_i$, is equal to the average kinetic energy of that particle. It is an important theorem in statistical mechanics. An analogue of the theorem is applicable in quantum mechanics, in which space averages are used instead of time averages.

VIRTUAL CATHODE. A region between the electrodes in an electron valve or tube where, owing to the space charge, the potential of the field is a minimum. This region can thus be considered to behave as a source of electrons.

VIRTUAL PROCESS. A process that can be represented by the emission of a sub-atomic particle and its absorption (or other interaction) within such a short time that the energy and momentum of the particle (a *virtual particle*) in the intermediate state are badly defined.

VIRTUAL STATE. Of an atomic nucleus: the state in which a nucleon can be spontaneously emitted by the nucleus. The existence of such a state does not imply the existence of a repulsive force, but rather that the kinetic energy of the nucleon is too great to allow the attractive force to hold the particle in the nucleus. See also: *Bound state.*

VIRTUAL WORK, PRINCIPLE OF. States that: a system with workless constraints is in equilibrium under applied forces if, and only if, zero (i.e. virtual) work is done by the applied forces in an arbitrary infinitesimal displacement that satisfies the constraints.

VISCOELASTIC SUBSTANCE. A solid or liquid substance which can both store and dissipate energy during mechanical deformation. See also: *Elastic liquid.*

VISCOMETER. An instrument for the measurement of viscosity. Most viscometers belong to one of the following four types: (a) *capillary*, which are based or Poiseuille's equation for the flow of liquid through a capillary tube; (b) *rotational*, which measure the viscous drag as a body is rotated in a liquid; (c) *oscillatory*, which measure the damping of a system in a liquid; and (d) *falling body*, which measure either the rate of fall of a body through a liquid or its terminal velocity. See also: *Terminal velocity.*

VISCOSITY. The resistance to fluid flow, set up by shear stresses within the flowing fluid.

VISCOSITY, COEFFICIENT OF. A coefficient (usually referred to simply as "the viscosity") relating the tangential force between neighbouring

layers in a flowing, non-turbulent fluid. It is defined as the tangential force per unit area required to maintain unit difference in relative velocity between two parallel planes situated at unit distance apart, i.e. it is the quantity η in the expression for the force per unit area: $F/A = \eta\dfrac{dv}{dx}$, where F is the tangential force, A the area of surface undergoing relative motion, v the relative velocity of the two planes and x their distance apart. The CGS unit is the *poise* (equal to 1 dyn s cm^{-2}, i.e. 10^{-1} N s m^{-2}), the SI unit being expressed in units of N s m^{-2} (i.e. Pa s). The reciprocal of the viscosity is known as the *fluidity.* See also: *Poise. Appendix II.*

VISCOSITY, KINEMATIC. The coefficient of viscosity of a fluid divided by its density. It measures the kinematic effect of the viscosity, i.e. the acceleration of the fluid. Typical values at 20°C are: for water 0·01 cm^2 s^{-1}, and for glycerine 6·8 cm^2 s^{-1}. The CGS unit of kinematic viscosity (cm^2 s^{-1}) is the *Stokes,* which is exactly equal to 10^{-4} m^2 s^{-1}, the SI unit being m^2 s^{-1}. See: *Appendix II.*

VISCOSITY, TEMPERATURE DEPENDENCE OF. (1) For a liquid: the relationship is given by $\eta = A \exp (B/RT)$, where η is the viscosity, A and B are temperature-independent parameters characteristic of the liquid, R is the gas constant and T the absolute temperature. (2) For a gas: the viscosity varies as $T^{\frac{1}{2}}$.

VISCOUS DRAG. The sum of the drag due to normal pressures and that due to skin friction. It is sometimes known as viscous resistance. For a spherical body it is given by *Stokes law.* See also: *Drag. Stokes law. Ship resistance. Terminal velocity.*

VISIBILITY. (1) A term formerly used to denote luminous efficiency. (2) The clarity with which objects can be seen. (3) The greatest distance at which objects can be seen.

VISION, PERSISTENCE OF. See: *Flicker.*

VISION, YOUNG–HELMHOLZ (OR HELM-HOLTZ) THEORY OF. See: *Colour vision, theories of.*

VISUAL ACUITY. The resolving power of the eye, expressed as the minimum detectable angular separation of point sources. See also: *Vernier acuity: Contour acuity.*

VISUAL PURPLE. A pigment (*rhodopsin*) occurring in the rods of the retina, which has a maximum absorption for light of wavelength about 5000 Å (0·5 μm), and therefore appears purple with white transmitted light. It is believed to be the pigment responsible for vision at low luminance levels.

VITREOUS HUMOUR. A fluid secreted in the eye, just behind the lens.

VITREOUS SOLID. A solid in the glassy state, its structure being that of a supercooled liquid of high viscosity. See also: *Glass.*

VLBI (VERY LONG BASELINE INTERFERO-METRY). Interferometry using a system of radio antennae spaced some hundreds or thousands of miles apart, used in radio astronomy.

VOICE MECHANISM. The system for the production of speech. It consists of the lungs and associated muscles, which maintain a steady flow of air; the larynx, which modulates this steady flow by the vibration of the vocal chords; and the vocal cavities of the pharynx, mouth and nose, which vary the harmonic content of the output of the larynx. The average frequency of the voice, produced by the vocal chords, lies between 90 and 160 Hz for the male and between 190 and 330 for the female, but extreme values of the singing voice range from 60 Hz (bass) to 2000 Hz (soprano). See also: *Formant. Speech spectrometry.*

VOID SWELLING. The swelling of metals and alloys as a result of irradiation by fast neutrons, typically in a fast reactor. It arises from the formation of small cavities about 100 Å (10^{-8}m) in diameter, and calls for specially designed components.

VOIGT EFFECT: MAGNETIC DOUBLE RE-FRACTION. The double refraction exhibited by light when it is passed through a vapour in a direction perpendicular to a strong magnetic field.

VOLT. The MKSA and SI unit of potential difference. It is defined as the difference of potential between two points of a conducting wire carrying a constant current of one ampere, when the power dissipated between these points is equal to one watt. See also: *Abvolt. Volt, international. Appendix II.*

VOLTAGE DIVIDER: POTENTIAL DIVIDER. A high resistance, provided with a fixed or adjustable tapping, used to provide a voltage between the tapping and one end terminal of the resistance, which is a known fraction of the applied voltage.

VOLTAGE, FORWARD. Voltage of that polarity which produces the larger current, e.g. at a p—n junction.

VOLTAGE, R.M.S. See: *Root-mean-square (r.m.s.) value.*

VOLTAGE TRANSFORMER: POTENTIAL TRANSFORMER. A transformer whose primary winding is connected to the main circuit and secondary winding to an instrument, e.g. a voltmeter. It permits instruments to be isolated from a high voltage supply and is used to extend their range.

VOLTAIC CELL. Another name for *electrochemical cell.*

VOLTAMETER: COULOMETER. An electrolytic cell used for measuring the quantity of electricity passing through a circuit in a given time by the determination of the amount of metal deposited, or gas liberated, due to the passage of the current. See also: *Silver voltameter.*

VOLTAMMETRY. Polarography in which any type of micropolarizable metallic electrode is used. See also: *Polarography.*

VOLTA PILE. The first electrochemical cell. It consisted of a pile of alternate disks of copper and zinc, each pair being separated from the next by paper moistened with brine.

VOLT, INTERNATIONAL. The former standard of electromotive force, defined as the steady potential difference which must be maintained across a conductor which has a resistance of one international ohm and which carries a steady current of one international ampere. One international volt is equal to 1·00034 absolute volts (abvolts). See also: *Abvolt.*

VOLTMETER. An instrument for measuring voltage. It commonly consists of a low-range ammeter with a high resistance in series.

VOLTMETER, ATTRACTED-DISK. A voltmeter, for the measurement of voltages in the range 10–500 kV, which consists of two parallel conducting plates, one fixed and one movable. Measurement of the movement of the latter, when attracted by the former, is the basis of the instrument. It is also known as the *attracted-disk electrometer.*

VOLUME. (1) The amount of space occupied by a body. See also: *Gallon. Litre.* (2) The loudness of a sound or the magnitude of the current giving rise to it.

VOLUMETRIC ANALYSIS. A form of chemical analysis in which a standard solution is used for determining the amount of a particular constituent present in another solution by measuring the volume of the standard solution that must be added to a given volume of the solution of interest to achieve chemical equivalence.

VOLUMETRIC GLASSWARE. Accurately graduated measuring flasks, cylinders, burettes and pipettes, mainly for use in chemical analysis.

VON MISES CRITERION. For yielding in an isotropic polycrystalline material: states that slip must be possible on at least five independent slip systems.

VORTEX. An intense spiral motion of a fluid in a limited region.

VORTEX SHEET. A surface separating two regions of fluid motion across which there is a discontinuous change in the velocity component tangential to the surface. The concept is of value in the study of trailing vortices.

VORTEX STREET. A regular arrangement of vortices situated alternately in two parallel rows as are street lamps. Such a vortex street resembles a double trail of vortices in the wake of a bluff cylinder in a certain range of Reynolds number, and was proposed by von Kármán as a mechanism for the drag of a solid body in the flow of a fluid of small viscosity.

VORTEX, TRAILING. One of a series of vortices passing from the main surfaces of an aeroplane and extending down-stream behind it.

VORTICITY. A measure of the rate of rotation in a fluid. At a point in a uniformly rotating fluid it is twice the mean angular velocity of a small element of fluid surrounding that point.

VULCANISM. A general term for all the phenomena associated with the expulsion of molten lava, lava fragments and gases from the deeper parts of the Earth, including the origin of the phenomena and their geological characteristics and locations.

VULCANIZATION. The treatment of rubber with sulphur or sulphur compounds resulting in a change in physical properties.

W

WADSWORTH SPECTROGRAPH. A spectrograph similar to the Littrow spectrograph, in which the mounting is composed of a prism and plane mirror and operates as a constant deviation prism. It is a common mounting for infrared spectrometers.

WAGNER EARTH. For an a.c. bridge: a subsidiary bridge which balances the main bridge with respect to earth, eliminating stray capacitances.

WAKE. The region behind a body which is moving in a fluid. It is characterized by turbulence, and the pressure is less than in the undisturbed portion of the fluid.

WALDEN RULE. For the conductance of a given electrolyte: states that the product of the equivalent conductance at zero concentration and the viscosity of the solvent is constant.

WALL EFFECT. In an ionization chamber or counter: the contribution to the ionization by electrons liberated from the walls.

WANKEL ENGINE. Proprietary name of a rotary internal combustion engine with a single sparking plug. It has a rotor of approximately triangular cross section, which is geared to a central driving shaft and rotates inside a close-fitting oval chamber so that, as each face of the rotor passes the sparking plug, a power stroke occurs.

WARM FRONT AND WARM OCCLUSION. See: *Front (meteorology).*

WASHBURN CORRECTIONS. In bomb calorimetry: the corrections that have to be made for the internal energy of the residual gas, water vapour, liquid water, and gases in solution such as carbon dioxide and nitric oxide. See also: *Calorimeter, bomb.*

WATCH. A portable timepiece relying on a coiled spring for power. See also: entries under *Clock or watch.*

WATCH, POSITION ERRORS OF. The changes in the rate of a watch produced by changes in its position relative to vertical. They may arise from lack of poise of the balance or changes in frictional conditions.

WATER EQUIVALENT. The product of specific heat and mass.

WATER HAMMER. A transient pressure wave occurring in a pipe filled with a fluid under pressure when a sudden change in the rate of flow occurs. It produces elastic waves which may travel along the pipe and give rise to a pulsating sound. The best known example is that afforded by water pipes (hence the name), but similar effects also occur in arterial blood.

WATER OF CRYSTALLIZATION. See: *Hydrated compound.*

WATERSPOUT. A tornado that occurs at sea, in which a column of water stretches from sea to cloud. See also: *Tornado.*

WATER, STRUCTURE OF. See: *Liquid structure.*

WATER TUNNEL: CAVITATION TUNNEL. Experimental equipment used for testing and research in which cavitation phenomena are reproduced.

WATT. The unit of power in the MKSA and SI systems. It is the power which gives rise to the production of energy at the rate of 1 J/s, and is equal to 10^7 erg/s. See also: *Volt. Appendix II.*

WATTLESS COMPONENT. See: *Reactive component.*

WAVE ANALYSER. An instrument for the analysis of a regular wave motion into its component frequencies.

WAVE DRAG. Drag associated with the presence of shock waves.

WAVE ENERGY DENSITY. Of a medium traversed by a wave: the total energy per unit volume.

WAVE EQUATION. (1) For wave motion in general: the equation $\nabla^2 \phi = \dfrac{1}{V^2}\left(\dfrac{\partial^2 \phi}{\partial t^2}\right)$, where

ϕ denotes the quantity characterizing the particular wave motion concerned, V is the velocity of the wave and t is time. (2) The Schrödinger equation.

WAVE FILTER. An electrical network whose insertion loss is low (and approximately constant) over a band or bands of frequencies and high over other bands, and can thus act as a filter.

WAVE FORM. The shape of the curve of the instantaneous values of a periodically varying quantity plotted against time.

WAVE FRONT. A surface at all points of which, at a given instant, a wave has the same phase. Thus, the radiation from a point source has spherical wave fronts.

WAVE FUNCTION. (1) For *electric charge:* a function giving the probability distribution of electric charge in space. (2) *Atomic:* the wave-mechanical description of the stationary states of atoms. See also: *Schrödinger equation.*

WAVE FUNCTION, ANTISYMMETRIC. See: *Fermi–Dirac statistics.*

WAVE FUNCTION, SYMMETRIC. See: *Bose–Einstein statistics.*

WAVEGUIDE. A system consisting of a metal tube, dielectric rod or tube, or a single wire, for the transmission of electromagnetic energy. Only waves of certain modes can be so transmitted.

WAVEGUIDE, ACOUSTIC. A waveguide for use with acoustic energy rather than electromagnetic. The simplest form is the old-fashioned speaking tube.

WAVEGUIDE JUNCTION. A point at which waveguides are coupled together.

WAVEGUIDE JUNCTION, HYBRID. A four-arm waveguide junction in which power fed into any one arm is divided equally between two other arms, when terminated by matched loads. No power flows into the remaining arm.

WAVEGUIDE LENS. A waveguide designed in such a way as to focus electromagnetic waves. Its shape is controlled by the fact that the refractive index is less than unity, the simplest being one in which power is fed in via an ellipsoidal surface and radiated from a plane one. See also: *X-ray lens.*

WAVEGUIDE, MATCHED. A waveguide having no reflected wave at any transverse section.

WAVEGUIDE, MODE. One of the possible configurations of the electromagnetic field of a travelling or standing wave in a uniform waveguide, characterized, for example, by the presence or absence of transverse and longitudinal components of the electric and magnetic fields. Such modes also exist in transmission lines and cavity resonators. See also: *Mode, dominant: Mode, fundamental.*

WAVE IMPEDANCE. At a point in a waveguide and for a given frequency: the ratio of the complex magnitude of the transverse electric vector to that of the transverse magnetic vector, chosen so that the real part is positive.

WAVE INTENSITY. The product of the wave energy density and the wave velocity. It is the rate of flow of energy per unit area of the wave front in the direction of propagation.

WAVELENGTH. The distance, in the direction of propagation of a wave motion, between two points in neighbouring cycles which have the same amplitude and phase.

WAVELENGTH CONSTANT. (1) See: *Wave number.* (2) Another name for phase-change coefficient.

WAVELENGTH STANDARDS. For electromagnetic radiation: (1) The *primary standard:* a standard to which all wavelengths are ultimately referred. The internationally agreed standard is that of the radiation corresponding to the transition between the levels $2p_{10}$ and $5d_5$ of the krypton isotope ^{86}Kr, taken (in vacuum) as $6057 \cdot 802\,10_5$ Å, where 1 Å is equal to 10^{-10} m exactly. See also: *Metre.* (2) *Class A secondary standards:* highly reproducible standards which have been directly compared with the primary standard with an accuracy comparable with the accuracy of that standard. The best sources of such standards are the noble gases. (3) *Class B secondary standards:* standards which are useful for routine measurements with a precision of better than 0·01 Å.

WAVE-MAKING RESISTANCE. See: *Ship resistance.*

WAVE MECHANICS. A form of quantum mechanics in which account is taken of the wave nature of sub-atomic and atomic particles. It stands in much the same relation to classical mechanics as does physical optics to geometrical optics. See also: *Matrix mechanics. Quantum mechanics. Schrödinger equation. Schrödinger equation, many body.*

WAVEMETER. A device for determining the frequency of an electric wave.

WAVE NORMAL. A direction at right angles to the wave front. In an isotropic medium the wave normal lies along the direction of propagation. See also: *Rays. Wave vector.*

WAVE NUMBER. Of a harmonic wave: the reciprocal of the wavelength. It is used in spectroscopy instead of frequency. Sometimes the wavelength constant $2\pi/\lambda$, where λ is the wavelength, is also called the wave number.

WAVE PACKET. A pulse, resulting from the superposition of waves of different wavelength, in which the amplitude is finite over only a limited region. If all the component waves move in the same direction the packet moves at the group velocity. See also: *Ehrenfest theorem.*

WAVE PLATE: RETARDATION PLATE. A plate of birefringent material cut so that a desired path difference is introduced between the ordinary and extraordinary vibrations of a ray of light travelling perpendicularly through it. See also: *Half-wave plate. Quarter-wave plate. Quartz wedge.*

WAVE POWER. The use of wave motion as a means of providing power. Various types of device have been proposed to convert the rise and fall of a floating object or system of objects into useful power. See also: *Tidal power.*

WAVE RESISTANCE. See: *Ship resistance.*

WAVES. (1) *Progressive:* a periodic disturbance propagated through a medium as a series of waves. At each point in the medium the physical quantity associated with the disturbance (e.g. mechanical displacement, electrical energy, electromagnetic energy) varies periodically with time about an equilibrium value but the medium itself suffers, in general, no translation. For harmonic oscillations about the equilibrium value the wave form takes the shape of a sine curve, the product of the wavelength and the frequency being equal to the phase velocity of the waves. (2) *Standing:* the waves in a "frozen" wave pattern which is formed by the interference of two progressive wave systems. Standing waves may occur in waveguides and transmission lines, may be optical or acoustic, may occur in streams of water, or generally wherever periodic waves are produced or scattered under suitable conditions. Standing waves are also known as *stationary waves.* (3) *In general:* any series of values of a periodically varying quantity which are a function of position as well as of time, e.g. a time-varying voltage or current in an electrical network. See also: *Waves of translation.*

WAVES, EPISODIC. Giant waves associated with ocean currents, turbulent seas and seabed topography, which are extremely dangerous even to the largest of ships. Their origin is poorly understood although the areas (often close to a continental shelf) and times of their occurrence are well known. They are sometimes called (incorrectly) *freak waves.* See also: *Storm surge: Storm tide. Tsunami.*

WAVES, GRAVITATIONAL. See: *Gravitational field.*

WAVES, GRAVITY. Surface waves whose motion is controlled by gravity and not by surface tension. They may be small wave systems in the atmosphere or surface waves in water. See also: *Waves, surface.*

WAVES OF TRANSLATION. Waves in shallow water, in which the height of the crests above the mean level becomes greater than the depth of the troughs below it, and in which water is carried forward.

WAVES, ROTATIONAL. A characteristic of surface waves on water whereby the surface elements undergo a circular motion in succession, if there is no interference from the bottom, i.e. where the water is sufficiently deep.

WAVES, SURFACE. Waves on the free surface separating two fluid phases (usually a liquid and a gas). If their motion is essentially controlled by gravity they are classed as *gravity waves*; if by surface tension they are classed as *ripples.*

WAVE TRAP. In a radio receiver: a network used to reject signals of unwanted frequency so as to reduce interference with wanted signals.

WAVE TUBE. See: *Electron wave tube. Travelling wave tube.*

WAVE VECTOR. A vector, usually denoted by **k**, which is normal to a wave front in a crystal, and whose length is $2\pi/\lambda$, where λ is the wavelength of the wave motion involved. It is a vector in reciprocal space (i.e. momentum space), and is widely used in problems involving crystal interactions, e.g. in X-ray diffraction and in the band theory of solids.

WAVE VELOCITY. See: *Phase velocity.*

WAVINESS. Of a surface: See: *Surface finish.*

WEAK INTERACTIONS. The weakest of the interactions, except for gravitational forces, between elementary particles. They are responsible for beta decay and the decay of some mesons and hyperons. Neither parity nor strangeness is conserved in these interactions. Particles which take part only in weak or electromagnetic interactions are known as *leptons.* See also: *Boson, intermediate. Interactions between elementary particles. Leptons.*

WEATHER. The totality of the atmospheric phenomena (principally temperature, precipitation and wind) existing at a given place either at a given time or over a short period, measured in days. It has been said that existing atmospheric phenomena constitute weather, while the average weather over long periods constitutes climate. See also: *Climate*.

WEATHER FORECASTING. The prediction of future weather on the basis of observations made at a network of places, on both land and sea, in which not only ground conditions are reported but conditions in the upper atmosphere. See also: the following entries.

WEATHER FORECASTING, LONG RANGE. Forecasting for periods beyond 2 or 3 days. It is based on analogy, particularly as regards conditions in the middle and upper layers of the troposphere, and the application of thermodynamics and the appropriate equations of motion to observations of wind, temperature and pressure at all levels in the troposphere.

WEATHER MAP: SYNOPTIC CHART. A map on which are marked synchronous observations of atmospheric pressure, temperature and wind velocity, together with observations of frontal regions, of the state of the weather, of the clouds, and of the visibility. Such charts or maps are used as a basis for forecasting the weather.

WEATHER SATELLITE. A satellite which orbits the Earth with the special function of relaying back meteorological observations, including television pictures. Such satellites may be in obvious orbit or their orbits may be such that the satellites appear to hover. See also: *Satellite*.

WEATHER SHIPS. Specially equipped ships situated in fixed positions and furnishing surface weather data and data from higher altitudes at specified hours.

WEBER. The MKSA and SI unit of magnetic flux. It is the flux which, linking a circuit of one turn, produces in it an electromotive force of 1V as it is reduced to zero at a uniform rate in 1 s. It is equal to 10^8 maxwell. See: *Appendix II*.

WEBER, BESSEL FUNCTIONS OF. See: *Bessel equation*.

WEBER LAW: FECHNER LAW. A general law of human sensation which states that the increase of stimulus necessary to cause an increase of sensation which is just perceptible, is a constant fraction of the whole stimulus. In the case of vision the *contrast sensitivity* (the smallest detectable difference in brightness divided by the background brightness) is known as the *Fechner fraction*.

WEIGHING. The determination of the mass of a body by comparing its weight with that of a known mass. In expressing the weight of the body, say in kilogrammes-weight, the word "weight" is usually omitted. See also: *Balance. Counterpoise weighing. Double weighing. Mass. Units, gravitational.*

WEIGHT. (1) Of a given body: the force exerted by gravity on the body. It is the product of the mass of the body and the acceleration due to gravity. It may be expressed in units of force or in gravitational units. (2) A body of known mass, used in weighing. (3) A factor used in weighting. See also: *Mass. Units, gravitational. Weighing. Weighting.*

WEIGHTED MEAN. See: *Mean: Average*.

WEIGHTING. Of a series of measurements: the operation of assigning a factor or "*weight*" to each of the measurements to allow for their relative reliability or importance. See also: *Mean: Average. Statistical weight.*

WEIGHTS, STANDARD. Weights which have been compared with a primary standard, almost always via a copy or a secondary or tertiary standard. See also: *Kilogram. Pound.*

WEIR. A dam or bulkhead over which water flows. It is usually employed to measure the volume of water flowing in a given time. The term is popularly applied to any small structure for impounding water.

WELDING. The joining of metallic components at faces which are made plastic or liquid by heat or pressure, with or without the use of additional material. In *resistance welding* the heat is obtained by passing an electric current through the joint. In *spot welding,* two sheets to be welded are pressed together between two electrodes. A resistance weld is then obtained between the sheets, of approximately the same area as the electrode tips. In *arc welding* an electric arc is struck between the two pieces to be joined and a rod of similar metal. In *butt welding* the surfaces to be joined are pressed closely together before welding begins. See also: *Electron-beam melting.*

WESTON CELL. See: *Standard cell*.

WET BULB TEMPERATURE. See: *Thermometer, wet-and-dry bulb*.

WETTING. See: *Contact angle*.

WHEATSTONE BRIDGE. See: *Bridge, electrical*.

WHIRLWIND. A miniature cyclone of some tens of metres in diameter.

WHISPERING GALLERY. A domed gallery in which feeble sounds can be heard at considerable distances, one of the most famous being in St. Paul's

Cathedral, London. A sound made at one point can be heard almost anywhere round the gallery. This is thought (following Rayleigh) to be due not to simple reflection but to the tendency for sound to creep over a concave surface without much attenuation. The behaviour of a whispering gallery is, however not, yet fully explained.

WHISTLER. In a lightning discharge: an electromagnetic disturbance which is propagated along a line of geomagnetic flux and gives rise to an audio signal of descending pitch.

WHITE DWARFS. See: *Stars, dwarf. Stellar evolution.*

WHITE HOLE. An explosion of energy from a singularity in space, which has been proposed as an explanation of the violent outbursts from the nuclei of certain galaxies and quasars, e.g. some Seyfert galaxies. As it would be a source of energy the term white hole is misleading, except as an opposite to black hole. See also: *Black hole. Stellar evolution.*

WHITE RADIATION. Radiation comprising a continuous range of wavelengths. It is also known as *continuous radiation* or *heterogeneous radiation*. See also: *Bremsstrahlung. X-rays.*

WIDMANSTÄTTEN STRUCTURE. A type of alloy structure resulting from the formation of a new phase within the body of the parent phase, the crystallographic orientations of the two phases being related.

WIEDEMANN–FRANZ LAW. States that the ratio of the thermal conductivity of any metal to its electrical conductivity is the same at the same temperature. The law holds well for many metals at ordinary temperatures but not so well at low temperatures. The same is true of the statement by Lorenz that the above ratio is proportional to the absolute temperature, i.e. that $k/(\sigma T)$ (the Lorentz number) is constant, where k is the thermal conductivity, σ the electrical conductivity and T the absolute temperature. See also: *Lorenz number: Lorenz constant.*

WIEN BRIDGE. See: *Bridge, electrical.*

WIEN RADIATION LAWS. (1) *Displacement law:* states that the wavelength at which the energy density of a black body is a maximum is given by $\lambda_{max} = \sigma/T$ where σ is *Wien's constant* and T the absolute temperature. (2) Usually itself known as the *Wien radiation law:* states that the distribution of energy in the spectrum of a black body is given by $E_\lambda \, d\lambda = (C_1/\lambda_5) \exp(-C_2/\lambda T)$, where $E_\lambda \, d\lambda$ is the energy radiated per unit area per unit time in the wavelength region λ and $\lambda + d\lambda$, C_1 and C_2

are constants, and T is the absolute temperature. It has been replaced by the Planck radiation formula.

WIGNER EFFECT. The change in the physical properties of graphite resulting from the displacement of lattice atoms by high-energy neutrons and other energetic particles in a nuclear reactor. It results inter alia in the building up of stored energy (*Wigner energy*), in the change of the dimensions of the crystal lattice, and hence in the change of overall bulk size. The term is sometimes extended by analogy to cover radiation damage in other materials.

WIGNER ENERGY. Energy stored as a result of the *Wigner effect.*

WIGNER NUCLEI: WIGNER NUCLIDES. See: *Mirror nuclei: Mirror nuclides.*

WIGNER–SEITZ METHOD. A method of calculating the energy levels of electrons in solid bodies, based on the assumption that each electron occupies a "cell" in which the potential energy of the electron has spherical symmetry.

WILSON CHAMBER. See: *Cloud chamber.*

WIMSHURST MACHINE. An electrical friction machine in which static electricity can be stored at relatively high potentials. It comprises two coaxial insulating disks revolving in opposite directions, each of which carries metallic foil sectors on which electric charge is induced, which make contact with fixed collecting combs as they revolve.

WIND. In general, any motion of atmospheric air, but the term usually refers to the horizontal component of such motion. See also: *Anticyclone. Antitrades. Cyclone. Depression, meteorological. Doldrums. Föhn. Gradient wind. Jet stream. Katabatic wind. Line squall. Monsoon. Squall. Tornado. Trade winds.*

WIND ROSE. A diagram indicating the frequency and velocity of winds at a given spot over specific periods.

WIND TUNNEL. An installation for the production of a controlled stream of air or other gas for experiments on fluid dynamics, typically on models of aircraft, missiles, etc. Various air speeds are used, from subsonic to hypersonic.

WIND TUNNEL, COMPRESSED-AIR (CAT). A wind tunnel in which compressed air is used, which makes it possible to reduce the size of the model or the wind speed of the test.

WIND TUNNEL, LOW-DENSITY. A wind tunnel designed for the study of the dynamics of rarefied gas, e.g. at supersonic or hypersonic speeds at high altitudes.

WING. A main lifting or supporting surface of an aircraft. See also: *Kutta–Joukowski hypothesis.* Entries under *Aerofoil.*

WITKA CIRCUIT. A voltage trebling circuit in which the voltages of two capacitors are added to that of a transformer secondary, producing a unidirectional voltage which varies from the peak voltage of the transformer to a value approaching, in alternate half cycles, three times that voltage.

W.K.B.J. OR W.K.B. APPROXIMATION. An approximate way of solving linear differential equations with slowly varying coefficients. The initials refer to Wentzel, Kramers, Brillouin and Jeffreys.

W MESON. See: *Boson, intermediate.*

WOLLASTON PRISM. See: *Polarizing prism.*

WOLLASTON WIRE. A thin (about $25\,\mu$m) platinum wire.

WORK. The transfer of energy from one system to another. The unit of work is thus the same as that of energy; in the MKS and SI systems it is the joule, and in the CGS system the erg. See also: *Energy. Erg. Joule.*

WORK FUNCTION. The energy, often expressed in electron volts, necessary to remove an electron from just inside to just outside the surface of a metal. It may be *thermionic* or *photoelectric*, and is a characteristic of the individual metal concerned. See also: *Einstein equation for photoelectric emission. Richardson equation: Richardson–Dushman equation.*

WORK FUNCTION, THERMODYNAMIC. The Helmholtz free energy. See: *Free energy.*

WORK HARDENING. The increase in the strength and hardness of a metal, produced by cold work. This may be explained in terms of the motion of dislocations, either out of the crystal altogether or in such a way as to impede or halt further motion within the metal crystals. See also: *Cold work.*

WORLD-. A prefix used to indicate that the quantity referred to is related to Lorentz transformations, i.e. is concerned with equations between four-dimensional vectors (*world-vectors*). Examples are: *World-force*, signifying a universal gravitational force in space-time, and *World-line*, signifying the path of a particle moving in space-time. See also: *Space-time.*

WRENCH. A combination of a couple, and a force which is parallel to the axis of the couple.

WRINGING. Of flat surfaces: the strong adherence between two highly finished flat surfaces when they are brought into contact with a trace of liquid between them.

WROUGHT ALLOY. An alloy shaped by any hot or cold working process. More usually it is restricted to an alloy subjected to hot working by forging, pressing, rolling, extrusion or similar means. Such hot working breaks up the cast structure of the alloy and removes porosity, as well as producing the required shape.

WULFF NET. Another name for a stereographic net.

WULFF THEOREM. A theorem concerning crystal growth: it states that, when a crystal is in its *equilibrium shape* (i.e. the shape which for a given mass minimizes the free energy), gravitational or other external forces being negligibly small, there exists within the crystal a point whose perpendicular distances from all the faces of the crystal are proportional to the surface free energies per unit area of these faces; and any other face, not belonging to the equilibrium shape, has such a surface-free energy that a plane drawn with the corresponding orientation and distance from the said point would lie entirely outside the crystal.

W VIRGINIS STARS. See: *Stars, variable, nomenclature of.*

X

X-BAND. Denotes that part of the radio-frequency spectrum lying between 7000 and 11000 MHz.

XEROGRAPHY. A photographic process in which a plate, coated typically with selenium, is electrically charged before exposure to light and then "developed" by dusting with electrically charged fine powder. Since those portions of the plate which have been exposed become conducting, the powder will delineate the exposed or unexposed regions according to the relative signs of the charge on the plate and powder, to an extent depending on the degree of exposure. A permanent record may then be obtained by transferring the powder to a supporting surface and giving it a suitable fixing treatment.

XERORADIOGRAPHY. The technique of Xeroradiography applied to X- and γ-ray radiography.

XI-PARTICLE. See: *Hyperon.*

X-RAY ABSORPTION ANALYSIS. The analysis of the atomic composition of materials from their X-ray absorption or emission spectra. See also: *X-ray spectrum.*

X-RAY ABSORPTION COEFFICIENT. The fraction of a beam of X-rays which is removed by absorption on passing through unit thickness of material, measured for very small thicknesses. It is less than the attenuation coefficient. See: *Absorption coefficient for radiation. Attenuation coefficient for radiation.*

X-RAY ABSORPTION EDGE. The X-ray energy or wavelength, corresponding to a given electron energy level in an atom of a specified element, at which a discontinuity occurs in the plot of absorption against wavelength or energy. The energy of the absorption edge (or *absorption limit* as it is also known) is the smallest X-ray energy that can excite characteristic X-radiation corresponding to the electron energy level concerned, and is labelled by the appropriate letter (K, L, M, etc.) of that energy level. See also: *Electron energy level. Electron shell.*

X-RAY ASTRONOMY. The study of celestial bodies, visible or invisible, by the observation and measurement of their X-ray emissions. Most observations are made from balloons, rockets or satellites, and the techniques employed include photography and the use of counters, ionization chambers and crystal spectrometers.

X-RAY CAMERA. A camera used in X-ray crystallography, by which a photographic record of an X-ray diffraction pattern may be obtained. See also: *Oscillating crystal method. Powder method. Rotating crystal method.*

X-RAY CRYSTALLOGRAPHY. The study of crystals by X-ray diffraction. See also: *Diffraction analysis.* Entries under *Crystal. Neutron diffraction. X-ray diffraction.*

X-RAY CRYSTALLOGRAPHY, DIVERGENT BEAM TECHNIQUE FOR. The use of a divergent beam of X-rays to produce Kossel lines. The technique can be used for the determination of accurate unit cell dimensions (to $\pm0.001\%$ for cubic crystals), the location of sub-structure boundaries and other variations in crystal texture, and deformation mechanisms in general. See also: *Kossel effect.*

X-RAY CRYSTALLOGRAPHY, INTERNATIONAL TABLES FOR. A series of volumes, published by the International Union of Crystallography, containing basic information on crystal symmetry, mathematical data, X-ray wavelengths, scattering factors and absorption coefficients, and so on, necessary for the determination of crystal structure by diffraction methods.

X-RAY DIFFRACTION. The diffraction of X-rays by a diffraction grating or, more commonly, by atoms in solids, liquids or gases. See also: *Dynamical theory of electron and X-ray diffraction. Electron diffraction. Neutron diffraction. X-ray crystallography.*

X-RAY DIFFRACTION TOPOGRAPHY. The study of the defect structure of a crystal by the detailed examination of the structure of X-ray diffraction images obtained from the crystal.

X-RAY DIFFRACTOMETER. See: *Diffractometer.*

X-RAY DISPERSION. The variation of the refractive index with wavelength. See also: *X-ray refraction.*

X-RAY DISPERSION, ANOMALOUS. The discontinuity in the curve of refractive index versus wavelength, in the neighbourhood of an absorption edge. It leads to a breakdown of Friedel's law and to a method for the determination of the phases of X-ray reflections from crystals having no centre of symmetry, and therefore to a method for obtaining a unique solution of the crystal structure by Fourier synthesis. Similar effects occur in neutron diffraction but only for a few isotopes so that there is much less scope for applications to structure determination. See also: *Fourier synthesis. Friedel law. Neutron dispersion, anomalous.*

X-RAY FILTER. See: *K-beta filter: Beta filter.*

X-RAY FLUORESCENCE ANALYSIS. The analysis of the atomic composition of materials from the spectra of the fluorescence X-rays emitted by them when irradiated by X- or γ-rays of appropriate energy. See also: *Auger emission spectroscopy. Electron-probe microanalysis. X-rays.*

X-RAY INTENSITY. At a given point in an X-ray beam: the amount of energy passing through unit area per unit time at that point, the area being normal to the direction of the beam. In X-ray crystallography, however, the intensity is often expressed in electron units, i.e. relative to the scattering by a single electron. See also: *Atomic scattering factor.*

X-RAY INTERFEROMETRY. The study of X-ray spectra by the use of interferometers in which Bragg reflecting crystals serve as beam amplitude splitters, mirrors and phase-sensitive detectors. A striking application is to the production of moiré fringes, from which lattice rotations and even lattice spacings can be directly measured. Other applications involve the examination of dislocations, stacking faults etc.

X-RAY LENS. A "lens" at which X-rays are reflected at a glancing angle, i.e. a mirror, since, owing to the nature of the refractive index, lenses of conventional shape are not feasible. It is concave and must be polished to a much higher degree of smoothness than is normal in optical technology. Such mirrors have been used in one form of X-ray microscope. See also: *X-ray microscopy. X-ray refraction.*

X-RAY MICROSCOPY. Any method of producing magnified X-ray images that reveal the structure of a material by contrast arising from differences in X-ray absorption or emission. In *reflection microscopy* X-rays are focused by reflection from mirrors at grazing incidence and used "optically" to provide a magnified image; in *contact microradiography,* an X-ray image is enlarged optically; in *projection microradiography* a projected image is obtained from a "point" source of X-rays; and in *X-ray diffraction topography,* an examination is made of X-ray diffraction images. See also: *Microradiography. X-ray diffraction topography. X-ray replica microscopy.*

X-RAY MONOCHROMATOR. See: *Monochromator.*

X-RAY POLARIZATION. See: *Polarization.*

X-RAY RADIOGRAPHY. Radiography by means of X-rays. The X-rays are usually produced by conventional X-ray tubes but, for high energy X-rays, particle accelerators may be employed, e.g. betatrons, cyclotrons, or Van de Graaff machines. See also: *Microradiography. X-ray microscopy.*

X-RAY REFLECTION. (1) True reflection of X-rays at glancing angles. (2) The apparent reflection of X-rays at a crystal surface or by parallel planes of atoms in a crystal, which arises from X-ray diffraction. See also: *Bragg equation: Bragg law. Integrated reflection.*

X-RAY REFLECTION CURVE. See: *Integrated reflection.*

X-RAY REFRACTION. The change in direction of X-rays in passing from one medium to another. For X-rays passing from air into a solid the *refractive index* is less than unity by a few parts per million for X-rays of 1 Å (10^{-10} m) wavelength and is given, in the absence of anomalous dispersion, by $1 - (n\lambda^2 e^2)/(2\pi mc^2)$, where n is the number of electrons per unit volume, λ is the wavelength, e is the charge on the electron, m the mass of the electron, and c the speed of light. Owing to the very small refractive index and the relatively high absorption of possible lens materials, refracting lenses for X-rays are not feasible and must be replaced by reflecting mirrors.

X-RAY REPLICA MICROSCOPY. A type of microscopy in which an X-ray image of a specimen is formed on a suitable sensitized material (typically methyl methacrylate) from which a replica is made, which constitutes a three dimensional absorption profile of the specimen. After being coated with a thin film of gold the replica is examined in the scanning electron microscope. The method is particularly suitable for biological materials since the replicas can be made at atmospheric pressure while the specimens are still moist. The resolution obtainable (about

10 nm, or 100 Å) compares well with that obtainable in the scanning electron microscope. See also: *Electron microscope, scanning. X-ray microscopy.*

X-RAYS. Electromagnetic radiation consisting of (1) *Continuous X-rays* (or *Bremsstrahlung*), arising from the retardation of charged particles, commonly electrons stopped by the target of an X-ray tube. The wavelength distribution is continuous with a maximum energy limit (the *quantum limit*) given by $\lambda = hc/eV$, where λ is the wavelength, h is Planck's constant, c is the speed of light and eV is the energy limit in electron volts. (2) *Characteristic X-rays*, arising from transitions involving electron transfer to the inner atomic shells. The wavelengths are confined to discrete groups, each group corresponding to the transfer of an outer electron to one of the inner shells, and denoted by the letter (K, L, M, ...) appropriate to the inner shell concerned. Characteristic X-rays arising from the absorption of X- or γ-rays are also known as *fluorescence X-rays.* See also: *Characteristic radiation. Electron shell. X-ray spectrum.*

X-RAY SCATTERING. See various entries under *Scattering.*

X-RAYS, CONTINUOUS. See: *X-rays.*

X-RAYS, HARD AND SOFT. Qualitative terms describing the penetrating power of a given beam of X-rays, "hard" X-rays being relatively highly penetrating and "soft" X-rays being relatively easily absorbed.

X-RAY SPECTROMETRY. (1) The measurement of the distribution of energy as a function of X-ray wavelength, the dispersion being effected by a crystal (often bent) or a diffraction grating, and the detector being an ionization chamber, counter, photographic plate or film or photomultiplier. (2) The study of X-ray spectra and their implications.

X-RAY SPECTROSCOPIC ANALYSIS. The analysis of the atomic composition of a material from the spectra emitted when the material serves as the target of an X-ray tube. In the electron-probe microanalyser the electron beam may be made to scan the target and thus provide a microanalysis of the material concerned or a reflection electron image of its surface. See also: *Electron probe microanalysis.*

X-RAY SPECTROSCOPY, SOFT. The production and examination of long wavelength X-ray spectra which, on account of their high absorption, require the use of a vacuum. One of the most important applications is to the study of the density of states in the valency band of a solid.

X-RAY SPECTRUM. Describes the combination of continuous and characteristic X-rays, consisting of the superposition of sharp lines on a continuous background. The sharp lines occur in groups characteristic of the transfer of outer electrons to each of the various inner shells (K, L, M, ...), each line being denoted by the letter appropriate to the inner shell concerned together with symbols (α, β, γ, ... and 1, 2, 3, ...) denoting the outer shells involved and the energy levels within shells. See also: *Electromagnetic spectrum. Moseley law. Term diagram. X-rays.*

X-RAYS, SECONDARY. See: *Secondary radiation.*

X-RAY STARS AND GALAXIES. Astronomical sources whose emission consists mostly of X-rays. About two-thirds of the X-ray stars lie within our Galaxy, but it appears that the majority of extragalactic X-ray sources are located in rich galaxy clusters. X-ray galaxies include the radio galaxy Virgo A, the large and small Magellanic clouds and one of the nearest quasars, 3C 273. There is a considerable amount of information available on the spectra and variability of X-ray stars but ideas on the underlying mechanism of their X-ray emission are still fluid. However it is considered that the X-ray emission from the Crab nebula is probably produced by the synchroton process since the X-rays emitted are plane-polarized.

X-RAY TUBE. A discharge tube designed for the production of X-rays. It involves essentially the production of an electron beam, which is accelerated by a suitable potential so as to strike a target (usually watercooled) of an appropriate material. The electrons are typically produced by an electron gun, with provision for focusing, and the potential may be provided by a transformer, Van de Graaff, betatron or linear accelerator, according to the voltage required. The target is commonly of high atomic number (e.g. tungsten) for radiography or radiotherapy, but of small atomic number (e.g. copper, iron, chromium) for X-ray crystallography and similar applications. A suitable window is incorporated (or sometimes more than one) to permit the X-rays to leave the tube without undue absorption. See also: *Flash radiography. Van de Graaff accelerator.*

X-UNIT. A unit of length used mostly in X-ray spectroscopy. It is defined by assigning the value 3029·45 X.U. to the spacing of the (200) planes of calcite at 18°C. The absolute value of the X-unit depends on the value of Avogadro's number and the conversion constant between X-units and Ångström units has been the subject of some discussion. The currently accepted value is 1 X.U. = $1\cdot00208 \times 10^{-3}$ Å. (*Note:* 1 Å = 10^{-10} m.) See: *Ångström unit.*

363

Y

YARD. The fundamental unit of length in the Imperial system of units. The *Imperial standard yard* was the former unit, expressed as the length of a bar of Baily's metal (a copper–tin–zinc alloy) but has been superseded by the *International yard* which is, by definition, 0·9144 m exactly. See also: *Length, units of. Metre. Wavelength standards. Appendix II.*

YAWING. Of a ship at sea or an aircraft in the air: the movement of the ship or aircraft about the vertical axis. See also: *Rolling. Pitching.*

Y-CONNECTION. See: *Delta connection.*

YEAR. (1) *Anomalistic year:* the interval between successive passages of the Sun through perigee, i.e. between successive passages of the Earth through perihelion. It is 365·2596 mean solar days. (2) *Besselian year:* that tropical year which begins when the mean longitude of the Sun is exactly 280° ($18^h 40^m$), this instant being near to the beginning of the *civil year* (0000^h on 1 Jan). (3) *Sidereal year:* the interval between two successive passages of the centre of the Sun past any one star situated in the ecliptic and devoid of proper motion. It is slightly longer than the tropical year and amounts to 365·2564 mean solar days decreasing by 10^{-7} mean solar days per century. (4) *Tropical year:* the interval between two successive passages of the Sun through the first point of Aries, or the interval between two similar equinoxes or solstices. It is equal to 365·242196 mean solar days. (5) *Civil year:* See: *Time.*

YERKES SYSTEM OF STELLAR CLASSIFICATION. See: *Stars, spectral classes of.*

YIELD POINT. Of a body subjected to stress: (1) the point at which the elastic limit is reached, i.e. at which plastic deformation first sets in, known as the *upper yield point.* (2) The point beyond the upper yield point, at which the increase of strain with no change of load (or often a reduction in load) ceases, owing to strain-hardening. It is known as the *lower yield point.*

YIELD STRAIN. The strain at the upper yield point.

YIELD STRESS. The stress at the upper yield point.

YLEM. The name proposed for a hypothetical substance of density about 10^{16} kg/m^3, consisting chiefly of neutrons, out of which all nuclei may have been formed. See also: *Neutron star. Nuclear matter.*

YOUNG–HELMHOLTZ THEORY OF COLOUR VISION. See: *Colour vision, theories of.*

YOUNG MODULUS. For an elastic material subjected to elongation: the stress which produces unit fractional elongation. See also: *Elastic modulus.*

YUKAWA PARTICLE. The particle, originally postulated by Yukawa, to explain the short-range nucleon–nucleon forces, which was subsequently discovered and termed a meson. See also: *Meson theory of nuclear forces.*

YUKAWA POTENTIAL. A potential function used to describe the meson field about a nucleon. See also: *Nuclear potential.*

Z

ZEEMAN EFFECT. The splitting of optical spectral lines into several components when the source of light is placed in a strong magnetic field. These components are polarized in a manner which varies with the direction from which the source is viewed relative to that of the magnetic field. The effect was predicted by Lorentz in a theory based on the behaviour of electrons as classical oscillators. According to this theory there should be two lines (circularly polarized in opposite directions) for each original line, when viewed parallel to the field direction; and three (one being undisplaced) when viewed at right angles to the field, the other displacements being symmetrical. See also: *Paschen–Back effect. Zeeman effect, anomalous.*

ZEEMAN EFFECT, ANOMALOUS. The splitting of optical spectral lines by the Zeeman effect into triplets displaying shifts which differ from those predicted by Lorentz, or into four or more lines. The effect is explained by the introduction of electron spin into the theory. Paradoxically, the "anomalous" effect is by far the most common, the "normal" effect being only a special case of this. One of the important applications is to the determination of the magnetic fields present in the solar and stellar atmospheres. See also: *Zeeman effect.*

ZENITH. That point on the celestial sphere which is vertically above the observer. See also: *Nadir.*

ZENITH ATTRACTION. The appearance of prominent meteor radiants near the zenith. See also: *Meteor radiant.*

ZENITH DISTANCE. Of a celestial body: the angular distance between the zenith and the body measured as the arc of the great circle passing through both. It is the complement of the altitude of the body.

ZEOLITES. Hydrous alumino-silicates of sodium, potassium, calcium and barium which are characterized by an open crystal structure and when dehydrated are honeycombed with regularly spaced cavities and interconnecting channels of molecular dimensions. They are used as molecular sieves and in ion exchange processes, including water softening. They can be regenerated after use without difficulty.

ZERNICKE–PRINS FORMULA. A formula relating the radial density function of a monatomic liquid to the intensity of the X-ray diffraction image (or sometimes the image obtained by electron or neutron diffraction). It may be written

$$4\pi r^2 \varrho(r) = 4\pi r^2 \varrho_0 + \frac{2r}{\pi} \int_0^\infty s \left(\frac{I}{Nf^2} - 1 \right) \sin rs \cdot ds$$

where $\varrho(r)$ is the radial density function at a distance r, ϱ_0 is the average value at large distances, $s = 4\pi \sin (\theta/2)/\lambda$ (where θ is the scattering angle and λ the X-ray or other appropriate wavelength), f is the atomic scattering factor, N is the effective number of atoms in the sample irradiated, and I is the diffracted intensity on a convenient scale. See also: *Radial density function.*

ZERO-POINT ENERGY. The energy remaining at the absolute zero of temperature. For any system it corresponds to the quantum state of lowest energy, i.e. it is the energy of the ground state, equal to one half quantum of energy. See also: *Absolute zero.*

ZERO SOUND. A wave motion in liquid ^{3}He at very low temperatures, whose velocity is a little higher than that of ordinary sound. It is observed under excitation at high frequency, when ordinary sound cannot be propagated.

ZETA POTENTIAL: ELECTROKINETIC POTENTIAL. At a solid–liquid interface in a colloid system: the potential existing across the diffuse portion of the electrical double layer. See: *Electrical double layer.*

ZITTERBEWEGUNG. The oscillatory motion (literally "trembling") postulated for a Dirac electron to overcome the paradox that the electron should move with the speed of light even in the absence of a field of force. The centre of mass of the electron cloud oscillates with a calculable amplitude and a speed equal to that of light, while the cloud as a whole has a translational speed which is generally much less than this.

ZODIAC. A region on the celestial sphere extending to about 9° on each side of the ecliptic, within which the Sun, Moon and planets appear to move. It is divided into twelve equal sections, each 30° of celestial longitude, bearing the names of the constellations that once corresponded to them (the *signs of the Zodiac*).

ZODIACAL LIGHT. Sunlight reaching the Earth after scattering from other bodies of the solar system, chiefly meteoritic dust. It is visible in the east just before dawn and in the west just after sunset.

ZONE, CRYSTALLOGRAPHIC. See: *Crystal zone.*

ZONE MELTING. A technique for the purification of metals (originally germanium) in which a bar of the metal is slowly passed through a furnace or induction coil so that a molten zone traverses the length of the bar. The method depends on the difference in solubility of the impurities in the solid and liquid phases.

ZONE PLATE. See: *Fresnel zones: Half-period zones.*

ZONES OF RADIATION, EXTRATERRESTRIAL. Two regions of high-energy charged particles at considerable heights above the Earth, commonly known as Van Allen belts, together with the equatorial ring current. See: *Equatorial ring current. Van Allen belts.*

ZWITTER ION. A molecule which contains separate centres of positive and negative charge. While it is electrically neutral overall it has a dipole moment: at one end it appears to be a positive ion and at the other a negative one. Most of the common amino-acids form such dipolar ions.

Quantity	SI unit	Symbol	Name and symbol of unit	SI equivalent		Reciprocal
energy	joule	J	erg	10^{-7}	J	10^7
			kilowatt hour (kWh)	$3\cdot6$	$J \times 10^6$	
			electron volt (eV)	$1\cdot602\ 189\ 2$	$J \times 10^{-19}$	$0\cdot624\ 146\ 10^{19}$
			IT calorie (cal)	$4\cdot1868$	J	$0\cdot238\ 846$
			British thermal unit (Btu)	$1\cdot055\ 06$	kJ	$0\cdot947\ 817$
force	newton	N	dyne	10^{-5}	N	10^5
			kg-force (kgf)	$9\cdot806\ 65$	N	$0\cdot101\ 972$
			poundal (pdl)	$0\cdot138\ 255$	N	$7\cdot233\ 01$
			pound-force (lbf)	$4\cdot448\ 22$	N	$0\cdot224\ 809$
			ton-force (tonf)	$9\cdot964\ 02$	N	$0\cdot100\ 361$
pressure	pascal	Pa	bar	10^5	Pa	10^{-5}
			standard atmosphere (atm)	$101\ 325$	Pa	$9\cdot869\ 23 \times 10^{-6}$
			torr / mm mercury	$133\cdot322$	Pa	$7\cdot500\ 62 \times 10^{-3}$
			inches mercury	$3386\cdot39$	Pa	$2\cdot9530 \times 10^{-4}$
power	watt	W	erg s^{-1}	10^{-7}	W	10^7
			Btu h^{-1}	$0\cdot293\ 071$	W	$3\cdot412\ 14$
			horse power (hp)	$745\cdot700$	W	$1\cdot341\ 02 \times 10^{-3}$
			cheval-vapeur (cv)	$735\cdot499$	W	$1\cdot359\ 62 \times 10^{-3}$
temperature	kelvin	K	degree Celsius (°C)	1	K	1
			t °C	$273\cdot15 + t$	K	
			degree Fahrenheit (°F)	$5/9$	K	$1\cdot8$
dynamic viscosity	pascal second	Pa s	poise (P)	10^{-1}	Pa s	10
kinematic viscosity	square metre per sec	m^2 s^{-1}	stokes (St)	10^{-4}	m^2 s^{-1}	10^4
magnetic flux	weber	Wb	maxwell (Mx)	10^{-8}	Wb	10^8
magnetic flux density, magnetic induction	tesla	T	gauss (Gs, G)	10^{-4}	T	10^4
			gamma (γ)	10^{-9}	T	10^9
magnetic field strength, magnetic intensity	ampere per metre	Am^{-1}	oersted (Oe)	$1/4\pi$	kA m^{-1}	4π
inductance	henry	H	centimetre (cm)	10^{-9}	H	10^9
luminance	candela per square metre	cd m^{-2}	stilb (sb)	10^4	cd m^{-2}	10^{-4}
illuminance	lux	lx	phot (ph)	10^4	lx	10^{-4}
activity (radioactive source)	becquerel	Bq	curie (Ci)	$3\cdot7$	$Bq \times 10^{10}$	$0\cdot270\ 270 \times 10^{-10}$
absorbed dose (ionizing radiation)	gray	Gy	rad	10^{-2}	Gy	10^2
exposure (ionizing radiation)	coulomb per kg	C kg^{-1}	röntgen (R)	$2\cdot58$	$C\ kg^{-1} \times 10^{-4}$	$0\cdot387\ 597 \times 10^4$

Appendix III

Values of Some Fundamental Physical Constants

		SI	CGS
Speed of light	$2\cdot997\ 924\ 580$ (12)	$\times\ 10^8$ m s^{-1}	$\times\ 10^{-10}$ cm s^{-1}
Charge on proton	$1\cdot602\ 189\ 2$ (46)	$\times\ 10^{-19}$ C	$\times\ 10^{-20}$ e.m.u.
Rest mass of electron	$9\cdot109\ 534$ (47)	$\times\ 10^{-31}$ kg	$\times\ 10^{-28}$ g
Planck constant	$6\cdot626\ 176$ (36)	$\times\ 10^{-34}$ J s	$\times\ 10^{-27}$ erg s
Boltzmann constant	$1\cdot380\ 662$ (44)	$\times\ 10^{-23}$ J K^{-1}	$\times\ 10^{16}$ erg K^{-1}
Bohr magneton	$9\cdot274\ 078$ (36)	$\times\ 10^{-24}$ J T^{-1}	$\times\ 10^{-21}$ erg gauss^{-1}
Stefan-Boltzmann constant	$5\cdot670\ 32$ (71)	$\times\ 10^{-8}$ W m^{-2} K^{-4}	$\times\ 10^{-5}$ erg cm^{-2} s^{-1} K^{-4}
Rydberg constant (∞)	$1\cdot097\ 373\ 177$ (83)	$\times 10^7$ m^{-1}	$\times\ 10^5$ cm^{-1}
Avogadro constant	$6\cdot022\ 045$ (31)	$\times\ 10^{23}$ mol^{-1}	$\times\ 10^{23}$ mol^{-1}
Faraday constant	$9\cdot648\ 456$ (27)	$\times\ 10^4$ C mol^{-1}	$\times\ 10^3$ e.m.u. mol^{-1}
Molar gas constant	$8\cdot314\ 41$ (26)	$\times$ J mol^{-1} K^{-1}	$\times\ 10^7$ erg mol^{-1} K^{-1}
Gravitational constant	$6\cdot672\ 0$ (41)	$\times\ 10^{-11}$ N m^2 kg^{-2}	$\times\ 10^{-8}$ dyn cm^2 g^{-2}

Note: The standard deviation is given below each value.

Appendix II

SI Units, Conversions and Equivalents

Base units			Derived units			
Quantity	*Name*	*Symbol*	*Quantity*	*Name*	*Symbol*	*Definition*
length	metre	m	energy	joule	J	$m^2\ kg\ s^{-2}$
mass	kilogram	kg	force	newton	N	$m\ kg\ s^{-2}$
time	second	s	pressure	pascal	Pa	$N\ m^{-2}$
electric current	ampere	A	power	watt	W	$J\ s^{-1}$
thermodynamic temperature	kelvin	K	electric charge	coulomb	C	$A\ s$
			electric potential difference	volt	V	$J\ C^{-1}$
amount of substance	mole	mol	electric resistance	ohm	Ω	$V\ A^{-1}$
luminous intensity	candela	cd	electric conductance	siemens	S	$A\ V^{-1}$
			electric capacitance	farad	F	$C\ V^{-1}$
Supplementary units			magnetic flux	weber	Wb	$V\ s$
plane angle	radian	rad	magnetic flux density	tesla	T	$Wb\ m^{-2}$
solid angle	steradian	sr	inductance	henry	H	$VA^{-1}\ s$
			luminous flux	lumen	lm	$cd\ sr$
			illuminance	lux	lx	$cd\ sr\ m^{-2}$
			frequency	hertz	Hz	s^{-1}
			activity (radioactive source)	becquerel	Bq	s^{-1}
			absorbed dose (ionizing radiation)	gray	Gy	$J\ kg^{-1}$

SI equivalents of some CGS, Imperial and other units (exact values are underlined)

Quantity	SI unit	Symbol	Name and symbol of unit	SI equivalent		Reciprocal
length	metre	m	ångstrom (Å)	$\underline{10^{-10}}$	m	$\underline{10^{10}}$
			micron (μm)	$\underline{10^{-6}}$	m	$\underline{10^{6}}$
			inch (in)	$\underline{2\cdot54}$	cm	$0\cdot393\ 701$
			foot (ft)	$\underline{30\cdot48}$	cm	$0\cdot032\ 808$
			yard (yd)	$\underline{0\cdot9144}$	m	$1\cdot093\ 61$
			statute mile	$\underline{1\cdot609\ 344}$	km	$0\cdot621\ 371$
			nautical mile	$\underline{1\cdot852}$	km	$0\cdot539\ 957$
			astronomical unit (AU)	$0\cdot149\ 600$	$m \times 10^{12}$	
			parsec (pc)	$30\ 857$	$m \times 10^{12}$	
mass	kilogram	kg	tonne (t)	$\underline{10^{3}}$	kg	$\underline{10^{-3}}$
			pound (lb)	$\underline{0\cdot453\ 592\ 37}$	kg	$2\cdot204\ 62$
			ton	$1016\cdot05$	kg	$0\cdot984\ 207 \times 10^{-3}$
			slug	$14\cdot5939$	kg	$0\cdot068\ 521$
			unified atomic mass unit (u)	$1\cdot660\ 565\ 5$	$kg \times 10^{-27}$	$0\cdot602\ 204 \times 10^{-27}$
area	square metre	m^2	barn (b)	$\underline{10^{-28}}$	m^2	$\underline{10^{28}}$
			hectare (ha)	$\underline{10^{4}}$	m^2	$\underline{10^{-4}}$
			square inch (in²)	$\underline{6\cdot4516}$	cm^2	$0\cdot155\ 000$
			square yard (yd²)	$0\cdot836\ 127$	m^2	$1\cdot195\ 99$
			area	$\begin{cases}4046\cdot86 \\ 0\cdot404\ 686\end{cases}$	m^2 hectare	$0\cdot000\ 247$ $2\cdot471\ 05$
volume	cubic metre	m^3	litre (l)	$\underline{10^{-3}}$	m^3	10^{3}
			pint (pt)	$0\cdot568\ 261$	litre	$1\cdot759\ 75$
			gallon, UK	$4\cdot546\ 09$	litre	$0\cdot219\ 969$
			gallon, US	$3\cdot785\ 41$	litre	$0\cdot264\ 172$
speed	metre per sec	ms^{-1}	foot per sec	$\begin{cases}\underline{0\cdot304\ 8} \\ 1\cdot097\ 28\end{cases}$	$m\ s^{-1}$ $km\ h^{-1}$	$3\cdot280\ 84$ $0\cdot911\ 344$
			mile per hour	$\underline{1\cdot609\ 344}$	$km\ h^{-1}$	$0\cdot621\ 371$
			knot	$\underline{1\cdot852}$	$km\ h^{-1}$	$0\cdot539\ 957$

Transition Eleme

Group	IA	IIA	IIIB	IVB	VB	VIB	VIIB	
Principal quantum number (Period) / Valence shell	s^1	s^2	$d^1s^2f^x$	d^2s^2	$(d^3s^2)^\dagger$	$(d^5s^1)^\dagger$	d^5s^2	$(d^6s^2)^\dagger$
$n=1$ — $1s$	1 H 1.008							
$n=2$ — $2s2p$	3 Li 6.94	4 Be 9.01						
$n=3$ — $3s3p$	11 Na 22.99	12 Mg 24.31						
$n=4$ — $4s3d4p$	19 K 39.10	20 Ca 40.08	21 Sc 44.96	22 Ti 47.90	23 V 50.94	24 Cr 51.996	25 Mn 54.94	26 Fe 55.85
$n=5$ — $5s4d5p$	37 Rb 85.47	38 Sr 87.62	39 Y 88.91	40 Zr 91.22	41 Nb 92.91	42 Mo 95.94	43 Tc 99	44 Ru 101.07
$n=6$ — $6s4f5d6p$	55 Cs 132.91	56 Ba 137.34	57* La 138.91	72 Hf 178.49	73 Ta 180.95	74 W 183.85	75 Re 186.2	76 Os 190.2
$n=7$ — $7s5f6d7p$	87 Fr	88 Ra 226	89 ⊙ Ac					

	58 Ce 140.12	59 Pr 140.91	60 Nd 144.24	61 Pm 145
* Lanthanide Series				
⊙ Actinide Series	90 Th 232.04	91 Pa	92 U 238.03	93 Np

†Variable valence shells

Notes Approximate atomic weights are shown below the symbols of th
elements except for those heavy elements where the term has litt
significance.
Although the actinides are shown as arising from the filling up
of the 5-f shell, there is so little difference between the 5-f and 6
electrons that it is often very difficult to distinguish between the

ts — *d*

VIII		IB	IIB	IIIA	IVA	VA	VIA	VIIA	0
$d^7s^2)^†$	$(d^8s^2)^†$	s^1d^{10}	s^2	s^2p^1	s^2p^2	s^2p^3	s^2p^4	s^2p^5	s^2p^6
									2 He 4.003
				5 B 10.81	6 C 12.01	7 N 14.01	8 O 15.999	9 F 18.99	10 Ne 20.18
				13 Al 26.98	14 Si 28.09	15 P 30.97	16 S 32.06	17 Cl 35.45	18 Ar 39.95
27 Co 58.93	28 Ni 58.71	29 Cu 63.54	30 Zn 65.37	31 Ga 69.72	32 Ge 72.59	33 As 74.92	34 Se 78.96	35 Br 79.91	36 Kr 83.80
45 Rh 102.91	46 Pd 106.4	47 Ag 107.87	48 Cd 112.40	49 In 114.82	50 Sn 118.69	51 Sb 121.75	52 Te 127.60	53 I 126.90	54 Xe 131.30
77 Ir 192.2	78 Pt 195.09	79 Au 196.97	80 Hg 200.59	81 Tl 204.37	82 Pb 207.19	83 Bi 208.98	84 Po	85 At	86 Rn

Inner Transition Elements — *f*

62 Sm 150.35	63 Eu 151.96	64 Gd 157.25	65 Tb 158.92	66 Dy 162.50	67 Ho 164.93	68 Er 167.26	69 Tm 168.93	70 Yb 173.04	71 Lu 174.97
94 Pu	95 Am	96 Cm	97 Bk	98 Cf	99 Es	100 Fm	101 Md	102 No	103 Lw

SYMBOLS OF THE ELEMENTS

Element	Symbol	Element	Symbol	Element	Symbol
Actinium	Ac	Gold	Au	Potassium	K
Aluminium	Al	Hafnium	Hf	Praseodymium	Pr
Americium	Am	Helium	He	Promethium	Pm
Antimony	Sb	Holmium	Ho	Protactinium	Pa
Argon	Ar	Hydrogen	H	Radium	Ra
Arsenic	As	Indium	In	Radon	Rn
Astatine	At	Iodine	I	Rhenium	Re
Barium	Ba	Iridium	Ir	Rhodium	Rh
Berkelium	Bk	Iron	Fe	Rubidium	Rb
Beryllium	Be	Krypton	Kr	Ruthenium	Ru
Bismuth	Bi	Lanthanum	La	Samarium	Sm
Boron	B	Lawrencium	Lr	Scandium	Sc
Bromine	Br	Lead	Pb	Selenium	Se
Cadmium	Cd	Lithium	Li	Silicon	Si
Caesium	Cs	Lutetium	Lu	Silver	Ag
Calcium	Ca	Magnesium	Mg	Sodium	Na
Californium	Cf	Manganese	Mn	Strontium	Sr
Carbon	C	Mendelevium	Md	Sulphur	S
Cerium	Ce	Mercury	Hg	Tantalum	Ta
Chlorine	Cl	Molybdenum	Mo	Technetium	Tc
Chromium	Cr	Neodymium	Nd	Tellurium	Te
Cobalt	Co	Neon	Ne	Terbium	Tb
Copper	Cu	Neptunium	Np	Thallium	Tl
Curium	Cm	Nickel	Ni	Thorium	Th
Dysprosium	Dy	Niobium	Nb	Thulium	Tm
Einsteinium	Es	Nitrogen	N	Tin	Sn
Erbium	Er	Nobelium	No	Titanium	Ti
Europium	Eu	Osmium	Os	Tungsten	W
Fermium	Fm	Oxygen	O	Uranium	U
Fluorine	F	Palladium	Pd	Vanadium	V
Francium	Fr	Phosphorus	P	Xenon	Xe
Gadolinium	Gd	Platinum	Pt	Ytterbium	Yb
Gallium	Ga	Plutonium	Pu	Yttrium	Y
Germanium	Ge	Polonium	Po	Zinc	Zn
				Zirconium	Zr

Group		IA	IIA	IIIB	IVB	VB	VIB	VIIB	
Principal quantum number (Period)	Valence shell	s^1	s^2	$d^1s^2f^x$	d^2s^2	$(d^3s^2)^\dagger$	$(d^5s^1)^\dagger$	d^5s^2	$(d^6s^2$
$n=1$	$1s$	1 H 1.008							
$n=2$	$2s\,2p$	3 Li 6.94	4 Be 9.01						
$n=3$	$3s\,3p$	11 Na 22.99	12 Mg 24.31						
$n=4$	$4s3d4p$	19 K 39.10	20 Ca 40.08	21 Sc 44.96	22 Ti 47.90	23 V 50.94	24 Cr 51.996	25 Mn 54.94	26 Fe 55.85
$n=5$	$5s4d5p$	37 Rb 85.47	38 Sr 87.62	39 Y 88.91	40 Zr 91.22	41 Nb 92.91	42 Mo 95.94	43 Tc 99	44 Ru 101.0
$n=6$	$6s4f5d6p$	55 Cs 132.91	56 Ba 137.34	57* La 138.91	72 Hf 178.49	73 Ta 180.95	74 W 183.85	75 Re 186.2	76 Os 190.2
$n=7$	$7s5f6d7p$	87 Fr	88 Ra 226	89 ⊙ Ac					

		58 Ce 140.12	59 Pr 140.91	60 Nd 144.24	61 Pm 145
*Lanthanide Series		58 Ce 140.12	59 Pr 140.91	60 Nd 144.24	61 Pm 145
⊙ Actinide Series		90 Th 232.04	91 Pa	92 U 238.03	93 Np

†Variable valence shells